Ceramics Before Farming

The Dispersal of Pottery Among Prehistoric Eurasian Hunter-Gatherers

Peter Jordan
Marek Zvelebil
Editors

Left Coast Press inc.
Walnut Creek, California

LEFT COAST PRESS, INC.
1630 North Main Street, #400
Walnut Creek, CA 94596
http://www.LCoastPress.com

ISBN 978-1-59874-245-9 hardcover

Library of Congress Cataloguing-in-Publication Data:

Ceramics before farming : the dispersal of pottery among prehistoric Eurasian hunter-gatherers / Peter Jordan, Marek Zvelebil, editors.
p. cm.—(Publications of the Institute of Archaeology, University College London)
Includes bibliographical references and index.
ISBN 978-1-59874-245-9 (hardcover : alk. paper)
1. Pottery, Prehistoric—Eurasia. 2. Eurasia—Antiquities. I. Jordan, Peter (Peter D.) II. Zvelebil, Marek.
GN849.C46 2009
666'.609012—dc22 2009028842

Printed in the United States of America

♾™The paper used in this publication meets the minimum requirements of American National Standard for Information Sciences—Permanence of Paper for Printed Library Materials, ANSI/NISO Z39.48–1992.

09 10 11 12 13 5 4 3 2 1

Dedication

We dedicate this volume to four scholars who, perhaps more than others, disregarded regional boundaries and ignored modern national frontiers in pursuit of a truly inter-regional Eurasian prehistory:

Gutorm Gjessing (1944) *The Circumpolar Stone Age*. Copenhagen: (Acta Arctica II).

Grahame Clark (1952) *Prehistoric Europe: The Economic Basis. London: Methuen;* (1971) *World Prehistory: A New Outline*. Cambridge: Cambridge University Press.

Tadeusz Sulimirski (1970) *Prehistoric Russia: An Outline*. London: John Baker.

Aleksei Pavlovich Okladnikov (1979) *Otkrytie Sibiri*. Novosibirsk: Nauka.

AHRC CECD AHRC CENTRE FOR THE EVOLUTION OF CULTURAL DIVERSITY

This volume is an output of the Centre for the Evolution of Cultural Diversity (CECD), which is funded by the UK's Arts and Humanities Research Council (AHRC). The project was co-managed by Peter Jordan (Principal Investigator, CECD) and Marek Zvelebil (Associate Member, CECD) and originally entitled 'Use of Ceramics by Old-World Hunter-Gatherers'.

Contents

List of Illustrations

Tables

Foreword

Brian Hayden

Readers will welcome this major advance in solving the enduring question of why pottery emerged in various parts of the world only towards the end of the prehistoric period. While there are many intriguing questions surrounding early pottery, including gender issues, dispersal mechanisms and symbolism (all of which are touched upon in the following chapters), the great question is *why* pottery technology developed initially. What was it used for? How were these uses dealt with before pottery, or were they? If cognitive factors were involved, what specific ideas or abilities changed and why? Why were there no pottery using hunter-gatherers before 16,000 BP, whereas 12% of the ethnographically known hunter-gatherers used pottery?

Since my contribution, some 13 years ago, to Barnett and Hoopes' volume on the origins of pottery (Hayden 1995), I have continued to probe the role and possible constraints of initial pottery. It has thus been a great pleasure to peruse the contributions of this volume and become acquainted with the most recent developments in the quest to understand the origins of pottery – a phenomenon, as Jordan and Zvelebil note, too long equated with the Neolithic revolution.

The basic properties of clay are not difficult to apprehend even by non-potters, especially when fires are built on clay-rich sediments such as the loess plains of Moravia, or when clay dries out from puddles, forming cupped and cracked clay features. A number of contributors (Jordan and Zvelebil, Kashina) have drawn attention to the fact that ceramic figurine technology was well known in some regions of Europe 26,000 years ago. However, this knowledge can probably be extended back much further, at least into the Middle Palaeolithic. At the Grotte du Prince, for example, seven slabs of hardened clay were arranged around a Middle Palaeolithic hearth, and postulated to have been used for grilling pieces of meat (Villeneuve 1906: 55–56). Thus, the cognitive awareness of ceramic technology has been around for a very long time, and it seems unlikely that either genetic changes in cognition or the emergence of new ideas can plausibly account for the appearance of the first pottery vessels, around 50,000 years later. On the other hand, the usually invoked physical constraints of mobility and humid climates hardly seem adequate to account for the lack of pottery during the first 2.5 million years either. As Jordan and Zvelebil note, while at least some degree of seasonal sedentism seems almost a pre-requisite for using heavy, coarse fabric, poorly fired pots like the earliest examples from the Far East, there is clearly more to the story.

If cognition was not the key explanatory factor, and physical constraints are not sufficient to account for the lack of pottery, it becomes essential, as Jordan and Zvelebil also enjoin us, to try to understand the motivations behind the initial use of pottery vessels. Necessity (resource stress or famine) is certainly a strong motivator, but one that has been with humanity since its inception, not just in the last 16,000 years. Moreover, Barnett observes that the major Pleistocene spread of pottery occurred around 13,000 BP during a period of climatic amelioration, not during a period of increasing stress.

Over a decade ago (Hayden 1990, 1995), I proposed that the development of feasting among complex hunter-gatherers constituted a new phenomenon that could have motivated people to develop a pottery technology. Feasting only emerged with other new technological advances that made complex hunting and gathering societies possible, beginning in some areas during the Upper Palaeolithic, but becoming much more widespread during the Mesolithic. Those technologies included not only new extractive devices (such as fishhooks, nets, leisters, sickles and weirs), but also new processing tools (such as grinding stones, boiling stones and earth ovens), and new storage technologies. In favourable areas, the cumulative effect of these innovations was to make the production of surpluses relatively reliable. These surpluses were owned by individuals or families and used by them in strategies to enhance their own survival and reproductive benefits. One of these strategies was the hosting of reciprocal feasts involving the gifting of food meant to impress guests so that they would enter into economic exchanges, marriage exchanges, and defensive/offensive alliances. These feasts also created the debt structures and hierarchies that characterize complex hunter-gatherers. Additional strategies that were undoubtedly used included the acquisition of mates in exchange for surpluses or wealth, the investment of wealth in children, the making and gifting of prestige objects (often of exotic origin), and ancestral worship (hence, burials and cemeteries).

Consequently, I suggested (Hayden 1995) that if pottery developed in order to enhance the impressiveness of feasts, then the initial use of pottery should either be:

(1) finely made serving vessels to impress guests (a prestige technology); OR (2) vessels used for producing specialty foods to impress guests (a new technology used to produce prestige materials).

The motivation would be the desire to favourably impress others through extra efforts to serve them sumptuously. Unfortunately, Pearson (2005) and others have focused on only the first of these possibilities (pottery as fine ware serving vessels) to critique the feasting model. And, in fact, early pottery characteristics in most areas do not seem to accord well with initial pottery as prestige serving vessels among most hunter-gatherers (although there are a number of examples in the Barnett and Hoopes volume [1995] of initial pottery developing among agriculturalists that fits this expectation very nicely).

However, if we turn to the second hypothesis (pottery used for preparing special dishes), there are a number of observations in the following chapters that seem to confirm early pottery being used to prepare specialty foods:

1 Initial pottery is generally a coarse ware, fired at low temperature, with thick walls and conical shapes. The presence of carbon crusts on interior walls confirms that these were primarily cooking pots, and some pots are even red ochred or have ritual types of elaborations seemingly for social display (Kashina, Pesonen and Leskinen, Skandfer).

2 Most pot volumes are relatively modest.

3 Almost everywhere, these pots first occur in very small numbers over extended periods of time as noted by Jordan and Zvelebil, McKenzie, Zhushchikhovskaya, Kaner, German, Skandfer, Stilborg and Holm. At Yuchanyan, perhaps the earliest-dated pottery assemblage, the remains of only two pots were recovered, representing semi-sedentary occupations spanning several thousand years (Yuan 2002: 161). As McKenzie, and Stilborg and Holm note, they served no critical need and must have played a very low role in overall subsistence processing. It seems that they were used for rare special events.

4 There appears to be a very strong association between sites where initial pottery is found and semi-sedentary sites where aquatic resources were important (Jordan and Zvelebil's 'estuarine model'; also Barnett, Budja, Chairkina and Kosinskaia, Cho and Ko, Crombé, Hallgren, Kaner, McKenzie, Papmehl-Dufay, Pesonen and Leskinen, Zhushchikhovskaya; also Aikens 1995).

In all, Stilborg and Holm conclude that early pottery was used to produce some sort of desirable product, while Budja views the context of some early pottery as feasting and playing a role in the control of power (via feasts and the reproduction of social relationships).

Given the above characteristics, what kinds of meant-to-impress foods could have been prepared in early pottery vessels? There are several possibilities, and specific kinds of luxury foods undoubtedly varied according to environments. But they all would have had the common denominator of being effort-demanding, unusual preparations. Some of the most likely foods would have included:

> Soups. As I have argued elsewhere (Hayden and Adams 2004: 98), soups would have been difficult to prepare for hunter-gatherers who only had baskets, especially if the extraction of oils for soups required prolonged boiling. Soups would have been relished in any seasonally cold climate, and they are ethnographically an essential part of many feasts, including potlatches and early Chinese feasts.
>
> Fish Oil. On the 'Pacific' Northwest Coast, the production of fish oil was, and still is, probably the most labour intensive kind of food. Large quantities of fish are required, there are several stages of fermentation in large box containers or canoes followed by boiling, skimming and refining in smaller containers similar in size to early pots. The major trade routes from the coast to the interior were known as 'grease trails' because of the volume of fish oil transported along them. Fish oil was highly prestigious in potlatches and it was served to guests in elaborately carved wooden bowls. Even today, with modern technology, a litre of fish oil costs at least US$120. Following Reid (1990) and others, numerous authors in this volume comment on the possible role of pottery in producing or storing fish oil (Budja, Crombé, Jordan and Zvelebil; see also Barich 1998: 115,122–123 for early African ceramics and the production of fish oil in pottery vessels by the Bozo of Nigeria). The return of groups to especially productive fishing sites on a yearly basis would ensure that fish oil pottery processing vessels could be cached and used again for producing fish oil for special occasions.
>
> Bone Grease and Marine Mammal Oil. In a similar fashion, the production of bone grease was a labour intensive undertaking resulting in a highly valued food especially in cold climates. Seal oil or other rendered mammal fats may have similarly been prepared and valued for use in feasts as indicated by several analyses (German, Jordan and Zvelebil, Papmehl-Dufay, Pesonen and Leskinen, Zhushchikhovskaya; see also Reid 1990).
>
> Nut Oil. Like fish oil, nut oil is highly labour intensive to produce, requiring 100 pounds of wild black walnuts to produce only 0.625 gallons of oil. The nuts must be crushed, boiled, and skimmed. Taché

and colleagues (2008) have recently examined early accounts of nut oil production, especially from Ozker (1982: 36), and concluded that nut oil must have been a prestige food, especially in northeastern North America. Pottery would have greatly facilitated the production and perhaps storage of all kinds of oils, including nut oils.

Alcohol. Several contributors (Jordan and Zvelebil, Pesonen and Leskinen, Stilborg and Holm) emphasize that initial pottery may have been used to produce fermented beers, just as it was by 9,000 BP in China and in later periods of prehistory (Roach 2005; Rojo-Guerra et al. 2006). Beer or alcohol is certainly one of the main constituents of most feasts throughout the world and may well be a tradition that began with complex hunter-gatherers in various areas.

Given this scenario, we are not constrained to think of initial pottery being used for the production of only one type of food, but expectations clearly focus on lipids from plant, animal and fish sources, as well as on fermented beverages. Quantities of pottery used for special feasting foods can be expected to be limited and infrequently used. These are all characteristics of most initial pottery traditions among hunter-gatherers documented in this volume. However, the same observations also apply to the initial adoption of pottery by hunter-gatherers in northeastern North America as well as the pre-colonial pottery-using hunter-gatherers of South Africa (Sadr 2004), and semi-sedentary groups around the Sahara and Nile where the aquatic adaptation extends back almost 20,000 years (Barich 1998: 34, 48).

The semi-sedentary nature of most sites (Budja, Chairkina and Kosinskaia, Kaner, Zhushchikhovskaya in this volume) indicates that all such groups were able to extract at least enough resources to make a modest degree of sedentism possible. While many people continue to refer to Mesolithic or earlier groups as 'egalitarian', there are usually enough indications of prestige goods circulating in the regions to make a case for some surplus production and, in some cases, feasting during these times. This is certainly true of the Chinese 'Mesolithic' where early pottery sites like Yuchanyan (18,000–14,000 BP) occur in exceptionally rich aquatic environments where people used 27 bird species (including 18 water birds), 28 mammal species, 33 shellfish species, 40 plant species, four fish species, and land molluscs (Yuan 2002: 160, 163, 165). A number of technological innovations occur during this period as well, possibly including watercraft that made the exploitation of aquatic environments more lucrative, just as canoes did on the 'Pacific' Northwest Coast, in Mesolithic Europe, and in Africa (Barich 1998: 48; Jiang and Liu 2005).

There are also a number of sites in the Chinese late Palaeolithic or 'Mesolithic' that have produced prestige items, including a shell bead necklace at

Donghulin imported from at least 290 km away (Hao et al. 2001), 125 perforated animal teeth, stone beads, shells and polished bone artefacts in the Upper Cave at Zhoukoudian (Kamminga and Wright 1988: 741–742, 755–756), a similar assemblage from Xiaogushan (Huang et al. 1986), ornamental animal canines at Yuchanyan together with small furbearers and birds (Yuan 2002: 159). Prepared intentional burials of some individuals also begin to occur in the Chinese Mesolithic. Thus, far from being the strict egalitarian cultures that some people describe them as, the pottery producing 'Mesolithic' sites in China appear instead to be more similar to transegalitarian societies elsewhere in the prehistoric world where feasting was part of the cultural repertory and special foods could be expected to be sought after.

It must also be recognized that pottery would have been only one solution to the need to impress feast guests with the food being given to them. On the 'Pacific' Northwest Coast and in Polynesia, serving vessels used in feasts were made of ornately carved wood, while fish oil, soups and kava were prepared in wooden containers. In other places with more suitable clay deposits and climates, such vessels were made of ceramics. Why Near Eastern pottery should appear at a time when societies were undergoing disintegrative stresses (Budja) is difficult to understand unless they were adopting an already perfected prestige technology from farther east and then applying it to more practical problems of existence (which is a general pattern of many prestige technologies including those based on metals, glass and domesticated animals).

If these suggestions eventually prove to be well-founded, a strong case may be made for initial pottery serving an important role in the reproduction of society (via its role in feasts), whether in hunting and gathering or agricultural societies. In like manner, any prestige serving vessel used at feasts could be expected to be imbued with symbolism at several levels (the symbolism of wealth, shared elite etiquette, group tradition, inherited status, and iconic meanings) just as the grease and serving bowls of the 'Pacific' Northwest Coast displayed masterly carving skills and totemic emblems. It is somewhat more problematic, however, as to whether common, undecorated coarse-ware cooking vessels would have conveyed similar connotations any more than modern banqueters are aware of the pots that their meals are prepared in. On the other hand, such pots did sometimes evolve into the elaborate symbolic wares such as the flame ware Jomon vessels which were undoubtedly used for serving some type of prestigious hot liquid while people gathered around a hearth. Despite the potential for symbolic communication, it seems unlikely to me that Larsson's proposal for the use of simple (or complex) motifs (such as triangles to decorate pots) could have constituted 'powerful mechanisms for maintaining social relations' or been the impetus behind the creation of 'new worlds', at least without important material and technological changes to support such use of symbols.

Thus, with this volume, we have achieved a new level in the dialectic between theory and data. It is a good time to take stock of this dialectic and to chart new avenues of exploration. Certainly, one of the most obvious areas to explore will be the analysis of residues from early pots. But equally important will be the careful documentation of subsistence staples, settlement patterns (particularly in relation to aquatic or nut resources), relative levels of seasonal sedentism, technological innovations, and intentional burials or prestige items. I suspect that they are all related to the development of the first pottery technologies among hunter-gatherers.

References

Aikens, C M (1995) 'First in the world: the Jomon pottery of Early Japan', in Barnett, W and Hoopes, J (eds), *The Emergence of Pottery*, Washington DC: Smithsonian Institution Press, pp 11–21

Barich, B (1998) *People, Water and Grain: The Beginnings of Domestication in the Sahara and the Nile Valley,* Amsterdam: L'Erma di Bretschneider

Barnett, W and Hoopes, J (eds) (1995) *The Emergence of Pottery,* Washington DC: Smithsonian Institution Press

Hao, S-G, Ma, X-P, Yuan, S-X, and Southon, J (2001) 'The Donghulin woman from western Beijing', 75 *Antiquity*, 517–522

Hayden, B (1990) 'Nimrods, piscators, pluckers, and planters: the emergence of food production', 9 *Journal of Anthropological Archaeology*, 31–69

——— (1995) 'The emergence of prestige technologies and pottery', in Barnett, W and Hoopes, J (eds), *The Emergence of Pottery,* Washington DC: Smithsonian Institution Press, pp 257–266

Hayden, B and Adams, R (2004) 'Ritual structures in transegalitarian communities', in Prentiss, W and Kuijt, I (eds), *Complex Hunter-Gatherers,* Salt Lake City: University of Utah Press, pp 84–102

Huang, W-W, Zang, Z, Fu, R, Chen, B, Liu, J, Zhu, M and Wu, H (1986) 'Bone artifacts and ornaments from Xiaogushan site of Haicheng', 5 *Acta Anthropologica Sinica*, 259–266

Jiang, L and Liu, L (2005) 'The discovery of an 8000-year-old dugout canoe at Kuahuqiao in the Lower Yangzi River, China', *Antiquity,* On-line 79(305), http://antiquity.ac.uk/ProjGall/305.html

Kamminga, J and Wright, R (1988) 'The upper cave at Zhoukoudian and the origins of the Mongoloids', 17 *Journal of Human Evolution*, 739–767

Ozker, D (1982) *An Early Woodland Community at the Schultz site 20SA2 in the Saginaw Valley and the Nature of the Early Woodland Adaptation in the Great Lakes Region*, Museum of Anthropology, University of Michigan Anthropological Papers 70, Ann Arbor: Museum of Anthropology, University of Michigan

Pearson, R (2005) 'The social context of early pottery in the Lingnan region of south China', 79 *Antiquity*, 819–828

Reid, K (1990) 'Simmering down: a second look at Ralph Linton's "North American Cooking Pots"', in Mack, J (ed), *Hunter-Gatherer Pottery of the Far West*. Nevada State Museum, 23 *Anthropological Papers*, 8–17

Roach, J (2005) '9,000-year-old beer recreated from Chinese recipe', *National Geographic News,* http://news.nationalgeographic.com/news/2005/07/0718_050718_ancientbeer.html

Rojo-Guerra, M A, Garrido-Pena, R, Garcia-Martinez-de-Lagran, I, Juan Treserras, J, Matamala, J C (2006) 'Beer and bell beakers: drinking rituals in Copper Age Inner Iberia', 72 *Proceedings of the Prehistoric Society*, 243–265

Sadr, K (2004) 'Feasting on Kasteelberg', 3 *Before Farming*, 1–17

Taché, K, White, D and Seelen, S (2008) 'Potential Functions of Vinette I Pottery. Complementary Use of Archaeological and Py-GC/MC Data', 36 *Archaeology of Eastern North America,* 63–90

Villeneuve, L de (ed) (1906) *Les Grottes de Grimaldi,* Volume 2, Monaco: Imprimerie de Monaco

Yuan, J (2002) 'Rice and pottery 10,000 yrs BP at Yuchanyan, Dao County, Hunan Province', in Yasuda, Y (ed), *Origins of Pottery and Agriculture*, New Delhi: Roli Books, pp 157–166

Preface

The emergence of ceramics and its early use amongst prehistoric hunting and gathering societies had occupied the research interests of one of us (Marek Zvelebil) for over twenty years, but the opportunity to produce a synthetic overview of early pottery began with a fortuitous chain of events that can be traced back to the 'Conference on Hunting and Gathering Societies (CHAGS)', which was held in Edinburgh in September 2002. A meeting with David Anderson (University of Aberdeen) at this conference eventually led to Peter Jordan hosting two Siberian archaeologists, Viktor M Vetrov and Evgeny M Ineshin, at the Institute of Archaeology, University College London, in early 2003. During this visit, Viktor Vetrov presented a fascinating lecture about his discovery of early hunter-gatherer ceramics at the site of Ust'Karenga on the Vitim River in Siberia (see McKenzie, this volume).

We happened to discuss this early material—not widely known outside Russia—a few weeks later at the University of Sheffield, and quickly realized that these early Siberian dates had the potential to bridge the chronologies between the earliest hunter-gatherer pottery in northern Europe, whose origins are imperfectly understood, and the very early, but apparently isolated, dates for the first pottery in East Asia. Suspecting that there may have been a continental-scale ceramic dispersal event, we decided to organize a preliminary session to discuss early pottery origins in northern Eurasia in order to locate regional specialists and gauge the quantity of available data.

The subsequent session formed part of the September 2003 meeting of the European Association of Archaeologists (EAA), which was held in the spectacular and particularly appropriate setting of St Petersburg. The panel was a big success and revealed an abundance of material on early ceramics, but also raised important questions about chronology and the social and practical motivations for the invention and adoption of early pottery vessels in hunter-gatherer settings. We decided to push on with the initiative via publication of a major synthetic volume, and this project was formally adopted into the portfolio of research conducted by the UK Arts and

Humanities Research Council (AHRC) Centre for the Evolution of Cultural Diversity (CECD) later that year. Publication contracts were secured and plans made for authors to present final versions of their chapters at a subsequent meeting of the Theoretical Archaeology Group (TAG) which was held in Sheffield in December 2005.

We would like to express an enormous debt to all the chapter authors for their excellent conference papers at both the EAA (2003) and TAG (2005) sessions, for their outstanding work on the individual chapters, and for their patience during the extended production process. Andrew Sherratt agreed to serve as discussant at the TAG session and contributed some stimulating reflections during the lengthy discussions which stemmed from the presentations. He had much to contribute to this Eurasian-scale initiative, and it is especially sad that his untimely departure—only a few weeks later—meant that he was unable to make a more significant contribution to our understanding of the wealth of materials, ideas and interpretations published in this volume.

The AHRC CECD provided travel and accommodation funds for several of the speakers, some of whom travelled to Sheffield from the shores of the Pacific Ocean during the middle of winter to take part in the conference. Both editors thank the AHRC for the financial support which made this research and publication possible. They would also like to express their gratitude to Stephen Shennan (AHRC CECD Phase 1 Director) and James Steele (AHRC CECD Phase 2 Director) for the encouragement and practical support which ensured that a bold research vision was transformed into a concrete reality.

Much encouragement, support, suggestions and inspiration came from a number of other valued colleagues and we would like to thank them all: Daniel Antoine, Andy Bevan, John Barrett, Charlotte Damm, Evgenii M Ineshin, Artur V Kharinskii, Patrick Quinn, Ken Sassaman, Viktor M Vetrov. Many thanks also to Yaroslav V Kuz'min, who has been instrumental in assembling and publishing the chronological frameworks which structure many of the interpretations presented in this volume.[1]

We would also like to thank the members of the Publications Committee at the Institute of Archaeology, University College London, all of whom have been very supportive of this research; thanks also to Marion Cutting. The insightful comments from two anonymous reviewers enabled us to significantly strengthen the collection of chapters and fill in some of the gaps in regional coverage, maps and associated illustrations. Many thanks to Mitch Allen and his team at Left Coast Press. We also benefited from editorial assistance from Jenny Moore and Heidi Robbins. Every effort has been made to standardize the transliteration of Russian words and terminology—we are grateful to Maria Nakhshina for undertaking this complex task.[2] Some of the illustrations were prepared or redrawn by

Jo Mincher, University of Sheffield, and the overview maps in the first chapter were drawn by Alison Sandison at the School of Geosciences, University of Aberdeen. Martin Brejka agreed to prepare the cartoon in Chapter 1 after we discussed V G Childe's comments about new technology and 'prudent housewives' with him over a few beers. Many thanks also to Rona Kennedy at the School of Geosciences, University of Aberdeen, for providing a last-minute helping hand as we went into final production.

The project was launched when Peter Jordan held a Leverhulme Trust Special Research Fellowship at University College London and later at the University of Sheffield. Many thanks to all these institutions, and in particular, to Peter Ucko and John Barrett for supporting the Fellowship. Marek Zvelebil would like to thank the Leverhulme Trust for grant F/00 M118/AP that enabled him to carry out research that contributed to the preparation of this volume.

Most of the introductory chapter was written over a wonderful week at Cabrespine in the south of France in September 2006. Intensive morning work on the volume was followed by late afternoon swimming in the Mediterranean and enjoyment of local seafood and fine wines. The editors would like to express a particular debt to Alena Haisová and Bibiána Jordan Horváthová for creating the material conditions that nourished body and soul and made such splendid intellectual seclusion possible. We also thank them for their long-standing encouragement during the extended work on the final volume.

As we stand on the brink of publication, our hope is that this body of work promotes the continued re-emergence of 'Eurasia' as an analytical concept, and breaks down some of the linguistic frontiers and narrow regionalism that continue to constrain our ability to research and understand some of the major transformations that have defined long-term human history.

Peter Jordan and Marek Zvelebil, Spring 2008

Notes

1. Radiocarbon Dates: Unless directly indicated, all dates are reported as: bp/bc (uncalibrated) and BP/BC (calibrated).
2. Russian Transliteration: For overall consistency Russian words have been transliterated according to the Library of Congress System, for example, when compiling references. This standard has also been applied to most Russian words, names, sites, locations or archaeological cultures mentioned in the text. In some limited cases, however, alternative spellings of personal names, geographic locations and archaeological sites or cultures have been retained and readers should be aware of potential discrepancies.

Part 1

Introduction

Chapter 1

Ex Oriente Lux: The Prehistory of Hunter-Gatherer Ceramic Dispersals

Peter Jordan and Marek Zvelebil

Introduction

The emergence of pottery remains one of the most important and longstanding issues in Old World archaeology. Since the inception of the discipline, the appearance of ceramics has provided archaeologists with primary data for constructing general chronologies of culture change. However, in European archaeology, the question of ceramic origins has been heavily influenced by broader debates surrounding the emergence of sedentary agricultural societies. To this day, there remains an implicit assumption that the earliest ceramic vessels were invented and used by the early farmers; moreover, if pots were a distinctive feature of farming societies, then the dispersal of ceramics must have been linked to the spread of agriculture. Through time, the arrival of ceramics and the spread of farming have become central themes in broader narratives about the transition to agriculture in a post-glacial world.

In contrast, the potential role of hunter-gatherer societies in the invention and dispersal of pottery technology has seen much less consideration. Where pottery was recorded in non-farming contexts, it was generally thought to represent partial adoption of the wider 'package' of traits that made up the Neolithic, part of a wider set of cumulative processes that would end with the inevitable replacement of foraging practices and populations by new forms of social life grounded in the farming economy.

These deeply established assumptions about the relationship between pottery, early agricultural dispersals and hunter-gatherers are now approaching a fundamental revision. Recent years have witnessed a rapid accumulation of new dates and material from across northern Eurasia, in particular, from Japan, China and the Russian Far East, which demonstrate long-term, extensive, and most importantly, 'independent' invention and use of pottery by hunter-gatherers long before the transition to agriculture in the early Holocene. These emerging insights challenge established histories of ceramic technology in three crucial ways: firstly, in some regions the earliest pottery is *older* than the start of the Holocene, making it clear that the first ceramic vessels must have been invented in the Pleistocene hunter-gatherer societies of the Upper Palaeolithic; secondly, the earliest pottery predates farming in *many* regions of northern Eurasia, indicating that knowledge of pottery technology spread without any association with agriculture, and that Upper Palaeolithic and early Holocene hunting and gathering societies were responsible for this dispersal process; thirdly, the growing evidence for early and very widespread use of ceramics by Eurasian hunting and gathering societies indicates that pottery can have an 'independent' technological history entirely distinct from that of the subsequent plant and animal domestications that characterized the Neolithic. Finally, the fresh materials and dates also contribute to the more general realization that the significance and diversity of this hunter-gatherer ceramic 'horizon' has yet to receive systematic attention from archaeologists. However, they also signal exactly where a new period of integrated research could begin.

The present volume initiates the process by undertaking the first continent-wide English language synthesis of pottery use amongst prehistoric hunter-gatherers, reconstructing the characteristics, chronologies and historical trajectories of early ceramic technologies as they were dispersed across northern Eurasia, subject to the data currently available. The volume also highlights significant gaps in current knowledge, and we suggest specific themes and regions where more work is required in order to investigate the local dynamics of the more general ceramic dispersal process.

The volume also attempts to identify new ways in which the invention, adoption, and use of pottery by hunter-gatherer societies might be researched and understood. We recognize a sense of creativity, contingency and technological improvisation by emphasizing the role of pottery as a cultural tradition. This would have been gradually incorporated into the social life of hunter-gatherer communities, and passed between generations and communities, resulting in innovations and transformations. By foregrounding the contingency of social tradition, we break from the constraints imposed by general schemes of social and technological progress, and we are also able to acknowledge the broader importance of ecological and environmental settings without assuming that pottery could function as no more than an adaptive 'tool'.

The present volume reflects the collective efforts of the two editors and contributing authors over recent years. The ambitious scale and international nature of the project have also meant that the volume has been a long time in production. The volume's more general origins can be traced back even further, with one of us (Marek Zvelebil) developing an interest in the hunter-gatherer pottery use and its origins in the 1980s (Zvelebil 1986c, d).

Despite long-standing recognition of widespread use of ceramics among north European hunter-gatherers, a recognition that generated a tradition of research among Finnish, Swedish and East Baltic archaeologists (Ailio 1909; Ayräpää 1952, 1955; Carpelan 1979; Edgren 1966, 1970, 1982; Halen 1994; Hulthén 1977, 1991; Larsson 2004, 2006; Meinander 1954, 1961, 1979; Nuñez 1975, 1990; Papmehl-Dufay 2006; Siiriäinen 1971, 1981, 1984; Welinder 1976; also Jaanits 1959; Kriiska 2004 in Estonia), the use of ceramics in hunter-gatherer communities continued to be regarded as a peripheral cultural practice in mainstream European prehistory that was more concerned with the development of farming societies and the emergence of cultural complexity. By the mid-1980s it was becoming increasingly apparent that the appearance of pottery in northern Europe may not have been connected to farming dispersals, but had its own historical trajectory, and that it was much more than a phenomenon marginal to the early spread of Neolithic farming societies across Europe—even though contacts and exchanges between foragers and farmers did exist (Zvelebil 1986d: 171, fig. 4; 1998a). The early use of ceramics among post-glacial hunter-gatherers extended well beyond the forager–farmer contact zones. This was puzzling, as pottery had previously been assumed to constitute an essential element of the Neolithic 'package' that was spreading slowly northwards through Europe. The presence of pottery in distant hunter-gatherer regions created a fundamental challenge to the established understandings of earlier prehistory. Even so, as many recent accounts indicate, hunter-gatherer ceramic-using societies continued to be regarded as a curious anomaly, resulting from contact with Neolithic farmers and requiring a special name within the social-evolutionary framework (such as sub-Neolithic, or para-Neolithic!), rather than warranting their own further research as distinct cultural traditions (i.e., Nuñez 1990; for an overview, see Werbart 1998; Zvelebil 1986c).

This conceptual problem did not exist for Soviet archaeologists, who were interpreting evidence for culture change within a Marxist-sanctioned evolutionary framework. In contrast to western notions of a Neolithic defined principally by farming, for Soviet archaeologists, it was the emergence of pottery that served as the key event defining the start of the Neolithic epoch. As a result, scholars of the Soviet era paid greater attention to ceramic technology in hunter-gatherer contexts. For example, in north-eastern Europe, Matiushin, Timofeev, Dolukhanov and Foniakov provided the first reliable chronological frameworks and systematic cultural descriptions of hunter-gatherer pottery (Dolukhanov 1979, 1986a, 1986b;

Dolukhanov and Foniakov 1984; Matiushin 1976, 1986; Timofeev 1987, 1990, 1998, 1999). Consequently, while in the West this phenomenon remained a marginal development, in the East it has become a central, epoch-defining event (see below).

These new materials and chronologies generated a series of important questions about the established history of ceramic origins. In particular, they indicated that either the local invention, or arrival, of ceramics in northern Europe had taken place at a much earlier date than previously thought, suggesting that the northern pottery was not linked to the dispersal of farming from the Near East. For example, at the Volga-Ural interfluve, early ceramics were eventually dated to c. 9,000 BP (8,000 BC), and in the Upper Volga Basin the first dated wares, which fall to 7,300 BP (6,000 BC), and also may have served as source areas for the first pottery in the eastern Baltic, where ceramics appear by 6,500 BP (5,400 BC) (Cyrek et al. 1986; Dolukhanov et al. 2005; Kempisty 1986; Nuñez 1990; Rimantienė 1992, 1998; Timofeev 1987, 1990, 1998). However, for a number of years these 'strange noises from the east' (Pavel Dolukhanov, TAG Conference Session, see below) did not attract significant attention.

Further crucial developments came in the later 1990s with publication of new dates for early pottery from eastern Eurasia, from China, Japan and the Amur. Some were so old that they caused initial disbelief, until validation confirmed that use of pottery amongst late Pleistocene hunter-gatherers was a rather widespread phenomenon, at least in East Asia. They also added another dimension to debates about the first pottery in northern Europe, which lay far outside the arrival zone of early farmers, and raised the possibility that there was an original source area for a distinct hunter-gatherer ceramic tradition that lay further to the east. The arrival of pottery into hunter-gatherer societies in northern Europe may have formed the most westerly evidence for a continent-wide dispersal event whose ultimate origins were located in the emergence of early pottery in the Far East of Asia (Kuz'min and Orlova 2000).

More recently, the idea of publishing the volume was triggered by a guest lecture given by Dr. Viktor Vetrov, Irkutsk State Pedagogical University, Russia, at the Institute of Archaeology, University College London, in the spring of 2003 (attended by Peter Jordan). The speaker described the Ust' Karenga ceramic assemblages from eastern Siberia which were dated to 10,000 BP (Vetrov 1985, and see McKenzie, this volume). During a later meeting at Sheffield University, the editors happened to discuss this new evidence, and in particular, its potential to 'bridge' the eastern and western chronologies of early pottery, and thereby shed new light on the wider history of ceramic dispersals. Plans for a volume exploring the earliest use of ceramics across northern Eurasia were developed rapidly, and activities began later that year with preliminary discussion papers invited for a special session on early ceramics at the St. Petersburg meeting of the European Association of Archaeologists (EAA) in September 2003. A wider suite of papers formed a full day session at the December 2005 meeting of the Theoretical Archaeology Group (TAG) which was held at Sheffield University.

Most chapters are revised versions of papers presented at EAA 2003, TAG 2005 or both conference sessions; the remaining papers were commissioned in the intervening periods in order to achieve coherent regional and topical coverage.

The volume is structured along the following lines: in his foreword, Hayden provides a stimulating reflection on some of the enduring questions that are central to understanding the emergence of early pottery. This introductory chapter (Jordan and Zvelebil) critiques current approaches and understandings of pottery use amongst hunter-gatherers, highlights the salient themes and questions, and draws out the overviews and insights, concluding with a summary of scope for further research. The 14 chapters forming the bulk of the volume comprise regional case-studies, and are organized geographically in order to track the emergence of early ceramic traditions in eastern Asia, their subsequent dispersal across Siberia, and subsequent arrival into the eastern and northern peripheries of Europe (papers by Kaner, Zhushchikhovskaya, Cho and Ko, McKenzie, Chairkina and Kosinskaia, Dolukhanov, Mazurkevich and Shukurov, German, Kashina, Pesonen and Leskinen, Stilborg and Holm, Skandfer, Hallgren, Larsson and Papmehl-Dufay) (see Figure 1.1).

Four further papers develop broader perspectives on the more complex issue of ceramic dispersals into forager–farmer interaction zones: Nowak discusses the introduction and use of pottery at the forager–farmer interface in present-day Poland; Crombé examines hunter-gatherer pottery along the western margins of early farming communities; Budja explores changing trajectories of ceramic tradition in Anatolia and south-east Europe; Gronenborn casts the widest net and explores ceramic traditions and inter-community contacts in north central Europe more generally. Barnett contributes a final discussion chapter which locates the history of Eurasian ceramic dispersals within a global context, and Hommel concludes the volume with the emerging chronology of early ceramic dispersals, a database that is certain to witness rapid expansion and revision over future years.

It is worth emphasizing that the chapters within the volume are characterized by differences in focus, detail and paradigmatic orientation, each reflecting in an interesting way the different research and scholarly traditions in different regions of Europe and Asia. In general, we have made no attempt to homogenize these differences, except for the sake of clarity and coherence. In general, and as far as possible, the volume editors have sought to use widely accepted English-language place names for sites and locations

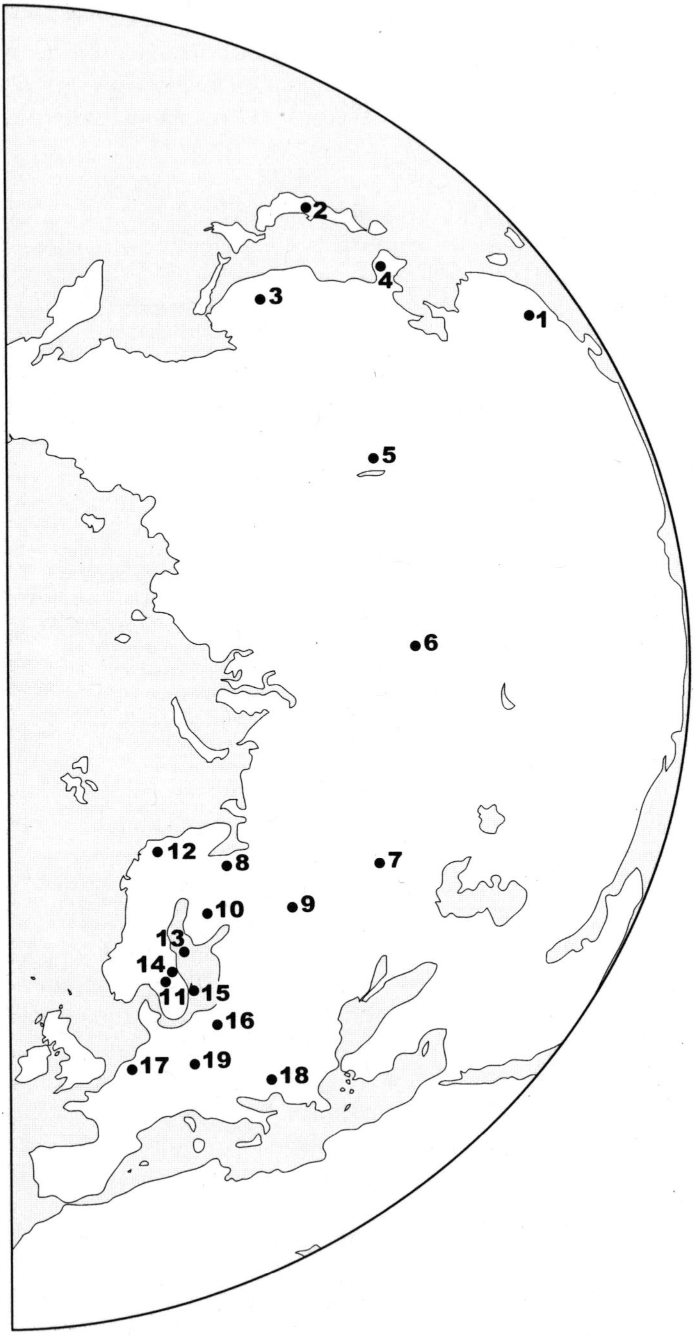

Figure 1.1 Early pottery regions and associated chapters.

mentioned in the text. Where such terms are absent—or politically contentious—the editors have followed the place names used by individual chapter authors.

Early Pottery in China

At this point we would like to address a significant gap in the regional case-studies. Very early pottery complexes have been reliably dated to 13,700–13,300 bp (about 17,200–14,700 BP) in Japan, the Russian Far East and China (Keally et al. 2004: 349). In the present volume, the first two regions are covered by Kaner (Japan) and Zhushchikhovskaya (Russian Far East). Unfortunately, a planned chapter on the earliest pottery in China failed to materialize in time for final publication, leaving a key centre of pottery innovation without a designated case-study. As Kuz'min (2003: 1) argues, comparative data from the Chinese sites are crucial for understanding the wider emergence of pottery, and we redress this gap in regional coverage by providing a brief overview of the Chinese materials.

Significant cultural differences were developing in China during the Upper Palaeolithic (Barker 2006; Chi 1999: 9). In the north, these were manifested by a cluster of microlith cultures identified in open sites across the North China Plain and adjacent regions. In southern China, along the northern and southern foothills of the Nanling Mountains, a 'cave-dweller' culture emerged, which was characterized by the use of chipped-pebble tools. These assemblages were found primarily in the deeply-stratified cave sites that are common along the river valleys of the Karst region (Chi 1999: 10, 2002: 30; Feng 2000). The northern tool tradition appears to have been linked to similar microlith cultures in adjacent areas of Northeast Asia, while the southern pebble lithic industry bore resemblances to the Hoabinhian Culture of adjacent continental Southeast Asia (Chi 1999: 9–10, 2002: 30).

In China, early pottery appears in both cultural traditions at different points prior to, and during, the late Pleistocene and early Holocene transition (Chi 2002: 30; Pearson 2005: 822; Zhao and Wu 2000). The oldest examples are the stroke- and cord-marked, round-based quartz-tempered pottery jars (Chi 1999: 10) recovered from cave sites in the south, now reliably dated to as far back as 13,700–13,300 bp (about 17,200–14,700 BP) (Keally et al. 2004: 349). In contrast, the flat-based pottery in northern China appears somewhat later than the earliest pottery in southern China, and also after the earliest pottery in Japan and the Russian Far East (Chi 2002: 29).

Zhao and Wu (2000) have published a full radiocarbon chronology for early pottery sites possessing what they argue to be pristine stratigraphic sequences, and these data appear to indicate that the earliest Chinese pottery is perhaps the oldest in East Asia (Figure 1.2). More recently, however, the accuracy of these dates has been questioned due to potential disturbance of the cultural layers (Keally et al. 2004; Kuz'min 2003: 1–2). At present, the most reliable—and widely accepted—14C age determinations for the earliest pottery complexes in southern China appear to fall in the range of approximately 13,700–13,300 bp (about 17,200–14,700 BP) (Keally et al. 2004: 349), and include the following sites: Yuchanyan, layer 3E, dated to c. 17,210–15,660 BP (13,680±270 bp [BA95058]); Miaoyan, layer 4 M, dated to c. 16,780–14,720 BP (13,320±270 bp [BA92034]), and Xianrendong, zone 3B1, dated to c. 15,530–14,130 BP (12,430±80 bp [UCR-3561]) (Keally et al. 2004: 347, table 1) (Figure 1.2). These dates place the emergence of the first Chinese pottery complexes at least at the very start of the Pleistocene–Holocene transitional period (c. 14,000–9000 years ago (Pearson 2005: 822), and possibly even earlier than the onset of climatic warming in the final Pleistocene (Barker 2006: 187–188, 228; Zhao and Wu 2000: 239).

More generally, the emergence of early pottery in China appears to form one element in a more general process of hunter-gatherer innovation, experimentation and transformation that marks the end of the Palaeolithic

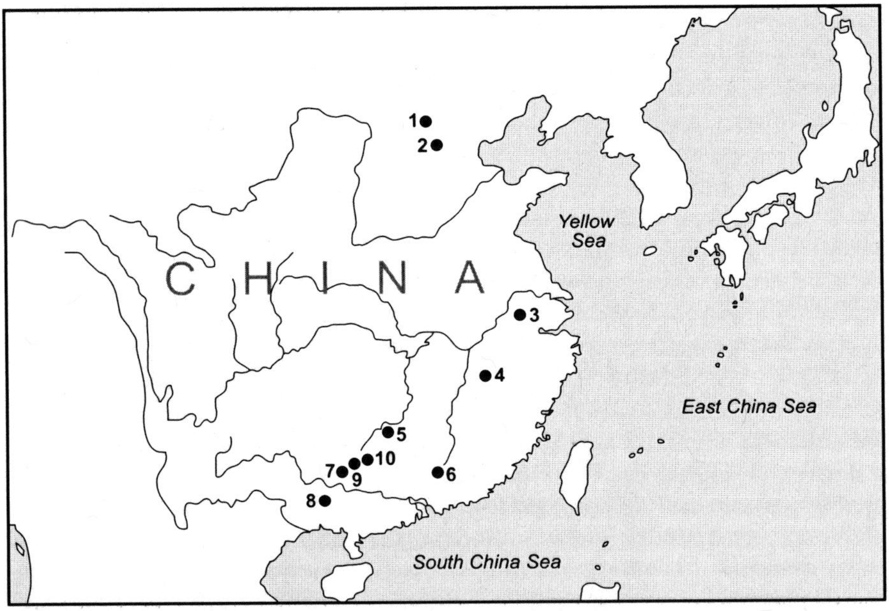

Figure 1.2 Selection of early pottery sites in China (after Zhao and Wu 2000: 234): northern sites: (1) Yujiagou; (2) Nanzhuangtou; southern sites: (3) Shenxiandong; (4) Xianrendong/Diaotonghuan; (5) Yuchanyan; (6) Quingtang; (7) Liyuzui; (8) Dingshishan; (9) Zengpiyan; (10) Miaoyan.

(Feng 2000: 230). This was characterized by population growth, increasingly settled life-ways and the diversification of subsistence beyond hunting to include a greater reliance on fishing and intensive shellfish collecting (Chi 1999; Feng 2000: 229). Finds of quern stones, stone rollers and spade-like implements point to the increasing importance of wild plant foods and resources in all areas, and, at some of the southern sites, this appears to have initiated the shift from gathering, through to cultivation and eventual domestication of wild rice by 7500 BC (Barker 2006: 225; Pearson 2005; Zhao and Wu 2000). In northern areas, the domestication of foxtail and broomcorn millet appears to have taken place only after 6500 BC (Barker 2006: 228; Chi 1999: 10).

In southern China, the largest assemblages have been found at Xianrendong cave site, with hundreds of sherds recovered from eight stratified layers (Chi 2002: 32) (see Figure 1.2). Most fragments are tempered with coarse-grained and poorly sorted quartzite grit, with particles ranging from 1 mm to over 5 mm. The sherds are brown in colour, occasionally dark brown, reddish brown or grey brown. Chi (2002: 33) describes two building methods: 'piece building' and coiling. There are two categories of the former, the first wiped or scored with a blunt fork-like object to create distinctive stripe-like decorations on the inner and outer surfaces of the vessel. Hand-smoothed plain wares form the second category, and all the 'piece built' pots are decorated with a single row of 'V' and 'U' shaped impressions along the inner and outer rims. On the outer surface and below this, there is usually a single row of puncture dots. Most vessels appear to have been round-based jars with a straight rim, but there is little indication of vessel size.

The coil-built pots are round-based, some with a flared rim. They were paddle-stamped to strengthen the pot walls, and fibres wrapped around the paddle left distinctive cord marks on the pots. Some fragments of the coiled pots had textile impressions of straw or cord-woven mats stamped on the exterior. According to their position in the stratigraphic sequence, the stripe-decorated pots are the oldest, followed by the plain and cord impressed wares, and later by the textile-impressed vessels (Chi 2002: 34).

The assemblage from Yuchanyan cave is much smaller, with only two piles of pot sherds recovered from the lower cultural layers (Chi 1999: 7, 2002: 34). Most are dark brown and tempered with quartz grit. The vessel walls are of uneven thickness, ranging from 5 mm up to 20 mm. One fully reconstructed vessel—a 'wide-mouthed cauldron' (Zhao and Wu 2000: 234)—had a slightly pointed round base with slanted sides that flare outwards to a rounded rim (Chi 2002: 33, fig. 12). The exterior has been paddled while cord-marks appear on both the exterior and interior, in a manner similar to the Xianrendong wares (Chi 1999, 2002).

The limited assemblages from Miaoyan (only five sherds) are similar to the hand-smoothed, quartz grit–tempered plain wares from Xianrendong

(Chi 2002: 34). Early pottery has also been recovered from Zengpiyan and Dayan (Zhao and Wu 2000), and the thick-walled, quartz grit–tempered, low-fired, cord-marked pottery bears resemblances to the Xianrendong cord-marked wares (Chi 2002: 34). At Dayan, the vessel forms are unclear, and at Zengpiyan there are only wide-mouthed, semi-hemispherical pots (Pearson 2005: 823).

Long-term excavations at a number of southern sites enable a detailed picture of subsistence to be reconstructed. Most sites share evidence for diversification of the initial hunter-gatherer economy, and similarities between the sites are so strong that they have been described as a single cultural horizon (Chi 2002: 31). Extensive faunal remains from the adjacent cave sites of Xianrendong and Diaotonghuan point to a heavy reliance on deer hunting, with boar, rabbit, fox, turtle and bird species playing lesser roles. At Yuchanyan, a range of deer species also dominated the terrestrial fauna, which included boar, cattle, rat and birds. There is widespread evidence for the exploitation of aquatic fauna, including fish, turtle, snails and molluscs (Chi 2002), a pattern of increasingly broad-based subsistence practices that embraces much of southern China (Feng 2000).

New relationships with wild resources also emerge during the Pleistocene–Holocene transition, albeit within a broader continuity in the emerging hunter-fisher-gatherer economy (Chi 1999: 10). There is possible evidence for selective 'harvesting' of young pigs at Zengpiyan (Pearson 2005: 823), and at sites across southern China, the substantial accumulations of tool debris, oysters, clams, mussels and other food remains (Feng 2000: 228) hint at more structured patterns of landscape mobility, with substantial periods spent at seasonal resource sites (Feng 2000: 228–229). Increasing evidence for deliberate and often ritualized human burials may also point to an increasingly settled population (Pearson 2005: 824), or perhaps repeated visits to specific places within the seasonal round.

Several sites now have reliable evidence for the rise of incipient rice cultivation (Chi 1999: 10, 2002: 30). Analysis of phytoliths at Diaotonghuan indicates that the population were exploiting changing proportions of wild, intermediate and domestic forms of rice at different points in the cultural sequence, possibly reflecting a shift from the intensive gathering of wild stands, through to cultivation and eventual domestication (Zhao 1998). The cultural sequence at Xianrendong reveals similar patterns (Chi 2002: 31), and widespread use of wild plant resources is also very much in evidence at Yuchanyan. Over 40 plant species were recovered through flotation, including four rice husks. Two of these came from the lower levels in the cultural sequence, perhaps also signalling the very early stages of rice domestication (Chi 2002: 31). However, in reviewing the wider subsistence evidence for southern China, Chi (1999: 8, 10) emphasizes long-term continuity in the underlying hunter-fisher-gatherer economy, in which deer and boar were the primary economic focus, supplemented by bird and aquatic resources. The gathering, cultivation and eventual domestication of wild rice is part of a wider process of economic diversification (Chi 1999: 8), although rice-based agriculture only comes to dominate subsistence significantly later, and possibly after 9000 years ago (Pluciennik and Zvelebil 2008: 475).

Contemporary with these southern 'cave-dweller' sites are over 100 'microlith culture' sites that extend across northern China, and have been dated to between 14,000 and 9,000 BP. Northern pottery sites include Yujiagou and Nanzhuangtou (see Figure 1.2), located north of the Huang He River, where the assemblages reflect the 'typical styles of North Chinese pottery production' (Zhao and Wu 2000: 233 and see Figure 1.1). The earliest northern pottery assemblages are found later than in the south, and are notably different from the round-based southern wares, these being largely flat-based jars (Chi 2002: 34; Zhao and Wu 2000: 235). Pots from Yujiagou are sand-tempered, reddish- or yellowish-brown in colour, slab-built and cord-marked along with what are probably fingernail impressions. Thermoluminescence dating of one sherd places its age around 11,000 bp (Chi 1999: 7). At Nanzhuangtou, the ceramic paste is tempered with quartz, mica or shell, and pots are coil-built and cord-paddled. Some have evidence of sooting, and some resemble small bowls (Chi 2002: 34; Zhao and Wu 2000: 235). This pottery may date to around 10,000 BP (Zhao and Wu 2000: 237). Chi (1999: 7) describes additional early northern pottery assemblages from Zhuannian and Zhejiangying (dated to c. 10,000–9,000 bp). These vessels consist of brown-coloured-quartz- or mica-tempered and low-fired flat-based jars and round-based pots, often found in association with stone mortars and pestles. At the early Holocene (c. 10,000 BP) river-terrace site of Donghulin, Zhao and colleagues (2003: 170–171) note the presence of pottery jars—these are red-brown in colour, sand-tempered, patch-built and unevenly fired plain wares, some with raised nodules around the rim.

During the transitional period from the Pleistocene to the Holocene, the northern sites are also characterized by evidence of increasing economic diversification, which includes hunting (primarily deer, boar, antelope, wild horse, cattle and donkey) and evidence for increasing reliance on aquatic resources. The presence of stone mortars, saddle querns and pestles points to widespread plant use (Chi 1999: 9), but the details of which plants were used, and whether these were cultivated or early domestic variants have yet to be established (Zhao et al. 2003: 173). To date, evidence for early agriculture has yet to be identified at northern microlithic sites (Chi 1999: 8).

There has been considerable debate as to the practical uses of early pottery in China, although there appears to be general consensus that the early mineral-tempered pottery was invented in hunter-gatherer societies (Zhao and Wu 2000: 236), either just prior to, during, or shortly after the onset of changing climatic conditions at the end of the Pleistocene (see dates above), and that it was primarily used for the preparation of food, perhaps over open fires, as evidenced by the sooting of several vessels (Zhao and Wu 2000: 237–238). Zhao and Wu suggest that the earliest southern pottery was used primarily for cooking wild or early forms of domesticated rice, but they also note that pottery appears alongside evidence for an expansion in the exploitation of snails, clams and other aquatic resources, all of which could have been boiled in ceramic vessels (Feng 2000: 228; Zhao and Wu 2000: 238). In addition, there is pottery but no evidence for plant domestication at the southern site of Zengpiyan (Pearson 2005: 823), and the early northern pottery complexes emerge in plant-using hunter-gatherer societies, but whether this extends to cultivation of domesticated variants remains as yet unclear (Chi 1999: 8; Zhao 2003).

Early pottery assemblages may best be understood as multi-purpose food-preparation vessels, perhaps forming a technology that improved the adaptive ability of prehistoric populations (Zhao and Wu 2000: 233). Kuz'min (2003: 2) also stresses the functional value of ceramic vessels, and ranks pottery alongside agriculture in its ability to transform subsistence and social practice. Chi (1999: 8) argues that pottery emerged during a period of environmental change, and was an essential element in the long-term process of economic diversification that characterized Upper Palaeolithic hunter-gatherer responses to the Pleistocene–Holocene transition. However, the accepted dates for the earliest round-based, cord-marked pottery complexes at Yuchanyan (17,210–15,660 BP) (Keally et al. 2004: 347) indicate the somewhat earlier invention of pottery, possibly within the last Pleistocene Glacial Episode of Eurasia (Zhao and Wu 2000: 239).

Others have attempted to reconstruct hunter-gatherer social dynamics in order to understand how pottery emerged. In a recent analysis of selected early pottery sites in southern China, Pearson (2005) rejects Hayden's argument that pottery was invented as a 'prestige technology' in hunter-gatherer societies where an egalitarian ethos was starting to break down as aggrandizers sought to increase their social status through competitive feasting (Hayden 1995; Rice 1999, and see discussions below). Instead, he argues that there is little evidence that early vessels served 'special purposes' in what appear to have been largely egalitarian hunter-gatherer societies (Pearson 2005: 826). Vessels have minimal decoration, occur in small quantities, and appear to have been used for food preparation in increasingly sedentary societies, where the main pottery-oriented activities 'centred around women and children and the hearth, such as cooking, firewood collection, water carrying, food consumption, and habitation' (Pearson 2005: 826–827). Hayden has recently countered these suggestions by pointing out that pottery may have arisen as an important new technology for producing prestigious and highly valued substances, including fats, lipids, grease or fermented beverages, all of which could have been central to ritualized feasting events (see Hayden's foreword to this volume).

Questions pertaining to the wider origins of pottery technology in East Asia remain unanswered (see the general models of pottery dispersal outlined below). Chi argues for a common source for all the early pottery complexes in East Asia, but is not yet able to identify the precise geographic point of origin (2002: 35). In contrast, Keally and colleagues (2004: 349), and more recently Kuz'min (2006), have highlighted the major differences in the forms and technologies of early pottery complexes in East Asia. Common origins resulting from migration or technological exchange are rejected, and they conclude that southern China formed one of three *independent* centres of pottery invention, along with Japan (Kaner, this volume) and the Russian Far East (Zhushchikhovskaya, this volume).

The very early emergence of pottery in China touches on a number of questions and debates that are central to the pursuit of more general understandings of ceramic innovation in hunter-gatherer societies. In the remaining sections of this introductory chapter, we examine the social, symbolic and practical factors that are likely to have affected the emergence and dispersal of pottery technology across northern Eurasia. We conclude with an initial attempt to synthesize the material presented in the individual chapters in the form of preliminary models, which we anticipate, will serve as a catalyst for further research on ceramic origins in northern Eurasia (Figures 1.3, 1.4).

Ceramic Innovation and Social Evolution

Our understandings of the history of technology and of the broader significance of technological change have long been conditioned by social evolutionary notions of progress that emerged in the middle of the 19th century and dominated archaeology and anthropology until the 1960s (Childe 1925, 1929, 1936, 1937, 1939, 1941, 1942, 1951, 1957, 1963; Engels 1884; Morgan 1877; Tylor 1871 and for critiques, see Pluciennik 2005 and references therein; Rudebeck 2000 and references therein).

Following the formative work of Tylor (1871), Henry Lewis Morgan included pottery making as a major defining criterion of his famous social evolutionary scheme (1877: 12–14) in which the presence of pottery was a distinguishing feature between lower barbarism and upper savagery. It is worth noting that Morgan did not make any specific links between pottery and agriculture; rather, the invention of pottery identified a separate stage in social and technological progress of mankind from savagery to barbarism. Agriculture was held to occur later, heralding the advance from middle

Figure 1.3 Ceramic innovation and prehistoric social context: Gordon Childe argued that 'pots were generally made by women and for women, and women are particularly suspicious of radical innovations'; many of the first ceramic vessels appeared to be deliberate copies of earlier basketry technologies, ensuring that the earliest pottery 'came to look less new-fangled and outlandish to the prudent housewife!' (Childe 1939: 105).

to upper barbarism (see Rudebeck 2000 and Pluciennik 2005 for further discussion of the significance of Tylor and Morgan's intellectual legacy in archaeology and prehistory).

It was, in fact, in *Pre-Historic Times* that Sir John Lubbock (1865) drew explicit links between growing crops, taming animals and the invention of pottery, all of which, he argued, formed defining features of the Neolithic (see Clark 1980; Daniel 1943: 64, and Dolukhanov et al., this volume). In western Europe, his argument for the coeval occurrence of ceramics and agriculture went on to enjoy enduring influence, thanks to them being taken up in the many writings of Gordon Childe. For Childe, pottery marked 'the earliest conscious utilization... of a chemical change' (1939: 101), embedded in the existing social traditions and sexual division of labour (Childe 1939: 101–105, and see Figure 1.3). Moreover, pottery was argued to have close links with agriculture, together forming a

Neolithic 'package', which defined a major advance in the march of technological and social progress, as well as the rise of new kinds of settlement and urbanism. The Neolithic was defined as a transformation of global significance, a 'veritable revolution' in human history. Moreover, it was a revolution that advanced outwards from its origins in the Near East, to encompass adjacent areas of Europe, part of the wider emergence of complex urban and agricultural societies. Defined through the concept of *ex oriente lux* (the light from the East), this process was argued to represent the gradual spread of new and progressive forms of social life (Childe 1925, 1929, 1957; see also Clark 1980; Daniel 1943, 1964; Zvelebil 1986a; see Budja, Dolukhanov et al., this volume, who discuss this issue further).

The powerful images of a Neolithic Revolution became so influential that archaeologists continue to find it difficult to disconnect the history of ceramic technology from notions of a Neolithic 'package', or from more general processes associated with the arrival of agriculture (e.g., Ammerman and Biagi 2003; Ammerman and Cavalli-Sforza 1984; Piggott 1965, etc; for a critique see Barker 2006; Zvelebil 1984, 1986b, 1989). The notion of a Neolithic 'package' has also proved so resilient that evidence of early pottery has even been taken as proxy data for the practice of agriculture. For example, Ammerman and Cavalli-Sforza (1984) developed their influential demic diffusion model for the origin of Neolithic dispersals into Europe (see also Ammerman and Biagi 2003) partly by classifying pottery-using sites in eastern and northern Europe as agricultural communities even where there was *no clear evidence* of farming (Zvelebil 1984, 1986d, 1989). Similar assumptions continue to influence research into the transition to agriculture (i.e., Ammermann and Biagi 2003; Gkiasta et al. 2003).

These biased treatments of the archaeological data generate several problems: first, they have skewed perceptions of the pace and pattern by which farming practices dispersed across Europe; second, they tend to reify and perpetuate idealized images of foragers and farmers, each epitomized by categorical differences in their material culture; third, they inject an implicit directionality into the adoption of new technologies by hunter-gatherer societies—even where use of pottery by hunter-gatherers is acknowledged, it is assumed to have originated in farming societies, and subsequently been adopted by foragers. Most importantly, however, the assumptions grounded in the standard definition of the Neolithic continue to obstruct reconsideration of the roles of hunter-gatherers in the 'independent' invention and dispersal of early pottery.

In summary, social evolutionary assumptions continue to linger at the core of current understandings of the transition to agriculture, where they reinforce the notion that farming emerged in core areas of early civilization in the Near East. At the same time, they also cast the hunting and gathering populations of the early post-glacial world as societies occupying a northern hinterland into which the innovations of farming and pottery were readily accepted as inevitable exemplars of social and technological progress.

In Russia, and other East European nations, archaeologists working in the Marxist Soviet tradition also drew on general models of social evolution but reached rather different conclusions (i.e., Gurina 1956, 1966; Pankrushev 1978; Ravdonikas 1940; for critique see Jacobs 1995; O'Shea and Zvelebil 1984; Rudebeck 2000). In his seminal work on *The Origin of the Family, Private Property, and the State*, Engels (1884) drew more heavily on Morgan's evolutionary scheme to outline his own vision of social progress, which would pass through a series of clearly elucidated formations that were explicitly defined in social terms, beginning with the primordial communal and egalitarian society of hunter-gatherers and ending with a communist society. This went on to form one of the foundational theories within Marxism, and once formally codified, was imported into archaeological research agendas where it shaped interpretation of diverse archaeological evidence, and remained largely unchallenged until the collapse of Communism in the late 1980s (Kohl and Fawcett 1995).

Liberated from the western social evolutionary notions that linked pottery to farming, but imprisoned within the Marxist canon, it has become standard practice to define the Neolithic in eastern Europe by the use of pottery, and not by the development of farming: a view that ensured greater attention was paid to the appearance of pottery among hunter-gatherers in the East than in the West of Europe (see Dolukhanov et al. in this volume for further discussion).

The Prehistory of Eurasian Hunter-Gatherer Ceramics

The deliberate firing of clay to create cultural artefacts has a much older ancestry than the first production of ceramic containers (Rice 1999: 4). Some of the best-known evidence for early ceramic objects has been recovered from the Upper Palaeolithic site of Dolní Věstonice, in the present-day Czech Republic. Remains of around 2,000 fired clay figurines including the 'Venus' figurine have been recovered. Production of ceramic images appears to have been a widespread practice amongst Ice Age hunter-gatherers. In Central Europe, figurines are regularly found at sites of the Pavlovian culture, and date between 28,000 and 22,000 BP (Oliva 2005; Soffer and Vandiver 1997; Soffer et al. 1993), and fired clay and figurines have been recorded in Greece (Pawlikowski et al. 2000). Moreover, this tradition appears to have continued well into the Holocene (Kashina, this volume).

There is general consensus that the earlier figurines were produced for ritual and symbolic purposes (Hayden 1990, 1995, 2003; Pettitt 2006; Verpoorte 2001; White 1992), perhaps manifesting evidence for the inception of social and symbolic behaviour in human cognitive evolution (Hayden 2003; Mithen 1990). Hayden (2003: 126), for example, locates the emergence of the first art and other forms of symbolic expression in the social context of emerging inter-personal tensions that characterized 'transegalitarian' hunter-gatherer societies who were able to store and redistribute increasing quantities of plant and animal resources, leading to social competition, the appropriation of food harvesting technology and of harvesting locales, and to the emergence of high-ranking individuals, a sort of incipient social elite.

The Upper Palaeolithic figurines—in fired, unfired, or dry-baked clay and acknowledged to have primarily ritual and prestige significance—were not associated with the production of fired clay pots. Within Upper Palaeolithic Europe, the lack of evidence for the transfer of fired clay technology from making clay figurines to making pottery poses an interesting question of function, knowledge or technology. One possible explanation may be that, rather than risk a shift to the development of a novel ceramic vessel technology (Rice 1999: 4–5), people preferred to continue with the well-established tradition of making baskets and plant fibre containers, the evidence for which was found in favourable preservation conditions at sites such as Dolní Věstonice and Pavlov (Andovasio et al. 1996; Oliva 2005; Soffer and Vandiver 1997).

Evidence for the first pottery starts to emerge in the late Upper Palaeolithic societies of East Asia, with fragments of early ceramic vessels recovered from sites in China, Japan and the Russian Far East. New dates now suggest that the Chinese material is the oldest in this region (see Keally et al. 2004, and above), and the limited evidence also points to localized emergence of distinct container traditions in each of these diverse areas (Kuz'min 2006), rather than the spread of one technological innovation from a single point of origin (see Budja, Cho and Ko, Hommel, this volume; and Akazawa 1986; Feng 2000; Kuz'min 2002; Kuz'min and Orlova 2000; MacNeish 1992; Zhao and Wu 2000; Zhushchikhovskaya 2005). In Korea (Cho and Ko, this volume), most of eastern Siberia (McKenzie, this volume), and in western Siberia (Chairkina and Kosinskaia, this volume), the dates fall into the early Holocene, after which there appears to be a rapid spread into northern and eastern Europe (see Dolukhanov et al. 2005; Hommel, this volume, and below).

These extensive data provide unequivocal evidence that hunter-gatherer societies of the Upper Palaeolithic invented the first pottery, and that they were also responsible for the early westward dispersals of ceramic technologies during the Pleistocene–Holocene transition. As such, the history of early hunter-gatherer ceramics in Eurasia contradicts Soviet Marxist definitions of the Neolithic, where pottery was regarded as an invention of post-glacial hunter-gatherers. Moreover, the Upper Palaeolithic dates for early first pottery generate fundamental problems of 'periodicity' by rendering obsolete the established form, sequence and content of key stages in human cultural development (see McKenzie, Zhushchikhovskaya, this volume; and see Feng 2000: 229–230 for related debates in Chinese archaeology).

The data also contradict the western notion of a 'Neolithic Revolution' in which pottery forms part of an inter-linked dispersal of agriculture and pottery. More interestingly, the westward spread of dates for early pottery also points to the existence of a larger east–west ceramic dispersal event, which originates in East Asia, extends across eastern and western Siberia, and eventually arrives into eastern and northern Europe. These emerging insights into the history of early ceramics in northern Eurasia therefore break open the categorical distinctions between hunter-gatherer and farmer societies that were established by European archaeologists in the 19th and 20th centuries.

To conclude, there is a fetching irony in the fact that Childe's concept of *ex oriente lux* ('the light from the East') still captures poetically the key features of this ceramic dispersal: the origins of pottery are undoubtedly *ex oriente,* but the notion of East now extends further in time and space to encompass the Palaeolithic hunter-gatherer societies of East Asia (present day China, Japan and the Russian Far East), rather than early Holocene farming communities of Anatolia.

Understanding Ceramic Technology as a Social 'Practice'

This history of material culture is one of the fields of enquiry to which archaeologists have made a particularly important contribution. Technologies have been explored from divergent perspectives: innovations have been regarded as the driving force of culture change, with discoveries marking milestones on humanity's general path of progress (Sherratt 1981, 1995, 1997). The form and content of new technological traditions, especially those of hunter-gatherers, have also been argued to reflect the more general functional requirements of populations as they seek out means to adapt most efficiently to the constraints of local environments (Torrence 2001). A third perspective on technology emphasizes the role of material culture as a historical phenomenon: new technologies become 'embedded' in social life. Their production and use eventually constitute cultural traditions whose form and content are conditioned as much by subjective 'technological choice' as the need to find 'optimal' solutions to specific adaptive problems (Lemonnier 1993).

Our attempt to understand the dispersal of ceramic technology among 'innovating' hunter-gatherers demands that we not only move beyond debating categories such as the Neolithic, but that we also develop ways in which we can navigate a balance between acknowledgement of the broader ecological and climatic changes at the Pleistocene–Holocene transition that provided the general environmental setting for the invention and early dispersal of pottery, and the spheres of ideas and performance that led to 'social construction' of ceramic technological traditions.

Our point of departure is to suggest that the hunter-gatherer lifestyle had its own characteristic features and dynamic relationships, and that these provide a shifting set of opportunities and constraints into which this new technology extended and expanded as 'active' elements of materiality. As such, the dispersal of ceramic technology can best be understood as a social tradition; a set of technological practices linked historically through time by inheritance based on social learning, and whose physical composition and patterns of take-up and use were historically contingent. We argue that the patterning of such transmissions needs to be researched and understood in its own terms.

In particular, when occupying the privileged position of being able to look back over the history of pottery, notions of 'technological choice' (Lemonnier 1993) are useful in reminding us that there was no doubt consciousness of alternative sets of pathways and traditions, rather than a deterministic and unilinear sequence in the invention and adoption of pottery. A sense of alternative technologies and choices also reminds us that the paths of technological change were local traditions grounded in social learning and passed from one generation to the next, and that we should not expect uniformitarian sequences, but be attuned to the role of human intentionality and habit in each of those stages. In particular, it is this sense of dynamic interplay between the techniques of pottery production, social practices and the wider material and environmental constraints facing prehistoric populations that we would like to draw out in the rest of this introductory chapter.

Understanding the role of pottery as a social practice starts with consideration of the production sequence. At its most basic level, the creation of ceramic vessels demands knowledge of how to obtain suitable fuels, clays and tempers, skills in mixing the clay paste, and in forming, decorating and drying the vessel, and its successful firing (see references in Rice 1999: 4). In the wider literature, emphasis has been placed on different aspects of this sequence of steps. Arnold's classic cross-cultural study explores the practical challenges and constraints that potters need to respond to and overcome in order to maintain a successful tradition of ceramic production (Arnold 1985, and see Zhushchikhovskaya 2005, and this volume). These external factors include suitable sources of clay and temper, appropriate climatic conditions for working and drying the clay, and the need for significant periods to be spent in one location in order to see the production sequence through to completion.

It is clear that several aspects of the ceramic tradition may be difficult to integrate alongside characteristic features of the 'classic' hunter-gatherer lifestyle (Arnold 1985), reinforcing social evolutionary notions. However, recent years have also witnessed an increasing appreciation amongst archaeologists and anthropologists of the broader range of variability that can be found within hunting and gathering societies (Arnold 1993; Ingold et al. 1988; Keeley 1988; Kent 1992; Layton 2001; Leacock and Lee 1982; Lee 1992; Panter-Brick et al. 2001; Schrire 1984; Shott 1992). In particular, there is a growing realization that the historically documented range of hunter-gatherer societies, and especially those that were often used as general models (San, Aborigines) of band societies past and present, probably represented only a small subset of the diverse hunter-gatherer societies in prehistory (Zvelebil 1998b). Many factors identified by Arnold (1985) and others may well have rendered pottery production difficult, though not impossible, even in some of the most mobile and egalitarian of hunter-gatherer groups.

However, at the other end of the 'foraging spectrum' (Kelly 1995) where hunter-gatherer populations occupy rich and seasonally predictable resource niches, for example, along rivers, lakes and shores, populations are often relatively sedentary for significant periods of the annual round. The classic ethnographic examples of 'complex' hunter-fisher-gatherer adaptations are found along the Pacific Northwest Coast, where kin-groups occupied substantial long houses over the winter months, and invested heavily in a diverse range of specialist technologies and highly-decorated material culture, the elaboration of which was spurred on by the politics of competitive feasting and ceremonial display (Aldenderfer 1993; Hayden 1990, 1995, 2002, 2003; Sahlins 1972).

On the Northwest Coast, serving vessels were generally made of ornately carved wood, but in other world regions—especially those with drier climates and better clay deposits—a similar intersection of sedentary living and increased social competition might easily have promoted the emergence of pottery and encouraged its integration into long-term community tradition. We can readily conclude that the pottery craft is *potentially* compatible with a wide range of hunter-gatherer adaptations, especially those that include some degree of seasonal sedentism.

Equipped with a better understanding of the considerable range of variability in hunter-gatherer adaptations, we began to appreciate the potential ease with which early pottery traditions might have been accepted and incorporated into local traditions. We also need to examine how pottery was integrated alongside a pre-existing repertoire of activities, where it may have formed an elaboration, transformation and eventual replacement of older container technologies. The wider literature on the emergence of pottery (Arnold 1985; Barnett and Hoopes 1995; Brown 1983; Brown 1989; Rice 1999; van der Leeuw and Pritchard 1984) outlines several antecedent forms of technological practice that may have been associated with the inception of pottery. These include the widespread Palaeolithic figurine tradition of dry and fired clay which also continued after the invention of pottery (Kashina, this volume). The coil-building of pots may have represented an elaboration in the rolling and joining of clay strips to produce vessels, rather than simple human forms. Slab-building methods used in some early pottery traditions may represent the extension of an 'architectural' tradition originating in adobe house construction, and later being transferred across to the production of containers (Rice 1999; Budja, this volume).

More often, there is widespread evidence for the more general 'ceramicization' of earlier textile technologies: basket, cord and net impressions are frequently found on early pottery, and may have been used as practical moulds into, or over which, wet clay was pressed (Rice 1999: 26; Zhushchikhovskaya 1997, 2005: 21–24, this volume). Textiles, baskets, boxes, skin bags and other soft technology containers may also have provided more general inspiration for the forms of early clay vessels (Sherratt 1997: 366–367; Wengrow 2001), the rhythmic and embodied movements of turning a basket and weaving in new lengths of grass or reed (Ingold 2000: 39–48) not dissimilar to the process of adding new coils to a pot. There may also have been more general cognitive associations between the production and decoration of basketry, and the imitation and incorporation of these practices into later pottery styles (see Ortman 2000).

The extensive literature on pottery performance characteristics has demonstrated significant variations in the ability of different kinds of pots to conduct heat, resist thermal and mechanical stress and perform a range of other practical roles requiring different degrees of porosity and permeability (Bronitsky and Hamer 1986; Reid 1989; Rice 1999: 26–31; Schiffer and Skibo 1987, 1997; Skibo et al. 1989; Tite and Kilikoglou 2002). For example, fibre-tempered pots conduct heat less than mineral-tempered vessels, but might be better suited as lighter, easily made, expedient containers (Schiffer and Skibo 1987) used for stone boiling rather than direct heating (Reid 1989).

Rice also notes the importance of shape on subsequent performance characteristics (1999: 30). Round, pointed-based and porous vessels can stand better the stresses of thermal shocking, while angular vessels are more liable to fracture when subjected to sudden direct heating. For this reason, flat-based vessels may have been used only for stone boiling, which subjects the clay vessel to a more gradually heating gradient, or for the dry toasting or parching of seeds or dough. In contrast, pointed-based wares may have initially been used for stone boiling, but served as more effective containers for use in direct heating and sustained boiling over open fires. There is also evidence that decorating vessels in ways that made the surface rougher (e.g., by adding cord impressions, incisions or punctuations) would have made handling heavy pots easier, but also improved both heat transfer and thermal shock resistance (Rice 1999: 30). The most important conclusion here is that the acquisition of practical knowledge about these variations in performance would have provided early potters with abundant scope for 'technological choice' through manipulation of raw material combinations, including specific tempers and clay, and the creation of particular forms and styles of decoration.

Pottery may also have been used by hunter-gatherers as a new storage technology for seal fat, animal foods and mixtures of vegetal carbohydrate foods and animal proteins. The latter may also have been subjected to deliberate fermentation processes and may have represented early attempts to develop alcoholic drinks (Edgren 1982; Halen 1994; Lindqvist 1987; Meinander 1954; Nuñez 1990 with references; Rowley-Conwy and Zvelebil 1989; Stilborg and Bergenstrahle 2000; for more general discussion see also Barnett and Hoopes 1995; Brown 1983; Brown 1989; Rice 1999: 26–31, 44–45; Sherratt 1995).

Pottery Adoption: Practical Implications for Hunter-Gatherers

The inherent flexibility of the ceramic craft, its ability to provide containers in various shapes and sizes, and with significantly different performance characteristics, means that the practical roles played by ceramics in hunter-gatherer societies could potentially have been very broad. However, the extensive literature on ceramics tends to stress the practical potentials of pottery as a new form of container technology (Barnett and Hoopes 1995; Brown D 1983; Brown J A 1989; Rice 1999: 8–12). In particular, pottery has been argued to confer some rather direct adaptive benefits on individuals and communities taking up its use, contributing to the steady dispersal of a superior technology. At the same time, related studies also demonstrate how the use of pottery brings with it some practical costs, including the challenge of integrating pottery production within local behavioural contexts, and alongside other container technologies.

Use of ceramic vessels has the potential to improve diet, nutrition and general community health in several crucial ways. The preparation of soups enables weaning infants to digest a wider range of detoxified foodstuffs, and provides gradual exposure to new and potentially dangerous foods (Barnett and Hoopes 1995; Meiklejohn and Zvelebil 1991). There are also benefits for adults as meat juices and lipids are not lost into the fire but are retained. Analyses of organic residues from hunter-gatherer pottery, such as the Ertebølle in Denmark, have revealed that fish and vegetable soup was the most likely content of some of the vessels (Andersen and Malmros 1985; Craig et al. 2007; Halen 1994).

Fish broths prepared in pots retain omega-3 lipids, regular consumption of which increases general female fertility and improves child survival rates

(Olsen and Secher 2002). Finally, when used for slow simmering, pots can be left unattended for lengthy periods, freeing up the time of short-ranging individuals from fireside food preparation duties. This would create opportunities to increase investment in child nutrition, as well as to conduct other food preparation and processing activities, which would further improve quality and quantity of the diet.

These factors provide strong grounds for suggesting that in prehistory, individuals, kin-groups and communities taking up use of the pottery craft would probably have gone on to enjoy better nutrition, general health and infant survival rates. These improvements would, in turn, have had major 'population consequences' (Shennan 2002: 100–123), including increased population densities, especially in areas rich in aquatic resources (Meiklejohn and Zvelebil 1991; Zvelebil 2000b), and may also have led to potential increases in community regionalization and territoriality (Shennan 2002: 123).

Security of the food reserves required through the leaner months of the year may also have been improved through the use of ceramic vessels, which are harder for rodents and other pests to chew into. Pottery might also be better for sealing and separating supplies (Nuñez 1990). For example, around the Baltic Sea basin, there is widespread evidence for the use of prehistoric pottery to store and transport seal fat and other marine products, with some of the larger vessels reaching capacities of 35–45 litres, and even up to 70 litres (Nuñez 1990 with references; Stilborg and Holm, this volume). However, the importance of storage in hunter-gatherer societies extends well beyond a means for surviving seasonal shortfalls, and has profound implications for social structure, territoriality and economy (Rowley Conwy and Zvelebil 1989). Strategies of storage present scope for political manipulation and for development of notions of ownership and redistribution, generating tensions within and between communities, and placing pressures on the egalitarian ideology of sharing that is believed to have been the common form of ideology among non-storing hunter-gatherers (Bird-David 1990, 1992; Ingold et al. 1988; Woodburn 1982).

While the adaptive advantages of pottery (Rice 1999: 20–25, 28–29) are rather self-evident, only populations that were able to integrate the lengthy ceramic production sequence into wider patterns of behaviour would have been able to benefit from pottery's practical advantages for improving diet and health. One of the biggest challenges in maintaining a viable pottery craft is the degree of community mobility: pottery is both fragile and heavy, and baskets may be lighter, durable and easier to carry. If we adopt the perspective that pottery production combines a series of sequential steps, then it is clear that several of these require a sustained and predictable presence in one location (see Eerkens et al. 2002 for further discussion).

It is clear that some ethnographically-known hunter-gatherers may move around so much that production and transport of pottery may simply be too difficult to accommodate alongside such mobile subsistence practices. The presumed absence of pottery-use among hunter-gatherers is also bolstered by social evolutionary thinking, which draws categorical distinctions between foragers and farmers. For example, Childe argued for a single and coeval emergence of farming and pottery within the context of settled village life, to which Hodder (1990) added the concept of *domus*, where the house, together with equipment such as pottery, were thought to be associated with the female sphere and with symbolic notions of taming and enculturation of the wild.

Closer reading of the technological literature indicates that in suitable environmental areas characterized by predictable periods of dry weather, the production, drying and firing of pots can be conducted within a few days or weeks, and does not require several months to be spent in the same location. Once considered to be no more than an 'ethnographic oddity' (Rice 1999: 28), evidence for the production and use of pottery by mobile, non-agricultural communities is widespread. For example, from an ethnographic sample of 862 societies, 103 (12%) were pot-making non-sedentary societies (Arnold 1985: 120). In addition, extensive variability in hunter-gatherer settlement and mobility patterns has been recorded in the literature. High mobility is one end of the spectrum; in contrast, most hunter-gatherers spend at least some periods in a single location, especially where resources permit. Interestingly, reviews of early pottery sites in a global perspective indicate a general lack of evidence for sedentary settlement:

> ...early pottery sites... typically lack evidence for full, year-round, permanent, sedentary settlement and housing, traits that were traditionally assumed to be necessary precursors to the widespread adoption of pottery. (Rice 1999: 24)

On the other hand, the archaeological record does include substantial evidence for the practice of more complex 'logistic' patterns of mobility, where task groups are sent out from base camps to specialist resource collection points, leaving significant sectors of the population residing in the same locations for lengthy periods. The pots would be heavy and breakable and so not easily transported from production sites to locations where the pots might need to be used. Instead they could be produced locally at the harvest site, and then cached for later use as either processing vessels or storage containers.

At this point it is worth revisiting notions of 'technological choice' to remind us that there would also have been knowledge of alternative traditions.

Barnett (1939, 1955), for example, provides ethnographic documentation of the extensive but 'perishable' processing, storage and cooking technologies used by winter-village Salish hunter-fisher-gatherers on the Pacific Northwest Coast. Aquatic resources played a major role in the local diet, and their storage and redistribution formed the central fabric of social life. Fish and shellfish were dried, smoked and roasted, fish oil was rendered in a variety of ways and stored in wooden boxes, as well as in sea lion and seal bladder containers. However, in the absence of pottery, most cooking involved stone boiling in bark containers, wooden 'canoe' shaped vessels, and in a variety of bentwood boxes whose sides were sewn or dowelled. A range of other carved and painted shell-inlaid dishes were used to serve food in feasting contexts, with more mundane wares for day-to-day consumption.

These ethnographic examples lead us to suspect that a range of similar 'organic' cooking, presentation and storage methods was widely used amongst the hunter-gatherer societies of the prehistoric past. If this was the case, how and why might pottery have come to form a rival technology to this range of organic alternatives? The more general point here is that individuals do not just jump automatically towards adoption of new technologies that are more adaptive or advantageous for reproductive success—they modify existing traditions and practices. What is it about pottery that makes it different from, indeed advantageous over, rival technological traditions? With many organic technologies able to perform the roles played by pottery, what, other than direct boiling ability, might have made pottery more attractive?

Brown (1989) has addressed this question (see Kaner, this volume), and argues that the most difficult and energy-demanding stage in the pottery production sequence is firing. The ability to fire large number of pots at any one time makes larger-scale production all the more worthwhile: in economic terms the per-unit labour costs of pots falls as larger numbers of vessels are fired together in a single production event, generating increasing economies of scale. In contrast, the per-unit labour costs of rival technologies like basketry, or the Salish boxes described above, remains the same whether a single item, or several hundred are made. Brown concludes that pots, though more fragile, would eventually become more desirable containers as they become relatively cheaper to make as larger numbers were produced. From this perspective, the adoption of pottery is attractive as a time-saving technology.

However, the argument for the inherent advantages of pottery also rests on the assumption that there is sufficient localized *demand* in the population for 'cheap', time-saving containers. This would require a population to be concentrated long enough in one place to take advantage of these potentials (see Eerkens et al. 2002), and would act against selection of pottery over rival technologies in low population density and higher mobility groups, where the key technological advantages of pottery over other rival technologies, would start to fall away. However, smaller and residentially mobile populations would be at a disadvantage, but may find ways of overcoming this by specialization and the development of regional distribution systems, which pool costs at larger social and spatial scales even in smaller populations (Eerkens et al. 2002).

In contrast, for hunter-gatherers practising more tethered patterns of mobility, perhaps revolving around exploitation of seasonal resources, the demand for 'cheap' processing and storage vessels would be heightened by seasonal population concentrations, as well as the need to process and store large quantities of resources quickly and efficiently in case of a later shortfall. In this sense, hunter-gatherer populations practising a delayed return economy would have some of the most urgent and seasonally and spatially concentrated demands for a cheap vessel technology, which may have hastened the uptake of a new technology that provided a range of practical advantages in addition to cheaper unit cost.

James Brown (1989) argues that certain features of the pottery craft award it with fundamental advantages over rival technologies, but only in the context of certain behavioural and ecological settings. Where might these features best be exemplified? Based on this practical argument, the most effective use of ceramics is predicted to be in ecotonal aquatic environments—along coasts, estuaries and in river valleys. In these settings, major resource concentrations of fish, water mammals, shellfish and birds can be more effectively exploited with the use of pottery in two ways: first, for the efficient and larger scale processing of small-sized animals such as shellfish; second, in extracting nutrients in larger quantities, and from a broader range of animals such as migratory fish and sea mammals, especially fats and oils, which were the most limiting resource in foraging economies and could be rendered in large quantities from fish (Speth and Spielman 1983; Spielman 1989; Zvelebil 2000b, in press).

Two further reasons reinforce the argument that pottery's adoption is best suited to aquatic zones: water-edge environments also have the greatest density of plant food in non-cultivated landscapes, and so the processing of plant foods in soups and stews within ceramic pots would be of additional value. Also, estuarine environments and water-edge locations have resources that are layered in time and space and so are likely to be characterized by resource scheduling conflicts (e.g., Brown 1989; Rowley-Conwy and Zvelebil 1989). Pots serve as a time-saving device of greatest value in the exploitation of estuarine and riverine environments (Rice 1999: 21).

This could be linked to literature on technology. Torrence (1983), Bleed (1986), Myers (1986) and Zvelebil (1984, 1986a) discuss ecological, economic and social contingencies of expedient and curated tools. If used for concentrated seasonal processing and storage, ceramics would fit the curated tool category very well: they are prepared in advance and employed at set times for specific needs, in order to increase efficiency and minimize risk in an unpredictable window of exploitation, especially when processing large amounts of food resources within a short time span.

If all this holds true then a strong case can be made that pots do provide key adaptive advantages to water-edge communities who take up their use. Moreover, pottery's emergence and spread may be implicated in the more widespread shift in the subsistence base, perhaps towards the more intensive processing of gathered vegetal resources and small-sized animal foods in relation to changing environmental conditions present at the start of the Holocene.

The arguments outlined above could be integrated into what might be termed the 'estuarine model' for the practical use of pottery, where estuaries in different aquatic environments—coasts, lakeshores or rivers—provided greatest advantages for the adoption and dispersal of pottery. This would suggest that pottery appeared at the start of the Holocene as part of a wider subsistence transition marked by a move away from the late Pleistocene economy of big game hunting, which was practised in small mobile bands, towards the local emergence of seasonally sedentary fishing, hunting and gathering economies, as deglaciation and sea-level changes created rich new ecotones in new, water-edge settings (Rice 1999: 10–11).

As some human populations shifted towards these new niches, economic and settlement patterns were refocused as part of new localized rises in population levels. In this setting, pottery may have emerged as a cheap new vessel technology that had implicit advantages over softer vessel technologies, and ensured efficient resource processing and increased general health and reproductive success, setting in train a circle of increasing resource exploitation, population concentration and growth at water-edge locations, in turn fuelling the demand for cheaper vessels. In addition, these lake- and river-edge settings may also have provided easy access to the essential raw materials of pottery production, namely fuel, water, temper and exposed quaternary clay deposits.

The wider dispersal of pottery would have been driven by the emergence of similar economic and demographic processes across northern Eurasia, creating 'corridors' of opportunity along which knowledge of the new technology could rapidly spread. Many contributors to this volume (for example, Chairkina and Kosinskaia, Cho and Ko, German, Kaner, McKenzie, Pesonen and Leskinen, Stilborg and Holm and Zhushchikhovskaya) do report sharp increases in ceramic production, or its dispersal into new areas, in exactly the period and environmental contexts indicated by the model.

The model also predicts that pottery should disperse more slowly and perhaps not at all, into areas not characterized by expanding sedentary populations occupying water-edge locations. In upland and high latitude zones, we can expect higher mobility and predation upon larger animals, representing a continuation of the Palaeolithic mobile game-hunting economy. The pattern of dispersal also seems to broadly fit this prediction: in western Finnmark and north western Fennoscandia (Skandfer, this volume), and within the uplands of the Central Siberian Plateau (McKenzie, this volume) and Chukotka (Zhushchikhovskaya 2005), we see a later arrival of pottery, perhaps compounded by the difficulties of working clay in colder climates (Arnold 1985; Zhushchikhovskaya 2005, this volume). In northern Fennoscandia, pottery traditions are adopted for about 1,000 years, but then fall out of use again (Skandfer, this volume), and there were significant delays in the adoption of pottery from the eastern to the western shores of the Baltic (Hallgren, this volume), despite the likely existence of inter-community contacts linking these areas.

Alongside this general pattern, many case studies drawn from eastern Eurasia indicate that the origins of ceramic vessel technology were actually much earlier than predicted by the 'estuarine model', and might not be a direct result of environmental changes at the opening of the Holocene. For example, Kaner (this volume) describes how the first pre-Jomon pots of the late glacial are limited in number and very small. However, at the Holocene transition in Japan, we do see a huge expansion in the scale of pottery production, as well as increases in vessel sizes, which suggests that although pottery use may have begun as a form of material culture that had narrow and ritualistic associations, its role changed as it was transformed into a more routine technology, crucial in enabling a shift towards intensified exploitation of a broader range of aquatic and other resources.

In the Russian Far East, Zhushchikhovskaya links the emergence of early pottery on the mainland to the general changes in environmental conditions marking the Pleistocene–Holocene transition, but notes that pottery use in these earlier periods was quite limited and it was not the dominant vessel technology. In contrast, the adoption of pottery in Sakhalin Island occurs much later, and in exactly the kinds of sedentary and aquatic-focused contexts that are predicted by the estuarine model. Cho and Ko (this volume), who discuss this issue, also favour the estuarine model, at least in Korea. In China, however, the earliest use of pottery dates back into the late Palaeolithic, and does not initially correlate with a shift to the use of aquatic resources—but could, arguably, reflect intensified exploitation of small-seeded vegetal foods and/or early use of ceramics for ritual reasons (Wu and Zhao 2003; Zhao and Wu 2000; and see review of the very early Chinese pottery, above). McKenzie (this volume) also notes that the arrival of widespread fishing technology appears to coincide with the occurrence of pottery in most areas of eastern Siberia.

Further to the west, Chairkina and Kosinskaia (this volume) situate the emergence of pottery within a more general shift towards a lakeside and riverside fishing and hunting economy along the Ob' River drainage in western Siberia. Similar arguments could be developed for the adoption and dispersal of pottery use in eastern Europe and the Baltic Sea basin, where the early centres of ceramic production appear to correlate with major river basins, and with coastal and estuarine areas (Dolukhanov 1986a, 1992, 1997; Dolukhanov and Mikliaev 1986; Fischer 1995, 2003; Fitzhugh 1975; Halen 1994; Nuñez 1990; Rimantiené 1992; Zvelebil 1987, 2008; Zvelebil and Dolukhanov 1991, and other authors in this volume).

These emerging insights suggest that pottery vessels may not have represented a *single* kind of technology, as revealed by the kind of focused economic and utilitarian analysis presented by Brown (1989). In contrast, the more complex patterns of dispersal, adoption and incorporation raise the possibility that the extended history of pottery use amongst earlier hunter-gatherers may also have been characterized by periods in which ceramic vessels performed broader social and symbolic roles that extended beyond the provision of an alternative 'cheap' vessel technology.

Hunter-Gatherer Ceramics: Social and Symbolic Roles

Diversity in current approaches to hunter-gatherer technology is neatly captured by Torrence's (2001) distinction between 'macro-' and 'micro'-scale perspectives. Studies falling into the former category tend to examine the composition of hunter-gatherer technologies in relation to success in energy capture, mediation of risk or other measurable currencies. The relative efficiency of different tool kits can then be assessed, enabling identification of general patterns in the way foragers employ technology in different environmental settings. In contrast, those seeking to develop micro-scale perspectives concern themselves with the social and symbolic constitution of technology, and it is to this aspect of hunter-gatherer pottery that we now turn.

Studies of diet and ecology have dominated research into hunter-gatherers, and led to detailed understandings of some of the general features of forager societies (Kelly 1995), but this emphasis on adaptation has also meant that the 'impressive materiality' of their social and symbolic worlds has often been overlooked (Conkey 2001). More recently, material culture studies have emerged across the humanities through a concern for understanding the 'active' roles played by objects and artefacts in the constitution of social life and the creative construction of identities (Tilley et al. 2006). Many meanings will be encoded into material culture due to the 'embedded' nature of production (Sillar 2000; Sillar and Tite 2000); other meaningful associations will be created as objects acquire a 'biography' in social contexts of use, exchange and deposition (Jones 2002).

The notion of 'technological practice' (Dobres and Robb 2000) emphasizes that performance of the pottery craft would have been central to the negotiation of identities, the acquisition of diverse personhoods, as well as forming a more general route of enculturation into specific social, symbolic and gender roles. For example, ethnographic studies of iron smelting in Africa highlight ways in which craft production is embedded in some of the most important aspects of social life, including identity, social status and ritual (Childs 1999). Pottery also has an extended production sequence, and this may have demanded extensive mobility across the landscape, as well as the ability to marshal and manipulate the knowledge and efforts of kin relations to secure assistance in the acquisition of fuel and raw materials. Along with iron smelting, pottery might also be classed as a 'transformative' technology, with firing the most risky stage in manipulating the chemical properties of the clay to produce 'artificial stone' (Childe 1939).

The firing event may have formed a more open and festive public spectacle, with the entire community invited in to observe and perhaps even participate, or else may have been held as a more restricted, hazardous and ritualized practice, with certain non-specialists or genders excluded from close contact. These attempts to broaden understandings of pottery production are useful because it is generally assumed that pots were invented, made and used by women, and this has been used to bolster some of the nutrition and dietary based accounts of pottery emergence reviewed above. In contrast, the *chaîne opératoire* approach adopted here suggests that simplistic gender categorizations may be too crude, especially if vessels are understood to represent the outcome of a complex production sequence, perhaps reflecting 'multiple authorship' (Finlay 2003).

The post-production life of pottery may also have been extensive, passing through many domains of social practice, and ensuring that vessels carried a range of meanings. Given these general observations, it is interesting to note that early ceramic traditions in northern Eurasia are characterized by major distinctions between decorated and undecorated wares (see Rice 1999: 13, 36). Major differences in the degree of decoration also contrast in all areas with the relative simplicity of the vessel form: these are mostly, but not exclusively rounded or pointed-based wares, although some regions also have flat-based traditions of ceramic production (Cho and Ko, Kaner, Zhushchikhovskaya, and other chapters in this volume), possibly reflecting the kinds of functional performance requirements noted above.

At one level, the deliberate addition of decoration on certain vessels may have reflected no more than a 'celebration of the material' (Bloch 1995:

215), and the addition of a further stage in the production sequence involving the careful and rhythmic pressing of design tools into the soft clay (see Kaner, this volume). These designs may have spoken of wider social themes, perhaps personal skill, or simply established forms of practice, without possessing any *specific* meaning. In contrast, decorative traditions may have represented a more *deliberate* investment of time and skill in certain kinds of vessel intended to advertise the status and prestige of its producers, owners and users (Hayden 1995). Whether intended to advertise prestige, or simply an embodied tradition of working, the choice to add specific motifs to only certain kinds of pottery might reflect attempts to arrange symbols in structured sequences, which would convey messages in ways akin to a textual or linguistic 'discourse' (e.g., Tilley 1991).

This discourse could have been participated in, and understood, in different ways by various sectors of the wider community, and perhaps beyond. The symbols may have expressed very general understandings about how the world came to be, and the role of humans and animals within it. More specific sets of meanings associated with the designs may only have been understood by certain age sets, genders or kin groups, and in certain cases, the meanings may have been known only to very exclusive groups, with the pots used only in exclusive ritualized settings where participation was restricted.

The individual chapters document a range of symbolic dimensions to the decoration of pottery. These include use of smaller fish and bird bones, and animal teeth, to make impressions into the wet clay (Chairkina and Kosinskaia, German, Papmehl-Dufay, Skandfer), widespread use of red ochre (Chairkina and Kosinskaia, German, Skandfer, Stilborg and Holm), explicit depictions of waterbird 'messenger' animals, trees and possibly ritual specialists (Chairkina and Kosinskaia, Kashina), potentially expressing central themes in circumpolar cosmology (see Zvelebil and Jordan 1999), as well as the long-standing ceramic figurine tradition (Budja, Chairkina and Kosinskaia, Kashina).

Pottery might also have expressed different forms of social identity, working at broader scales to signal conceptual distinctions between insiders and outsiders. In a classic study, David Clarke linked level of attribute affinity between assemblages to shared identities (more than 60% affinity), contact networks (30–60%), and 'techno-complex' entities (5–30%), in other words, ecologically circumscribed social traditions, expressed by shared technological traits (Clarke 1968, 1979; Hodder 1978, 1982; Hodder and Orton 1976).

A number of chapters in this book examine these nested notions of identity, including the more general roles of pottery in social practice (Kaner), and pottery as a general form of cultural tradition (Papmehl-Dufay, Skandfer) as well as the potential for pottery to serve as a more explicit or negotiated expression of group identities (Larsson). Other chapters examine variation in pottery styles in relation to regional forms of social interaction and exchange, including community exogamy as a means of ensuring dispersal (Pesonen and Leskinen) and/or resistance to adoption of the pottery innovation (Hallgren, Skandfer).

Hunter-gatherer pottery also enters the archaeological record in the form of grave goods (Budja, Chairkina and Kosinskaia, Kashina, McKenzie, this volume). These may have been vessels used in everyday contexts, perhaps with close personal associations, which were then placed with the dead as gifts for some form of afterlife. Conversely, they could have been pots that were made in distinctive ways and intended from the outset to be included in funerary contexts. Halen, for example, notes the special decoration of small-sized pots found in burials at Lillberget in northern Sweden, dated to c. 3,900 BC. Decoration on such pottery was 63.3% more common than that found at the nearby settlement, and included decoration on the interior surface of the pot, not just on the outside (Halen 1994). In this case it seems that mortuary ceramics had different production sequences than pottery made for living persons (see Papmehl-Dufay, this volume).

In summary, many theoretical approaches treating material culture as being 'meaningfully constituted' examine how objects carry significance for those who produce and use them, but they tend not to address or explain how novel technologies are invented, or explore how they are drawn into established forms of social life and disperse further through extended social networks. In contrast, alternative bodies of theory have provided a more explicit set of mechanisms for addressing these issues, and make explicit links between the emergence, elaboration and dispersal of new innovations, and the roles these objects perform as 'prestige' technologies that are open to manipulation by individuals aspiring to enhance their social status (Hayden 1995; Rice 1999: 12, 36, 42; and see Budja, this volume).

There are a complex bundle of 'prestige' arguments, and these can be broken down into different themes. First, Hayden (1995) has argued that early ceramic technology was appropriated or sponsored by a restricted number of aggrandizers within communities. These individuals were always present in human groups but had the opportunity to develop political control/status as part of emerging rank distinctions within the new ecological situations at the start of the Holocene. The archaeology of these practices should include special purpose vessels for preparation and storage, and for the serving of either new special foods, or older special foods in new containers. For storage capacity should be large and visible, food vessels should be decorated in new media (Hayden 1995; Rice 1999).

One flaw with this model is the fact that the pottery production sequence has many stages, and these may have involved many more people (and locations) than could be controlled by one or two individuals. However, the production, use and exchange of pottery could have provided more general opportunities for the negotiation of access, authority and other resources within a wider ritually sanctioned discourse, in which production and conspicuous use of pottery came to enjoy a central place. Those having the skills and practical knowledge would then be able to manipulate that discourse and acquire prestige.

Obtaining decorated vessels could have conveyed status if the items were used to display abundant food stores, or to serve valued food and liquids. Pots might also have served as items which were transacted between individuals or groups, either as a general token, reflecting kin relations, or as a socially sanctioned form of material currency in a more established network of exchange: when displayed openly, decorated vessels would advertise the owner's skill and status in the manipulation of local or regional alliances.

These pottery exchanges may also have facilitated development of social storage in the strong seasonal regimes of northern Eurasia, where alliances provided a means for coping with bad years and unforeseen shortfalls (Halstead and O'Shea 1989). Exchanges may have been made in either balanced reciprocity, or generosity may have been used to undertake political manipulation. Locally made pots would have carried symbols of identity as they were exchanged further afield (see Clarke 1968; Hodder 1978, 1982; Knutsson 2004; Rice 1999; Sahlins 1972; but see Eerkens et al. 2002 for a critical appraisal). In these ways, pots might have served as a *general* prestige good, whose acquisition was open to all with aspirations, and forming a cultural resource which could be acquired, used and exchanged more widely, rather than a more restricted technology controlled by small numbers of aggrandizers.

However, the desire to accumulate prestige need not have played a role in the dispersal of pottery. The ceramic craft might also have spread gradually via routine practice and the cumulative exchange of members between communities possessing knowledge of different forms of vessel technology (Moore 2001; Pesonen and Leskinen, this volume). These contacts could have facilitated the dispersal of new technologies, without the need for individuals to perform aggrandizer roles. Knowledge of pottery would have spread into communities practising 'soft' container technology; contacts may also have led to 'hybridization' of distinct ceramic traditions. For example, ceramic traditions at the frontier between La Hoguette and the LBK are highly distinctive: the technology of pots is clearly derived from the LBK tradition which was dispersed with Neolithic farming populations; however, the pottery tradition was then 'enculturated' into pre-existing hunter-gatherer practices through addition of 'Mesolithic' decorations that had a wider currency on bone and antler artifacts, and which predated the arrival of pottery (see Gronenborn, this volume).

Hunter-Gatherer Ceramics: Understanding Diversity in Technological Traditions

We have argued that the history of pottery can be understood as a form of 'technological tradition'. After their invention in East Asia, these practices were gradually incorporated into the social life of hunter-gatherer communities at different times, and in different ways (Figure 1.4), culminating in a general dispersal event that eventually spanned the entire Eurasian continent. At a general scale, the technology tradition is united by a common

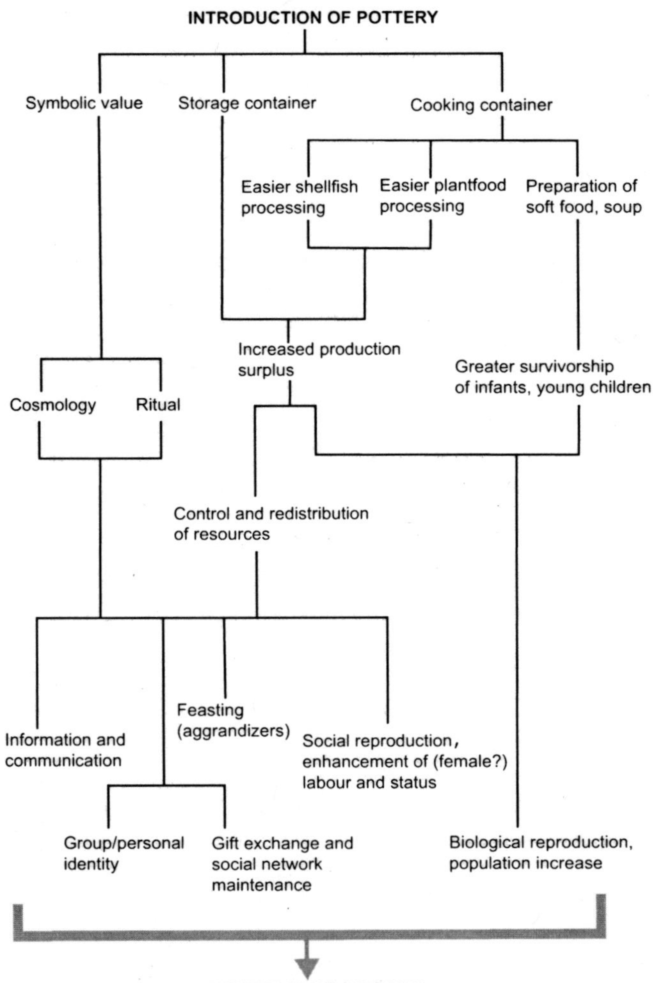

Figure 1.4 An emerging hunter-gatherer ceramic dispersal model: potential practical, social and symbolic dimensions to the adoption of pottery.

production sequence, starting with the acquisition and mixing of moist clay, its shaping, decoration and eventual firing into a durable vessel. In this sense, the history of pottery technology constitutes the transmission of a distinct body of cultural knowledge. In more local contexts, these inherited practices would have been subjected to cumulative modification, either via deliberate attempts at innovation, or via learning error. Local features of pottery traditions, for example, vessel size, degree and composition of decoration, as well as concepts of 'appropriate biographies' (their modes of production, use and deposition) would all have involved a negotiated accommodation between personal creativity, the weight of community tradition, and practical performance requirements.

The process of inheritance might be explored at different scales, and from diverse theoretical perspectives. In the shorter term, ceramic production and use might be viewed as the social 'performance' of an embodied technological tradition (Dobres and Hoffman 1994; Dobres and Robb 2000; Gosselain 1998; Killick 2004; Lemonnier 1993; Sillar 2000), thereby stressing its role in the day-to-day reproduction of authority, identity and social life (Hallgren, Kaner, Larsson, Papmehl-Dufay and Skandfer, this volume). Longer-term perspectives might also be developed through deployment of 'structuration' theory (Barrett 2000, 2001; Giddens 1984) in which the dynamics of pottery craft could be understood as a specific form of structured knowledge, transmitted through the routines of daily practice, and modified by the strategic interventions of group or individual agency (i.e., Larsson and Rzepecki 2002–2003; Zvelebil 2005).

Alternatively, perspectives drawing on 'cultural transmission' theory would involve defining pottery traditions in terms of cultural traits that were passed through social learning and subjected to cultural 'descent with modification' (Boyd and Richerson 1985; Cavalli-Sforza and Feldman 1981; Jordan 2007; Lipo et al. 2006; Mace et al. 2005; Shennan 2002, 2004, 2009). As the transmission process unfolds, new traits might catch on and increase in frequency; other features of the pottery—or perhaps the entire ceramic container tradition—might gradually decline or eventually drop out of use entirely (as pottery appears to have done in northern Finnmark, see Skandfer, this volume).

This 'descent with modification' perspective also specifies how cultural traits might be subjected to various forms of selection. In addition to standard forms of natural selection, which operate on humans in the same way as other biological organisms via their genetically inherited variations, there is another special form of natural selection that acts on individuals via their cultural traditions (Bettinger 1991; Boyd and Richerson 1985). For example, the widespread adoption of new technology such as pottery might result in general improvements to diet and nutrition, ensuring demographic expansion and the successful reproduction of the tradition. Finally, the traditions themselves are also subject to 'cultural selection': after learning one technique from a parent, the offspring may modify it through personal experimentation; at other times they may switch entirely to another method because this is what they have seen others do (Shennan 2002, 2004).

Several examples of 'selection' are pertinent to the discussion of ceramic dispersals: individuals might acquire the ability to coil boiling baskets from their parents, but then switch to the production of pottery boiling vessels either because they find out for themselves that they perform better, or perhaps because pot makers form the local majority, or because those individuals thought to be more prestigious and powerful have begun to practise pottery making. Likewise, individuals might switch from making slab-built to coil-built pottery, or from flat-based to pointed-based wares, for the same kinds of reason. In northern Eurasia, the concept of cultural 'selection' is a potentially useful means to understand how pottery would need to be consciously *accepted* by individuals and communities as they encountered it, especially since the first pottery would almost certainly have had to disperse into social settings occupied by potentially 'rival' organic container technologies like boxes and baskets.

As Hallgren discusses in the present volume, this social acceptance of new innovations may not have been forthcoming in all areas. For example, the new technology may have challenged older stocks of knowledge, practice and authority, and could only be taken up after protracted social negotiation. In addition to the practical challenges of incorporating the production sequence into the hunter-gatherer seasonal round, the roles of sanction and imitation—particularly by local leaders—would have provided powerful tools for accelerating or blocking the spread of pottery into new areas. Recent studies tracking the spread of innovations explore the roles played by innovators, and early and late adopters, but also stress the crucial role of opinion leaders in ensuring that the new technology either succeeds or fails to catch on, despite the fact that it may possess important practical benefits. The risks, costs and complex social dynamics associated with the process, and in particular the central role of imitation, may also result in effective rather than optimal technologies being acquired (Layton 1989, 2000). In summary, inter-community and inter-personal networks of contact are essential mechanisms for the dispersal of new technologies. However, while contacts facilitate uptake, they do not *guarantee* that novel technologies—or optimal variants—will succeed in being selected, accepted and incorporated into pre-existing social traditions.

General Models of Ceramic Dispersal in Northern Eurasia

Understanding of contextual factors likely to affect the uptake of pottery (Figure 1.4), along with the development of a 'cultural transmission' perspective outlined above, enable us to attempt an initial elucidation of the ceramic prehistory of northern Eurasia at several spatial scales. In tracking the continent-wide dispersal of ceramic technologies we can identify several trajectories (Figure 1.5, and see Hommel, this volume, for initial attempts to elucidate an Eurasian ceramic dispersal chronology):

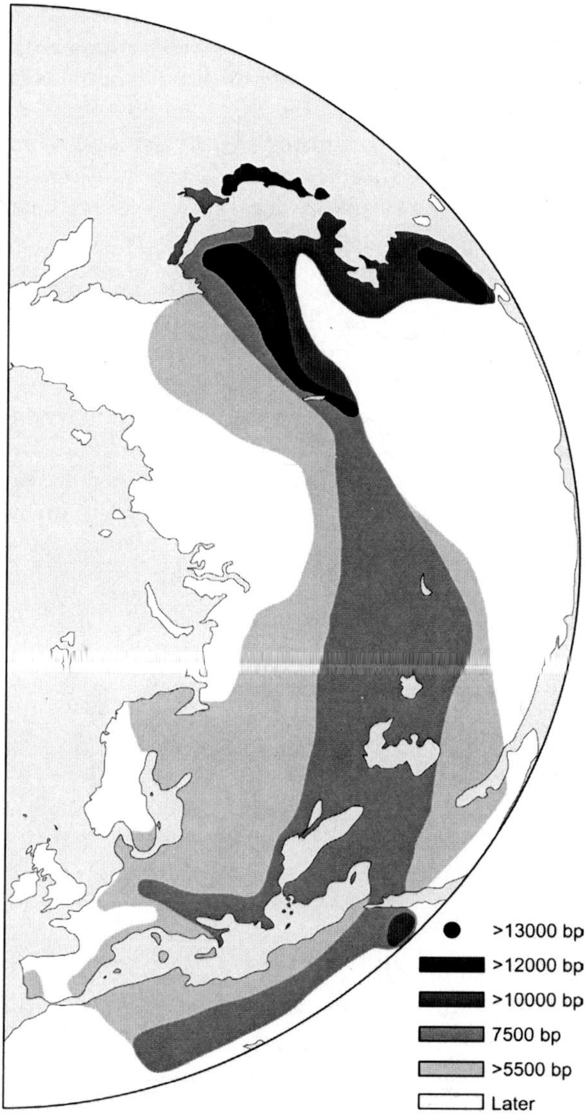

Figure 1.5 Periodization of the spread of early pottery in northern Eurasia: a preliminary model (data from Hommel, this volume).

1a. Single late Palaeolithic origin hypothesis: The '*ex oriente lux*' East Asian dispersal marked by a localized emergence and spread of pottery amongst hunter-gatherer societies, beginning in China around 17,000–15,000 BP, thereafter spreading north to the Amur River (16,500–14,900 BP) and east to Japan (16,750–15,700 BP) (Keally et al. 2004: 349). From there, ceramic production may have reached Korea by 12,000 BP, although this early date has been rejected by some (Kuz'min 2006). The gradual dispersal west across the forest-steppe regions of southern Siberia occurs at the beginning of the post-glacial period c. 11,000—9,000 years ago, and then along the northern shores of the Caspian Sea. Eventually, in the early Holocene, pottery reaches the southern Urals and spreads into Eastern Europe, a very slow process, lasting some 8,000 years.

1b. Multiple origin hypothesis at the Pleistocene–Holocene transition: a case has been made by Keally and colleagues (2004: 349) and Kuz'min (2006) for three independent centres of pottery invention in East Asia. This argument is based on differences in ceramic traditions between the three putative areas of origin: south-central China, the Amur river valley, and Japan; and on the relatively short time-span within which ceramics occurred within all three regions, even though they are geographically distant. A parsimonious interpretation of the C14 dates would data the origin of ceramics in all three regions within the 14th millennium BP (13,700–13,300 uncal BP, or c. 17,300–15,000 cal BP). While the differences in ceramic production, form and decoration may be a result of local, historically situational use of a technological innovation, only further dating results can provide reliable information about the true age and pace of dispersal of ceramic technology. At a slightly later time, another possibly independent centre of ceramic production was emerging in northern Africa, where the production of pottery is attested from about 10,000 BP (Close 1995). North African ceramic traditions may have influenced the further development of the ceramic craft both in the Near East, and in Mediterranean Europe.

2. After c. 7,500 BC (9,500 BP), in the context of early post-glacial environmental conditions, pottery is dispersed further to the northwest, via the northerly route through central Russia, the Upper Volga, into Karelia and beyond, forming various local traditions of pointed-based pitted and combed wares, such as the Sperrings pottery of Finland, and entering the East Baltic and northern Scandinavia by about 5,000 BC (Carpelan 2002; Dolukhanov et al.

2005; Hallgren 2004; Nuñez 2004; Skandfer 2005; and authors in this volume).

3. More tentatively, it is possible that as a part of this general process of ceramic dispersal, the production of pottery also spreads via a more southerly route from Central Asia along the eastern shore of the Caspian south, into northwest Iran and northern Syria. It is clear that contacts between these regions already existed in the early post-glacial period, and according to Matiushin (1976, 1986; see also Clark 1980 for further discussion), the operation of these social networks ensured the introduction of microlithic technology from the Near East into western parts of Central Asia and the southern Urals region. These inter-community cultural borrowings between foragers and farmers may have then been responsible for the introduction of pottery from the Asian North into the Near East—crucially some 1,000 to 1,500 years *after* Neolithic farming practices were developed in the Near East, which generated the early farming period of the 'aceramic Neolithic' in the Near East between about 8,000 and 6,500 BC (e.g., Barker 2006; Moore 1995). North African ceramic traditions may also have been another source of inspiration for the development of early ceramic technology in the Near East.

4. The subsequent spread of farming from the Near East into Europe, from c. 6,500 BC onwards, is already marked by the presence of both Neolithic farming practices and the production of pottery. The spread of these 'entangled' traditions across Europe equates with the classic western definition of the Neolithic as a social tradition (and for some, also as a demographic dispersal, e.g., Ammerman and Biagi 2003; Ammermann and Cavalli-Sforza 1984).

5. Early farming cultural traditions such as the Cardial Ware in south-west Europe, or the LBK culture in Central Europe come subsequently into contact with hunter-gatherer and other early farming communities in the north temperate and northwest parts of Europe, leading to a complex situation of inter-communal contacts and cultural borrowing between people of at least three different traditions: farming Cardial Ware tradition, farming LBK and derived traditions (such as Stroke-Ornamented Ware, or Lengyel Ware), and hunter-gatherer communities of northern France, the Low Countries, northwest Germany and southern Baltic lowlands, some of whom may have already adopted ceramics earlier from the cognate communities in eastern Europe (see, for example, Cahen and Otte 1990; Dolukhanov et al. 2005; Fischer and Kristiansen

2002; Jeunesse 2000; Zvelebil et al. 1998). Philipe Crombé and Detlef Gronenborn address these complex issues in the later sections of this volume.

6. Around the edges of the hunter-gatherer ceramic dispersal, a number of areas remain without pottery for significant periods. Reasons may include the high mobility in upland areas where small water bodies mean low seasonal concentrations of population. This would include upland and interior Scandinavia, the Urals and much of Arctic Siberia. Moreover, these areas may have other material culture media for the articulation of prestige, for example, developed stone/amber working industries and associated 'discourses' regarding these prestige systems (Knutsson 2004, see also Armit and Finlayson 1995 for Britain). For example, the Pitted Ware tradition of coastal southern and central Scandinavia did not disperse far from its coastal core areas because it never seems to have become a prestige technology. Rather, as a low decorated ware, it served as a practical cooking tool, or, when more extensively decorated, it was associated with burial or ritual contexts (Larsson 2003, 2006; Papmehl-Dufay 2003, 2006, see also Papmehl-Dufay in this volume) and so it was linked to ancestry and identity rather than prestige.

What were the key stages in this dispersal? The following general model of ceramic dispersals can be suggested:

Stage (1): Experimentation: We see an extended phase in Japan, China and the Amur River basin of very occasional use. Pottery is a new tradition that becomes embedded into everyday practice very gradually. These vessels tend to be either plain and coarse wares used largely for practical tasks, as, for example, in China, or the Amur region, but may also have been used for ritual and symbolic purposes, for example, in pre-Jomon communities in Japan. In this format, the tradition was subjected to modes of cultural selection and transmission ensuring that it remained a minor practice over many generations and for several hundred years, often as a rough, low-fired and experimental technology. In this period, which falls within the late Palaeolithic (c.16,000–10,000 BC), we tend to see the use of organic tempers, making pots lighter and more portable, perhaps also improving their performance as stone boiling vessels. Frequent textile impressions on pottery surfaces also suggest that basketry was being used for moulding wet clay, hinting at the deliberate fusion of older organic container technologies with new strategies for the deliberate firing of ceramic vessels. Zhushchikhovskaya (this volume) notes the range of very different forming methods used at early pottery sites in the Russian Far East, which suggest localized experimentation, and the partial 'ceramicisation' of older textile practices. It may also be the case that several established traditions converged in the emergence of ceramic technology as a new form of social practice: the 'soft technology' of basket making, the practical benefits of using pottery in areas where small-sized, bulk resources (such as fish, shellfish or grains) were widely available, as well as the incorporation of pottery in aggrandizing displays, social competition or in ritual use.

Stage (2) Intensification: A combination of changing climatic and ecological conditions at the beginning of the Holocene, and the associated shift amongst human populations towards increasing exploitation of rich estuarine ecozones, would have generated exactly the kind of behavioural niche that Brown (1989) identifies as being ideal for the ceramic craft to expand and develop its technological potential as a cheap and effective means of producing processing and storage vessels in seasonally sedentary communities. The increasing selection of pottery traditions unfolds within the context of a benign cycle of population growth along the major water transport routes, the shift to shell processing, and also refinement of the craft towards higher-fired, mineral-tempered vessels of greater functional utility and durability in direct heating. Improvements in technological praxis, and emphasis upon practical use, are then the key features of the stage.

Stage (3) Integration: Pottery making becomes deeply embedded in the social practices of water-edge, delayed-return hunter-fisher-gatherers. We see the ceramic craft begin to take on the identity of a prestige technology whose production sequence and subsequent biographies are incorporated into the communities in different ways. The craft is generally present throughout the population, rather than being sponsored by a few aggrandizers for use in competitive feasting. There is also a move away from incidental basketry impressions that marked an experimental stage towards the heavy use of specific design tools and decoration formats which refer to wider notions of cosmology, identity and belief. As an impressive and central element of hunter-gatherer materiality, the pot traditions begin to convey powerful signals relating to differences in identity and social status. Pottery spreads in the form of a heavily decorated technology that conveys general status on users and their life biographies, but also expresses group identities and refers to broader cosmological principles. The key feature of this stage, then, is added emphasis on social and symbolic aspects of ceramic use as a social practice.

Stage (4) Dispersal and Differentiation: Pottery spreads from its core areas of origin in East Asia and southern Siberia as a finished technology which conveys status, prestige and practical benefits on all who take it up. This process accelerates from about 7,500 BC, with another even more rapid episode of spread occurring at 5,000–4,500 BC. Pottery persists as a heavily decorated prestige technology until other rival forms of material culture take up this role, and at this point, pots return to rough and functional forms. This can be seen, for example, in the abandonment of heavily decorated Pit-Comb Ware ceramics in favour of plain wares of the late Neolithic Pitted Ware culture and the ceramics of the Early Bronze Age, that coincides with the incorporation of non-ceramic prestige items in northern Europe (including early metal forms or their polished stone replicas) and with a greater use of pottery as a storage container for seal fat and other products, and for domestic food preparation (see Edgren 1982; Halen 1994; Lindqvist 1987; Nuñez 1990 with references; Nuñez and Okkonen 1999; Stilborg and Bergenstråhle 2000; Zhushchikhovskaya 2005 for Sakhalin Island). We also see further improvements and refinements, including the emergence of flat-based vessels for parching, roasting seeds and grains and so on, perhaps linked to wider diversification in the resources exploited.

As noted elsewhere, several factors could influence the characteristics and directions of pottery dispersals. First, the uplands and the Arctic—areas of high mobility—conferred less practical advantage upon pottery adoption. Second, if the pots were poorly decorated, they would not present an inviting medium for prestige technology or for communicating social messages. Hence, they would not spread into hinterlands. The rate of adoption of ceramic technology would depend then, on the condition and social traditions of ceramic production in neighbouring regions where pottery was already in use. For example, the limited decorative load of the Pitted Ware culture in coastal Scandinavia may have acted as a disincentive for the uptake of pottery in the Scandinavian hinterland. Third, if there are rival prestige technologies, they can replace the use of ceramics as a prestige and symbolic practice. These factors contribute to complex regional and historically contingent patterns of dispersal and eventual use. Bearing this in mind, the scheme presented above provides merely a broad outline of events, and remains to some extent tentative. Finally, several authors treat another important topic: intercultural contacts and cultural transmission between hunter-gatherer communities and the first farming cultures in Europe. The social context of these contacts was conditioned by the structure of agricultural frontier zones, some of considerable duration, that extended along ecological boundaries between south-east and central Europe, and between central and north-west (Atlantic) and northern (Baltic) margins of northern Europe (Zvelebil 1998a, 1998b, 2000a). Within ceramics as a social practice, the currency of such intercultural exchanges rested in the decorative design, production techniques and pottery forms, to which both foragers and farmers contributed from their own ancestral styles and traditions (see Budja, Crombé, Gronenborn, Nowak, this volume).

Hunter-Gatherer Ceramics in Eurasia: Conclusions and Outlook

We opened this volume by arguing that the origin of pottery was one of the most important questions in Old World archaeology. We went on to stress that fired clay was a Palaeolithic hunter-gatherer invention, and noted how legacies of the social evolutionary thinking of the 19th century, as well as the western notions of the 'Neolithic Package' and *ex oriente lux*, were hindering our ability to understand the full implications of the systematically older dates for early ceramics in East Asia and across Siberia. In turn, these biases hindered appreciation of the crucial role played by hunter-gatherer societies in the subsequent use and long-distance dispersal of ceramic technology across northern Eurasia and into Europe.

The wealth and diversity of evidence throughout this volume also points to the range of very different roles performed by ceramic vessel and figurine traditions. Despite a sense of general unity in the use of firing clay to create a cultural artefact, the chapters outline a remarkable diversity and flexibility in production, decoration and firing sequences, as well as in the potential practical, social and symbolic 'biographies' and roles played by the pottery vessels and figurines: what emerges are vivid glimpses of the creative and multi-contextual incorporation of the pottery craft into very different settings, not the spread of a single novelty or 'tool' possessing strictly prescribed modes of production, use and deposition. Clearly, the pottery craft was not a uniform tradition, but one that possessed an inherent capacity for performing a range of very different social and symbolic roles.

Given this inherent flexibility and diversity, can we trace out a general trajectory of pottery adoption by hunter-gatherers? The evidence presented in the volume tentatively suggests that the earliest vessel technologies went through initial and experimental phases. Here, the preliminary use of pottery was socially contextualized towards either practical or ritual activities depending on local conditions, although the most early vessels tend to be relatively simple in form and poorly decorated. This stage is followed by their deeper incorporation into the fabric of social life which stimulates increasingly intensive decoration and leads on to rapid dispersals across communities generating similar styles over large areas. Eventually, this dispersal is followed by a subsequent break down into diverse regional styles often with more simple decorations. For example, the early pottery phases in Japan suggest the presence of initially very conservative traditions, with pots made as small ritualistic vessels; Phase 4 Jomon witnesses an explosion in their more functional usage and by the Mid to Late Jomon we see the florescence of symbolism and high decoration. Comb Ware and Pitted Ware Cultures also appear to follow this dynamic: early phases tend to be plain wares; we then see a shift to elaborately decorated wares. Pit Comb Ware of this phase is interesting in that it is decorated in a very structured way over large areas, perhaps corresponding to the kind of 'spread horizon' traditionally associated with a population dispersal, and possibly associated with the movement of Finno-Ugric populations from the Urals and eastern margins of eastern Europe westwards to northern Russia and the East Baltic (Dolukhanov 1998; Hajdu 1977; Künnap 2000; Nuñez 1998, 2000; Zvelebil 2001). This regional-scale uniformity gradually breaks down and is replaced by more localized wares. Rather than seeing these later shifts in terms of population dispersals, we might interpret these changes in terms of shifting emphasis in the uses of pottery between more social and symbolic domains on one hand, and more functional storage and food preparation roles on the other.

In conclusion, a new appreciation of hunter-gatherer ceramics provokes an entire rethink of later hunter-gatherer prehistory, and highlights once again the extent to which the modern ethnographic record of hunter-gatherers describes such a small section of former diversity, moreover, one which is impoverished materially and culturally and forced into marginal areas by modern historical developments (Pluciennik 2002, 2004, 2005; Sahlins 1972; Woodburn 1988). In contrast, the material and technological diversity of the Late Palaeolithic and Mesolithic hunter-gatherers was probably far greater than we can even start to imagine. Once again, arising from this rethink, we need to move on from regarding hunter-gatherers as socially and technologically simple, capable only of reacting adaptively to ecological change, in contrast to the developed social and cognitive models of behaviour that we regularly use to explore societies of the farming Neolithic and Metal Ages. While we cannot reject the role of adaptation outright as being an important theme in hunter-gatherer prehistory, we must also be aware that it is such a partial dimension of the human story, and one which does not address issues of motivation and individual intent. In contrast, a new appreciation of the impressive materiality of the hunter-gatherer world can provide us with opportunities to apprehend better the social contexts to human action. We need to address more effectively the potential breadth of evidence for hunter-gatherer symbolic behaviour, and consider how belief, identity, prestige and competition might have been made possible through the 'active' roles played by pottery—as well as stone tools, bone and antler artefacts and other technologies—in prehistoric forager communities.

To address these issues fully, much more work will be required across Eurasia to move studies of ceramic technology beyond the construction of general chronologies of cultural history, and away from social evolutionary schema and the origins of agriculture, to focus more resolutely on the important question of how cumulative choices led to pottery traditions becoming embedded in hunter-gatherer social life.

This programme of research will establish the study of hunter-gatherer pottery as a distinct field of enquiry, and will require greater attention to regional sequences, and the means to attend to the social roles played by pottery through reconstruction of contextual life histories of vessels. We need more studies of organic residue analysis, which appears to hold enormous promise, as well as provenance and production techniques; we need to develop a more detailed understanding of how pots were produced, used and moved around hunter-gatherer landscapes.

References

Ailio, J (1909) *Die Steinzeitliche Wohnplatzfunde in Finland*, Finska Fornminnesföreningen volumes 1 and 2, Helsinki: Helsingförs

Akazawa, T (1986) 'Hunter-gatherer adaptions and the transition to food production in Japan', in Zvelebil, M (ed), *Hunters in Transition*, Cambridge: Cambridge University Press, pp 151–166

Aldenderfer, M (1993) 'Ritual, hierarchy, and change in foraging societies', 12 *Journal of Anthropological Archaeology*, 1–40

Ammerman, A J and Biagi, P (2003) *The Widening Harvest. The Neolithic Transition in Europe: Looking Back, Looking Forward*, Boston: Archaeological Institute of America

Ammerman, A J and Cavalli-Sforza, L L (1984) *The Neolithic Transition and the Genetics of Population in Europe*, Princeton: Princeton University Press

Andersen, S H and Malmros, K (1985) '"Madscorpe" pa Ertebollekar fra Tybrind Vig', 1984 *Aarboger for nordisk Oldkyndighed og Historie*, 78–95

Andovasio, J M, Soffer, O and Klima, B (1996) 'Upper Palaeolithic fibre technology: interlaced woven finds from Pavlov 1, Czech Republic, c. 26 000 years ago', 70 *Antiquity*, 526–534

Arnold, D (1985) *Ceramic Theory and Cultural Process*, Cambridge: Cambridge University Press

Arnold, J E (1993) 'Labor and the rise of complex hunter-gatherers', 12 *Journal of Anthropological Archaeology*, 75–119

Armit, I and Finlayson, B (1995) 'Social strategies and economic change. Pottery in context', in Barnett, W and Hoopes, J (eds), *The Emergence of Pottery*, Washington DC: Smithsonian Institution Press, pp 267–275

Ayräpää, A (1952) 'The settlement of prehistoric age', 72 *Fennia*, 285–300

——— (1955) 'Den yngre stenalderns kronologi I Finland och Sverige', *Finskt Museum*, 5–52

Barker, G (2006) *The Agricultural Revolution in Prehistory*, Oxford: Oxford University Press

Barnett, H G (1939) 'Culture element distributions: IX Gulf of Georgia Salish', 1 *Anthropological Records*, 5

——— (1955) *The Coast Salish of British Colombia*, Studies in Anthropology 4, University of Oregon Monographs, Eugene: University of Oregon

Barnett, W and Hoopes, J (eds) (1995) *The Emergence of Pottery*, Washington DC: Smithsonian Institution Press

Barrett, J (2000) 'A thesis on agency', in Dobres, M-A and Robb, J E (eds), *Agency in Archaeology*, London and New York: Routledge, pp 61–68

——— (2001) 'Agency, the duality of structure and the problem of the archaeological record', in Hodder, I (ed), *Archaeological Theory Today*, Cambridge: Polity Press, pp 141–164

Bettinger, R (1991) *Hunter-Gatherers: Archaeological and Evolutionary Theory*, New York: Plenum

Bird-David, N (1990) 'The giving environment: another perspective on the economic system of gatherer-hunters', 31(2) *Current Anthropology*, 189–196

——— (1992) 'Beyond the original affluent society: a culturalist reformation', 33(1) *Current Anthropology*, 25–47

Bleed, P (1986) 'The optimal design of hunting weapons: maintainability or reliability', 51 *American Antiquity*, 737–747

Bloch, M (1995) 'Questions not to ask of Malagasy carvings', in Hodder, I, Shanks, M, Alexandri, A, Büchli, V, Carman, J, Last, J and Lucas, G (eds), *Interpreting Archaeology: Finding Meaning in the Past*, London: Routledge, pp 212–215

Boyd, R and Richerson, P (1985) *Culture and the Evolutionary Process*, Chicago: University of Chicago Press

Bronitsky, G and Hamer, R (1986) 'Experiments in ceramic technology: the effects of various tempering materials on impact and thermal-shock resistance', 51 *American Antiquity*, 89–101

Brown, D (1983) 'Pots as tools', in Moore, J A and Keene, A S (eds), *Archaeological Hammers and Theories*, New York: Academic Press, pp 108–134

Brown, J A (1985) 'Long-term trends to sedentism and the emergence of complexity in the American Midwest', in Price, T and Brown, J (eds), *Prehistoric Hunter Gatherers: The Emergence of Cultural Complexity*, Orlando: Academic Press, pp 201–231

——— (1989) 'The beginnings of pottery as an economic process', in van der Leeuw, S and Torrence, R (eds), *What's New? A Closer Look at the Process of Innovation*, London: Unwin Hyman, pp 203–224

Cahen, D and Otte, M (eds) (1990) *Rubane et Cardial. Etudes et Recherches*, Archéologique de l'Université de Liège 39, Liège: Université de Liège

Carpelan, C (1979) 'Om asbestkeramikens historia i Fennoskandien', *Finskt Museum*, 5–25

——— (2002) 'Inarilaisten arkeologiset vaiheet', in Leholaq, V-P (ed), *Inari-Aanar: Inarin historia jaäkaudesta nykypäivän*, Inari: Regional Municipal Press, pp 28–95

Cavalli-Sforza, L L and Feldman, M W (1981) *Cultural Transmission and Evolution: A Quantitative Approach*, Princeton: Princeton University Press

Chi, Z (1999) 'The Mesolithic and the Neolithic in China', 26 *Documenta Praehistorica*, 1–13

——— (2002) 'The discovery of early pottery in China', 29 *Documenta Praehistorica*, 29–35

Childe, V G (1925) *The Dawn of European Civilisation*, London: Kegan Paul

——— (1929) *Danube in Prehistory*, Oxford: Clarendon Press

——— (1936), (1937), (1939), (1941) *Man Makes Himself*, 2nd edn 1937; 3rd edn 1939, 4th revd edn 1941, London: Watts and Co

——— (1942) *What Happened in History*, Harmondsworth: Penguin

——— (1951) *Man Makes Himself: Man's Progress through the Ages*, American printing of a revised English edn (1941); New York: New American Library

——— (1957) *The Dawn of European Civilisation*, 6th edn, London: Kegan Paul

——— (1963) *Social Evolution*, Harmondsworth: Penguin

Childs, S T (1999) '"After all, a hoe bought a wife": the social dimensions of ironworking among the Toro of East Africa', in Dobres, M-A and Hoffman, C R (eds), *The Social Dynamics of Technology: Practice, Politics, and World Views*, Washington DC: Smithsonian Institution Press, pp 23–45

Clarke, D L (1968) *Analytical Archaeology*, London: Methuen

——— (1979) 'Towards analytical archaeology—new directions in the interpretative thinking of British archaeologists', in Colleagues of D L Clarke (eds), *Analytical Archaeologist. Collected Papers of David L. Clarke*, London: Academic Press, pp 145–181

Clark, J G D (1980) *Mesolithic Prelude*, Edinburgh: Edinburgh University Press

Close, A (1995) 'Few and far between: early ceramics in north Africa', in Barnett, W and Hoopes, J (eds), *The Emergence of Pottery*, Washington DC: Smithsonian Institution Press, pp 23–38

Conkey, M (2001) 'Hunting for images, gathering up meanings: art for life in hunting-gathering societies', in Panter-Brick, C, Layton, R H and Rowley-Conwy, P (eds), *Hunter-Gatherers: An Interdisciplinary Perspective*, Cambridge: Cambridge University Press, pp 267–291

Craig, O E, Forster, M, Andersen, S H, Koch, E, Crombé, P, Milner, N J, Stern, B, Bailey, G N and Heron, C P (2007) 'Molecular and isotopic demonstration of the processing of aquatic products in northern European prehistoric pottery', 49 *Archaeometry*, 135–152

Cyrek, K, Grygiel, R and Nowak, K (1986) 'The basis for distinguishing the ceramic Mesolithic in the Polish lowland', in Malinowski, T (ed), *Problems of the Stone Age in Pomerania*, Warsaw: Warsaw University, pp 95–126

Daniel, G E (1943) *The Three Ages: An Essay in Archaeological Method*, Cambridge: Cambridge University Press

——— (1964) *The Idea of Prehistory*, Harmondsworth: Penguin

Dobres, M-A and Hoffman, C R (1994) 'Social agency and the dynamics of prehistoric technology', 1(3) *Journal of Archaeological Method and Theory*, 211–258

Dobres, M-A and Robb, J (eds) (2000) *Agency in Archaeology*, London: Routledge

Dolukhanov, P M (1979) *Ecology and Economy in Neolithic Eastern Europe*, London: Duckworth

——— (1986a) 'Natural environment and the Holocene settlement pattern in the northwestern part of the USSR', 3 *Fennoscandia archaeologica*, 3–16

——— (1986b) 'The late Mesolithic and the transition to food production in Eastern Europe', in Zvelebil, M (ed), *Hunters in Transition*, Cambridge: Cambridge University Press, pp 109–120

——— (1992) 'Evolution of lakes and prehistoric settlement in northwestern Russia', in Coles, B (ed), *The Wetland Revolution in Prehistory*, Exeter: University of Exeter Press, pp 93–98

——— (1997) 'The Pleistocene-Holocene transition in northern Eurasia: environmental changes and human adaptations', 41/42 *Quaternary International*, 181–191

——— (1998) 'The most ancient north Europeans. Consensus in sight?' in Julku, K and Wiik, K (eds), *The Roots of Peoples and Languages of Northern Eurasia* I, Turku: Societas Historiae Fenno-Ugricae, pp 9–27

Dolukhanov, P M and Foniakov, D I (1984) 'Modelirovanie kul'turno-istoricheskikh protsessov', in *Kompleksnye metody izucheniia istorii s drevneishikh vremen do nashikh dnei*, Moscow: Nauka, pp 33–35

Dolukhanov, P M and Mikliaev, A N (1986) 'Prehistoric lacustrine pine dwellings in the northwestern part of the USSR', III *Fennoscandia Archaeologica*, 81–90

Dolukhanov, P M, Shukurov, A, Gronenborn, D, Sokoloff, D, Timofeev, V and Zaitseva, G (2005) 'The chronology of Neolithic dispersal in central and eastern Europe', 32 *Journal of Archaeological Science*, 1441–1458

Edgren, T (1966) *Die Jäkärlä Gruppen*. Finska Fornminnesföreningens Tidskrift 68, Helsinki: Oy Weilin and Goos AB

——— (1970) 'Studier over den snörkeramiska kulturens i Finland', *Finska Fornminnesföreningens Tidskrift* 72, Helsinki: Oy Weilin and Goos AB, pp 1–118

——— (1982) 'Formgivning och function, en kamkeeramisk studie', 3 *Iskos*, 26–41

Engels, F (1884) (1972) *The Origin of the Family, Private Property, and the State*, New York: Pathfinder Press

Eerkens, J W, Neff, H and Glascock, M D (2002) 'Ceramic production among small-scale and mobile hunters and gatherers: a case study from the southwestern Great Basin', 21(2) *Journal of Anthropological Archaeology*, 200–229

Feng, Z (2000) 'The Mesolithic in South China', 27 *Documenta Praehistorica*, 225–231

Finlay, N (2003) 'Microliths and multiple authorship', in Larsson, L, Kindgren, H, Knutsson, K, Loeffler, D and Åkerlund, A (eds), *Mesolithic on the Move*, Oxford: Oxbow Books, pp 169–176

Fischer, A (ed) (1995) *Man and Sea in the Mesolithic. Proceedings of the International Symposium, Kalundborg, Denmark 1993*, Oxbow Monograph 53, Oxford: Oxbow Books

——— (2003) 'Exchange: artefacts, people and ideas on the move in Mesolithic Europe', in Larsson, L, Kindgren, H, Knutsson, K, Loeffler, D and Åkerlund, A (eds), *Mesolithic on the Move*, Oxford: Oxbow Books, pp 385–387

Fischer, A and Kristiansen, K (eds) (2002) *The Neolithisation of Denmark—150 years of debate*, Sheffield Archaeological Monographs 12, Sheffield: J R Collis Publications

Fitzhugh, W (ed) (1975) *Prehistoric Maritime Adaptations of the Circumpolar Zone*, The Hague: Mouton

Giddens, A (1984) *The Constitution of Society: Outline of a Theory of Structuration*, Cambridge: Polity Press

Gkiasta, M, Russell, T, Shennan, S and Steele, J (2003) 'Neolithic transition in Europe: the radiocarbon record revisited', 77 *Antiquity*, 45–63

Gosselain, O P (1998) 'Social and technical identity in a clay crystal ball', in Stark, M T (ed), *The Archaeology of Social Boundaries*, Washington DC: Smithsonian Institution Press, pp 78–106

Gurina, N N (1956) 'Oleneostrovskii Mogil'nik', *Materialy i Issledovaniia po Arkheologii SSSR* 47, Izdatel'stvo Akademii Nauk SSSR, Moskva

——— (1966) 'Sources of ancient culture (Mesolithic epoch)', *Materialy i Issledovaniia po Arkheologii SSSR* 126, Izdatel'stvo Akademii Nauk SSSR, Moskva

Hajdu, P (1977) *Finno-Ugrian Languages and Peoples*, London: Thames and Hudson

Halen, O (1994) *Sedentariness during the Stone Age of Northern Sweden*, Acta Archaeological Lundensia 20, Stockholm: Gotab

Hallgren, F (2004) 'The introduction of ceramic technology around the Baltic Sea basin in the 6th millennium', in Knutsson, H (ed), *Coast to Coast—Arrival*, Uppsala: Wikstroms, pp 123–143

Halstead, P and O'Shea, J (eds) (1989) *Bad Year Economics: Cultural Responses to Risk and Uncertainty*, Cambridge: Cambridge University Press

Hayden, B (1990) 'Nimrods, piscators, pluckers, and planters: the emergence of food production', 9 *Journal of Anthropological Archaeology*, 31–69

——— (1995) 'The emergence of prestige technologies and pottery', in Barnett, W and Hoopes, J (eds), *The Emergence of Pottery*, Washington DC: Smithsonian Institution Press

——— (2002) 'Hunting and feasting: health and demographic consequences', 2002, issue 3/4 *Before Farming* (electronic journal) http://www.waspress.co.uk/journals/beforefarming/

(2003) *Shamans, Sorcerers and Saints*, Washington DC: Smithsonian Institution Press

Hodder, I (ed) (1978) *The Spatial Organisation of Culture*, London: Duckworth

——— (1982) *Symbols in Action*, Cambridge: Cambridge University Press

——— (1990) *The Domestication of Europe*, Oxford: Blackwell

Hodder, I and Orton, C (1976) *Spatial Analysis in Archaeology*, Cambridge: Cambridge University Press

Hulthén, B (1977) *On Ceramic Technology during the Scanian Neolithic and Bronze Age*, Theses and Papers in North-European Archaeology 6, Stockholm, Department of Archaeology, University of Stockholm

——— (1991) *On the Ceramic Ware in Northern Scandinavia during the Neolithic, Bronze and Early Iron Age. A Ceramic-Ecological Study*, Archaeology and Environment 8, Umeå: Department of Archaeology, University of Umeå

Ingold, T (1988) 'Notes on the foraging mode of production', in Ingold, T, Riches, D and Woodburn, J (eds), *Hunters and Gatherers, 1: History, Evolution and Social Change*, Oxford: Berg, pp 269–285

——— (2000) *The Perception of the Environment. Essays in Livelihood, Dwelling and Skill*, London: Routledge

Ingold, T, Riches, D and Woodburn, J (eds) (1988) *Hunters and Gatherers*, Oxford: Berg

Jaanits, L (1959) *Poseleniia Epokhi Neolita i Rannego Metalla v Priust'e Reki Emaigi.* Tallinn: AN ESSR

Jacobs, K (1995) 'Return to Olenii Ostrov: social, economic and skeletal dimensions of a boreal forest Mesolithic cemetery', 14 *Journal of Anthropological Archaeology*, 359–403

Jeunesse, C (2000) 'Les Composantes Autochtone et Danubienne en Europe Centrale et Occidentale entre 5,500 et 4,000 av. J.-C.: Contacts, Transfers, Acculturations', in Richard, A, Cupillard, C, Richard, H and Thevenin, A (eds), *Les Dernières Chasseurs-Cueilleurs d'Europe occidentale*, Besançon: Presses Universitaires Franc-Contoises, pp 361–379

Jones, A (2002) *Archaeological Theory and Scientific Practice.* Cambridge: Cambridge University Press

——— (2005) 'Archaeometry and materiality: materials-based analysis in theory and practice', 46(3) *Archaeometry*, 327–338

Jordan, P (2007) 'Continuity and change in different domains of culture: an emerging approach to understanding diversity in technological traditions', in Kohler, T A and van der Leeuw, S E (eds), *The Model-Based Archaeology of Socionatural Systems*, Santa Fe: School for Advanced Research Press, pp 13–39

Keally, C T, Taniguchi, Y, Kuzmin, Y V and Shewkomud, I Y (2004) 'Chronology of the beginning of pottery manufacture in East Asia', 46(1) *Radiocarbon*, 345–351

Keeley, I (1988) 'Hunter-gatherer economic complexity and population pressure: a cross-cultural analysis', 7 *Journal of Anthropological Archaeology*, 343–411

Kelly, R L (1995) *The Foraging Spectrum: Diversity in Hunter-Gatherer Lifeways*, Washington DC: Smithsonian Institution Press

Kempisty, E (1986) 'Neolithic cultures of the forest zone in northern Poland', in Malinowski, T (ed), *Problems of the Stone Age in Pomerania,* Warsaw: Warsaw University, pp 187–214

Kent, S (1992) 'The current forager controversy: real versus ideal views of hunter-gatherers', 27 *Man* (N.S.), 45–70

Killick, D (2004) 'Social constructionist approaches to the study of technology', 36(4) *World Archaeology*, 571–578

Knutsson, K (2004) 'The historical construction of "Norland"', in Knutsson, H (ed), *Coast to Coast—Arrival*, Uppsala: Wikstroms, 45–71

Kohl, P L and Fawcett, C (eds) (1995) *Nationalism, Politics and the Practice of Archaeology*, Cambridge: Cambridge University Press

Kriiska, A (2004) *Aegade Alguses*, Tallinn: T T Raamatutrukikojas

Künnap, A (ed) (2000) *The Roots of Peoples and Languages of Northern Eurasia II and III*, Tartu: Societas Historiae Finno-Ugricae

Kuz'min, Y V (2002) 'The earliest centers of pottery origin in the Russian Far East and Siberia: review of chronology for the oldest Neolithic cultures', 29 *Documenta Praehistorica*, 37–46

——— (2003) 'Introduction: changing the paradigm', 19(2) *The Review of Archaeology*, 1–2

——— (2006) 'Chronology of the earliest pottery in East Asia: progress and pitfalls', 80 *Antiquity*, 362–371

Kuz'min, Y V and Orlova, L A (2000) 'The Neolithization of Siberia and the Russian Far East radiocarbon evidence', 74 *Antiquity*, 284, 356–364

Larsson, M (2003) 'People and sherds. The pitted ware site Aby in Ostergotland, Eastern Sweden', in Samuelsson, C and Ytterberg, N (eds), *Uniting Sea*, OPIA 3, Uppsala: Department of Archaeology and Ancient History, pp 117–131

——— (2004) 'Living in cultural diversity. The Pitted Ware Culture and its relatives', 14 *Journal of Nordic Archaeological Science*, 61–69

——— (2006) *A Tale of Strange People. The Pitted Ware Culture in Southern Sweden.* Kalmar Studies in Archaeology 2, Report Series no. 96, Lund: University of Lund, Department of Archaeology and Ancient History

Larsson, M and Rzepecki, S (2002–2003) 'Pottery, houses and graves. The early Funnel Beaker Culture in southern Sweden and central Poland', 8–9 *Lund Archaeological Review*, 1–21

Layton, R (1989) 'Pellaport', in van der Leeuw, S E and Torrence, R (eds), *What's New? A Closer Look at the Process of Innovation*, London: Unwin Hyman, pp 33–53

——— (2000) *Anthropology and History in Franche-Comté: A Critique of Social Theory*, Oxford: Oxford University Press

——— (2001) 'Hunter-gatherers, their neighbours and the nation state', in Panter-Brick, C, Layton, R H and Rowley-Conwy, P (eds), *Hunter-Gatherers. An Interdisciplinary Perspective*, Cambridge: Cambridge University Press, pp 292–321

Leacock, E and Lee, R (1982) *Politics and History in Band Societies*, Cambridge: Cambridge University Press

Lee, R B (1992) 'Art, science, or politics? The crisis in hunter-gatherer studies', 94(1) *American Anthropology*, 31–54

Lemonnier, P (ed) (1993) *Technological Choices. Transformation in Material Cultures since the Neolithic*, London: Routledge

Lindqvist, C (1987) 'A carbonised cereal grain (Hordeum sp) and faunal remains from Kolsvidja Upper Stone Age habitation sites on Aland', *Finskt Museum*, 5–40

Lipo, C P, O'Brien, M J, Shennan, S J and Collard, M H (eds) (2006) *Mapping Our Ancestors: Phylogenetic Methods in Anthropology and Prehistory*, New York: Aldine/de Gruyter, pp 149–168

Lubbock, J (1865) *Pre-historic Times, as Illustrated by Ancient Remains, and the Manners and Customs of Modern Savages*, London: Williams and Norgate

Mace, R, Holden, C J and Shennan, S (eds) (2005) *The Evolution of Cultural Diversity: A Phylogenetic Approach*, London: UCL Press

MacNeish, R S (1992) *The Origins of Agriculture and Settled Life*, Norman: University of Oklahoma Press

Matiushin, G I (1976) *Mezolit Iuzhnogo Urala*, Moscow: Nauka

——— (1986) 'The Mesolithic and Neolithic in the southern Urals and central Asia', in Zvelebil, M (ed), *Hunters in Transition: Mesolithic Societies of Temperate Eurasia and their Transition to Farming*, Cambridge: Cambridge University Press, 133–150

Meiklejohn, C and Zvelebil, M (1991) 'Health status of European populations at the agricultural transition and the implications for the adoption of farming', in Bush, H and Zvelebil, M (eds), *Health in Past Societies: Biocultural Interpretations of Human Skeletal Remains in Archaeological Contexts,* British Archaeological Reports international series 567, Oxford: Tempus Reparatum, pp 129–145

Meinander, C F (1954) 'Die Kiukais-Ku1tur', *Suomen Aluinasimuis-toyhdistyksen Aikakavskirja/Finska Förnminnes-Foreningens Tidskrift* 53, Helsinki: K.F.Puromiehen Kirjapaino O.Y., pp 1–191

——— (1961) 'De subneolitiska kulturgrupperna I norra Europa', XXXIXX B 4 *Societas Scientiarum Fennica Arsbok,* 61–71

——— (1979) 'Innovation centres in the Arctic', 2 *Fenno-Ugrica Suecana*, 85–94

Mithen, S J (1990) *Thoughtful Foragers: A Study of Prehistoric Decision Making*, Cambridge: Cambridge University Press

Moore, A M T (1995) 'The inception of potting in western Asia and its impact on economy and society', in Barnett, W K and Hoopes, J W (eds), *The Emergence of Pottery*, Washington DC: Smithsonian Institution Press, pp 39–55

Moore, J H (2001) 'Ethnogenetic patterns in Native North America', in Terrell, J E (ed), *Archaeology, Language and History: Essays on Culture and Ethnicity*, Westport: Bergin and Garvey, pp 30–56

Morgan, H L (1877) (1964) *Ancient Society*, reprinted 1964, Cambridge, MA: Harvard University Press

Myers, A M (1986) 'Reliable and maintainable technological strategies in the Mesolithic of mainland Britain', in Torrence, R (ed), *Time, Energy and Stone Tools*, Cambridge: Cambridge University Press, pp 78–92

Nuñez, M (1975) *Report on a Study of Finnish Prehistoric Pottery*, Helsinki: Finnish Academy

——— (1990) 'On subneolithic pottery and its adoption in late Mesolithic Finland', VII *Fennoscandia archaeologica*, 27–50

——— (1998) 'Old and new ideas about the origins of the Finns and Saami', in Julku, K and Wiik, K (eds), *The Roots of Peoples and Languages of Northern Eurasia* I, Turku: Societas Historiae Fenno-Ugricae, pp 151–160

——— (2000) 'Problems with the search for the ancestral Finns', in Künnap, A (ed), *The Roots of Peoples and Languages of Northern Eurasia* II *and* III, Turku: Societas Historiae Fenno-Ugricae, pp 60–68

——— (2004) 'All quiet on the eastern front?' in Knutsson, H (ed), *Coast to Coast—Arrival*, Uppsala: Wikstroms, pp 345–367

Nuñez, M and Okkonen, J (1999) 'Environmental background for the rise and fall of villages and megastructures in north Ostrobothnia 4000–2000 cal BC', in Huurre, M (ed), *Dig It All. Papers Dedicated to Aari Siiriäinen*, Helsinki: Finnish Antiquarian Society and Archaeological Society of Finland, pp 105–115

O'Shea, J and Zvelebil, M (1984) 'Oleneostrovskii Mogil'nik: reconstructing social and economic organisation of prehistoric hunter-fishers in Northern Russia', 3 *Journal of Anthropological Archaeology*, 1–40

Oliva, M (2005) *Palaeolithic and Mesolithic of Moravia*, Brno: Moravske Zemske Muzeum

Olsen, S F and Secher, N J (2002) 'Low consumption of seafood in early pregnancy as a risk factor for preterm delivery: prospective cohort study', 324 (7335) *British Medical Journal*, 447–457

Ortman, S (2000) 'Conceptual metaphor in the archaeological record: methods and an example from the American southwest', 65 *American Antiquity*, 613–645

Panter-Brick, C, Layton, R H and Rowley-Conwy, P (eds) (2001) *Hunter-Gatherers. An Interdisciplinary Perspective*, Cambridge: Cambridge University Press

Papmehl-Dufay, L (2003) 'Stone Age island archaeology', in Samuelsson, C and Ytterberg, N (eds), *Uniting Sea. Stone Age Societies in the Baltic Sea Region*, OPIA 33, Uppsala: Department of Archaeology and Ancient History, pp 180–203

——— (2006) *Shaping an Identity. Pitted Ware Pottery and Potters in Southeast Sweden*, Thesis and Papers in Scientific Archaeology 7, Stockholm: Archaeological Research Laboratory

Pawlikowski, M, Koukouzelis, M, Ginter, B and Kozlowski, J K (2000) 'Emerging ceramic technology in structured Aurignacian hearts at Klisoura Cave 1 in Greece', 4 *Archaeology, Ethnology and Anthropology of Russia*, 19–30

Pankrushev, G (1978) *Mezolit i Neolit Karelii, Tom 1: Mezolit*. Leningrad: Nauka

Pearson, R (2005) 'The social context of early pottery in the Lingnan region of south China', 79 *Antiquity*, 819–828

Pettitt, P B (2006) 'The living dead and the dead living: burials, figurines and social performance in the European mid Upper Palaeolithic', in Knusel, C and Gowland, R (eds), *The Social Archaeology of Funerary Remains*, Oxford: Oxbow Books, pp 292–308

Piggott, S (1965) *Ancient Europe*, Edinburgh: Edinburgh University Press

Pluciennik, M (2002) 'The invention of hunter-gatherers in seventeenth-century Europe', 9(2) *Archaeological Dialogues*, 98–151

——— (2004) 'The meaning of "hunter-gatherers" and modes of subsistence: a comparative historical perspective', in Barnard, A (ed), *Hunter-Gatherers in History, Archaeology and Anthropology*, Oxford: Berg, pp 17–19

——— (2005) *Social Evolution*, London: Duckworth

Pluciennik, M and Zvelebil, M (2008) 'The origins and spread of agriculture', in Bentley, R A, Maschner, H D G, and Chippendale, C (eds), *Handbook of Archaeological Theories*, Lanham: Rowman and Littlefield Publishing Group, pp 467–486

Ravdonikas, V I (1940) 'Neoliticheskii mogil'nik na Onezhskom ozere', VI *Sovetskaia arkheologiia*, 16–62

Reid, K C (1989) 'A material science perspective on hunter-gatherer pottery', in Bronitsky, G (ed), *Pottery Technology: Ideas and Approaches*, Boulder: Westview Press, pp 167–180

Rice, P (1999) 'On the origins of pottery', 6(1) *Journal of Archaeological Method and Theory*, 1–54

Rimantienė, R K (1992) 'The Neolithic in the eastern Baltic', 6 *Journal of World Prehistory*, 97–143

——— (1998) 'The first Narva culture farmers in Lithuania', in Zvelebil, M, Dennell, R and Domańska, L (eds), *Harvesting the Sea, Farming the Forest*, Sheffield: Sheffield University Press, pp 213–218

Rowley-Conwy, P A and Zvelebil, M (1989) 'Saving it for later: storage by prehistoric hunter-gatherers in Europe', in Halstead, P and O'Shea, J (eds), *Bad Year Economics*, Cambridge: Cambridge University Press, pp 40–56

Rudebeck, E (2000) *Tilling Nature, Harvesting Culture. Exploring Images of the Human Being in the Transition to Agriculture*, Acta Archaeologica Lundensia, 8th series, 32, Stockholm: Almqvist and Wiksell International

Sahlins, M (1972) *Stone Age Economics*, London: Tavistock

Schiffer, M B and Skibo, J M (1987) 'Theory and experiment in the study of technological change', 28 *Current Anthropology*, 595–622

——— (1997) 'The explanation of artifact variability', 62 *American Antiquity*, 27–50

Schrire, C (ed) (1984) *Past and Present in Hunter Gatherer Studies*, Orlando: Academic Press

Shennan, S J (2002) *Genes, Memes and Human History*, London: Thames and Hudson

——— (2004) 'An evolutionary perspective in agency in archaeology', in Gardner, A (ed), *Agency Uncovered: Archaeological Perspectives on Social Agency, Power and Being Human*, London: UCL Press, pp 33–50

——— (ed) (2009) *Pattern and Process in Cultural Evolution*. Berkeley: University of California Press

Sherratt, A G (1976) 'Resources, technology and trade, an essay in early European metallurgy', in Sieveking, G, Longworth, I J and Wilson, K E (eds), *Problems in Economic and Social Archaeology*, London: Duckworth, pp 557–581

——— (1981) 'Plough and pastoralism: aspects of the secondary products revolution', in Hodder, I, Isaac, G and Hammond, N (eds), *Pattern of the Past: Studies in Honour of David Clarke*, Cambridge: Cambridge University Press, pp 261–305

——— (1995) 'Alcohol and its alternatives: symbol and substance in pre-industrial cultures', in Goodman, J, Lovejoy, P E and Sherratt, A (eds), *Consuming Habits: Drugs in History and Anthropology*, London: Routledge, pp 11–46

——— (1997) *Economy and Society in Prehistoric Europe: Changing Perspectives*, Edinburgh: Edinburgh University Press

Shott, M J (1992) 'On recent trends in the anthropology of foragers: Kalahari revisionism and its archaeological implications', 27 *Man* (N.S.), 843–871

Siiriäinen, A (1971) 'Shoreline dating of the Saarsniemi 1 ceramics in Finland', 1971 *Suomen Museo*, 9–19

——— (1981) 'On the cultural ecology of the Finnish Stone Age', 87 (1980) *Suomen Museo*, 5–40

——— (1984) 'On the late Stone Age Asbestos Ware Culture of Northern and Eastern Finland', Fenno-Ugri et Slavi 1983, 4 *Iskos*, xx

Sillar, B (2000) 'Dung by preference: the choice of fuel as an examplar of how Andean pottery production is embedded within wider technical, social and economic practices', 42(1) *Archaeometry*, 43–60

Sillar, B and Tite, M (2000) 'The challenge of "technological choices" for materials science approaches in archaeology', 42(1) *Archaeometry*, 2–20

Skandfer, M (2005) 'Early Northern Comb Ware in Finnark: the concept of Saraisniemi 1 reconsidered', 22 *Fennoscandian Archaeological*, 3–28

Skibo, J M, Schiffer, M B and Reid, K C (1989) 'Organic-tempered pottery: an experimental study', 54 *American Antiquity*, 122–146

Soffer, O and Vandiver, P (1997) 'The ceramics from Pavlov 1—1957 excavation', in Svoboda, J (ed), *Pavlov 1: 1957 Excavations*, Brno: Archeologicky Ustav CR, pp 383–402

Soffer, O, Vandiver, P, Oliva, M and Seitl, L (1993) 'Fiery Venus. Stone Age art made to explode', January/February 1993 *Archaeology*, 36–39

Speth, J D and Spielman, K A (1983) 'Energy source, protein metabolism and hunter-gatherer strategies', 2 *Journal of Anthropology Archaeology*, 1–31

Spielmann, K A (1989) 'A review: dietary restrictions on hunter-gatherer women and the implications for fertility and infant mortality', 17 *Human Ecology*, 321–345

Stilborg, O and Bergenstråhle, I (2000) 'Traditions in transition. A comparative study of the Patterns of Ertebølle lithic and pottery changes in the Late Mesolithic ceramic phase at Skateholm I, III, and Soldattorpet in Scania, Sweden', 6 *Lund Archaeological Review*, 23–41

Tilley, C (1991) *Material Culture and Text: The Art of Ambiguity*, London: Routledge

Tilley, C, Keane, W, Küchler, S, Rowlands, M and Spyer, P (eds) (2006) *Handbook of Material Culture*, London: Sage

Timofeev, V I (1987) 'On the problem of the early Neolithic of the East Baltic area', 58 *Acta Archaeologica*, 207–212

——— (1990) 'On the links of the East Baltic Neolithic and the Funnel Beaker Culture', in Jankowska, D (ed), *Die Trichterbeckerkultur*, Poznan: Instytut Prahistorii Universytetu im. Adama Mickiewicza w Poznaniu Zaklad Archeologii Wielkoposki IHKM PAN w Poznaiu, pp 135–149

——— (1998) 'The beginning of the Neolithic in the Eastern Baltic', in Zvelebil, M, Dennell, R and Domańska, L (eds), *Harvesting the Sea, Farming the Forest,* Sheffield: Sheffield Academic Press, pp 225–236

——— (1999) 'The East-West relations in the late Mesolithic and Neolithic Baltic region', in Kosko, A (ed), *Beyond Balkanization,* Baltic Pontic Studies vol. 5, pp 44–58

Tite, M S and Kilikoglou, V (2002) 'Do we understand cooking pots and is there an ideal cooking pot?' in Kilikoglou, V, Hein, A and Maniatis, Y (eds), *Modern Trends in Scientific Analysis of Ancient Ceramics*, BAR International Series 1011, Oxford: British Archaeological Reports, pp 1–8

Torrence, R (1983) 'Time budgeting and hunter-gatherer technology', in Bailey, G (ed), *Hunter-Gatherer Economy in Prehistory: A European Perspective*, Cambridge: Cambridge University Press, pp 11–22

———(2001) 'Hunter-gatherer technology: macro and microscale approaches', in Panter-Brick, C, Layton, R H and Rowley-Conwy, P (eds), *Hunter-Gatherers: An Interdisciplinary Perspective*, Cambridge: Cambridge University Press, pp 73–95

Tylor, E B (1871) *Primitive Culture*, London: J Murray

van der Leeuw, S E and Pritchard, A C (eds) (1984) *The Many Dimensions of Pottery*, Cingula VII, Universiteit van Amsterdam

Verpoorte, A (2001) *Places of Art, Traces of Fire*, Leiden: Faculty of Archaeology, University of Leiden

Vetrov, V M (1985) 'Keramika Ust'-Karengskoi kul'tury na Vitime', in Konovalov, P B (ed), *Drevnee Zabaikal'e i Ego Kul'turnyi Sviazi*, Novosibirsk: Nauka, pp 123–130

Welinder, S (1976) 'The economy of the Pitted Ware Culture in eastern Sweden', 1975–76, new series, l *Meddelanden Fran Lunds Universitets Historika Museum*, 20–30

Wengrow, D (2001) 'The evolution of simplicity: aesthetic labour and social change in the Neolithic Near East', 33 *World Archaeology*, 168–188

Werbart, B (1998) 'Subneolithic: what is it? Subneolithic societies and the conservative economies of the circum-Baltic region', in Zvelebil, M, Dennell, R and Domańska, L (eds), *Harvesting the Sea, Farming the Forest*, Sheffield: Sheffield Academic Press, pp 37–45

White, R (1992) 'Beyond art: towards an understanding of the origins of material representations in Europe', 21 *Annual Review of Anthropology*, 537–564

Woodburn, J (1982) 'Egalitarian societies', 17 *Man* (N.S.), 431–451

——— (1988) 'African hunter-gatherer social organization: is it best understood as a product of enscapulation?' in Ingold, T, Riches, D and Woodburn, J (eds), *Hunters and Gatherers* 1, Oxford: Berg, pp 31–64

Wu, X and Zhao, C (2003) 'Chronology of the transition from Palaeolithic to Neolithic in China', 24(2) *The Review of Archaeology*, 15–20

Zhao, Z (1998) 'The middle Yangtze region in China is one place where rice was domesticated: phytolith evidence from the Diaotonghuan Cave, Northern Jiangxi', 72 *Antiquity*, 885–897

Zhao, C and Wu, X (2000) 'The dating of Chinese Early Pottery and a discussion of some related problems', 27 *Documenta Praehistorica*, 233–239

Zhao, C, Yu, J, Wang, T, Wu, X, Hao, S, Ma, X and Xia, Z (2003) 'A study on an early Neolithic site in North China', 30 *Documenta Praehistorica,* 169–173

Zhushchikhovskaya, I S (1997) 'On early pottery making in the Russian Far East', 36(2) *Asian Perspectives*, 159–174

——— (2005) *Prehistoric Pottery Making of the Russian Far East*, translated and edited by R L Bland and C M Aikens, BAR International Series 1434, Oxford: Hadrian Books

Zvelebil, M (1984) 'Clues to recent human evolution from specialised technologies?' 307 *Nature*, 314–315

——— (1986a) 'Postglacial foraging in the forests of Europe', 254(5) *Scientific American*, 104–115

——— (ed) (1986b) *Hunters in Transition. Mesolithic Societies of Temperate Eurasia and their Transition to Farming*, Cambridge: Cambridge University Press

———(1986c) 'Mesolithic prelude and Neolithic revolution', in Zvelebil, M (ed), *Hunters in Transition*, Cambridge: Cambridge University Press, pp 5–15

——— (1986d) 'Mesolithic societies and the transition to farming: problems of time, scale and organisation', in Zvelebil, M (ed), *Hunters in Transition*, Cambridge: Cambridge University Press, pp 167–187

——— (1987) 'Wetland settlements in Eastern Europe', in Coles, J M and Lawson, A J (eds), *European Wetlands in Prehistory*, Oxford: Clarendon Press, pp 94–116

——— (1989) 'Economic intensification and postglacial hunter-gatherers in north temperate Europe', in Bonsall, C (ed), *The Mesolithic in Europe*, Edinburgh: John Donald, pp 80–88

——— (1998a) 'Agricultural frontiers, Neolithic origins, and the transition to farming in the Baltic basin', in Zvelebil, M, Dennell, R and Domańska, L (eds), *Harvesting the Sea, Farming the Forest*, Sheffield: Sheffield Academic Press, pp 9–28

——— (1998b) 'What's in a name? The Mesolithic, the Neolithic and social change at the Mesolithic-Neolithic transition', in Edmonds, M R and Richards, C (eds), *Understanding the Neolithic of North-Western Europe*, Glasgow: Cruithne Press, pp 1–35

——— (2000a) 'The social context of the agricultural transition in Europe', in Renfrew, C and Boyle, K (eds), *Archaeogenetics: DNA and the Population Prehistory of Europe,* Cambridge: McDonald Institute Monographs, pp 57–79

——— (2000b) 'Fat is a feminist issue: on ideology, diet and health in hunter-gatherer societies', in Donald, M and Hurcombe, L (eds), *Gender and Material Culture in Archaeological Perspective*, Basingstoke: Macmillan, pp 209–221

Zvelebil, M (2001) 'Revisiting Indreko's culture historical model: "origin and area of settlement of the Fenno-Ugrian peoples"', No 1, Vol 5(55/50) *Trames Journal of the Humanities and Social Sciences*, 27–47

——— (2005) '*Homo habitus*: agency, structure and the transformation of tradition in the constitution of the TRB foraging-farming communities in the North European plain (ca 4500–2000 BC)', XXXII *Documenta Praehistorica*, 87–101

——— (2006) 'Mobility, contact and exchange in the Baltic Sea Basin 6000–2000 BC', in Lovis, W, Donahue, R and Whallon, R (eds), *Mobility, Contact and Exchange in Mesolithic Europe*. Journal of Anthropological Archaeology, Special Issue

——— (2008) 'Innovating hunter-gatherers. The Mesolithic in the Baltic', in Bailey, G and Spikins, P (eds), *Mesolithic Europe*, Cambridge: Cambridge University Press, pp 18–60

Zvelebil, M, Dennell, R and Domańska, L (eds) (1998) *Harvesting the Sea, Farming the Forest*, Sheffield: Sheffield Academic Press

Zvelebil, M and Dolukhanov, P (1991) 'Transition to farming in eastern and northern Europe', 5(3) *Journal of World Prehistory*, 233–278

Zvelebil, M and Jordan, P (1999) 'Hunter fisher gatherer ritual landscapes: questions of time, space and representation', in Goldhahn, J (ed), *Rock Art as Social Representation*, BAR International Series 794, Oxford: Hadrian Books, pp 101–127

Part 2

Early Ceramic Innovations and Dispersals

Chapter 2

Long-Term Innovation: Appearance and Spread of Pottery in the Japanese Archipelago

Simon Kaner

Introduction

The early appearance of pottery in the Japanese archipelago can be understood as a process of technological innovation involving accumulative changes in small-scale practices, each of which potentially had social meaning. The making of ceramic vessels took place in a social milieu, in which potters' decisions including the selection of clay, the choice of temper, the forming of the pot and the application of decoration were all constrained by their experience and knowledge of the medium in which they were working, offering archaeologists the possibility of investigating the development of what Marcia Anne Dobres has termed the social values of artefact production (Dobres 2000: 116–117). The creation of each pottery vessel offered the potential to reproduce or transform the social significance of each of these small decisions and resulting actions, and can therefore be considered an expression of the social agency of the potter involved. This chapter will argue that pottery production in the Japanese archipelago developed in the context of the clash of a series of what Margaret Boden terms 'conceptual spaces' (Boden 1998), including new subsistence practices, new ways of structuring social space, and new engagement with the soil, which engendered the conditions for creativity at the end of the Pleistocene. Pottery production took several thousand years to become embodied, or normalized, in the routinized activity of foraging communities of the prehistoric Japanese archipelago (Figures 2.1 and 2.2). These innovations took place, embedded in a series of changing social and ecological landscapes which

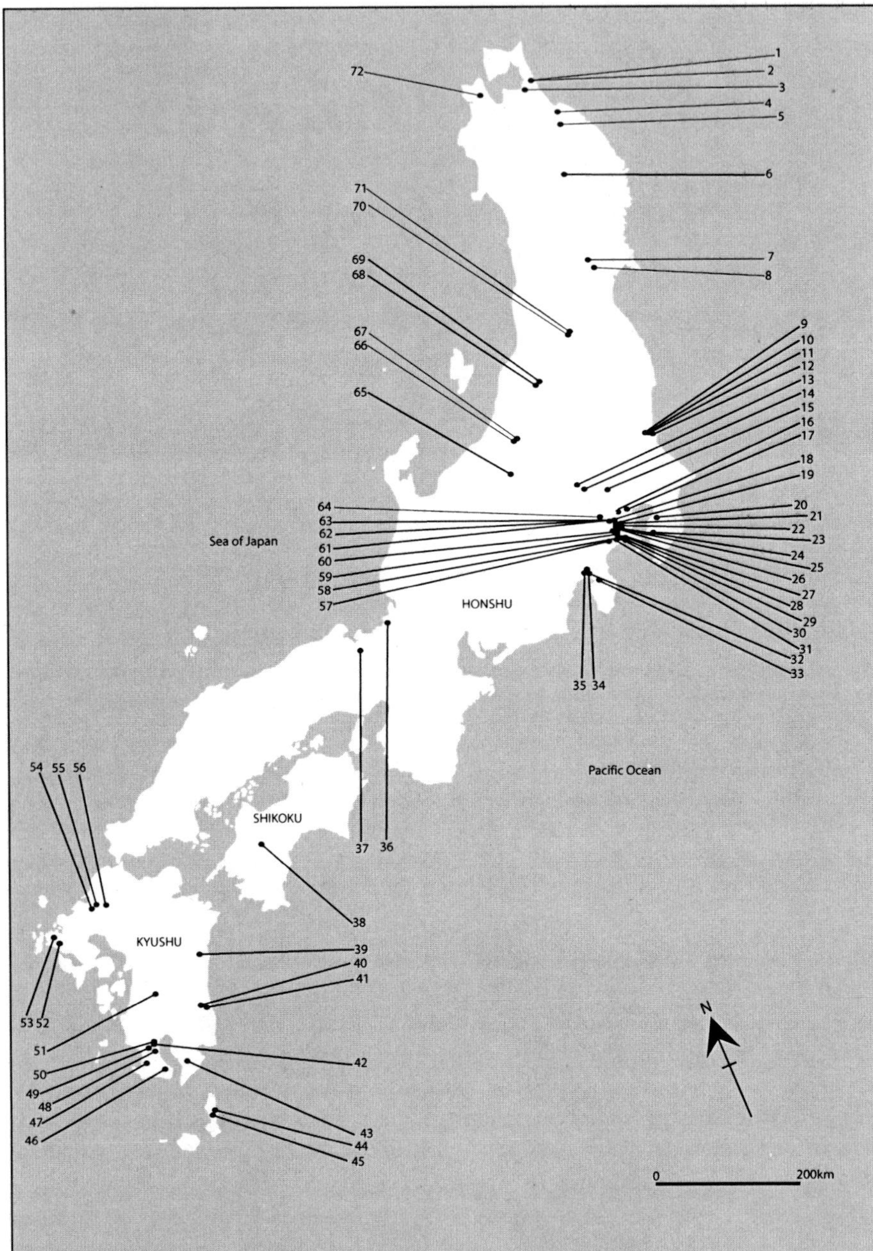

Figure 2.1 Important Incipient Jomon sites (adapted from Yokohama-shi 1996). (1) Omotedate; (2) Hatsuchazawa; (3) Chojakubo; (4) Kamotai; (5) Babano II; (6) Daishincho; (7) Zazaragi; (8) Ohara B; (9) Toyabami; (10) Higashi Ishikawa Niihori;

(11) Ushirono; (12) Haranotera; (13) Nishiya and Mizukubo; (14) Miyabayashi; (15) Junibankochi; (16) Mukianohara; (17) Hyakunincho; (18) Kurokawa Higashi; (19) Nasunahara; (20) Minamihara; (21) Hanamiyama; (22) Tsukidematsu and Nokendo; (23) Maemifunedai; (24) Ueno Location 2 and Sagamino Site 149; (25) Nagahori Kita; (26) Fukami Suwayama; (27) Minami Kuzuno; (28) Imada; (29) Karasawa; (30) Minami Kajiyama; (31) Keio Gijuku Daigaku Shonan Fujisawa Campus; (32) Onoue Iraune Kita; (33) Nakamichi A; (34) Tsuijibana Kita; (35) Kuzuharazawa IV, Ojigakubo and Hatsunegahara; (36) Torihama; (37) Mushagatani; (38) Kamikuroiwa rockshelter; (39) Kuruta; (40) Shiiyakata I; (41) Dochi Nishi; (42) Kakuriyama; (43) Ishiki and Higashi Kurotsuchida; (44) Okunonita; (45) Sankakuyama; (46) Iwamoto; (47) Kakoinohara; (48) Sojiyama; (49) Yokoi-Takenoyama; (50) Kagiyazono; (51) Shiratoribira B; (52) Sempukuji Cave; (53) Fukui Cave; (54) Kashihara; (55) Monden; (56) Hara; (57) Sannomiya, Miyanomae and Kirigaoka; (58) Daikanyama; (59) Terao; (60) Katsusaka; (61) Kawashimadani; (62) Tama New Town site 57; (63) Tama New Town sites 116, 125, 496, 796; (64) Maeda Kochi; (64) Ishigoya Cave; (65) Motonoki; (66) Tazawa and Jin; (67) Muroya Cave; (68) Kosegasawa Cave; (69) Hinata Cave West; (70) Hinata Cave; (71) Ichinosawa rockshelter; (72) Odai Yamamoto I.

were relatively unstable amid the regular fracturing of networks of social meanings and associations. This process eventually led to the flourishing of one of the most impressive traditions of prehistoric pottery manufacture, the Jomon (Habu 2004; Kobayashi 2004a; Figure 2.3).

This chapter will (1) survey the context for the appearance of pottery in the Japanese archipelago and provide accounts of pottery and associated assemblages from the late Palaeolithic/Incipient Jomon (c.16,000–c.10,000 BP); (2) consider the history of research into this topic; (3) explore the epistemological issues raised by the classification of chronological periods and technological traditions in the archipelago; and (4) discuss ideas as to why the technological innovation represented by the adoption of pottery took so long to take off and relate this to theoretical models concerning the adoption and innovation of pottery.

The Adoption of Pottery: Current Models and Environmental Conditions

The calibrated AMS dates for pottery fragments from Odai Yamamoto in Aomori Prefecture suggest that pottery was being manufactured as early as 16,000 years ago in the late Palaeolithic. It was not until considerably later, however, well into the Jomon tradition that pottery bearing the eponymous cord-marked ('Jomon' in Japanese) decoration appeared (Figure 2.3).

The earliest dates for pottery in the Japanese archipelago are from Odai Yamamoto I at the northern tip of the largest island in the archipelago, Honshu (Odai Yamamoto 1999; Figure 2.4). Forty-six potsherds of a flat-based vessel in the form of either a deep bowl or pot were discovered in July

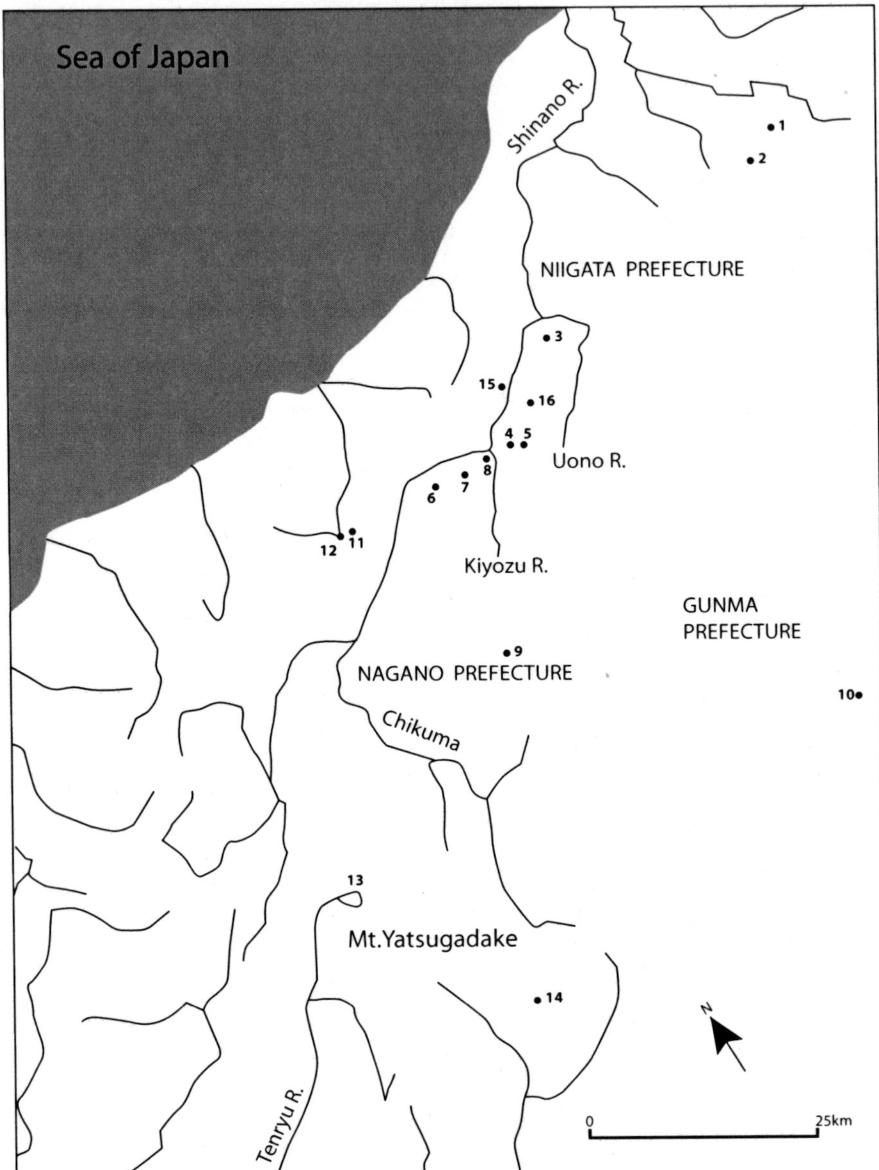

Figure 2.2 Sites in Central Honshu (adapted from Sato 2001b, Figure 1). (1) Kosegasawa Cave; (2) Muroya Cave; (3) Araya; (4) Jin and Tazawa; (5) Nakabayashi; (6) Kamiyama; (7) Motonoki; (8) Unoki and Unoki Minami; (9) Okarino; (10) Iwajuku; (11) Lake Nojiri; (12) Chausuyama; (13) Lake Suwa and Uenodaira; (14) Yadegawa; (15) Kubodera; (16) Hizono.

1998 by a team from Kokugakuin University in Tokyo led by Taniguchi Yasuhiro. The sherds are reported to be 7.6 mm thick, with no fibre tempering and no clear evidence for decoration, although nine of the sherds did bear narrow engraved lines. Petrographic analysis of the clay indicated that the vessel was made of locally available materials. Thirty of the 46

Figure 2.3 Incipient Jomon cord-marked pots from Muroya Cave, Niigata Prefecture. Height of pot on left 20 cm (from Kobayashi 2004a, figure 3.8, courtesy Nagaoka City Museum of Science).

Figure 2.4 Fragments of a pottery container from Odai Yamamoto I, Niigata Prefecture (from Odai Yamamoto I Site Excavation Team 1999).

sherds had carbonized accretions, some of which formed a line suggesting the vessel had been used to boil its contents. Five samples of these carbon adhesions and three samples of charred wood associated with the artefacts were submitted for Accelerator Mass Spectrometry dating, the results of which, when calibrated using the MacCALIB 3.0 program gave dates of 16,500–16,000 BP. The pottery sherds were associated with stone tools of the Mikoshiba-Chojakubo culture, comprising unifacial blade axes, a large bifacial projectile point (spearpoint), small projectile points (arrowheads), blades, scrapers, gravers and a core. The dates are supported by previous dates for the Hachinohe tephra, which provides a *terminus post quem* for sites with Mikoshiba-Chojakubo assemblages. Analysis of the carbonized accretions at Odai Yamamoto using C13 isotope analysis suggests that the remains represent C3-type plant foods. However, this cannot be regarded as proven since it is possible that plant charcoal got into the cooking process. It was not until considerably later, however, well into the Holocene, that pottery appears in the large quantities that are so distinctive of the Jomon tradition.

This discovery has encouraged Japanese archaeologists to situate the Palaeolithic–Jomon transition at around 16,000 BP. Pottery appeared with what were previously considered late Palaeolithic stone tools, adding to a long-standing debate regarding the beginnings of the Jomon period, a debate which has in certain respects structured the questions that have been asked about the archaeological material. The Incipient Jomon period, which is now thought to have lasted for 6,000 years, from Odai Yamamoto until just over 10,000 BP, saw the establishment of subsequent Jomon cultural traditions which continued in the archipelago until the introduction and adoption of rice agriculture from the East Asian continent during the 1st millennium BC. The foundations were laid for significant changes in both subsistence and settlement patterns, with the appearance and decline of some 70 major regional pottery styles and over 400 local variants now recognized by Japanese archaeologists (Kobayashi 2004a: 30–31). Developments included the increased use of plant foods and marine resources, and what Imamura terms 'an intensification in the scale of settlements, all of which imply a stable life' (1996: 56).

The environmental conditions during the Incipient Jomon changed considerably across the Japanese archipelago. Major shifts occurred in the distribution of different forest zones. As the temperatures fluctuated at the end of the Pleistocene, the warm temperate evergreen broadleaf forest that came to dominate most of the western part of the archipelago from about 10,000 BP was pushing the cool temperate deciduous broadleaf forest further northeast (Yasuda and Miyoshi 1999). The fluctuations were significant, however, and suggest that the adoption of pottery cannot be regarded as a simple functional response to environmental change.

Incipient Jomon communities are thought to have comprised bands of fishers, hunters and gatherers, made up of relatively amorphous and open extended family groups, although social relations are usually seen as having no causal significance with regard to the major cultural developments and innovations of the period, such as the adoption of pottery, archery and water transportation (Ikawa-Smith 1986). Population density is estimated to have been low, with the implication that there were no social constraints on where these groups could wander in search of game, game that was becoming increasingly infrequent due to a combination of hunting pressure from these human groups and changes in the natural habitat. Little consideration has been given to the agency of these foragers, their knowledge and the effect this may have had on the developmental histories of their communities, which are regarded as insignificant given the magnitude of the environmental changes occurring around them. These communities are not thought of as being culturally or socially complex: they left no great artistry, no evidence for permanent settlements. Indeed, it is even suggested that these communities were considerably less complex than their late Palaeolithic ancestors. The ways in which the Incipient Jomon inhabitants of the archipelago reacted and responded to the situations are considered to have been dictated by their environmental circumstances. They are perceived as having had no choice:

> 'Conditions in Kyushu [one of the regions in which pottery manufacture originated] at that time offered only marginal chances of survival. The inhabitants who stayed on were driven to find alternative sources of nourishment [to the large game animals that were becoming extinct]. Boiling otherwise indigestible plants, especially certain grasses and ferns, made them edible and sometimes even palatable.' (Kidder 1993: 58, present author's additions in square brackets)

According to this view, innovation was the result of necessity and the fight for survival in an increasingly hostile natural world (Aikens 1995; Aikens and Akazawa 1996; Tsutsumi 2002).

In his interesting and theoretically innovative survey of Japanese prehistory, Mizoguchi (2003) argues that pottery appeared in the Japanese archipelago as part of a major transition in the construction of identity. In line with many Japanese archaeologists, he notes an apparent shift in mobility strategies towards the end of the Pleistocene, with a shift from 'circulating high mobility' to 'radiating mobility with fixed stations to call at regularly'. This change is associated with a series of related changes in spatio-temporal organization: (1) a shift from unfixed resource locales (i.e., chasing herds of large mammals) to fixed resource locales (i.e., shell beds, patches of rich nut bearing trees); (2) a shift from a situation of high uncertainty and low predictability to one of decreasing uncertainty and increasing predictability, resulting in (3) a shift from a situation in which the projection of past experiences to the future is difficult or risky, to one in which it becomes increasingly possible and important. He argues that the hoarding practices characteristic of the Mikoshiba-Chojakubo culture, to which the Odai Yamamoto assemblage belongs, is related to an 'emerging sense of attachment to particular locales'. In what could be described as a re-casting of the ideas set out in Lewis Binford's 'Post-Pleistocene adaptations' (Binford 1968), the major causative factor behind this shift in what Mizoguchi terms the 'topography of identities', however, remains the new environmental conditions and disbursement of resources at the end of the Pleistocene, leading to less residential mobility and more sedentism. Mizoguchi does not fully explore either the process whereby this transition in identity formation took place, nor the workings of the agency of the communities and individuals involved.

Richard Pearson (2006) has attempted to bring an agency perspective to bear on the early appearance of pottery and apparently relatively sedentary village communities in southern Kyushu, following on from an earlier study of the origins of pottery in southern China (Pearson 2005). These interesting accounts, however, do not consider the specificity of the adoption of pottery as a process of socially embedded technological innovation.

Kobayashi Tatsuo (2004b), on the other hand, in reflecting on the origins of pottery in the Japanese archipelago is explicit about the specific implications and significance of pottery in terms of the subsequent development of Jomon culture. He usefully differentiates between the reductive technology of stone tool manufacture and the additive, accumulative technology of pottery. Kobayashi considered that the two essential prerequisites for the subsequent development of the Jomon pottery tradition were the presence of the technological basis for pottery manufacture and the perceived need to make pottery (2004b: 20). The need for pottery, in Kobayashi's opinion, was, however, still driven by new foodstuffs, in particular nuts, and subsequently shellfish.

One of the most innovative interpretations to have been proposed in recent years for the early appearance of pottery in the Japanese archipelago has been proposed by James Brown and transcends direct environmental causality (Brown 1989). He argues that the invention of pottery was an economic process, and proposed a model for the adoption of pottery which was based on an understanding of time-budgeting or scheduling within the societies adopting pottery technology. He made a number of references to the appearance of pottery in the Jomon which makes a review of this paper of particular relevance here.

Brown argued that in order to go beyond the previous mentalist or adaptationist models, an economic model is needed which takes into account supply and demand. He suggested that pottery production should reflect pressure on the overall demand for vessels of all kinds (1989: 210). He took up Dean Arnold's correlation between the development of sedentism and the adoption of pottery (Arnold 1985), noting that in sedentary societies, pottery making could easily be subsumed into the women's schedule of activities, based as it was around the house. Perhaps most interestingly, he noted that by scheduling the various stages involved in the manufacture of pottery, it would be possible to increase the numbers of pots being produced without a matching increase in the amount of time invested in pottery manufacture. This sets pottery manufacture apart from the production of containers made from other materials, such as basketry and stone. Brown made a series of other interesting observations. He observed that the production of pottery vessels took a long time to become a well-established facet of material production in the Japanese archipelago, and that it was important to understand how pottery production would fit into the overall scheduling of activities within any society.

Brown's essay retains its interest in part because of its resonance with more recent writings on the Jomon which have focused on the reconstruction of subsistence-settlement systems. These emphasize how human groups respond through their spatio-temporal behaviour to the changing availability of resources in the landscape (Fujiyama 2005; Habu 2004), Mizoguchi's (2003) time-geography influenced approach also creates a framework for understanding a series of 'topographies of identity', suggesting that a more cultural motor has generated the patterning perceived in the archaeological traces of a series of cultures in the archipelago. Brown's model is of further interest as it provides a series of propositions against which more social approaches might be examined (Hayden 1995; Pearson 2005), suggesting that the innovation and adoption of pottery might be the result of specific social strategies.

Considering the approaches presented above, a number of questions remain. Why are so few sites from this period producing pottery? Imamura states that only five of the 'few dozen' known Mikoshiba-Chojakubo sites have pottery 'and even in these five sites the amount of pottery is very small. ... It is more likely that pottery was used in very small quantities and on rare occasions, so that only a few sites have it' (Imamura 1996: 52). Furthermore, the manufacture of pottery containers started before the environmental changes that have been assumed as a cause for social, economic or culinary changes as argued by the models presented above.

The Appearance of Pottery in Central Honshu

As a case study, I will now turn to the appearance of pottery in central Honshu, with particular attention being given to the area including the modern prefectures of Niigata and Nagano (Nagano-ken 1992; Niigata-ken 1986). This region is adjacent to the major geological and environmental divide of the archipelago, the *fossa magna* which runs from the Sea of Japan coast to the Pacific coast. The geomorphology of the region comprises major mountain massifs including the Japan Alps, coastal plains and inland mountain basins. Major river systems include the Shinano and Chikuma drainages leading out to the Japan Sea coast. In northern Nagano, Lake Nojiri has provided important environmental evidence from the lake sediments, as well as one of the most dramatic of all the Palaeolithic sites in Japan, which produced the well-preserved remains of a Nauman's elephant and other palaeofauna (Aso and Jitsukawa 1983).

The appearance and early development of pottery in central Honshu makes a useful case study as it allows us to explore: firstly, the appearance of pottery in conjunction with the collapse of the late Palaeolithic stone procurement systems; secondly, the early proliferation of pottery manufacturing techniques; and thirdly, why it takes so long for pottery to be adopted in large quantities in relationship to the relatively slow and limited take-up of building practices, with which pottery later becomes closely related. In addition, the region contains a number of key sites in the history of Incipient Jomon archaeology.

There are a series of major Incipient Jomon sites in central Honshu, including caves such as Muroya and Kosegasawa, rockshelters and open sites, notably a cluster in the upper reaches of the Shinano, including Motonoki, Unoki, Tazawa and Jin (Nagaoka Shiritsu Kagaku Hakubutsukan 2001; Niigata-ken 1986: 68–72, 78–92). Little has been written about these sites in English, and research is ongoing, which will clarify the nature of Incipient Jomon activity in the upper reaches of the Shinano River (Sato 2001a). The Shinano was a major salmon river (Matsui 1996), and recent work in Japan and in the Russian Far East (Tabarev 2006) is building on earlier hypotheses that salmon was being exploited on a large scale in the Incipient Jomon in this region. A particularly high density cluster of Incipient Jomon sites, including Motonoki, Tazawa and Jin, are known from the broad riverine terraces of the upper reaches of the Shinano River. It is possible to begin to reconstruct the occupational history of the Incipient Jomon landscape, which reflects finely attuned site locations tailored to shifts and changes in the river itself, suggesting that the exploitation of riverine resources was of great significance in the seasonal round of these early Holocene foragers (Sato 2001a). The predominant natural forest type is currently deciduous broadleaf trees including beech and deciduous oak, which were firmly established in the region by the end of the Incipient Jomon.

Many archaeologists now divide the Incipient Jomon, and the Pleistocene–Holocene transition, or the late Palaeolithic–Jomon transition, into a number of phases. For present purposes, I divide the sequences into five stages.

Before Pottery

Since the confirmation of the Palaeolithic occupation of the archipelago at Iwajuku in 1949 (Aikens and Higuchi 1982: 42–46), some 4,500 Pleistocene archaeological sites have been identified (Ono et al. 2002). Central Honshu has a number of important Palaeolithic sites, with evidence for human occupation dating back to before 30,000 years ago. One of the best known of these is a cluster of locations around Lake Nojiri in the northern part of Nagano Prefecture, where bones and tusks of Nauman's elephant and other large Pleistocene fauna were recovered from the muds around the fringes of the lake in 1948 and again in the early 1960s (Aso and Jitsukawa 1983; Nagano-ken 1992: 53–61). One of these sites, Tategahana, produced bone items thought to have been worked. From about 30,000 years ago, a series of knife-blade cultures are recognized from sites such as Chausuyama near Lake Suwa, in the southern part of Nagano Prefecture. Points start to appear about 18,000 years ago, again evidencing stylistic differences, for example the distinctive assemblages from Omegura and Shibukawa in Nagano Prefecture (Nagano-ken 1989: 64–68).

Stone tools were made at Omegura. Some 1,500 worked stone pieces along with 15,000 flakes and other debitage were recovered from an ovate shallow pit surrounded by larger rocks. The materials being used were obsidian, derived from sources in the central mountains of Honshu, including Hoshikuso Pass and Tsumetayama in the Yatsugadake volcanic massif. Particularly good assemblages of stone points are known from sites such as Uenodaira in Suwa City, where over 180 spear points were recovered, over half of which were obsidian, the rest being andesite and chert.

Around 14,000 years ago, stone tools became much smaller with the appearance of microliths in central Honshu. The first microliths discovered in the archipelago were found at Yadegawa in the eastern part of Nagano Prefecture, located on the western edge of a plateau on the southeastern flanks of Yatsugadake (Tsutsumi 2004). The site was first reported in 1853, but was only excavated for three years from 1979, during which time some 10 individual sites and 18 artefact scatters were investigated, proving that Yadegawa was an important site over a considerable period of time. Interestingly, these sites in central Honshu appear to be on a border between two major microlithic traditions. The first links western Honshu and Kyushu with the Korean peninsula and China. The second has closer affiliations with the north, suggesting connections between northern Honshu and Hokkaido, the Russian Far East and Mongolia. This liminality may also have contributed to the creative context within which early pottery was shortly to appear in central Honshu and across much of the archipelago.

One of the most famous of the microlithic sites in central Honshu is Araya (Aikens and Higuchi 1982: 74–78). Located near the confluence of the Shinano River and one of its tributaries, the artefact assemblage, in particular the distinctive Araya burins, link it to the northern microlithic tradition that extends into northeast Asia and Alaska. Sato and colleagues (1994: 178) consider that the 16 features which were discovered during the second excavation (an area of 64 sq m) represent the remains of pit houses which had burnt features resembling interior hearths. These were sealed beneath silt layers, indicative of flooding episodes at this location. The low position of this site, close to the river bed, suggests exploitation of salmon spawning grounds at this early stage. Palaeolithic fishing thus appears to have been an important subsistence activity even before the widely reported evidence from the Incipient Jomon site of Maeda Kochi (Miyazaki and Keally 1986). Some of the stone tools at Araya were being made in very particular ways, notably the famous Araya burins. It is suggested here that the manner of making these stone tools, and the activities involved in fishing, were subject to social choices and constraints as were most other aspects of technology both at this time and later.

Therefore, in the period before the appearance of pottery in central Honshu, big game hunters had been exploiting the rich lithic sources in the central mountains and the region had already taken on something of a liminal character, being the border area between northern and southern traditions of blade and then microlithic manufacture. The production of stone tools was an area of activity, a conceptual space in Boden's terms, in which specific ways of doing things were already important, producing artefacts essential for everyday existence and resulting in the distinctive types of stone tools which are recognized by modern archaeologists. There was already an awareness of the significance of certain locations in the landscape, including sources of obsidian, and places such as Yadegawa and Araya, to which Palaeolithic groups returned again and again. Certain locations within these sites were used for living and everyday activities, as demonstrated by the clusters of artefacts at sites such as Kamiyama (Niigata-ken 1986). Activities already had a certain spatial structuring, suggesting the existence of routinized ways of doing things which took on specific social significances. Tentative steps were being made towards modifying dwelling places through creating scoops in the ground, above which were perhaps tents or other temporary shelters (Sato 1997). These sites were places of considerable information flow, where new ideas could be introduced, discussed and tried out. The scene was set for innovations which would have far-reaching consequences.

Pottery Appears

While the earliest known pottery was being made in the north of Honshu at Odai Yamamoto and other sites across the archipelago including Ushirono,

Terao, Maeda Kochi and Senpukiji, the occupation of central Honshu is represented by sites such as Jin, Okarino and Uenohara. This first phase of pottery production is thought to have lasted some 1,200 years from c. 16,750–15,700 BP to c. 15,750–14,350 BP. The most important tree species in central Honshu were spruce (*Picea*), fir (*Abies*) and hemlock (*Tsuga*), making up a sub-boreal conifer forest, mixed with some deciduous oak (*Quercus [Lepidobalanus]*) (Keally et al. 2003: 9). The presence of the deciduous oak at this early stage is significant, as nut exploitation is an important component of the Incipient Jomon economy, and boiling and leaching nuts to remove toxic tannins became an important subsistence activity for much of the Jomon. This phase is usually correlated with the Older Dryas cold period. Even during the Pleistocene, nut consumption featured in meals that formed the focus of special social events, which became expressions of the social conceptualization of space.

During this first phase, pottery is associated with the Chojakubo-Mikoshiba cultural complex, the main attributes of which we encountered at Odai Yamamoto. The Mikoshiba site itself is located on the right bank of the Tenryu River in Nagano Prefecture (Hayashi 1983; Miyashita 1994). Discovered in 1958, artefacts recovered during subsequent excavations secured its position as the typesite for the Chojakubo-Mikoshiba assemblage. Location A produced a lithic assemblage of over 90 artefacts including 16 points, 14 Mikoshiba-style stone axes, 12 blades, 3 scrapers, 7 cores and flakes. Both the points and the axes were large: the largest point being 25.2 cm in length, the shale and obsidian points taking slightly different shapes. The axes, the largest of which was 23.6 cm long, were partly polished and included large bifacial and unifacial examples, as well as smaller discoidal axes. The edges of some of the axes had been polished using distinctive 'cucumber' shaped whetstones, which are only found in these types of assemblage. The distribution of the artefacts at Location A has been used to suggest that the site was a dwelling, a grave, or a hoard. The objects were arranged in an arc-shape within an oval area measuring 5 m east to west and 3 m north to south, with the cores to the north, axes and whetstones to the northwest, and smaller points, axes and scrapers to the southwest, all clustered around a central space about 1 m in diameter which contained only ash and small pieces of obsidian debitage. Artefacts of a similar nature accompanied the small pottery fragments from Odai Yamamoto, with the exception of arrowheads which do not usually feature in Chojakubo-Mikoshiba assemblages other than at Odai Yamamoto.

The Chojakubo site in Aomori Prefecture was excavated by Sato Tatsuo in 1962, and the discoidal axes discovered there were taken as proof for continental influences on Japan at this time (Tozawa 1994: 544–545). The 'Chojakubo culture' was considered by Sato and Yamanouchi Sugao to be an immediate forerunner of the Jomon, and indeed no pottery was recovered from the site. Recent research, however, has placed the Chojakubo assemblage with those from Mikoshiba in Nagano, and Mosanru in Hokkaido, and they are now considered to be part of the Incipient Jomon.

In the Shinano River region, this phase is represented by Okarino site (Sato 1988). Okarino is located at the confluence of the Uono and Tsunagi Rivers. Layer IV produced plain pottery associated with large points of the Mikoshiba-Chojakubo type, along with microcores and microblades, all sealed beneath the As-YPk tephra. This suggests that either the microblade assemblages were contemporary with the Mikoshiba-Chojakubo assemblage, or at least that the change took place relatively rapidly. Obsidian disk-shaped microcores were also found at this site, reminiscent of those from Uenohara in Nagano Prefecture. At Uenohara, on the western side of Lake Nojiri, a series of five stone-built hearths were discovered in a line at Location C. Artefacts included microcores, microblades, partly polished stone axes, scrapers and hammer stones. A short distance away, forming another 'block' were three cobble arrangements associated with large points and gravers. Again, microcores and partly polished stone axes were found in association with each other. A number of interpretations have been put forward: (1) different groups of people living close to each other; (2) different functional adaptations within one group; (3) a change in technology over a very short time.

Although sites from this phase remain relatively small and few in number, and those with pottery even fewer, several significant observations can be made. Firstly, there is evidence for the deliberate and careful spatial structuring of activities within sites. Although there is no evidence for length of occupation of particular locales, it does seem that certain activities were associated with particular places within the occupational space. Careful spatial structuring of activities could be indicative of social meanings being assigned to those activities. Secondly, the presence of plant fibres in some pottery fragments indicates an early technological association between plants and pottery production. The use of plant fibres for inclusion within clays suggests the recognition of the properties of such tempers and would have required a series of decisions to be taken in the production stages of the life histories of the pots. Thirdly, Fujiyama (2005) has argued that people at this time in the southwestern Kanto area, a region where intensive archaeological activity permits the reconstruction of residential mobility patterns on the basis of detailed site distribution data, led a 'middle-range migratory life in which they moved within a large area'. He detects a significant change in lifestyles between this phase and the following phase (Phase 2), in which 'people led a short-range circling life where their movement was mostly limited within a small area' (Fujiyama 2005: 23). Fujiyama explains these changes in terms of 'environmental factors such as the development of deciduous broadleaf forests. In Phase 2 where new plant resources were developed, intensive resource utilization within a small area became possible, and living behaviours of Phase 1 were drastically changed' (Fujiyama 2005: 23). Pottery was therefore being experimented with by people who were still very mobile, and were thus not yet challenged by the new conceptual space that comprised dealing with the constraints introduced by staying in one location for extended periods of time, although repeated visits to the same location as part of an annual round was quite likely. Lastly, the Shinano-Chikuma region is once again on a boundary at this stage, with sites producing both the Mikoshiba-Chojakubo assemblages, and microlithic assemblages, suggesting a continued intensive flow of information, and fluidity of social networks. It is in this context that the first pottery is being made. The middle-range circulatory lifestyle suggests that social networks were relatively unstable and subject to the regular fracturing of networks of social meanings and associations. The conditions for further innovation were thus set.

Establishment of Decorated Pottery

What happened next represents a major intensification of the innovation process. During the slightly warmer conditions of this second phase of the development of pottery, a stage which is considered to have lasted about 1,700 years from c. 15,750–14,350 BP to c. 13,500–13,150 BP, people began not only to make more pottery, but decorate them as well. Keally and colleagues report from their sample of 17 sites from around Japan that the average number of potsherds increased dramatically from 28 to 494 (Keally et al. 2003).

Why did the quantity and nature of ceramic production change so markedly at this time? One possibility is that the climatic warming at this time, correlated with the Bølling and Allerød warm phases in the European sequence, saw a gradual spread of the warmer species such as deciduous oaks, birch and alder. And yet this development could not have occurred without the first steps which were taken in the previous phase.

The most representative form of pottery decoration was the application of strips of clay to the outer surface of the pottery vessels, the so-called Linear Relief pottery, a form which in the Shinano region is found at sites such as Jin and Kosegasawa. There are, however, only a few sites in the central Honshu which have Linear Relief pottery, although Sato (forthcoming) is able to suggest that there are four stages in the linear relief pottery of the Shinano River. Tazawa, Kubodera Minami, Hizono and Yashikita III all produced artefacts associated with this phase (Sato 2001a).

Tazawa is located on a river terrace close to the confluence of the Shinano River and the Kiyozu River in Niigata Prefecture at a height of 210 m above current sea levels, and some 20 m above the current river bed (Watanabe 1994b). It forms part of an important cluster of sites, with Nakabayashi site some 100 m to the southeast, and Jin approximately 50 m to the west. The site was excavated in 1968 by archaeologists from Tohoku University, and Linear Relief pottery was found in association with points and bifacial axes. The majority of the artefacts were recovered from Layer II, on the level summit of a low natural levee. Over 200 sherds of plain and Linear Relief ware were recovered, but it was thought that the undecorated sherds were from plain parts of the linear relief decorated pots. The sherds were quite thick, 4–5 mm thick, and well fired. A series of quite broad bands of clay had been applied to the outer surface of the vessels, the eponymous Linear Relief design. The lithic assemblage comprised points, unifacial axes, awls, scrapers and cores, and large quantities of flakes and other debitage. Most of the points were the characteristic leaf shapes, with a few tanged points among them. A strong association between quite small chipped unifacial axes and Linear Relief pottery was established at this site.

The Nakabayashi site has a very similar location to Tazawa (Watanabe 1994a). A further 1.1 km away from the Kiyozu River, on the opposite bank, is the famous point production site at Motonoki. Nakabayashi was excavated in 1965 by Serizawa Chosuke, and many points were found in association with pottery. The site spanned some fairly uneven surfaces within the sands and gravels of the terraces, and four layers were discerned within the depressions. Artefacts in Layer III were thought to be in their primary context, indicating a living floor immediately above the gravels. Layers I and II contained points and some later pottery sherds. The lithic assemblage included points, scrapers, cores, and some other indeterminate tools and a large number of flakes. There was no clear association of points with pottery sherds at this site, and Serizawa placed it right at the end of the pre-ceramic period, although Yamanouchi Sugao took this absence of a relationship to support his contention that Serizawa was mistaken to associate pottery and stone points at Motonoki as well.

One of the most important sites for this stage is the cave at Kosegasawa in Niigata Prefecture (Aikens and Higuchi 1982: 107–110). Over 1,000 sherds of pottery were recovered during excavations in 1958 and 1959 undertaken by Nakamura Kozaburo and the Nagaoka Municipal Science Museum (Nagaoka Shiritsu Kagaku Hakubutsukan 2001). These sherds were associated with an extensive lithic assemblage which included over 500 large bifacial projectile points, another 500 smaller points considered arrowheads, over 100 adzes, thought to indicate a lot of woodworking at the site, and blades, awls, burins, gravers, sidescrapers and endscrapers, some of which were tanged, a distinctive artefact type that became very common throughout the Jomon period. Both the lithic and ceramic assemblages at Kosegasawa were reappraised in the early 1990s. This was reported in the proceedings of the meeting of the Japanese Archaeological Association, held in Niigata (Sato et al. 1994). An important result of this reanalysis was the redefinition of the pottery styles into some 22 styles, which were then clustered into six broader groups. This is the earliest evidence for such variability within a ceramic assemblage of the Incipient Jomon, and has important implications for our understanding of the process of ceramic innovation in the archipelago.

It can be suggested that the ceramic assemblage from Kosegasawa represents a new stage in the adoption of pottery. The assemblage indicates a new degree of confidence with working in the medium of clay. It is interesting that this occurs at a time of environmental amelioration, with greater abundance of reliable food resources than would have been available in the previous, colder phase. Yet this is not to say that warmer climatic conditions *per se* were the causal factor in the increase in pottery diversity. Instead, the exploitation of new foodstuffs, using new technologies such as archery, was one of the conceptual spaces, the clash and interaction of which gave rise to conditions for increased creativity, out of which arose the increasing diversity of pottery forms. The same can be argued for the appearance of the tanged scrapers which were to become an essential part of the Jomon lithic toolkit. This heightened creativity in the ceramic conceptual space was perhaps further encouraged by developments in the social networks, which were becoming increasingly regularized and routinized. Sites such as Kosegasawa were probably not occupied all year around, and yet their long-term usage does imply, as Fujiyama (2005) suggests for areas further to the south, that human groups were beginning to range less widely. As social networks became increasingly embedded, it suited the potters to express identity and differentiation in their medium of choice, namely ceramic decoration. The conditions for creativity facilitated new expressions of agency, represented in the sequences of small decisions required to produce increasingly diverse assemblages. And yet none of this would have been possible without the previous centuries of experimentation and working of clay, which provided the expertise to make the enhanced range of ceramic forms found at Kosegasawa. Therefore, while improving environmental conditions facilitated the further development of pottery witnessed in this stage, it was the preceding history of pottery making that underpinned the developments seen at Kosegasawa.

More Pottery in a Cooler Climate

With the return to the colder conditions of the Younger Dryas, correlated with the third phase of the development of pottery production in the Japanese archipelago, the average number of pottery sherds from sites doubled, from 494 in the preceding phase to just over 1,000. The 2,100 years of this phase (c. 13,500–13,150 BP to 11,230–11,180 BP) are reported as witnessing lower sea levels, and therefore a greater exposure of the coastal plain around the archipelago Coniferous forests also regained ground at the expense of deciduous broadleafed forests during this period (Keally et al. 2003: 10). The diversity in pottery styles continued to develop, with new forms including the so-called Nail-Impressed wares, punctate wares and the early appearance of certain types of cord-marking, associated with lithic assemblages comprising arrowheads, querns, grinding stones, scrapers and thumbnail scrapers. Taniguchi Yasuhiro (Odai Yamamoto I 1999) argues for five stages in the pottery from this phase, all exhibiting considerable variation, although Keally and colleagues (2003) note that there is still uncertainty over the details of the sequence.

Unoki Minami is a representative site from this period (Niigata-ken 1986) on a natural levee in the Shinano River, close to the confluence with the Kiyozu River. Two blocks of artefacts were discovered, and refitting of lithic materials suggests they were occupied at about the same time. Flask-shaped pits were found, which contained nail-marked pottery, points, arrowheads and shaft straighteners. It is possible that one of these pits was a burial.

Motonoki is one of the most famous of the Incipient Jomon stone spear production sites. It is located on the third of five terraces on the right bank of the Shinano River in Tsunan Town, Niigata Prefecture, at a height of 220 m above sea level, approximately 50 m above the river bed. There are a number of important sites in the vicinity, including an early Initial Jomon site, Unoki, which produced a significant assemblage of rouletted ware pottery. The first excavation at Motonoki took place in 1956, undertaken by Serizawa Chosuke, with the second excavation being undertaken by Yamanouchi Sugao in 1957, when large quantities of artefacts and features such as pits were discovered. Most of the artefacts were discovered in the layers which comprised the lower parts of the topsoil and the upper parts of the underlying loam. The majority of the artefacts were bifacially worked points, which totalled over 1,500 in all. There were many partly finished points as well as flakes and debitage, and gravers which are thought to have been used to fashion hafts for the spears, indicating that this was an important production site. Other items in the normal Incipient Jomon toolbox, notably stone axes and arrowheads, were conspicuous by their absence. Small quantities of pottery were discovered, mainly with impressions of cordage, but also some undecorated sherds and fingernail impressed wares. This pottery assemblage was termed the Motonoki style, and is located in the later part of the Incipient Jomon.

Serizawa attributed the large number of points to the end of the preceramic age, and that they had subsequently become mixed with the later pottery sherds. Yamanouchi's investigation, however, indicated that the sherds and the points were part of a contemporary original assemblage, and that therefore the site should be considered a Jomon spear production site. These two views formed the opposing sides in what was to become known as the Motonoki Debate. Further investigations undertaken in 2006 were intended to clarify the situation (Motonoki Iseki Chosadan 2006).

If early pottery usage in the Japanese archipelago was a response to changes in the environment, then it might be expected that the return to cold conditions in central Japan would have resulted in less pottery use and increased reliance on the hunting technologies of the Palaeolithic. Instead, however, there followed the development of large-scale production facilities for stone tools, as at Motonoki, and increasing sophistication of pottery production, as evidenced by the new pottery designs and assemblages from sites such as Unoki Minami. By this stage, pottery production and use had become embedded in the set of normal productive activities of these foragers, although the unpredictability of environmental conditions at this time may have contributed to the increasingly regular fracturing of social networks and the need to establish new ones, perhaps involving longer distance contacts. At this stage, while the inhabitants of Motonoki and Unoki Minami were facing environmental instability, early village communities at places like Sojiyama and Kakoinohara were thriving within the deciduous forests of southern Kyushu at the southern tip of the archipelago (Pearson 2006). The foragers of the Younger Dryas were armed with new knowledge based on developments in the earlier phases, which allowed them to intensify their exploitation of the environment, and build on the achievements of earlier generations of potters.

Into the Holocene: Pottery Ascendant

With the end of the Younger Dryas, and the onset of the Holocene, once again warmer climatic conditions encouraged the eastward spread of the deciduous broadleaf forests through most of central and eastern Honshu, while in the southwest of the archipelago, the evergreen oak forests predominated. The first 1,000 years of this ameliorated environment, from 11,230–11,180 cal BP to 10,220–10,160 cal BP, represents the fourth and final phase of the Incipient Jomon, by the end of which the ecological and cultural patterns which were to structure human occupation of the Japanese archipelago for the following eight millennia were firmly established.

During this period, the production of pottery underwent an exponential increase, with an average of over 15,600 pottery sherds from the 10 sites surveyed by Keally and colleagues, a fifteen-fold increase over the previous period. The eponymous distinctive cord-marked decoration (*jomon*) became the predominant decorative style, with the development of the early cord marked, including tajomon series of types (Imamura 1996). Pottery production had become a normal part of the schedule of activities for Jomon communities, a thoroughly embedded technology, and part of what Dobres (2000) terms the 'embodied knowledge' of the Jomon communities.

A key site from this phase is the Muroya Cave, which was originally occupied in the preceding phase. Excavated between 1960 and 1962, as in the case of Kosegasawa, by Nakamura Kozaburo and the Nagaoka Municipal Science Museum (Nakamura 1964), some 5,800 pottery sherds were recovered from the lower layers of the 3 m deep deposits, in association with small arrowheads, scrapers and gravers. The occupation of Muroya continued into the first half of the Early Jomon (c. 7,000 BP). The south-facing mouth of the cave is located about 40 m above the Muroya River, 218 m above the current sea level. The cave is up to 3 m high, 7 m wide, and 8 m deep, and the entrance comprises a 2 m wide terrace. The Muroya Lower assemblage was characterized by flat-bottomed vessels with a variety of decorative motifs employed around a thick band of clay along the outer rim (Nagaoka Shiritsu Kagaku Hakubutsukan 2001; Nakamura 1964; Watanabe 1994c: 623).

A further development was the widespread appearance of pit dwellings associated with early cord-marked pottery assemblages (Suzuki 2001). In the Shinano region, the Hizono site (Sato 1994) produced seven such structures, while key locations such as Muroya and Motonoki also continued to be utilized. The Tajomon phase is important as it saw the first full-scale pit dwellings being constructed in central Japan and the development of village communities comparable to those which had flourished earlier in southern Kyushu. The construction of these buildings, and the new forms of engagement with the soil they represented, that is, digging, itself represents a new 'conceptual space', as well as creating new forms of space for inhabitation. As with the creation of ceramic containers, which required the repetition of a sequence of actions, each of which were predicated on prior knowledge and specific decisions affecting each segment of the *chaîne operatoire*, so with house construction, each stage of which also resulted from decisions constrained and enabled by already established ways of operating.

One aspect of Jomon house building which is of particular note in the present context is the way in which houses were often built overlapping with the remains of previous structures, as seen at sites such as Oshikakubo in Shizuoka Prefecture (Kaner 2003). In different contexts, Jones (2001) has identified a homology between making pots and building houses. It is significant that in the fourth phase of the Incipient Jomon both house construction and pottery manufacture become major routinized practices within Jomon activity systems.

The storage pit from Higashi Kurotsuchida in Kyushu is often regarded as evidence for significant storage activities in the Incipient Jomon, and an accompanying shift from what James Woodburn identified as immediate return systems to delayed return systems. As Richard Pearson (2006) has noted, however, there are very few such pits in the Incipient Jomon, in contrast to the later Jomon period where pits become commonplace. Pit digging, representing a particular form of material engagement with the surrounding world, was not a new phenomenon in the Incipient Jomon. Already by 20,000 years ago, as noted by Takeshi Tsutsumi (2002), late Palaeolithic hunters were employing pit-digging technology to construct traps for animals at sites such as Hatsunegahara in Shizuoka Prefecture. The digging of pit traps became relatively common in some areas in the Initial Jomon, at sites such as Kirigaoka (Imamura 1996). In addition to this, the digging of certain resources out of the ground—notably obsidian from locations such as the Wada Pass in Nagano Prefecture, where a series of scooped out depressions remain from Jomon exploitative practices—probably formed part of the Incipient Jomon repertoire of actions (Kokuyoseki 2004).

Why Early Pottery in the Japanese Archipelago?

Current understanding of the development of pottery from central Japan suggests a rather different scenario from its appearance in southern Kyushu (Pearson 2006). In the upper reaches of the Shinano River, foragers were exploiting fish from very early on, as attested to by the location of Araya. They were also making use of the well-developed lithic exploitation networks that took them into the higher mountain areas. The relatively unstable environmental conditions which characterized the Incipient Jomon brought new possibilities for subsistence, as the large fauna represented by sequences at Lake Nojiri vanished. People started to eat different things. They occupied good fishing locations and cave sites, probably on a seasonal basis but over long periods of time, and they created pottery vessels for their use at those sites. They made pots which exhibited considerable variability, even within established traditions of decoration and form. I suggest that this variability is a reflection of creativity at work in the negotiation and renegotiation of what Dobres (2000) terms the 'social values of artefact production'. The minute details of making each pot, the pushing, pulling, stabbing, dragging, twisting and puncturing of the pottery surface each required small, repeated and deliberate actions on the part of these early potters. Observing how this was done, remembering and then reproducing these acts of creation, comprises the establishment of 'embodied knowledge', drawn on to create new objects through the performance of a series of often complex sequences of actions, each of which had particular cultural and social meanings. This embodied knowledge of ceramic production complemented other forms of embodied knowledge, each of which can be identified as having its own conceptual space, whether that be the manufacture of stone tools, or perhaps most significantly for these Incipient Jomon communities, the exploitation and manipulation of riverine and plant resources, in particular nuts. It was the clash of these different 'conceptual spaces', as argued by Margaret Boden (1998), that gave rise to the conditions for creativity within the Incipient Jomon. This creativity was expressed through pottery in the first three phases of the Incipient Jomon, but only in a limited way. People were experimenting with ceramic manufacture, but within the constraints of a mobile way of life. It was only in the later stages of the Incipient Jomon, at the dawn of the Holocene, that Jomon groups began to engage with ceramics on a much larger scale, at the same time that they began to build more substantial structures.

The elements of this story have been known for some time. What has been lacking until now is an appropriate interpretative framework within which to place them. We can now begin to postulate about why it took such a long time for pottery manufacture to really take off in the archipelago. Not only did the environmental conditions have to be right, but people needed to develop the traditions of embodied knowledge that gave them the ontological security to invest in larger scale ceramic production, the confidence in the embodied knowledge of ceramic manufacture. The difference between the Palaeolithic fishers who were catching and eating salmon from the Shinano River at Araya, and their successors who were enjoying salmon feasts at Maeda Kochi, was the accumulated experience of 5,000 years of experimentation in pottery making. These skills were gradually embedded into their normal, routinized, social technologies, and included the ability to find suitable clays, form vessels and decorate the leather-hard ceramic bodies with twisted plant fibres. It was this embodied knowledge that underpinned the economic needs for pottery production, the rigours of supply and demand, suggested by James Brown.

Conclusion: The Social Embodiment of Early Pottery

In this chapter, I have considered how pottery production during the early stages of pottery usage created a new materiality for the foragers of the central area of the prehistoric Japanese archipelago, and how this new materiality was shaped by a recursive relationship between the foragers and their environment, a relationship mediated by their own, albeit imperfect, knowledge of the material conditions of their existence.

The past several years have witnessed a dramatic increase in the quantity and quality of data available for the earliest pottery-using societies in the Japanese archipelago. Developments in radiocarbon dating have been matched by increasing sophistication in models of social and cultural development, but these have not yet achieved an explanatory or interpretive level which really addresses how and why pottery appeared as early as it did, and why it then took so long to take hold in the archipelago.

I have reviewed the evidence for early pottery usage in the central part of the Japanese archipelago. The interpretations and explanations for the appearance of pottery have been constrained by the nature of archaeological discourse on the Palaeolithic–Jomon, or Pleistocene–Holocene transition. I have argued that a more contextual approach is now feasible, one in which earlier work on the processes of innovation can usefully be revisited, so as to be integrated with an understanding of the conditions for creativity that structured the long process by which pottery making became an established part of social practice in the Jomon period.

The detailed work on early ceramics in the Japanese archipelago available in Japanese literature, and ongoing research into Jomon ceramic production processes, mean that we have available a good understanding of the life histories of these vessels. With contemporary studies on social agency and technology to hand, there is also no doubt that Jomon archaeology can provide rich contextual case studies. In order to fulfil the potential offered by this fascinating material, however, we must recall Dobres' contention that: 'the detailed materials analysis of artifact life histories should be undertaken not only to identify and describe decision-making sequences of artifact making, use and repair activities, but also to understand the social contexts and organisational dynamics structuring and giving meaning to them' (Dobres 2000: 164).

Acknowledgements

This paper was inspired by discussions at an international conference on 'Contexts for the Earliest Pottery in the World' held at the Universities of Cambridge and East Anglia in November 2001, sponsored by the Japan Foundation, the Great Britain Sasakawa Foundation, the Daiwa Anglo-Japanese Foundation and Japan 2001, organized by the author in conjunction with an exhibition of Jomon archaeology at the Fitzwilliam Museum in Cambridge. I am grateful to the participants for sharing their knowledge on that occasion, in particular to Professor Kobayashi Tatsuo, Miyao Toru, Nakamura Oki and Sato Masaichi who subsequently introduced me to the sites of central Honshu, under the auspices of the British Academy funded Shinano River Project. Thanks are due to the editors of the present volume for their patience and to Dr. Liliana Janik for invaluable advice on earlier drafts.

References

Aikens, C M (1995) 'First in the world: the Jomon pottery of early Japan', in Barnett, W K and Hoopes, J W (eds), *The Emergence of Pottery: Technology and Innovation in Ancient Societies*, Washington DC: Smithsonian Institution Press, pp 11–22

Aikens, C M and Akazawa, T (1996) 'The Pleistocene–Holocene transition in Japan and adjacent northeast Asia: climatic change, broad-spectrum diet, pottery and sedentism', in Straus, L G, Eriksen, B V, Erlandson, J M and Yesner, D R (eds), *Humans at the End of the Ice Age: The Archaeology of the Pleistocene–Holocene Transition*, New York: Plenum, pp 215–228

Aikens, C M and Higuchi, T (1982) *Prehistory of Japan*, London and New York: Academic Press

Arnold, D E (1985) *Ceramic Theory and Cultural Process*, Cambridge: Cambridge University Press

Aso, M and Jitsukawa, J (1983) 'Nagano-ken Nojiri-ko no isekigun [The site group at Lake Nojiri, Nagano Prefecture]', in Tozawa, M and Ambiru, M (eds), *Sendoki Jidai no Iseki [Preceramic Sites]*, Tokyo: Yuzankaku, pp 303–309

Binford, L R (1968) 'Post-Pleistocene adaptations' in Binford, S R and Binford, L R (eds), *New Perspectives in Archaeology*, Chicago: Aldine, pp 313–341

Boden, M A (1998) 'What is creativity?' in Mithen, S (ed), *Creativity in Human Evolution and Prehistory*, London: Routledge, pp 22–60

Brown, J A (1989) 'The beginnings of pottery as an economic process', in van der Leeuw, S and Torrence, R (eds), *What's New? A Closer Look at the Process of Innovation*, London: Unwin Hyman, pp 203–221

Dobres, M A (2000) *Technology and Social Agency*, Oxford: Blackwell

Fujiyama, R (2005) 'Hyoga jidai shumatsuki no kyoju kodo ron [Living behaviours at the end of the Ice Age]', 20 *Nihon Kokogaku,* 1–25 (in Japanese)

Habu, J (2004) *Ancient Jomon of Japan*, Cambridge: Cambridge University Press

Hayashi, S (1983) 'Nagano-ken Mikoshiba iseki [The Mikoshiba site, Nagano Prefecture]', in Tozawa, M and Ambiru, M (eds), *Sendoki Jidai no Iseki [Preceramic Sites]*, Tokyo: Yuzankaku, pp 286–290

Hayden, B (1995) 'The emergence of prestige economies and pottery', in Barnett, W K and Hoopes, J W (eds), *The Emergence of Pottery: Technology and Innovation in Ancient Societies*, Washington DC: Smithsonian Institution Press, pp 257–266

Ikawa-Smith, F (1986) 'Late Palaeolithic and Early Holocene technologies', in Pearson, R, Barnes, G L and Hutterer, K (eds), *Windows on the Japanese Past: Studies in Archaeology and Prehistory*, Michigan: Center for Japanese Studies, pp 199–216

Imamura, K (1996) *Prehistoric Japan: New Perspectives on Insular East Asia*, London: UCL Press

Jones, A (2001) *Archaeological Theory and Scientific Practice*, Cambridge: Cambridge University Press

Kaner, S (2003) 'The oldest pottery in the world', 1 *Current World Archaeology*, 44–49

Keally, C T, Taniguchi, Y and Kuz'min, Y V (2003) 'Understanding the beginning of pottery technology in Japan and neighbouring East Asia', 24(2) *The Review of Archaeology*, 3–14

Kidder, J E (1993) 'The earliest societies in Japan', in Brown, D M (ed), *Cambridge History of Japan*, Volume 1, *Ancient Japan*, Cambridge: Cambridge University Press, pp 80–107

Kobayashi, T (2004a) *Jomon Reflections* [edited and adapted by S Kaner and O Nakamura], Oxford: Oxbow Books

——— (2004b) 'Re-thinking the first pottery use and the concept of the Incipient Jomon in Japan', in *Recent Palaeolithic Studies in Japan: Proceedings for Tainted Evidence and the Restoration of Confidence in the Pleistocene Archaeology of the Japanese Archipelago*, Tokyo: The Japanese Archaeological Association, pp 61–67

Kokuyoseki Taiken Myujiamu (2004) *Kokuyoseki no Gensanchi o Fukuru: Takayama Iseki Gun [Investigating the Obsidian Source: the Takayama Site Group]*, Tokyo: Shinsensha

Matsui, A (1996) 'Archaeological investigations of anadromous salmonid fishing in Japan', 27(3) *World Archaeology*, 444–460

Miyashita, K (1994) 'Mikoshiba iseki [The Mikoshiba site]', in Tozawa M (ed), *Jomon Jidai Kenkyu Jiten [Dictionary of Jomon Period Research]*, Tokyo: Tokyodo, p 611

Miyazaki, H and Keally, C T (1986) 'A terminal Pleistocene salmon-fishing and lithic worksite at Maeda Kochi, Tokyo, Japan', 3 *Current Research in the Pleistocene*, 96–97

Mizoguchi, K (2003) *An Archaeological History of Japan 30,000 BC–AD 700*, Philadelphia: Pennsylvania University Press

Motonoki Iseki Chosadan (2006) Niigata-ken Naka-Uono-gun Tsunan-machi Motonoki iseki hakkutsu chosa panfuretto [Pamphlet produced for the excavation of the Motonoki site, Tsunan Town, Naka-Uono County, Niigata Prefecture], unpublished photocopy held on file

Nagaoka Shiritsu Kagaku Hakubutsukan (2001) *Juyo Bunkazai Koko Shiryo Ten: Kaen Doki to Kosegasawa, Muroya Doketsu Shutsudohin [Exhibition of Important Archaeological Cultural Properties: Flame Pottery and Excavated Objects from the Kosegasawa and Muroya Caves]*, Nagaoka: Nagaoka Municipal Museum of Science

Nagano-ken (1989) *Nagano-ken Shi, Tsushi Ron 1: Genshi Kodai [History of Nagano Prefecture. Narrative Volume 1: Prehistoric and Ancient Periods]*, Nagano: Nagano-ken Kankokai

——— (1992) *Nagano-ken shi [History of Nagano Prefecture]*, Nagano City: Nagano ken Kankokai

Nakamura, K (1964) *Muroya Doketsu [Muroya Cave]*, Nagaoka: Nagaoka Municipal Museum of Science

Niigata-ken (1986) *Niigata-ken Shi. Tsushi Ron 1. Genshi kodai [History of Niigata Prefecture. Narrative Volume 1: Prehistoric and Ancient Times]*, Niigata: Kyokukosha

Odai Yamamoto I Site Excavation Team (eds) (1999) *Odai Yamamoto I Iseki no Kokogaku Chosa [Archaeological Research at the Odai Yamamoto I Site]*, Tokyo: Kokugakuin University

Oguma, H, Suzuki, T, Kitamura, R, Ushino, T, Dobashi, Y, Tachiki, H, Suganuma, W, Yoshii, M, Sawada, A, Maeyama, K and Ono, A (1994) 'Niigata-ken Kosegasawa doketsu iseki shutsudo ibutsu no saikento [A re-examination of the artefacts from Kosegasawa Cave, Niigata Prefecture]', in Ono, A and Suzuki, T (eds), *Kan Nihonkai Chiiki no Doki Shutsugenki no Yoso: 1993 Nendo Nihon Kokogaku Kyokai Shinpojiumu [Aspects of the Period of the Development of Pottery in the Circum-Japan Sea Region: 1993 Japanese Archaeological Association Symposium]*, Tokyo: Yuzankaku, pp 77–174

Ono, A, Sato, H, Tsutsumi, T and Kudo, Y (2002) 'Radiocarbon dates and archaeology of the Late Pleistocene in the Japanese islands', 44(2) *Radiocarbon*, 477–494

Ono, A and Suzuki, T (1994) *Kan Nihonkai Chiiki no Doki Shutsugenki no Yoso: 1993 Nendo Nihon Kokogaku Kyokai Shinpojiumu [Aspects of the Period of the Development of Pottery in the Circum-Japan Sea Region: 1993 Japanese Archaeological Association Symposium]*, Tokyo: Yuzankaku

Pearson, R (2005) 'The social context of early pottery in the Lingnan region of south China', 79 *Antiquity*, 819–828

——— (2006) 'Jomon hot spot: increasing sedentism in south-western Japan in the Incipient Jomon (14,000–9,250 cal BC) and Earliest Jomon (9,250–5,300 cal BC)', 38(2) *World Archaeology*, 239–258

Sato, H (1997) 'Nihon kyusekki jidai kenkyu to kyoju keitai ron [Japanese Palaeolithic research and theories of dwelling form]', in *Sumai no Kokogaku [The Archaeology of Dwelling]*, Tokyo: Doseisha, pp 2–12

Sato, M (1988) *Okarino Iseki [The Okarino Site]*, Yuzawa: Yuzawa Town Board of Education

——— (1994) *Hizono Iseki [The Hizono Site]*, Nakazato: Nakazato Village Board of Education

——— (2001a) The Incipient Jomon in the Upper Reaches of the Shinano River, Niigata Prefecture. Paper presented at the conference, Contexts for the Oldest Pottery in the World, Cambridge and Norwich, November

——— (2001b) 'Shinanogawa joryuiki no iseki to ibutsu no yoso [Aspects of artefacts and sites from the upper drainage of the Shinano River]' in *Nagaoka Shiritsu Kagaku Hakubutsukan (ed), Juyo Bunkazai Koko Shiryo Ten: Kaen Doki to Kosegasawa Muroya Dokutsu Shutsudo Ten [Exhibition of Archaeological Data including Important Cultural Materials: Exhibition of Artefacts Excavated from the Kosegasawa and Muroya Caves and Flame pots]*, Nagaoka: Nagaoka Municipal Science Museum, pp 113–125

Sato, M, Hoshino, Y, Ishizaka, K and Oka, S (1994) 'Shinanogawa suikei ni okeru Jomon jidai sosoki iseki no yoso [Aspects of Incipient Jomon sites in the Shinano River drainage]', in Ono, A and Suzuki, T (eds), *Kan Nihonkai Chiiki no Doki Shutsugenki no Yoso: 1993 Nendo Nihon Kokogaku Kyokai Shinpojiumu [Aspects of the Period of the Development of Pottery in the Circum-Japan Sea Region: 1993 Japanese Archaeological Association Symposium]*, Tokyo: Yuzankaku, pp 175–200

Sempukuji Cave Research Publication Committee (eds) (2002) *Sempukuji Doketsu Kenkyu Ron [Research on Sempukuji Cave]*, Sasebo: Showado

Shiraishi, H (1994) 'Jomon jidai sosoki no shudan kozo e no sekkin [An approach to the social structure of the Incipient Jomon period]', 5 *Jomon Jidai*, 1–28

——— (2004) 'Approaches to Late Paleolithic settlement system in Japan', in *Recent Palaeolithic Studies in Japan: Proceedings for Tainted Evidence and the Restoration of Confidence in the Pleistocene Archaeology of the Japanese Archipelago*, Tokyo: The Japanese Archaeological Association, pp 47–60

Suzuki, C (2001) 'Jomon sosoki jukyoshi to jukyogata no ato [Incipient Jomon houses and house-shaped features]', 12 *Jomon Jidai*, 1–26

Tabarev, A V (2006) 'People of salmon: technology, art and ritual of stone age cultures, Russian Far East', in *Archaeological Elucidation of the Japanese Fundamental Culture in East Asia, Proceedings of the Kokugakuin University 21st Century COE Program International Symposium*, Tokyo: Kokugakuin University, pp 111–120

Taniguchi, Y (2004) 'Calibrated radiocarbon dates for the earliest pottery assemblages in the Japanese archipelago', 519 *Kokogaku Janaru*, 4–10

Tozawa, M (ed) (1994) *Jomon Jidai Kenkyu Jiten [Dictionary of Jomon Period Research]*, Tokyo: Tokyodo Shuppan

Tsutsumi, T (2002) 'Origins of pottery and human strategies for adaptation during the termination of the Last-Glacial period in the Japanese archipelago', in Yasuda, Y (ed), *The Origins of Pottery and Agriculture*, New Delhi: Roli Books and Lustre Press, pp 241–262

——— (2004) *Hyoga ki o Ikinuita Kariudo [Native Hunters of the Glacial Age: Yadegawa Site]*, Tokyo: Shinsensha

Watanabe, H (1994a) 'Nakabayashi iseki [Nakabayashi site]', in Tozawa, M (ed), *Jomon Jidai Kenkyu Jiten [Dictionary of Jomon Period Research]*, Tokyo: Tokyodo Shuppan, pp 567–568

——— (1994b) 'Tazawa iseki [Tazawa site]', in Tozawa, M (ed), *Jomon Jidai Kenkyu Jiten [Dictionary of Jomon Period Research]*, Tokyo: Tokyodo Shuppan, p 537

——— (1994c) 'Muroya dokutsu [Muroya Cave]', in Tozawa, M (ed), *Jomon Jidai Kenkyu Jiten [Dictionary of Jomon Period Research]*, Tokyo: Tokyodo Shuppan, p 623

Yasuda, Y and Miyoshi, N (1999) *Nihon Retto Shokuseishi [History of Vegetation in the Japanese Archipelago]*, Tokyo: Asakura Shoten

Yokohama-shi Rekishi Hakubutsukan and Yokohama-shi Furusato Rekishi Zaidan Maizo Bunkazai Senta (1996) *Jomon Jidai Sosoki [The Incipient Jomon]*, Yokohama: Yokohama City Historical Museum and Yokohama City Archaeological Research Centre

Chapter 3

Pottery Making in Prehistoric Cultures of the Russian Far East

Irina S Zhushchikhovskaya

Introduction

In this chapter I will examine evidence for the emergence of pottery making traditions amongst hunter-fisher-gatherer cultures of the Russian Far East. Systematic programmes of archaeological excavation conducted from the 1950s to the present have led to the discovery of a large number of sites and the recovery of immense quantities of archaeological material. Abundant pottery assemblages are a characteristic feature of most archaeological sites, especially in the southern regions of the Russian Far East. Analysis of these ceramic traditions has generated important insights into questions of chronological and cultural patterns of prehistoric society (Andreev 1957, 1960; Andreeva 1991; Andreeva et al. 1986; Brodianskii 1987; Derevianko 1973, 1976; Myl'nikova 1999; Okladnikov 1970, 1984; Vasil'evskii and Golubev 1976; Zhushchikhovskaya 1998, 2001, 2005; Zhushchikhovskaya and Shubina 2006). During the last two decades, the tendency towards social, economic and palaeoecological reconstruction in Far Eastern archaeology had to be strengthened (Derevianko and Kononenko 2003; Kuz'min 2005; Popov et al. 1997; Vasil'evskii et al. 1997; Vostretsov 1986, 1998; Vostretsov et al. 2003). The analytical approaches are also applied to the study of archaeological ceramic assemblages and prehistoric pottery making (Grebenshchikov and Derevianko 2001; Myl'nikova 1999; Zhushchikhovskaya 2001, 2005).

I will examine here the links between broader natural, social and economic factors and the development of distinct ceramic traditions in different regions of the southern Russian Far East. The problem of emergence, or invention of pottery technology and the problem of transformation of pottery making traditions in the course of their history are accented separately in the chapter. The research is based on methods and approaches accepted in current archaeology (Arnold 1985; Hally 1986; Rice 1987; Shepard 1985).

The research location includes two mainland regions—Primor'e and Priamur'e, or the lower reaches of the Amur river—and the insular region of Sakhalin Island (Figure 3.1). All three territories fall within moderate latitudes and lie alongside the northern parts of the Sea of Japan.

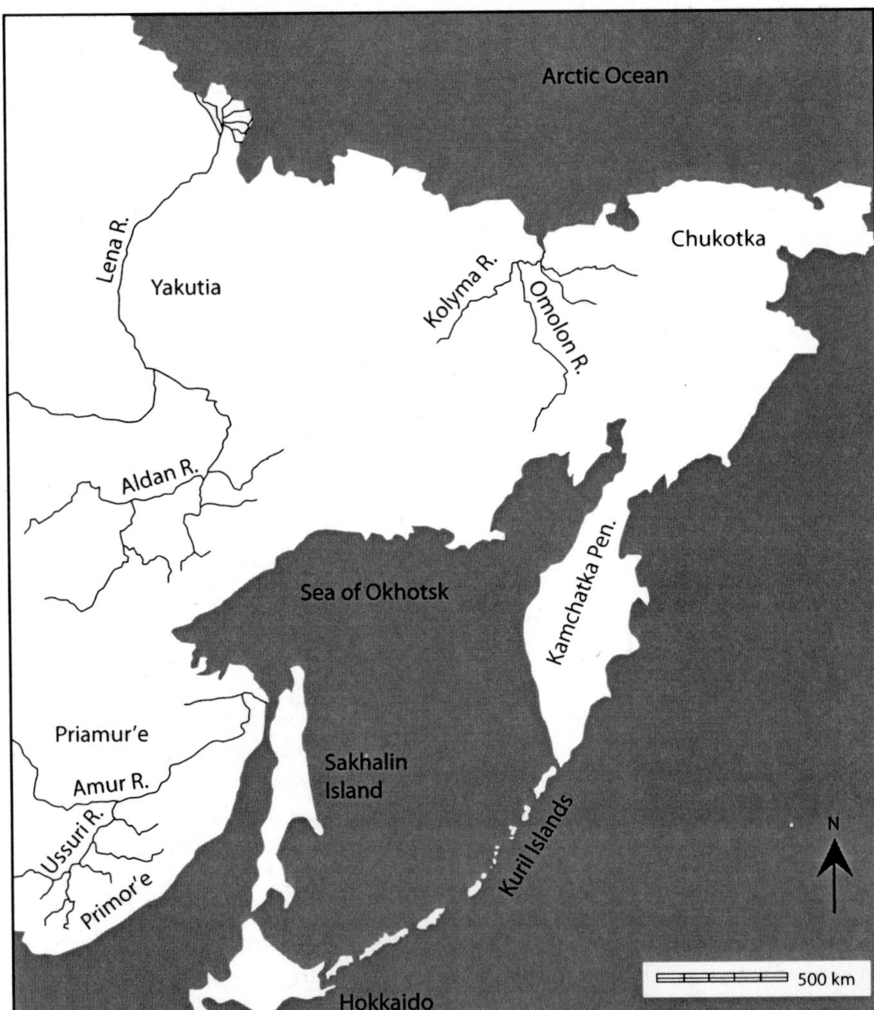

Figure 3.1 Research area: southern regions of the Russian Far East.

Chronological frames extend from about 13,000–10,000 bp to 1200–1300 AD. The oldest date correlates with the late Palaeolithic–Neolithic transition period when most early pottery was made in both the Russian Far East and East Asia as a whole. The later date corresponds to the last stage of prehistoric society represented by Okhotsk culture in Sakhalin Island.

The basic culture-historical divisions determining the dynamics of the historical process within the outlined temporal frameworks are the late Palaeolithic–Neolithic Transition, the Neolithic, and the Palaeometal period[1]. These definitions have certain archaeological contents and temporal limits which will be explained in the course of this discussion.

These locations are characterized by divergent economic and cultural histories (Andreeva et al. 1986; Kuz'min 2005; Vasilevskii 2000a, 2000b, 2003; Vasil'evskii and Golubev 1976; Vasil'evskii et al. 1997; Vostretsov 1986, 1996, 1998; Vostretsov et al. 2003). A hunter-gatherer mode of subsistence dominated in the Primor'e and Priamur'e regions during the late Palaeolithic–Neolithic transition period and through most of the Neolithic. Crops and animal domestication developed later during the late Neolithic and Palaeometal period. On Sakhalin Island the prehistoric cultures subsisted through hunting-gathering; domesticated crops and animals are not found in this territory (Table 3.1).

These economic differences caused divergent social developments. The first millennium AD witnessed the emergence of early political states in

Table 3.1 Economic transformations in prehistoric cultures of the Russian Far East.

Region Period	Southern Mainland Part (Priamur'e, Primor'e)	Insular Part (Sakhalin Island)
Palaeolithic–Neolithic transition (Early Neolithic)	***11,000–7,000 uncalibrated BC:*** Land hunting, fishing, plant gathering.	***7,000–6,000 uncalibrated BC:*** Land hunting, fishing, plant gathering.
Neolithic	***6,000–1,500 bc:*** *Gathering Branches:* Fishing, land hunting, plant and marine gathering. *Producing Branches:* Initial agriculture ***(from 3–2 Mil. BC)***	***5,000–1,000 uncalibrated BC:*** *Gathering Branches:* Land hunting, fishing, plant gathering.
Palaeometal	*2,000 bc to c. AD 500 (uncalibrated):* *Gathering Branches:* Fishing, land hunting, plant and marine gathering. *Producing Branches:* Primitive agriculture + domestication.	*From around AD 1 to AD 1500 (uncalibrated)* Sea-mammal hunting, fishing, plant and marine gathering.

Priamur'e and Primor'e' while the hunting-fishing-gathering society continued in Sakhalin into the 17th and 18th centuries when European explorers discovered the region (Krushanov 1989).

I present two case studies here. Case study I concerns the temporal dynamics of prehistoric pottery making in the regions of the southern mainland Russian Far East, Primor'e and Priamur'e (the basin of Amur river) (Figure 3.1). These regions have similar natural conditions as well as a shared economic and social history. Therefore, I will consider summarized materials of both regions.

Case study II concerns pottery making in the hunter-gatherer cultures of the isolated and insular Sakhalin Island.

These cases show resemblances and differences in pottery making reflecting regional patterns of prehistory.

Case Study I: Pottery Making in the Southern Mainland Regions of the Russian Far East

1. Palaeolithic–Neolithic Transition Period: The Beginning of Pottery

The late Palaeolithic–Neolithic transition period in the southern Russian Far Eastern area corresponds to the late Pleistocene–early Holocene period and dates to about 13,000–9,000 bp. The main features of this period are the combining of archaic Palaeolithic traditions and some new elements of Neolithic context. Abrasive techniques in stone processing and new types of stone tools were introduced gradually. Alongside hunting-gathering, fishing began to play an important role in subsistence. Salmon fishing marked a shift to a sedentary or semi-sedentary mode of life in the most favourable zones (Derevianko and Kononenko 2003; Vasil'evskii et al. 1997). A significant innovation of the late Palaeolithic–early Neolithic period was the invention of pottery making technology. Based on this, some researchers suggest an alternative definition of this culture-historical period—the Early or Incipient Neolithic (Derevianko and Medvedev 1995; Kuz'min 2005).

The earliest pottery sites include Gasya, Khummy and Gosyan in the Lower Amur river basin, Gromatukha in Middle Amur river basin and Chernigovka 1, Gorny Khutor and Ustinovka 3 in Primor'e (Figure 3.2; Table 3.2) (see Derevianko and Kononenko 2003; Derevianko and Medvedev 1995; Garkovik 2002; Jull et al. 1998; Lapshina 1999; Medvedev 1993, 2003a; Zhushchikhovskaya 1997, 2005: 12–20).

Pottery emerged in the late Pleistocene–early Holocene era in the southern mainland of the Russian Far East and the larger East Asian region including Japan, eastern and south-eastern China and the southern Korean

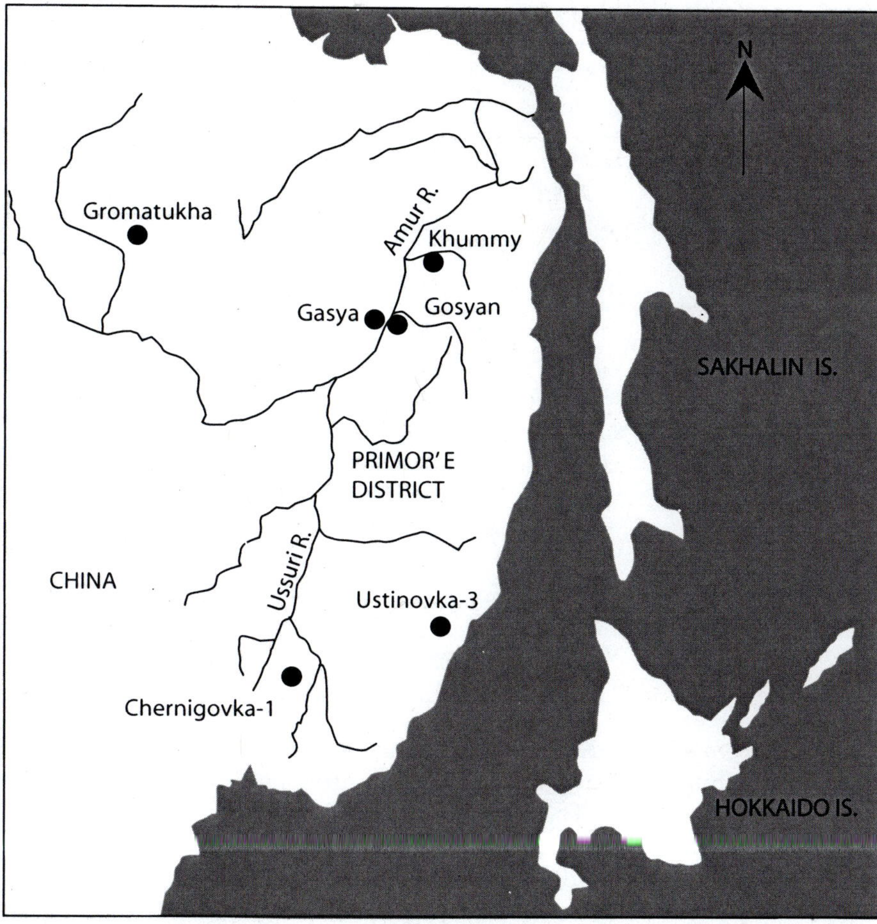

Figure 3.2 Archaeological sites containing earliest pottery (Primor'e and Priamur'e regions).

peninsula. Important traits shared by both Far Eastern sites and sites with the oldest pottery from other East Asian regions are the appearance of progressive elements in the lithic industry and the beginning of a tendency to a sedentary mode of life (Kajiwara and Yokoyama 1995; Kobayashi 2004: 7–17; Myl'nikova and Nesterov 2008; Underhill 1997).

Detailed descriptions of early pottery assemblages have been presented in several recent publications (Lapshina 1999; Medvedev 2003a; Zhushchikhovskaya 1997, 2005: 12–20). In general, early pottery tends to be represented by limited quantities of highly fragmented sherds, which limit the scope of insights and conclusions that can be drawn from this material. Despite these constraints, interesting ceramic assemblages have been

Table 3.2 Archaeological sites with early pottery (Russian Far East).

Site Name	Location	Type of Site	Referenced Cultural Layer	Date
Gasya	Lower Amur area	Multi-layered settlement	Lower layer	12,480±120 bp (charcoal)—lower part of layer, 10,875±90 bp (charcoal)—upper part of layer.
Khummy	Lower Amur area	Multi-layered settlement	Lower layer	13,260±100–10,345±110 bp (charcoal)
Goncharka 1	Lower Amur area	Multi-layered settlement	Lower layer	10,000–8,000 BP—provisionally.
Gromatukha	Middle Amur area	Multi-layered settlement	Lower layer	14,510±240–8,770±60 bp (micro-testing of organic matter from ceramic samples)
Chernigovka-1	Western Primor'e	Multi-layered settlement	Lower layer	10,770±75–7,475±65 bp (micro-testing of organic matter from ceramic samples)
Ustinovka-3	Eastern Primor'e	Multi-layered settlement	Lower layer	9,360±30 bp (dating of ceramic samples)

recovered from Gasya, Khummy, Gosyan, Gromatukha, Chernigovka 1, and Gorny Khutor, all of which have been attributed to the same broad technological tradition. These technologies share similar paste recipes, which include the addition of plant tempers. Traces of carbonized grass have been detected within the ceramic body, and in certain cases, the species of plant can be determined; for example, samples from Chernigovka 1 and Khummy include *Cyperaceae*.

The earliest pottery is characterized by simple forms, with open mouths and straight walls. It is difficult to deduce the shape of the vessel base from the remains except in Gasya, where a flat-bottomed vessel has been recovered. Surface treatments like rubbing, smoothing and slipping were not practised. Firing temperatures were low, from 450 to 600°C, suggesting that vessels were fired on open hearths. These features are common to early pottery in various regions of the world—Near East, Central America, East Asia and some others (Amiran 1965; Barnett and Hoopes 1995; Hoopes 1994; Rice 1999; Zhushchikhovskaya 2005: 24–26).

The most intriguing problem is the recognition of the forming method used by the earliest Far Eastern potters. Any direct evidence is limited but some assumptions can be made from the zig-zag or net-like impressions on the surfaces of vessels (Figure 3.3). These impressions were largely on the

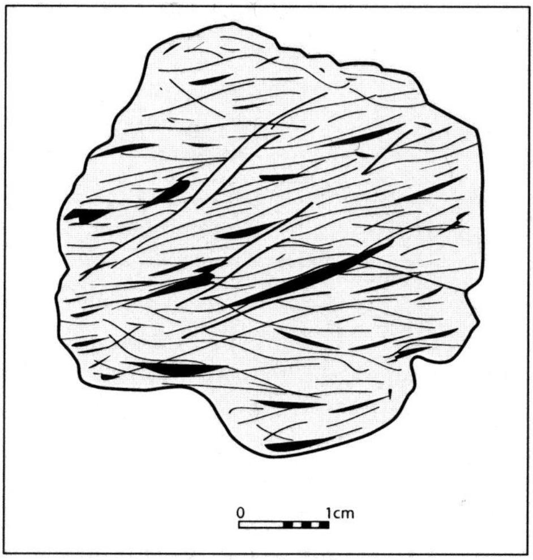

Figure 3.3 Inner surface of pottery fragment from Chernigovka-1 (Primor'e Region), with net-like impressions.

outer surfaces, and in some cases, on the inner surfaces as well. After detailed analysis these impressions were explained as the negatives of plaited or woven items. Experimental studies show that baskets or cord bags might have been used as moulds during pottery making (Zhushchikhovskaya 1997, 2005: 21–24).

It is difficult to surmise the functions of these old vessels. Certain sherds recovered from Gasya show a thick layer of carbonized residues on both outer and inner surfaces indicating that they might have been used for cooking (Medvedev 2003a). Residue analysis to reconstruct practical uses of these vessels has yet to be carried out.

Several important points emerge concerning the production and function of the earliest pottery of the Russian Far East. First, pottery production in the communities of the late Pleistocene–early Holocene period was at an incipient stage. Pottery was not manufactured on a large scale judging by the frugal recovery of sherds at the sites. Technological and morphological characteristics correspond with early pottery in various world regions and reflect the lowest level of pottery making.

Second, a distinctive feature of early Far Eastern pottery making seems to be the usage of basketry and plaited containers as shape moulds. The first pottery samples were based upon the older and more established basketry and plaiting crafts (Adovasio 1977; Adovasio and Lynch 1973; Crowfoot 1954). Containers made of plant material served as models for early pottery which obviously had a different function because their production included the heating process. In general the archaeological record of the southern

Russian Far East confirms the well-known work of H Wormington and A Neal, who uncovered the links between basketry and early pottery by examining ethnographic data from Native Americans (Wormington and Neal 1951).

Third, the specific use and function of early pottery in the Russian Far East has yet to be determined. A systematic study of organic residues in these early ceramics could throw light upon the actual usage of early vessels.

Important areas of research are the reasons and conditions for the early emergence of pottery in the Far East and East Asia. According to C Serizawa and F Ikawa-Smith, the invention of pottery technology in East Asia was a result of the changing ecological situation in the late Pleistocene–early Holocene period. The appearance of new kinds of plants and animal food resources caused by global warming (c. 13th–10th millennium bp) induced new methods of food preparation, in particular, thermal processing in ceramic containers (Ikawa-Smith 1976; Serizawa 1976). Researchers discussing the earliest pottery from Odai-Yamamoto-1 on Honshu island dated around the thirteenth millennium BC, found several fragments with a residual layer on the inside, and suggest that these containers served as cooking vessels (Odai Yamamoto 1 Site Excavation Team 1999: 139). Discoveries of the earliest pottery sites in China, dated around 12th–9th millennium BP, also point to their use as cooking vessels (Underhill 1997). Thus, the first clay vessels seem to have resulted from the adaptive needs of people in the late Pleistocene–early Holocene.

Besides the functional need for new ways and means of food preparation, it is possible to determine the natural and socio-economic factors and conditions supporting the possibilities for ceramic technology invention (Zhushchikhovskaya 2005: 26–29).

1. Global warming of the climate around 13th–10th millennium bp. Warmer climatic conditions favoured working with clay which is sensitive to air temperature. Research in traditional pottery making proves that warm climate is more conducive to all stages of ceramic production than cold climate (Arnold 1985: 61–98).
2. Availability of good-quality raw materials. Quaternary clay deposits formed in East Asia (Japanese islands, China, Far East) throughout the middle and upper Pleistocene till the end of the Pleistocene period. Global climatic changes in East Asia resulted in rich deposits of various high-quality clays suitable for ceramic production.
3. Early development of fishing in prehistoric cultures of the Far East and East Asia. Archaeological records indicate the development of fishing in the late Pleistocene–early Holocene period in the Far East (Table 3.1), Japanese islands, east and south-east China. Fishing led

to much time being spent on river, lake and sea shores which were rich in clay deposits. This provided the opportunity to these communities to observe and experiment with clay.

4. Early tendency to sedentary mode of life. The tendency to sedentary or semi-sedentary mode of life began in the Far East (Table 3.2), the Japanese islands, east and south-east China during the late Pleistocene–early Holocene period. Fishing brought the onset of a sedentary mode of life in the Far East and the Japanese islands. This mode of living facilitated the use of local resources and the management of the cycles of pottery production.

These factors created favourable conditions for the exploitation of a new raw material, clay, with the help of new technologies. Thus, the earliest pottery making efforts could be viewed as a phenomenon of the adaptation of the prehistoric populations of certain East Asian regions to the new natural and socio-economic conditions during Late Pleistocene–early Holocene.

2. Neolithic Period: Further Developments in Pottery Making

The Neolithic period in Primor'e and Priamur'e regions is dated generally to 8,000–3,500 bp (Figure 3.4). The technological context of the Neolithic is characterized by a developed stone industry including abrasive and bifacial retouching techniques, widespread pottery making technology, and the appearance of primitive textile technology.

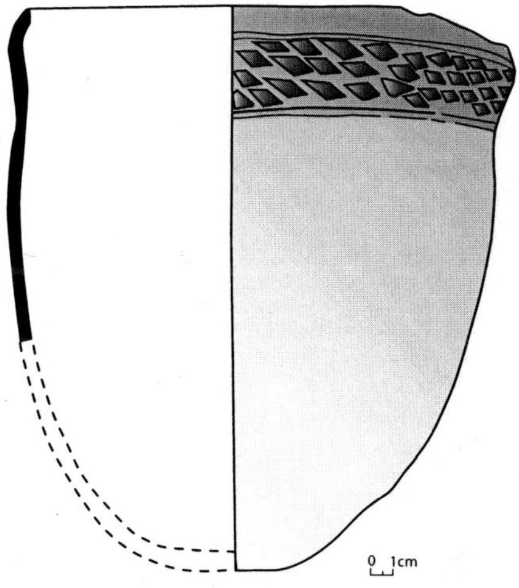

Figure 3.4 Typical Neolithic vessel from the Rudninskaia Culture, Primor'e Region.

In contrast to some of the areas of Eurasia, the introduction of agriculture is not a principal Neolithic indicator for prehistoric Far Eastern cultures. Hunting-gathering was the principal subsistence activity of the Neolithic period. However, intensive development of fishing caused a high degree of sedentism. Agriculture appeared at a late stage of the Neolithic in the 3rd millennium BC, and did not play an important role in the economy until the Palaeometal period (Andreeva 1991; Brodianskii 1987; D'iakov 1992; Medvedev 2003b; Okladnikov 1984; Okladnikov and Derevianko 1973; Popov et al. 1997; Sergusheva 2008; Shevkomud 2004; Vostretsov 1998). The dating and subsistence patterns for the Primor'e and Priamur'e archaeological cultures are summarized in Table 3.3.

In contrast with the late Pleistocene–early Holocene sites, the mid-Holocene sites of Primor'e and Priamur'e regions contain large assemblages of pottery including some intact vessels from each excavated site. It is clear that the use of ceramic vessels was widespread and intensive in these later hunter-fisher-gatherer communities. This was made possible by

Table 3.3 Subsistence patterns and lifestyles of the Neolithic in southern mainland Far East.

Region	Culture, Date	Subsistence Pattern	Lifestyle
	Boismanskaia, 6,355±60 BP to 5,030±140 BP (or 5,420–5,150 BC to 4,230–3,520 BC)	Marine fishing, gathering, hunting. Land hunting, gathering.	Seasonal sedentary or all-year sedentary
Primor'e		Land hunting, gathering, river fishing.	Seasonal sedentary
	Rudninskaia, *end. of* 8,380±60 BP to 5,890±45 BP (or 7,540–7,130 BC to 4,932–4,682 BC)	Land hunting, gathering, river fishing. Marine fishing, gathering. Initial plant cultivation.	Seasonal sedentary and all-year sedentary
	Zaisanovskaia, 5,780±60 BP to 3,090±35 BP (or 4,780–4,510 BC to 1,430–1,270 BC)		
Priamur'e	Kondonskaia, approximately between 5,000 and 4,000 BP (no radiocarbon dating as yet)	River fishing, land hunting, gathering.	All-year sedentary as main tendency
	Voznesenovskaia, *about second half of* 4,520±25 BP to 3,160±50 BP (or 3,350–3,100 BC to 1,520–1,310 BC)	River fishing, land hunting, gathering. Initial plant cultivation.	All-year sedentary as main tendency

the favourable natural conditions in the southern mainland regions of the Russian Far East (Zhushchikhovskaya and Ponkratova 2000). For example, the Holocene climate of this region was characterized by a rather long and relatively dry summer season lasting 4 to 5 months which was suitable for pottery production. The Primor'e and Priamur'e regions also possessed good quality clay deposits along with sources of mineral tempers (sands, rocks) and mineral pigments (ochre) (Grebenshchikov and Derevianko 2001; Zhushchikhovskaya 2001).

Neolithic pottery technology progressed way beyond early pottery. Different stages of the production cycle changed drastically. Neolithic pottery making technology is characterized by the mineral tempering of ceramic paste, the ring forming method, gradual development of surface treatment methods—from rubbing to slipping and polishing—and an open firing method with average temperatures of 600–650°C. The most popular model was a simple-shaped vessel with an open orifice, straight or slightly profiled walls, and a flat bottom (Figure 3.5).

Variation in size of Neolithic pottery is limited. Most vessels are about 15–20 cm high with a few large ones being around 35–40 cm.

In general, technological, morphological and size characteristics of the vessels give no clear indication of likely pottery functions. Residue analysis has yet to be conducted. In some cases it is possible to recognize the probable usage of certain vessels for cooking where a carbonized organic layer is present on the inner or outer surface. However, it is difficult to distinguish between cooking, table and storage wares.

Cross-cultural contacts between the southern Far Eastern population and the more western territories of north-eastern China, Central Asia, southern and western Siberia during late Neolithic (3rd–2nd millennium

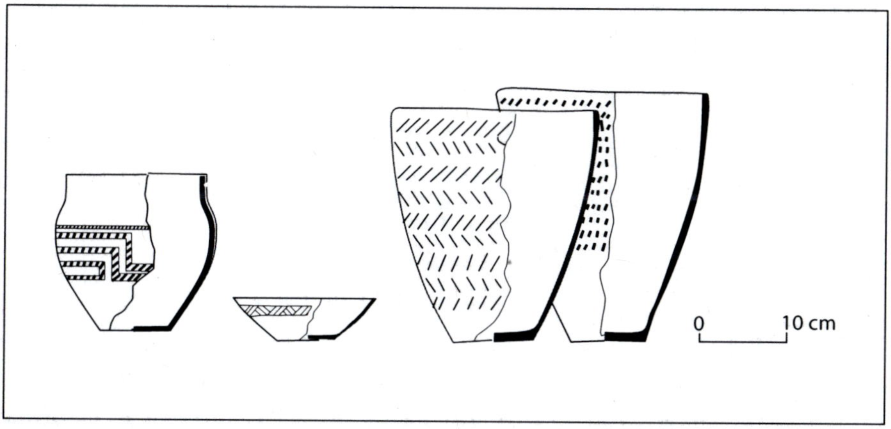

Figure 3.5 Coarse- and fine-wares from the (Neolithic) Zaisanovskaia Culture, Primor'e Region.

BC) resulted in the introduction of early agriculture and some changes in material culture (Andreev 1957; D'iakov 1992; Kuz'min 2005; Popov et al. 1997; Vostretsov et al. 2003; Zhushchikhovskaya 2006). These processes influenced pottery production and use as well. I shall focus on the western and south-western Primor'e region to study this process.

Pottery assemblages from the Neolithic sites of western and south-western Primor'e represent the most developed and probably later variant of the Zaisanovskaia culture and show improved firing techniques, the widening of morphological variability and larger-scale pottery production and function differentiation. Pottery assemblages can be classified into 'coarse' and 'fine' ware. The former is simpler in terms of technique and design. 'Fine' ware is characterized by more elaborate technology, developed shapes, and complicated geometric decorations (Figure 3.6). Obviously 'coarse' and 'fine' ceramic ware served different purposes. While 'coarse' ware served cooking and storage functions, 'fine' ware was used for eating and as prestige items.

3. Palaeometal Period: Principal Achievements in Pottery Making

The Palaeometal period lasted from the end of 2nd millennium BC to the beginning of the 1st millennium AD and was marked by new phenomena and changes in the life and culture of prehistoric Far Eastern populations (Figure 3.7; Table 3.4). The period is characterized by the appearance of metals—bronze and iron—and some metal-working skills, the further development of agriculture (crops and domestic animals), and the integration of complex subsistence patterns including gathering and food producing. Early Iron Age

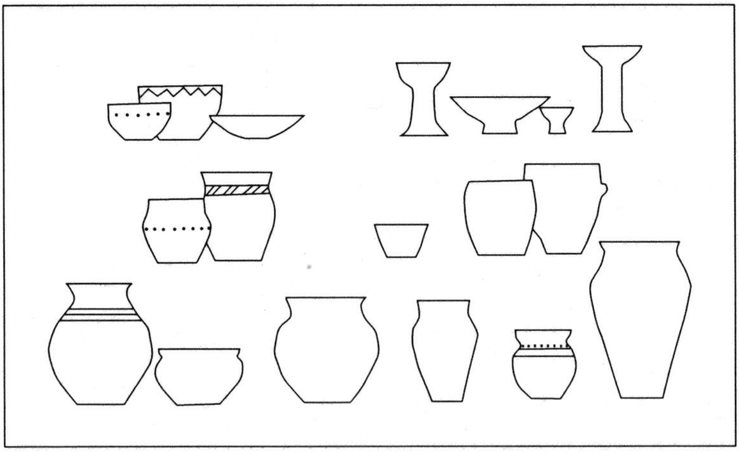

Figure 3.6 Typical pottery vessel forms from the Palaeometal Period, Primor'e Region.

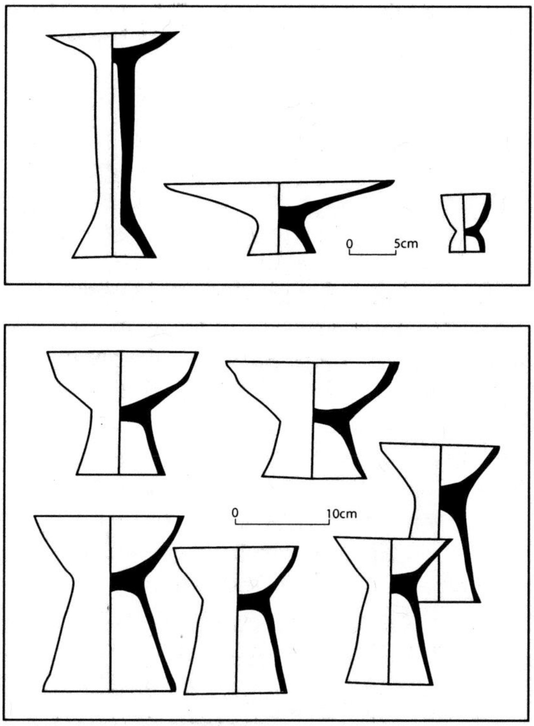

Figure 3.7 Non-utilitarian (ritual/prestige) vessels from sites of (A) Yankovskaya Culture and (B) Krounovskaya Culture, Palaeometal Period, Primor'e Region.

Table 3.4 Subsistence patterns and lifestyles of Early Iron Age in southern mainland Russian Far East.

Region	Culture, Date	Subsistence Pattern	Lifestyle
Primor'e	Yankovskaya, 900 bc to 400–300 bc	Marine fishing, gathering, hunting. Land hunting, gathering. Plant cultivation, animal domestication.	All-year sedentary, Seasonal sedentary
	Krounovskaya, 400 bc to AD 200–300 (uncalibrated)	Land hunting, river fishing, gathering. Plant cultivation, animal domestication.	All-year sedentary
Priamur'e	Uril'skaya, 1,100–1,000 bc to 500–400 bc	River fishing, land hunting, gathering. Plant cultivation, animal domestication.	All-year sedentary
	Pol'tsevskaya, 400–300 bc to AD 100 (uncalibrated)	River fishing, land hunting, gathering. Plant cultivation, animal domestication.	All-year sedentary

cultures corresponding to the last stage of the Primitive Society epoch in the southern Far East are characterized by the highest level of economic and technological development. The economic situation was more stable and productive than in the Neolithic, stimulating the growth of population, increasing settlement areas, a more complex way of life, and intensification of craft activity (Andreeva et al. 1986; Brodianskii 1987; Derevianko 1973, 1976; D'iakov 1989; Kon'kova 1989; Vostretsov 1996; Zhushchikhovskaya 2005).

A complex process of cultural movement and exchange took place in the southern Far East during the Palaeometal period. The Sinegaisko-Lidovskaya cultural community owed its ethnic lineage to migrants from the more western territories of Eurasia who introduced bronze into the Primor'e region. The Yankovskaya and Krounovskaya cultures of Early Iron Age did not share ethnic links. The former probably descended from the previous Neolithic Zaisanovskaia culture. The latter migrated to Primor'e from neighbouring areas of China and co-existed with the Yankovskaya culture for a while. We have a more definite understanding of the ethnic lineage of people in the Priamur'e region.

The Palaeometal period saw significant advancement in pottery making and functional variability. Hand-building was still the preferred forming method as in the Neolithic period. However, surface treatment and firing technology improved dramatically. Operations of rubbing, slipping, polishing and painting were applied to post-forming processing of the clay vessels. Potters used simple kilns. The remains of firing constructions were discovered at one of the Yankovskaya sites. The average index of firing temperature was 700–750°C, sometimes 800–850°C. Some researchers estimated firing temperatures of around 900–1,000°C for pottery in the Priamur'e region (Grebenshchikov and Derevianko 2001: 36). While the oxidizing regime was a basic one in firing practice, the technology of 'smudging' was used intentionally for producing black-surfaced vessels. Technological specialization was a distinctive feature of pottery production during the Early Iron Age in the Yankovskaya, Krounovskaya, Uril'skaya and Pol'tsevskaya cultures.

In terms of forms and size, the Palaeometallic pottery were much broader than the Neolithic (Figure 3.8). This is especially notable in the cultures of the Early Iron Age (Grebenshchikov and Derevianko 2001; Zhushchikhovskaya 2005: 83–118). Ceramic assemblages of the Yankovskaya, Krounovskaya, Uril'skaya, and Pol'tsevskaya cultures consist of two morphological classes—restricted vessels (with restricted orifice) and unrestricted vessels (with unrestricted orifice). The orifice and body are clearly separated in restricted vessels providing new functional possibilities, for example, in cooking and storage spheres. The sizes of restricted vessels vary from small (10–15 cm high) to large (35–40 cm high) to very large (50–60 cm high). Unrestricted vessels have a broad range of shapes from a flattened dish to

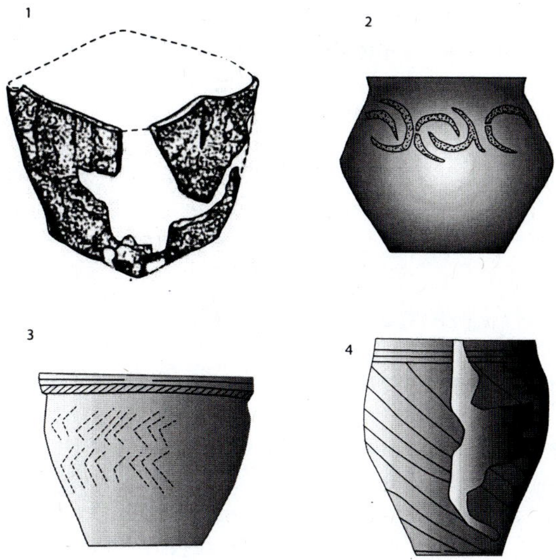

Figure 3.8 Typical Neolithic pottery from Sakhalin Island.

a tall conical-shaped container. A specific type of Yankovskaya and Krounovskaya pottery is the footed vessel. Their morphological structure consists of two parts—a dish-like or bowl-like unrestricted reservoir and a foot of conical or cylindrical shape (Figure 3.9). The footed vessels are of small and medium size.

Certain pottery forms show a different degree of technological complexity and perfection; this observation is important for indicating a ceramic vessel's function. Much time and labour was invested in the production of footed vessels in order to craft them for specific functions. This

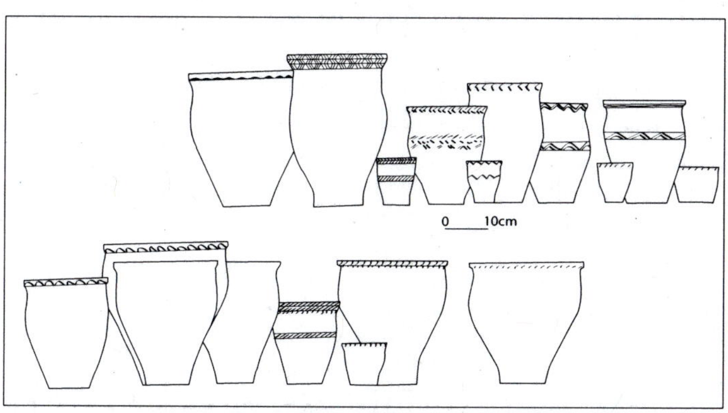

Figure 3.9 Pottery from the site of Starodubskoye 3, Okhotskaya Culture, Neolithic-Palaeometal Transitional Period, Sakhalin Island.

ware is characterized by the careful processing of ceramic paste, refined surface treatment and a complex firing method. In contrast, the production of cooking pots involved basic technology. Several medium–sized vessels with restricted orifice have a carbonized residual layer on inner and sometimes outer surfaces. Storage containers of large and very large size usually have slipped and polished walls minimizing water permeability which is important for holding liquids.

II. Case Study 2: Prehistoric Pottery Making in Sakhalin Island

1. The Neolithic: Pottery Making of the Hunter-Gatherers

Most recent investigations in Sakhalin Island indicate the appearance of the first ceramics in the area around 7,000–5,500 bc. Some pottery fragments were discovered at sites attributed to the earliest stage of the Neolithic (Vasilevskii 2003: 30–31; Zhushchikhovskaya and Shubina 2006). However, nothing much can be concluded from these finds.

There is greater evidence of pottery making on the Sakhalin Island at the beginning of the later Neolithic (Zhushchikhovskaya and Shubina 2006). Neolithic sites contain both abundant lithics and pottery, which suggests that ceramics had an established role from the outset amongst the island's prehistoric hunter-gatherer populations. Natural conditions for this craft were less favourable than on the mainland southern Far Eastern regions. The Holocene climate allowed for only two to three summer months a year, a significant part of which was marked by rains, mist and winds. These conditions were not conducive to pottery, especially the drying and firing stages (Zhushchikhovskaya 2005: 31–58).

Neolithic pottery developed within different cultural contexts but under similar socio-economic conditions (Table 3.5).

Table 3.5 Subsistence patterns and lifestyles of the Neolithic in Sakhalin.

Culture, Date	Subsistence Pattern	Lifestyle
Yuzhno-Sakhalinskaya, 6,740±150 BP to 5,648±490 BP (or 5,626±345 BC to 4,495±525 BC)	Land hunting, gathering. Marine fishing, possible marine hunting.	Seasonal sedentism, possible full sedentism.
Imchinskaya, 5,810±90 BP to 2,570±110 BP (or 4,680±150 BC to 643±172 BC)	Land hunting, gathering, river fishing. Marine gathering, possible marine hunting.	Seasonal sedentary.
Anivskaya, second half of 3,005±125 BP to 2,360±110 BP (or 1,520–920 BC to 790–200 BC)	Land hunting, gathering. Marine fishing.	Seasonal sedent ism, possible full sedentism.

Pottery assemblages of various Neolithic cultural units are characterized by both common and specific traits (Table 3.6).

Ceramic traditions of certain cultures had different origins that produced the differences in their specification. Pottery of the Yuzhno-Sakhalinskaya and Anivskaya cultures located in southern Sakhalin seem to be influenced by cultural impulses from the Japanese islands where the Jomon pottery tradition flourished between 12–10th millennium BC and the first half of 1st millennium BC (Kobayashi 2004). The ceramic assemblages of the Imchinskaya culture were the result of close contacts with the Neolithic Proto-Voznesenovskaya culture of the lower Amur river basin. In spite of particular differences, Neolithic pottery assemblages of Sakhalin have common regional, temporal and economic traits. They are: technological simplicity and primitiveness, morphological invariability, dominance of the unrestricted vessel model, and small and medium-sized containers. These characteristics probably indicate a low degree of functional differentiation among the Neolithic pottery of Sakhalin Island. It is not possible to classify the pottery assemblages functionally (Zhushchikhovskaya and Shubina 2006).

2. Palaeometal period: Innovations in Economy and Pottery Making

Significant changes in economy and material culture occured in the period between the middle of 1st millennium BC and the middle of 1st millennium

Table 3.6 Pottery characteristics of early ceramics of Sakhalin Island

Culture	Pottery Characteristics					
	Paste	Forming	Surface Treatment	Firing	Morphological Model	Vessels' Sizes
Yuzhno-Sakhalinskaya	Clay+ Grass Temper	Slab Construction	Smoothing	Open Fire, 500–600°C	Unrestricted, flat-bottomed, straight-walled, rectangular horizontal section– "Box-like" vessels.	Height: 8–18 cm
Imchinskaya	Clay + Shell Temper	Coiling	Smoothing	Open Fire, 600–650°C	Unrestricted, flat-bottomed, slightly profiled walls.	Height: 8–20 cm
Anivskaya	Clay+ Sand Temper	Coiling	Smoothing Slipping	Open Fire 600–650°C	Unrestricted, flat-bottomed, Slightly profiled walls	Height: 10–25 cm

AD, which corresponds with the Susuyskaya archaeological culture of southern Sakhalin. This period is interpreted as a transitional link between the Neolithic and later Palaeometal period. Subsistence patterns changed from a hunter-gatherer economy based on land-mammal hunting to sea-mammal hunting and gathering of other marine resources (Table 3.7). In this period, settlements were concentrated in coastal areas. This shift is also reflected in new specialized complexes of bone and stone tools. Sites are characterized increasingly by marine faunal remains (Vasilevskii 2000a, 2000b; Vasil'evskii and Golubev 1976).

Ceramic technology underwent changes during this period. Technological standards were not markedly different from earlier Anivskaya pottery. However, size and form varied in this later pottery. In the ceramic assemblages of the Susuyskaya culture, the vessels with slightly restricted orifices appear together with unrestricted vessels. The large vessels, up to 30 cm high and with a body diameter up to 25–30 cm, have mostly unrestricted or slightly restricted orifices. The medium- and large-sized vessels have a conical-shaped contour, wide mouth, narrow bottom and slightly profiled walls. Small vessels which are not so frequently represented are largely rough, straight-walled, wide-mouthed and wide-bottomed.

One remarkable feature of the Susuyskaya culture pottery is the presence of a thick (1–2 mm) residual carbonized layer on the inner and outer surfaces of the pots. This was found mostly in large and medium containers and ocassionally in small ones. The phosphate chemical analysis of the residual layer detects the high content of phosphorous (Ph) that indicates the origin (land or sea animal) of the organic matter. Clearly, the main function of much of the pottery was cooking, particularly animal products rich in fat. Some of the large containers might have been used for storage, but there is a notable absence of serving vessels and 'fine wares'.

The current periodization of Sakhalin Island's prehistory dates the Palaeometal period from about the middle of 1st millennium to the first half of 2nd millennium AD and correlates it with the sites of the Okhotskaya archaeological culture which is considered to be a direct descendant of the Susuyskaya culture (Vasilevskii 1995, 2000a, 2000b).

Table 3.7 Subsistence patterns and lifestyles of transitional Neolithic–Palaeometal and Palaeometal periods in Sakhalin.

Culture, Date	Subsistence Pattern	Lifestyle
Susuyskaya, 500 bc to AD 500 (uncalibrated)	Sea-mammal hunting, sea gathering, land gathering.	All-year sedentary
Okhotskaya, 500 bc to AD 1,200–1,300 (uncalibrated)	Sea-mammal hunting, sea gathering, land gathering.	All-year sedentary

Archaeological records provide evidence of the first appearance of metal in the settlements of the 1st millennium AD. Iron artefacts were imported to Sakhalin from the Japanese islands and probably also from the mainland Far Eastern regions. In the sites of 2nd millennium AD, not only iron implements but also fragments of iron cauldrons were unearthed (Shubin 1973; Shubin and Shubina 1984). No traces of local metal production were discovered on the island. However, it may be supposed that some operations of metalworking were known to the peoples of the Okhotskaya culture.

The economic base of Okhotskaya culture continued and developed the traditions of the previous Susuyskaya culture. Highly specialized sea-mammal hunting, marine fishing and gathering were the main subsistence activities. This type of economy necessitated a stable sedentary mode of life.

The Okhotskaya culture sites yielded large pottery assemblages consisting of fragments as well as whole vessels. Pottery making of the period preserved basic traditions of the previous periods, but also saw some innovations. There is the restricted usage of polishing for surface treatment and the increasing of firing temperatures up to 700–750°C. Most popular containers were slightly restricted and unrestricted conical-shaped vessels of medium and large sizes, up to 35–40 cm high (Figure 3.9). The pottery of this period is also characterized by a thick residual layer on the vessels' surface. Chemical phosphate analysis revealed residues of animal products. It can be safely concluded that the vessels were used for the thermal processing of sea mammals. Ethnographic records of some Northwest American populations which practised sea-mammal hunting for their basic subsistence, reveal the use of ceramic pots for thermal processing of animal products (Reid 1990). Results of isotope analysis on human bones from Okhotsk sites on Hokkaido Island suggest that the diet of this culture's population was based on marine products (Ono 2002).

The sites of the latest stages of the Okhotskaya culture are dated to around the middle of the 2nd millennium AD and provide evidence of some changes in pottery production and function. Ceramic assemblages in dwellings on sites Anfel'tsevo-2 and Ozersk show the increasing role of small-sized vessels with straight walls, wide mouths and wide bottoms. This simple shape was a continuation from the previous Susuyskaya culture. On the other hand, the number of large cooking containers decreased. Probably this was caused by the introduction of iron cauldrons which were more effective and long-lasting than ceramic containers. The making of pottery in the suboptimal natural climatic conditions on the island was not an easy and productive activity. With the appearance of metallic cooking containers, the ceramic ones began to lose their practical importance for the marine hunter-gatherer communities (Zhushchikhovskaya 2005: 57–58).

Early Ceramics in the Russian Far East: Overview and Conclusions

This chapter has examined the emergence and long-term history of pottery production in two regions of the southern Russian Far East, each characterized by different economic histories and natural and climatic conditions. Several conclusions can be drawn.

1. The southern mainland Russian Far East has a special place in the earliest history of world pottery making. Late Pleistocene–early Holocene communities of the Priamur'e (Amur river basin) and Primor'e areas represent a case of an independent local origin of ceramic technology within the East Asian region. The process of pottery invention may be considered a part of wider responses to climatic amelioration and transformations in society and economy around the 13th to 10th millennium BP, which included an increasing reliance on fishing and more sedenatary lifestyles.
2. Archaeological records of the Neolithic period show some common principal traits in the development of pottery in the mainland southern Far East and on Sakhalin Island. Both regions possessed similar economic social and technological conditions. Neolithic pottery assemblages from Primor'e, Priamur'e and Sakhalin Island are characterized mainly by technological primitiveness, simplicity, and a small degree of morphological and size diversity. Neolithic ceramic assemblages of mainland and insular regions of the Russian Far East do not present evidence of functional differentiation. Based on the use-wear traces, it is possible to recognize the use of ceramic vessels for cooking. However, one cannot conclude about the other functions of these ceramic vessels.
3. Certain progressive changes in pottery production and function patterns in the considered regions are connected with the latest stage of the Neolithic. These changes are synchronic and due in some measure to important innovations in the economic life of prehistoric peoples: the invention of agriculture in the mainland region and the beginning of a subsistence mode based on sea-mammal hunting in the Sakhalin Island. Newer developments are more significant and representative in the mainland areas: the improvement of pottery techniques, the increase in morphological diversity, and the development of functional differentiation. In Sakhalin's communities corresponding to the Late Neolithic or transitional Neolithic–Palaeometal stage, the changes were restricted to some development of pottery morphology (the invention of a slightly restricted vessel model) and an increase in ceramic container sizes.

4. The contrast between the mainland and insular regions in pottery production and functions is quite remarkable for the Palaeometal period. An increasing reliance on agriculture in the mainland region caused a more complex social structure and community life, and the intensification of various activities, particularly pottery making during the 1st millennium BC. Case studies from around the world testify that the advent of agriculture resulted in more complex pottery production methods and functions (Dow 1985; Gebauer 1995; Khlopin 1964; Nelson 1993; Underhill 1991b). Together with economic and social factors determining the development of pottery making and ceramic ware functions, favourable natural conditions also stimulate pottery making. Climate and the availability of raw materials also played a role in the use of pottery by the community.

Pottery assemblages of the Palaeometal period from Sakhalin Island provide evidence of only a few progressive technological changes and no evidence of increased functional differentiation in comparison with earlier periods. This hunter-gatherer community principally used pottery for cooking, particularly the thermal processing of animal products. Preheistoric pottery making on Sakhalin Island made slow progress due not only to social and economic conditions, but also to unfavourable natural conditions which prevented the improvement of pottery production methods.

Note

1. This chapter employs the 'eastern' definition of Neolithic, i.e., pottery-using hunter-gatherers. See the introductory chapter (Jordan and Zvelebil) for further discussion of the various definitions of Neolithic.

References

Adovasio, J M (1977) *Basketry Technology*, Chicago: Aldine

Adovasio, J M and Lynch, T F (1973) 'Preceramic textile and cordage from Guittarrero Cave, Peru', 38(1) *American Antiquity*, 84–90

Amiran, R (1965) 'The beginning of pottery making in the Near East', in Matson, F R (ed), *Ceramics and Man*, Chicago: Aldine, pp 240–247

Andreev, G I (1957) 'Poselenie Zaisanovka-1 v Primor'e [The Zaisanovka 1 site in Primor'e]', 2 *Sovetskaia Arkheologiia*, 121–145

——— (1960) 'Nekotorye voprosy kul'tur Yuzhnogo Primor'a III-I tysiacheletii do n.e. [Some questions on the cultures of southern Primor'e of the third to first millennium BC]', 86 *Materialy i Issledovaniia po Arkheologii SSSR*, 136–161

Andreeva, Z V (1991) *Neolit Iuga Dal'nego Vostoka (Drevnee Poselenie v Peshchere Chertovy Vorota) [The Neolithic of the Southern Far East (An Early Site in Chertovy Vorota Cave)]*, Moscow: Nauka

Andreeva, Z V, Zhushchikhovskaya, I S and Kononenko, N A (1986) *Yankovskaya Kul'tura [The Yankovskaya Culture]*, Moscow: Nauka

Arnold, D E (1985) *Ceramics Theory and Cultural Process*, Cambridge: Cambridge University Press

Barnett, W K and Hoopes, J W (eds) (1995) *The Emergence of Pottery*, Washington DC: Smithsonian Institution Press

Brodianskii, D L (1987) *Vvedenie v Dal'nevostochnuiu Arkheologiiu [Introduction to Far Eastern Archaeology]*, Vladivostok: Far Eastern State University Press

Crowfoot, G M (1954) 'Textiles, basketry, and mats', in Singer, C, Holmyard, E, and Hall, A (eds), *A History of Technology* 1, Oxford: Oxford University Press, pp 413–455

Derevianko, A P (1973) *Rannii Zheleznyi vek Priamur'a [The Early Iron Age of Priamur'e]*, Novosibirsk: Nauka

——— (1976) *Priamur'e. 1 Tysiacheletie do n. e. [Priamur'e. The First Millennium BC]*, Novosibirsk: Nauka

Derevianko, A P, and Kononenko, N A (eds) (2003) *Foraging Population of the Sea of Japan during the Late Pleistocene–Early Holocene*, Novosibirsk: Institute of Archaeology and Ethnology Press [in Russian and English]

Derevianko, A P and Medvedev, V E (1995) 'The Amur River basin as one of the earliest centers of ceramics in the Far East', in Kajiwara, H and Yokoyama, Y (eds), *Toadzia Kekutonodoku no Kigaen (The Origin of Ceramics in Eastern Asia). Proceedings of the International Symposium*, Sendai: Tohoku Fukushi University Press, pp 11–26

Derevianko, A P and Petrin, V T (1995) 'Neolithic Age in the south of Russian Far East: on chronology', in Kajiwara, H and Yokoyama, Y (eds), *Toadzia Kekutonodoku no Kigaen (The Origin of Ceramics in Eastern Asia). Proceedings of the International Symposium*, Sendai: Tohoku Fukushi University Press, pp 51–53

D'iakov, V I (1989) *Primor'e v Epokhu Bronzy [Primor'e in the Bronze Age]*, Vladivostok: Far Eastern State University Press

——— (1992) *Mnogosloinoe Poselenie Rudnaya Pristan i Periodizatsiia Neoliticheskikh Kul'tur Primor'a [The Multi-Layered Rudnaya Pristan Site and Periodization of Neolithic Cultures of Primor'e]*, Vladivostok: Far Eastern State University Press

Dow, M M (1985) 'Agricultural intensification and craft specialization: a non-recursive model', 24(2) *Ethnology*, 137–152

Garkovik, A V (2002) 'Ancient Ceramics from East Regions of Russia' in Guo Jingkun (ed), *The Proceedings of the International Symposium of Asian Ceramics 2002*, Shanghai: Shanghai Institute of Ceramics, pp 366–377

Gebauer, A B (1995) 'Pottery production and the introduction of agriculture in Southern Scandinavia', in Barnett, W K and Hoopes, J W (eds), *The Emergence of Pottery*, Washington DC: Smithsonian Institution Press, pp 99–114

Golubev, V A and Zhushchikhovskaya, I S (1987) 'Neoliticheskaia kul'tura iuzhnogo Sakhalina v svete alaniza keramicheskikh kompleksov [The Neolithic culture of southern Sakhalin in light of analysis of ceramic complexes]', *Voprosy Arkheologii Dal'nego Vostoka SSSR*, Vladivostok: Institute of History, Archaeology and Ethnography of Peoples of Far East Press, pp 25–33

Grebenshchikov, A V and Derevianko, E I (2001) *Goncharstvo Drevnikh Plemen Priamur'a [Pottery Making of Prehistoric Population of Priamur'e]*, Novosibirsk: Institute of Archaeology and Ethnology Press

Hally, D J (1986) 'The identification of vessel function: a case study of Northwest Georgia', 51(2) *American Antiquity*, 267–295

Hoopes, J W (1994) 'Ford revisited: a critical review of the chronology and relationships of the earliest ceramic complexes in the New World, 6,000–1,500 BC', 8 *Journal of World Prehistory,* 1–49

Ikawa-Smith, F (1976) 'On ceramic technology in East Asia', 17 *Current Anthropology*, 513–515

Jull, E, Malli, Z, Biddul'f, D, Derevianko, A P, Kuz'min, Y V, Medvedev, V E, Tabarev, A V, Zenin, V N, Vetrov, V M, Lapshina, Z S, Garkovik, A V and Zhushchikhovskaya, I S (1998) 'Radiouglerodnaia khronologiia drevneishikh neoliticheskikh kul'tur iuga Dal'nego Vostoka Rossii i Zabaikal'ya po rezul'tatam priamogo datirovaniia keramiki metodom uskoritel'noi mass-spektrometrii [The radiocarbon chronology of the earliest Neolithic cultures of the southern Russian Far East and Zabaikalye based on the results of the direct dating of ceramics by the method of accelerator mass-spectrometry]', in Derevianko, A P (ed), *Paleoekologiia Pleistotsena i Kul'tury Kamennogo Veka Severnoi Azii i Sopredel'nykh Territorii* V.2, Novosibirsk: Institute of Archaeology and Ethnology Press, pp 63–68

Kajiwara, H and Yokoyama, Y (eds) (1995) *Toadzia Kekutonodoku no Kigaen [The Origin of Ceramics in Eastern Asia]. Proceedings of International Symposium*, Sendai: Tohoku Fukushi University Press

Kashina, T I (1977) *Keramika Kul'tury Yanshao [The Pottery of Yanshao Culture]*, Novosibirsk: Nauka

Khlopin, I N (1964) *Geoksyurskaya Gruppa Poselenii Epokhi Eneolita [The Geoksyur Group of Sites of the Eneolithic Period]*, Moscow, Leningrad: Nauka

Kobayashi, T (ed) (1989) *Jomon Doki Talkan [Comprehensive Book of Jomon Pottery]*, Tokyo: Shyogakkan Press [in Japanese]

Kobayashi, T (2004) *Jomon Reflections*, Oxford: Oxbow Books

Kon'kova, L V (1989) *Bronzoliteinoe Proizvodstvo na Iuge Dal'nego Vostoka SSSR [Bronze Metallurgical Production in the Southern Far East of the USSR]*, Moscow

——— (1996) 'The earliest bronze artifacts from the Soviet Far East', 26 *Bulletin of the Museum of Metals*, 1–20

Kononenko, N A (ed) (1996) *Pozdnii Paleolit—Rannii Neolit Vostochnoi Azii i Severnoi Ameriki (Materialy Mezhdunarodnoi Konferentsii) [The Late Paleolithic–Early Neolithic of Eastern Asia and North America (Materials from the International Conference)]*, Vladivostok: Institute of History, Archaeology and Ethnography of Peoples of Far East Press

Krushanov, A I (ed) (1989) *Istoriia Dal'nego Vostoka SSSR s Drevneishikh Vremen do XVII Veka [The History of Far East of USSR from Old Past to XVII Century]*, Moscow: Nauka

Kuz'min, Y V (2005) *Geochronology and Paleoenvironment in the Late Paleolithic and Neolithic of Temperate East Asia*, Vladivostok: Pacific Institute of Geography

Lapshina, Z S (1999) *Drevnosti Ozera Khummi [The Antiquity of Lake Khummi]*, Khabarovsk: Priamur'e Geographic Society Press

Loehr, M (1968) *Ritual Vessels of Bronze Age China*, New York: Asia Society

Medvedev, V E (1993) 'The Gasya site and the problem of the appearance of ceramics in the east of the Russian part of Asia', in *The Aspects of Initial Period of Pottery Invention in the Areas Surrounding Sea of Japan*, Niigata: Niigata University press, pp 1–11 [in Japanese]

——— (2003a) 'Kogda i kak byla otkryta na Dal 'nem Vostoke drevneishaia keramika? [When and how was the oldest pottery discovered in the Far East?]', in Derevianko, A P (ed), *Problemy Arkheologii i Paleoecologii Severnoi, Vostochnoi i Tsentral'noi Azii [The Problems of Archaeology and Palaeoecology of North, East and Central Asia]*, Proceedings of the International Conference, Novosibirsk: Institute of Archaeology and Ethnography, Siberian Branch of the Russian Academy of Sciences, pp 38–43

——— (2003b) 'Akademik A P Okladnikov i neolit Nizhnego Amura: razvitie idei [Academician A P Okladnikov and the Neolithic of Lower Amur: evolutions of conceptions]', in Derevianko, A P (ed), *Problemy Arkheologii i Paleoecologii Severnoi, Vostochnoi i Tsentral'noi Azii [The Problems of Archaeology and Palaeoecology of North, East and Central Asia]*, Proceedings of the International Conference, Novosibirsk: Institute of Archaeology and Ethnography, Siberian Branch of the Russian Academy of Sciences, pp 164–171

Myl'nikova, L N (1999) *Goncharstvo Neoliticheskikh Plemen Nizhnego Amura [Pottery Making of the Neolithic Tribes of the Lower Amur]*, Novosibirsk: Institute of Archaeology and Ethnology Press

Myl'nikova, L N and Nesterov, S P (2008) 'Fiziko-chimicheskoe issledovanie keramiki pamiatkina Kosanii: k probleme proiskhozhdeniia goncharstva', in Popov, A N (ed), *Neolithic and Neolitization of the Japanese Sea Basin: Individual and the Historical Landscape*, International Conference proceedings, Vladivostok: Far Eastern State University Press, pp 161–170

Nelson, S M (1993) *The Archaeology of Korea*, Cambridge: Cambridge University Press

Odai Yamamoto 1 Site Excavation Team (eds) (1999) *Archaeological Research at the Odai Yamamoto 1 Site (Inquiry into the Question of the End of the Paleolithic Culture and the Beginning of the Jomon Culture)*, Tokyo: Kokugakuin University

Okladnikov, A P (1970) 'Neolit Sibiri i Dal'nego Vostoka [The Neolithic of Siberia and the Far East]', in *Kamennyi vek na Territorii SSSR,* Moscow, Leningrad: Nauka, pp 172–193

——— (1984) *Keramika Drevnego Poseleniia Kondon [Ceramics of the Early Kondon Site]*, Novosibirsk: Nauka

Okladnikov, A P and Derevianko, A P (1973) *Dalekoe Proshloe Primor'a i Priamur'a [The Distant Past of Primor'e and Priamur'e]*, Vladivostok: Primorskii Book Press

Ono, H. (2002) Micro Wear, Heavy Abrasion and Degeneration of Tempomandibular joint among the Okhotsk People: From the Viewpoint of Archaeology, in *Okhotsk Culture Formation, Metamorphosis and Ending*. The Proceedings of Japan and Russia Cooperative Symposium, Sapporo: Hokkaido University Museum press, pp 139–145

Popov, A N, Chikisheva, T A and Shpakova, E G (1997) *Boismanskaia Arkheologicheskaia Kul'tura Yuzhnogo Primor'a [The Boisman Archaeological Culture of Southern Primor'e]*, Novosibirsk: Nauka

Reid, K (1990) 'Simmering down: a second look at Ralph Linton's "North American Cooking Pots"', 23 *Anthropological Papers*, 8–17

Rice, P M (1987) *Pottery Analysis: A Sourcebook*, Chicago: University of Chicago Press

——— (1999) 'On the origins of pottery', 6(1) *Journal of Archaeological Method and Theory*, 1–54

Saiko, E V (1982) *Tekhnika i Tekhnologiia Keramicheskogo Proizvodstva Srednei Azii v Istoricheskom Razvitii [The Technique and Technology of Ceramic Production of Middle Asia in Historical Development]*, Moscow: Nauka

Serizawa, C (1976) 'The stone age of Japan', 19 *Asian Perspectives*, 1–14

Sergusheva, E (2008) 'Poiavlenie zemledeliia na territorii iuzhnogo Primor'a [The origin of agriculture in southern Primor'e]', in *Neolithic and Neolithization of the Japanese Sea Basin: Individual and the Historical Landscape*, The Proceedings of International Conference', Vladivostok: Far Eastern State University press, pp 205–217

Shepard, A O (1985) *Ceramics for the Archaeologist*, Washington DC: Carnegie Institution of Washington

Shevkomud, I Y (2004) *Pozdnii Neolit Nizhnego Amura [Late Neolithic of Lower Amur Region]*, Novosibirsk: Institute of Archaeology and Ethnology Press

Shubin, V O (1973) 'Raskopki mnogosloinogo poseleniia Ozersk [Excavations of the multi-component Ozersk site]', in *Arkheologiia Amuro-Sakhalinskogo Regiona [Archaeology of the Amur-Sakhalin Region]*, Vladivostok, pp 5–29

Shubin, V O and Shubina, O A (1984) *Novye Radiouglerodnye Datirovki po Arkheologicheskim Pamiatnikam Sakhalinskoi Oblasti [New Radiocarbon Dates of the Archaeological Sites of the Sakhalin District]*, Yuzhno-Sakhalinsk

Underhill, A P (1991a) 'A guide to understanding ceramic change', in Weeder, E (ed), *The Rise of Great Tradition: Japanese Archaeological Ceramics from the Jomon through Heian Periods (10,500 B.C.–A.D. 1185)*, New York: Agency for Cultural Affairs, Government of Japan, Japan Society, pp 10–14

——— (1991b) 'Pottery production in chiefdoms: the Longshan period in Northeast China', 23(1) *World Archaeology*, 12–27

——— (1997) 'Current issues in Chinese Neolithic archaeology', 11(2) *Journal of World Prehistory*, 103–160

Vasilevskii, A A (1992) 'Neoliticheskie kul'tury ostrova Sakhalin (kratkii ocherk) [Neolithic cultures of Sakhalin Island (a brief outline)]', 28 *Arkheologiia Hokkaido*, 115–136 [in Japanese]

——— (1995) 'Korrektsiia radiouglerodnykh datirovok i absoliutnaia khronologiia arkheologicheskikh kul'tur o. Sakhalin [The correction of radiocarbon dates and the absolute chronology of the archaeological cultures of Sakhalin Island]', 2 *Kraevedcheskii Biulleten'*, 93–110

——— (2000a) 'K poniatiiu "neolit" i ego periodizatsii na o-ve Sakhalin [On understanding the "Neolithic" and its periodization on Sakhalin Island]', in Vostretsov, Y E and Kliuev, N A (eds), *Vpered . . . v proshloe*, Vladivostok: Dal'nauka, pp 150–160

——— (2000b) 'Okhotskaia problema v svete sovremennykh issledovanii (Sakhalin, Hokkaido, Kuril'skie ostrova, 1980–90-e gg.) [Okhotsk problem in light of modern investigations (Sakhalin, Hokkaido, Kurile Islands, 1980s and 1990s)]', 1 *Uchenye Zapiski Sakhalinskogo Gosudarstvennogo Universiteta* [Scientific Notes of Sakhalin State University], 6–12

——— (2003) 'Kamennyi vek ostrova Sakhalin [The Stone Age of Sakhalin Island]', Avtoreferat dissertatsii na soiskanie uchenoi stepeni doktora istoricheskikh nauk

[Thesis for the dissertation of Doctor of Historical Sciences], Novosibirsk: Institute of Archaeology and Ethnology

Vasil'evskii, R S and Golubev, V A (1976) *Drevnie Poseleniia na Ostrove Sakhalin (Susuiskaya Stoianka) [Early Sites on Sakhalin Island (The Susuya Site)]*, Novosibirsk: Nauka

Vasil'evskii, R S, Krupianko, A A and Tabarev, A V (1997) *Genezis Neolita na Iuge Dal'nego Vostoka Rossii (Kamennaia Industriia i Problema Rannei Osedlosti) [The Genesis of the Neolithic in the South of the Russian Far East (The Stone Industry and the Problem of Early Sedentism)]*, Vladivostok: Far Eastern State University Press

Vitelly, K (1993) 'Pots, potters, and the shaping of Greek Neolithic society', in Barnett, W K and Hoopes, J W (eds), *The Emergence of Pottery*, Washington DC: Smithsonian Institution Press, pp 55–64

Vostretsov, Y E (1986) *Metod Landshaftnogo Analiza (na Primere Poselenii Krounovskoi Kul'tury Zheleznogo Veka v Primor'e [The Method of Landscape Analysis (with Krounovskaya Culture's Sites in Primor'e as a Case Study)]*, Problemy Arheologicheskih Issledovanii na Dal'nem Vostoke SSSR, Vladivostok: Far Eastern Scientific Centre press, pp 135–147

——— (1996) 'Vzaimodeistvie morskikh i zemledel'cheskikh adaptatsii v basseine Yaponskogo moria [The Interaction of marine and agricultural adaptation systems in Japan Sea basin]', in *Primor'e v Drevnosti i Srednevekov'e,* Ussuryisk: Ussuryisk Pedagogical College Press, pp 17–23

——— (1998) *Pervye Rybolovy v Zalive Petra Velikogo (Chelovek i Priroda v Bukhte Boismana)* [The First Fishermen in Peter the Great Gulf (Man and Nature in Boisman Bay)], Vladivostok: Dal'nauka

Vostretsov, Y E, Sergusheva, E A, Masayuki, K, Kazuo, M and Hiroki, O (2003) 'Novye dannye o rannem zemledelii v Primor'e: neoliticheskii kompleks poseleniia Krounovka-1 [Recent data on early agriculture in Primor'e region: Neolithic complex of site Krounovka-1]', in Derevianko, A P (ed), *Problemy Arkheologii i Paleoecologii Severnoi, Vostochnoi i Tsentral'noi Azii*, Novosibirsk: Institute of Archaeology and Ethnology Press, pp 373–378

Wormington, H and Neal, A (1951) *The Story of Pueblo Pottery*, Museum Pictorial N 2, Denver: Museum of Natural History

Zhushchikhovskaya, I S (1997) 'On early pottery making in the Russian Far East', 36(2) *Asian Perspectives,* 159–174

——— (1998) 'Keramika poseleniia Boismana-1 [Ceramics of the Boisman 1 Site]', in Vostretsov, Y E (ed), *Pervye Rybolovy v Zalive Petra Velikogo (Priroda i Drevnii Chelovek v Bukhte Boismana) [The First Fishermen in Peter the Great Gulf (Man and Nature in Boisman Bay)]*, Vladivostok: Nauka, pp 123–196

——— (1999) 'Dynamics of prehistoric pottery making of the Russian Far East', in Guo Jingkun (ed), *The Proceedings of the International Symposium of Asian Ceramics 1999*, Shanghai: Shanghai Institute of Ceramics, pp 471–484

——— (2001) 'Prehistoric and ancient pottery making of Northern Japan Sea Basin: spatio-temporal dynamics of ceramic pastes', in Millard, A (ed), *Archaeological Sciences, Proceedings of conference held at the University of Durham, 2–4 September 1997*, BAR International Series 939, Oxford: Archaeopress, pp 31–45

——— (2005) *Prehistoric Pottery Making of the Russian Far East*, BAR International Series 1434, Oxford: Archaeopress

Zhushchikhovskaya, I S (2006) 'The Russian Far East and cultural interaction in Prehistoric Eurasia in the light of ceramic studies', in Peterson, D, Popova, L and Smith, A (eds), *Beyond the Steppe and the Snow,* Leiden-Boston: Brill, pp 399–414

Zhushchikhovskaya, I S and Ponkratova, I Y (2000) 'Syr'evaia baza, klimat i traditsii drevnego goncharstva (po materialam kul'tur Vostochnoi i Severo-Vostochnoi Azii) [Source of raw materials, climate, and traditions of early pottery making (based on materials from the cultures of Eastern and Northeastern Asia)]', in Vostretsov, Y E and Kliuev, N A (eds), *Vpered . . . v Proshloe [Forwards . . . into the Past]*, Vladivostok: Dal'nauka, pp 103–149

Zhushchikhovskaya, I S and Shubina, O A (2006) 'Pottery making and the culture history of Neolithic Sakhalin', in Dumond, D and Bland, R (eds), *Archaeology in Northeast Asia,* University of Oregon Anthropological Papers 65, University of Oregon Press, pp 91–128

Chapter 4

Hunter-Gatherer Ceramics of Neolithic Korea

Daeyoun Cho and Ilhong Ko[1]

Introduction

The post-glacial period in the Korean peninsula saw the production of ceramics and ground stone tools in a hunting-gathering context. These pottery making hunter-gatherers are considered to be 'Neolithic', following a usage of the term specific to East Asia, in which the Neolithic refers to the presence of ceramics and ground stone tools, rather than agriculture (Barnes 1993; Kuz'min 2006; and see Jordan and Zvelebil, this volume, for wider discussion of the various definitions of 'Neolithic'). This Korean Neolithic, spanning from 8,000 bc to around 1,500 bc, is also at times referred to as the '*Chulmun* (comb marked) period' (e.g., Nelson 1993), in reference to the comb-pattern decoration considered emblematic of Korean Neolithic ceramics.

Korean Neolithic pottery was first identified by the Japanese archaeologist Hujita Ryosaku in 1930 at the Dongsamdong shell midden. However, as the Japanese colonial understanding of Korean prehistory had maintained that a Korean presence in the peninsula began with the 'Chalcolithic period', it was only following liberation that these ceramics came to dated to an earlier Neolithic period (Ahn 2002). It was through the stratigraphic sequences identified at the sites of Jitabri and Gumtanni in North Korea in the 1960s that the existence of pottery preceding the Bronze Age could at last be proved. Indeed, subsequent studies of Neolithic ceramics must be understood within this post-colonial context of Korean archaeology, in which 'identifying' this newly confirmed Neolithic period was the key issue.

Therefore, ceramic typology and distribution have had primacy over other avenues of research, as they provide the means by which spatial and temporal parameters of the Korean Neolithic are defined. The low resolution of Neolithic sites, mainly comprising campsites and shell middens, and the difficulty in obtaining radiocarbon dates from these sites, has also contributed to the way in which Neolithic ceramics have been divorced from their social and economic context, and approached merely as 'fossil records', representing the temporal position and spatial distribution of Neolithic cultures.

The past decade in Korean archaeology has witnessed significant advancements in the study of Neolithic pottery. A marked increase in the number of Neolithic sites excavated, and the recent discoveries of seminally important sites, such as Gosanni and Munamri, which yielded the earliest examples of Neolithic pottery, have facilitated lively debates concerning pottery chronology (e.g., Kim 2003; Lim 2003), and have led to a clearer understanding of the regional variation of Korean Neolithic ceramics (e.g., Lee, D J 2001; Lim 2001). More importantly, the accumulation of data in the western and southern coastal regions of the peninsula has at last made it possible to examine issues of subsistence economy and residence patterns for Late Neolithic populations (e.g., Kim 2006; Lee, J J 2001, 2002a, 2002b; Lim 1998), allowing us to situate our understanding of Neolithic ceramics within a social and economic context. This chapter therefore presents a brief introduction to Korean Neolithic pottery and examines some of the recent work that has been done relating ceramic change to changes in the subsistence economy and residence mobility of hunter-gatherers in the Korean peninsula.

The Broader Regional Background

Studies regarding the emergence of pottery have traditionally focused on the Japanese Jomon, which has yielded pottery dating to c. 12,700 bp (Aikens 1995). However, new evidence of Final Pleistocene pottery from regions neighbouring Japan has made it possible to suggest that the emergence of pottery around this time was not limited to the Japanese archipelago, but was in fact a wider East Asian phenomenon. Based on a critical reevaluation of radiocarbon dates, Kuz'min (2006), for example, has proposed that pottery in East Asia emerged simultaneously in three separate regions (southern China, Japan and the Russian Far East) during the Last Glacial (c. 13,700–13,300 BP).

As with the Korean Neolithic ceramics, the earliest ceramics of Japan and the Russian Far East were produced by hunter-gatherer communities. The specific nature of these neighbouring ceramics and the social and economic contexts in which they were used are discussed elsewhere in this volume. However, it is of interest to note here that the earliest examples of far eastern Russian pottery, from the Osipovka and Gromatukha complexes, are fibre-tempered, with flat bases, thick walls and almost no surface decoration (Kuz'min 2006)—characteristics which are also observed in the earliest pottery from Korea.

As for the Chinese mainland, two different cultural traditions have been identified for this period of transition from the Upper Pleistocene to the Holocene. The cave-dwelling communities of southern China produced a lithic industry related to the contemporary pebble lithic industry of Southeast Asia (the Hoabinhian culture). It is from this southern region that the earliest Chinese ceramics have been identified, at the sites of Yuchanyan, in Hunan Province (two ceramic vessels dated to 14,810±230 bp–12,320±120 bp), and at Xianrendong and Diaotonghuan, in Jiangxi Province (cultural layers dated to between 17,640±60 bp and 12,430±80 bp) (Zhang 2002). In contrast to this, the earliest ceramic dates obtained for sites from northern China, which are represented by a microlithic industry associated with the lithic industries of the Russian Far East and Japan, fall broadly around 10,000 bp (Zhang 2002).

The Korean Neolithic Sequence

The earliest example of Korean Neolithic pottery was discovered in a secure cultural context at Gosanni on the island of Jeju, off the southern coast of Korea. A possible date of 10,500–10,000 bp was suggested for the Gosanni pottery (Im 1995; Kang 2002), which acted to push back the beginning of the Korean Neolithic by around 2,000 years, as until then, raised-design ware (dated to around 8,000 bp) had been considered to be the earliest pottery tradition in Korea. The discovery of this earliest pottery in association with a late Palaeolithic microlithic assemblage at Gosanni also acted to bridge the chronological gap that had existed between the end of the Korean Palaeolithic, dated to around 12,000 bp, and the beginning of the Neolithic. The end of the Korean Neolithic, on the other hand, is marked by the appearance of a new set of Bronze Age material culture, comprised of coarse plain ware, longhouses and rectangular houses, semi-lunar reaping knives and dolmen burials.

Ceramic chronologies for the Korean Neolithic have mainly focused on regional sequences. Sequences for the southern coastal region of the peninsula were constructed around the ceramics from the Dongsamdong and Sugari shell middens (e.g., PNUM 1981) while sequences for the central-western region were structured around the stylistic changes observed in comb-pattern ware (e.g., Lim 1983). Attempts have recently been made to combine these regional chronologies to produce a general ceramic sequence that may be applied to the whole of the peninsula. The chronological framework set out by Ahn (2002) represents a tentative attempt at such a pan-peninsular sequence, taking into account the various chronologies that have been outlined for the different regions in the peninsula, as well as incorporating more recently obtained radiocarbon dates. Our discussion of Neolithic ceramics in this chapter will therefore follow the basics of the chronological scheme proposed by Ahn, beginning with an introduction of the earliest Korean Neolithic pottery, which will then be followed by an examination of the pottery traditions of the Initial, Early, Middle and Late Neolithic periods.

Beginning of the Korean Neolithic

Archaic plain ware marks the beginning of the Korean Neolithic tradition. While examples of archaic plain ware[2] had previously been found at the Dongsamdong shell midden (Kim 1986), it was through excavations at Gosanni (JNUM 1998; Kang 2002) that this pottery could securely be attributed to the earliest period of the Neolithic. The lack of charcoal at this site meant that radiocarbon dates could only be obtained though the direct dating of pottery. Unfortunately, the datasets received have given very different dates (10,180±65 BP, 4,480±45 BP, 6,230±320 BP, 6,910±60 BP), indicating the need for additional dating (Kuz'min 2006: 368). However, a microlithic assemblage consisting of over 10,000 stone tools, including arrow points, scrapers, points, bifaces, blades and microblade cores, was found in association with archaic plain ware under the K-Ah volcanic ash layer at Gosanni (JNUM 1998; Kang 2002), and similarities between it and the microlithic assemblage from layer 6 of the Kamikuroiwa rockshelter site in Japan, dated to around 10,000 BP, have been noted (Ahn 2002). Thus, it may tentatively be suggested that archaic plain ware, representing the earliest form of Korean Neolithic pottery, made its appearance at around the 9th millennium bc.

Although the pottery identified at Gosanni has been termed 'archaic plain ware', this ceramic type includes vessels decorated with narrow zigzag motifs and those that have holes punched along the rim, in addition to the plain, undecorated specimens (Figure 4.1). A round-bottomed variant of archaic plain ware, called 'thumbmark ware', can be observed slightly later in the archaeological record. Both types were made using organic temper, possibly plant fibre. It has been observed that this use of plant fibre as a temper is part of a wider Northeast Asian tradition that can be identified throughout the Amur River basin and Eastern Siberia (Lee 2002). Archaic plain ware has been identified at other coastal locations in the peninsula, such as Gumyoungri (Kang 2002) on the western tip of Jeju Island and

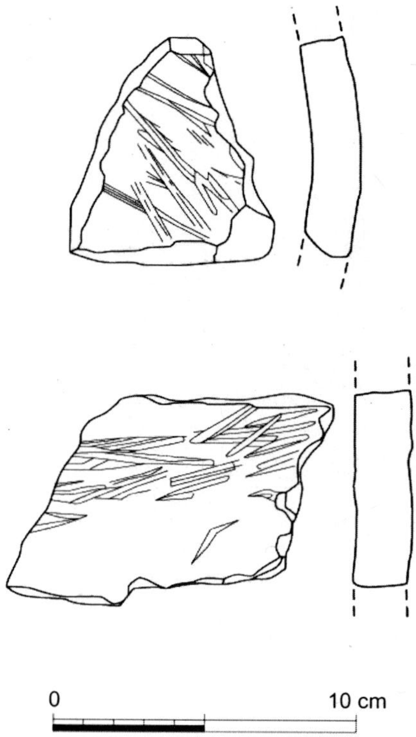

Figure 4.1 Undecorated example of archaic plain ware (after JNUM 1998: 69, figure 25; redrawn by author).

Munamri (NRICP 2004) on the eastern coast of Korea. Thumbmark ware has been found at both inland coastal locations, the Ojinri rockshelter at Chungdo (PNUM 1994) being an example of an inland location site, and the shell midden at Bumbang (PCM 1993, 1997) an example of a coastal location site.

Initial Neolithic

This period, from about 8,000 to 5,000 BC, is represented by raised-design ware. These vessels, which are generally beaker-shaped with relatively narrow, flat bases, are decorated with appliqué bands around the rim and body (Figure 4.2). It has been suggested that the origins of raised-design ware can be traced to the middle reaches of the Amur River (Ha 2002; Lee 2002). However, as similar traditions producing raised-band decorated pottery can be found throughout the wider region of Northeast Asia (such as

Figure 4.2 Raised-design ware from Osanni (after SNUM 1984: 92, plate 35-1; with permission from the Seoul National University Museum).

the linear relief ware of the Jomon period), further analysis must be carried out before any such links can be proposed. In Korea, this pottery tradition is prevalent along the eastern and southern coastal areas. Key sites yielding raised-design ware are Yangyang Osanni (SNUM 1984, 1985b, 1988a) and Gosung Munamri (NRICP 2004) on the eastern coast, and Shinamri (NMK 1975, 1989), Dolsan Songdo (NMKKJ 1989, 1990) and Sangnodaedo (DUM 1984) on the southern coast. Raised-design ware has also been identified further inland, at Geochang Imbulri (Ahn 1988) and Damyang Sangsi (YUM 1984).

Gosung Munamri is a settlement and burial site that has been of central importance in understanding the ceramic sequence for the earlier stages of the Korean Neolithic. It was at this site that the chronological relationship between archaic plain ware and raised-design ware could be confirmed in a clear stratigraphic sequence (NRICP 2004). The large numbers of composite fishhooks and harpoons found in association with raised-design ware at this site provide important clues regarding the subsistence strategy adopted by coastal communities in the Initial Neolithic. In addition, the discovery of obsidian tools in association with raised-design ware at the nearby site of Osanni suggests that long-distance trade networks may have been in place in this period, as obsidian does not occur locally (SNUM 1988a).

In contrast to these eastern coastal locations, the raised-design ware of the southern coastal region contains a more diverse range of appliqué designs. It appears that raised-design ware was in use for a longer period in this region, as evidenced by the radiocarbon dates from Youndaedo, Bumbang shell midden and Songdo, which are relatively later than that obtained for Osanni (Ahn 2002). In addition, raised-design ware is often found in association with incised ware at many of these southern coastal sites, leading some to suggest that the two ceramic traditions may have co-existed for a significant period of time in the southern coastal areas (Han 1995).

Early Neolithic

The Early Neolithic, dating to between 5,000 and 4,000 bc, is dominated by the beaker-shaped vessels that are decorated around the vessel rim with impressed motifs (Figure 4.3). Although the literal translation of the Korean term describing this pottery type—*Abinmun*—is 'nail impressed', many of the decorations were in fact impressed using sharp incising tools. The distribution of impressed ware overlaps with that of raised-design ware, although an increase in the number of impressed ware sites in the western

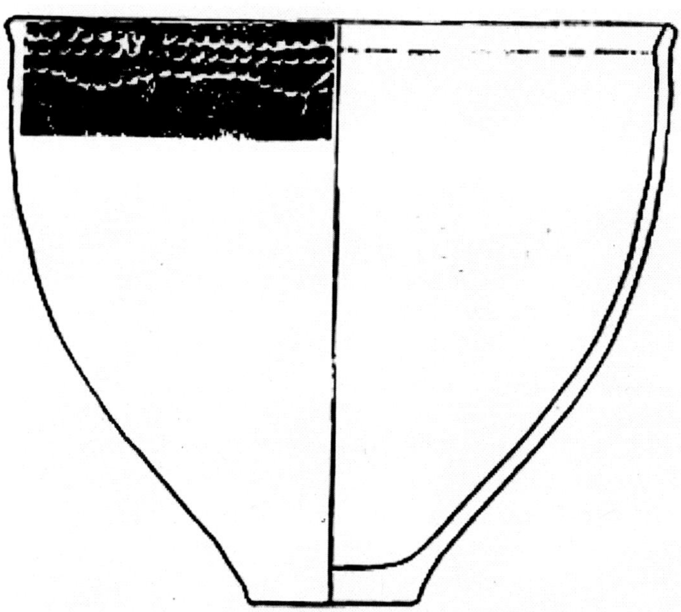

Figure 4.3 Impressed ware from Osanni (after SNUM 1984: 24, figure 11-1; with permission from the Seoul National University Museum).

coastal region can be noted. In the eastern coastal region, impressed ware has been identified at Osanni and Munamri. In the southern coastal region, the ceramic tradition of this period is represented by the 'Youngsundong type' vessels, which are decorated with a band of short horizontal herringbone impressions below the vessel rim.

The co-existence of raised-design ware and impressed ware observed at several sites has led to the suggestion that a continuity of pottery tradition may have existed between the Initial and Early Neolithic (e.g., Lim 1993). However, the opposite can also be proposed, in particular when focusing on the different decorative schemes of the two ceramic types (e.g., Ha 1993). In fact, although raised-design ware and impressed ware may be found in similar locations (generally coastal) and in association with similar stone tool assemblages (composite fishhooks and harpoons), it has also been observed that impressed ware vessels tend to be smaller than raised-design ware vessels (Song 2002). Thus, it appears that further research on the ceramics themselves, and on the economic and social context of the Initial and Early Neolithic, must take place before we may begin to consider the relationship between raised-design ware and impressed ware.

Middle Neolithic

It is in the Middle Neolithic period, dating from 4,000 to 1,500 bc, that comb-pattern pottery—generally considered as *the* representative artefact of the Korean Neolithic—with its pointed, torpedo-shaped bases and elaborate surface decorations, came to be used throughout the peninsula.[3] This period also witnessed a marked increase in the number of Neolithic sites found inland, along the middle and upper reaches of main river ways (Kim 2004). Comb-pattern ware shows marked regional variation. Evidence will now be examined according to the four broad ceramic regions that have been established for this period (Figure 4.4).

Central-western Region

In this region, comb-pattern ware is found along the Han River, at sites such as Amsadong (NMK 1994, 1995; SNUM 1985a) and Misari (MPSET and KPPDB 1994), along the Daedong River, at the Gungsan shell midden (Do and Hwang 1957a) and Jitabri (Do and Hwang 1957b), and along the western coast, at locations such as the island site of Oido (SNUM 1988b). Comb-pattern ware vessels from this region have up to three different zones of decoration, a typical example consisting of a band of short slanting lines along the mouth, a wide horizontal herringbone pattern around the body, and a wide vertical herringbone pattern for the base (Figure 4.5). This

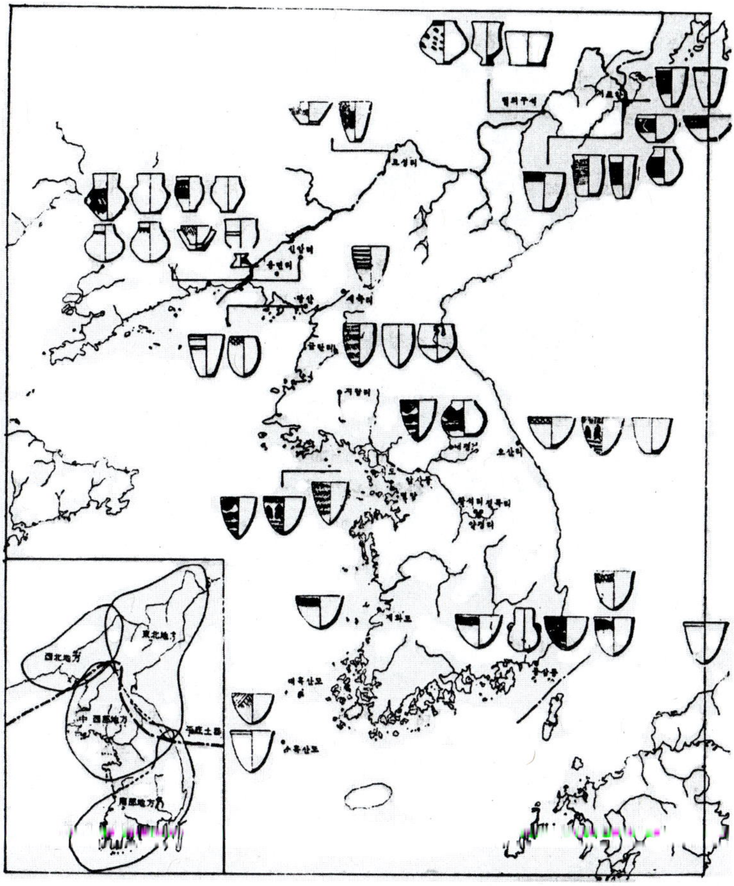

Figure 4.4 The four regional zones of Korean Middle Neolithic pottery (after Han 1983: 483, figure 2; with permission from the estate of Dr. Han Young-hee).

tripartite division of surface decoration characterizes the earliest examples of comb-pattern ware. In later examples, the separate design zones gradually merge into one, as observed for the Daedong River area (Han 1983), or the lower decoration zone disappears altogether, and the horizontal herringbones of the vessel body are replaced with vertical herringbones, as observed for the Han River area (Lim and Park 1988).

The chronological sequence for comb-pattern ware outside the Daedong River area has been based primarily on the stylistic analysis of its decorative motifs. This is due to the fact that much of the late Middle Neolithic material from the central-western region comes from midden sites and campsites, both of which have poor stratigraphic resolution. Settlement sites in this area are similarly unable to produce useful sequences due to

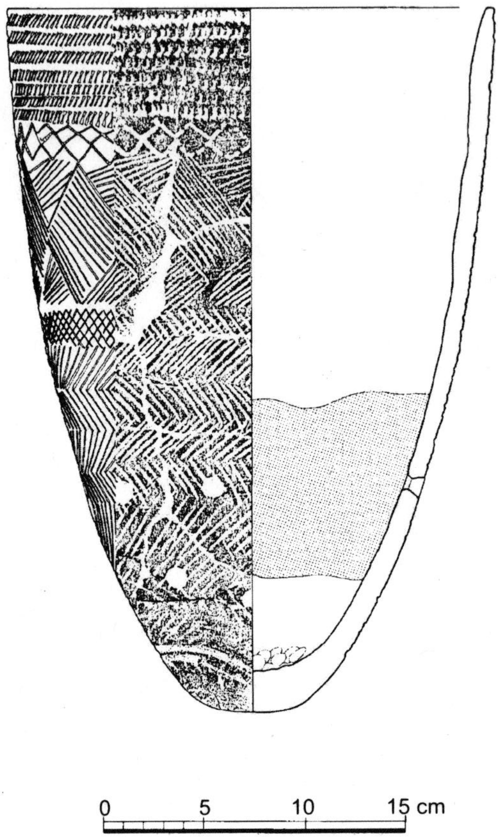

Figure 4.5 Comb-pattern ware from Amsadong (after NMK 1994: 298, figure 116; with permission from the National Museum of Korea).

their riverine locations. Consequently, as there is no direct evidence to suggest that the different motif combinations used in comb-pattern ware are representative of temporal variation, it has alternatively been proposed by Kim and Yang (2001) that the different styles of comb-pattern ware found at midden sites along the western coastline may in fact represent the presence of different communities that converged at these marine resource sites at certain times of the year. However, this is not to say that the proponents of this scheme ignore the influence of temporal factors in stylistic variation; they too acknowledge that the tripartite surface decoration of comb-pattern ware became simplified with time.

It remains to be seen if the specific styles of comb-pattern ware found at coastal midden sites represent regional community identity. However, it does appear that the Middle Neolithic in this region saw the seasonal movement of communities to coastal areas. This was the result of a new inland subsistence strategy adopted at the time, which also included hunting, river fishing using nets (as evidenced by the stone fishnet sinkers found at inland river sites), and the intensive use of plant resources (Lim 2006). Evidence of such seasonal forays into coastal locations where marine resource acquisition and processing took place can be seen in the faunal remains of the shell midden at So-Yeonpyoung Island (Kim 2006).

The Eastern Coastal and Northeastern Region

The Middle Neolithic of the eastern coastal region is represented by the site of Jikyungri (KNUM 2002), where comb-pattern ware was found in association with stone fishnet sinkers. The presence of the latter, which replaced the composite fishhooks of the previous period, is seen to represent a shift in the Neolithic subsistence strategy. This adoption of a new subsistence strategy is further suggested by the discovery of a saddle quern and handstone at the site, which have been interpreted as indicating the intensification of plant resource use (Song 2002).

The pottery tradition for the northeastern region in this period is represented by the flat-bottomed comb-pattern ware discovered at Osanni. As observed at Jikyungri, these Osanni flat-bottomed vessels were later replaced by comb-pattern ware vessels with pointed bases (Lim 2004). Contemporary to this flat-bottomed comb-pattern ware of the Middle Neolithic is the pottery assemblage from Seopohang layer 3 (Kim and Seo 1972).[4] The site of Seopohang is located on the northeastern coastline of the peninsula, near the border with Russian Siberia. Its pottery assemblage represents a deviation from the ubiquitous comb-pattern ware of this period, and comprises a wide array of vessel forms, such as deep bowls, bowls, jars and cups (forms not observed in comb-pattern ware), which were decorated with a variety of motifs, including squared spirals (the 'lightning' motif), triangles and buttons, etc.

Flat-bottomed comb-pattern ware and the Seopohang pottery assemblage have both been associated with the Neolithic ceramic traditions of the Russian Far East and Siberia (Ahn 2002). In particular, similarities have been noted between the pottery and lithic assemblages of Seopohang and Boisman-2 of southern Primor'e (Popov et al. 1997), although in this case, it is the artefact assemblage of Seopohang[5] layer 1 and 2 which provides the closest links with the Boisman culture.

The Southern Coastal Region

The comb-pattern ware of this region is distinctive in its use of thick, deep lines, called *taesunmun* (broad line designs), which begin and end with clear, rounded impressions. The use of *taesunmun* grouped lines and lattice motifs in decorating the vessel rim of comb-pattern ware is thought to illustrate the strong continuation of the local ceramic tradition, as similar motifs were widely used in the earlier raised-design ware of this region (Song 2002). These *taesunmun* decorations represent a significant deviation from the short, slanting lines observed in vessels from the central-western region of Korea, from which comb-pattern ware originated. The presence of *taesunmun* comb-pattern ware in association with impressed ware at several Middle Neolithic sites is similarly thought to represent the existence of a strong, regional ceramic tradition.

This strong continuation of the regional ceramic tradition has been understood in association with the regional subsistence economy. Song (2002) has suggested that the Tsushima warm current, which heavily influences the climate of this region, provided optimal conditions for the subsistence strategy of the previous period (i.e., marine resource exploitation) to be maintained. Hence, as comb-pattern ware is thought to have spread throughout the peninsula in conjunction with a new subsistence strategy utilizing a wider spectrum of resources, it is therefore possible that in the southern coastal regions of the peninsula, where, compared to other regions, a broadening of the resource spectrum may not have been as necessary, a more selective adoption of comb-pattern ware took place in the Middle Neolithic (Song 2002). Evidence of this regional ceramic tradition and regional subsistence strategy can be observed at the Sugari shell midden (PNUM 1981).

The Northwestern Region

Pottery tradition of the northwestern region of the peninsula is represented by the 10 sherds of pottery discovered in and around the limestone cave of Uiju Misongri. Eight of these sherds contained the typical herringbone patterns of comb-pattern ware, but the other two sherds were decorated with a distinctive zig-zag motif (Kim 1963). A large number of sites yielding pottery decorated with a similar zig-zag motif have recently been excavated in the Liao River area of southern Manchuria. It therefore appears that links may exist between the northeastern region of China and the northwestern region of Korea (Kim 2001).

Late Neolithic

This period from around 1,500 bc witnessed the gradual disappearance of decoration from Neolithic ceramics. In the northern region of the peninsula, this transition can be observed at the site of Seopohang (Kim and Seo 1972) and at Shinamri (NMK 1975, 1989), located at the mouth of the Abnok River. In the central region of Korea, herringbone patterns disappeared from comb-pattern ware, leaving the vessels bereft of decoration, apart from the short slanting lines surrounding the vessel rim. In other examples, the horizontal herringbone and crisscross patterns covering the vessel mouth and body came to be executed with less care (NMKKH 2005). Finally, it has been observed that in the southern region of Korea, decorations with thick deep lines were replaced with faint lattice patterns, parallel lines or short slanting lines, all of which are consigned to the area of the vessel mouth. The most striking decorative feature of these faintly decorated, beaker-shaped vessels, instead, becomes their double rim (NMKKH 2005). The distribution of these double-rimmed vessels was originally thought to have been limited to the southern coastal areas, as represented by sites such as the Sugari shell midden and Yulri (PNUM 1980), but recent discoveries of double-rimmed vessels at Kunsan Gado (CNUM 2001), the Noraesum shell midden (WUMPRI 2002) and Jinan Galmuri (HCPI 2003) indicate that this pottery type may have been more widespread than originally thought (NMKKH 2005). The reason for this drastic disintegration of the Korean Neolithic pottery tradition is generally thought to be associated with changes taking place in the subsistence economy, as will now be discussed.

The Emergence, Spread and Dissolution of Neolithic Ceramics in a Hunting-Gathering Context

The appearance of pottery in a hunting-gathering context in the Korean peninsula around the Early Holocene is an issue that has not been explored much in Korean archaeology. This is because the discipline has been more concerned with identifying and establishing a date for this earliest Neolithic pottery. However, it appears possible to suggest that what Aikens (1995) proposed for the appearance of Japanese Jomon pottery in a hunting-gathering context—that the emergence and use of pottery may have been directly connected to the adoption of an economy based on the hunting of marine and terrestrial animals, fishing and gathering—may also be applicable to Korea. Indeed, it is highly likely that the earlier Neolithic communities of Korea produced and used archaic plain ware, raised-design ware and impressed ware within a similar hunting, fishing and gathering economy.

An examination of the artefactual, faunal and floral evidence suggests a broadening of the subsistence spectrum taking place at around 4,000 BC, which goes in hand with an increase in inland site occupations. It has been argued that environmental change and/or population increase may have led to a certain degree of resource stress for Middle Neolithic communities, with the inability of shellfishing to provide sufficient food during periods of seasonal stress leading to the adoption of alternative subsistence strategies

(Lee, J J 2001). In addition to river fishing using nets, it appears that Middle Neolithic communities in the central-western region of Korea began to experiment with other strategies of food procurement, namely the use of domesticated plants. Although direct correlations between plant domestication and environmental/population pressure must be regarded with caution (the lack of conclusive evidence has been pointed out by Song [2002] and Ahn [2005]), it is generally accepted that the domestication of cultigens took place from this period onwards (Ahn 2005; Choe and Bale 2002; Nelson 2000). Crawford and Lee (2003) have used AMS dating to confirm that broomcorn and foxtail millet were established no later than 3,400 cal. BC in the Korean peninsula. However, this does not indicate a transition from a hunting-gathering-fishing economy to a farming economy. Rather, it is argued that these cultivated plants played an extremely limited role in the Neolithic subsistence economy (Ahn 1998; Kim 2002), and that their use should be considered as one of many strategies that were adopted to broaden the subsistence spectrum in this period (Lim 2006). Consequently, it was within this context of diversifying subsistence strategies that comb-pattern ware came to be made, used and spread throughout the peninsula. Unfortunately, however, the specific ways in which this pottery type may have been associated with the newly emerging social and economic conditions of the Middle Neolithic has yet to be examined in depth.

The dissolution of Neolithic ceramics has been interpreted as going hand-in-hand with the demise of the Neolithic subsistence economy (Lim 2006). Farming became established as the main subsistence strategy in the Bronze Age, and in particular, it has been noted that the coastal and island resources exploited extensively in the Neolithic were no longer used in the Bronze Age (Kim 2002). This, in conjunction with the transition from hunting-gathering-fishing to farming, brought an end to Neolithic ceramics.

Conclusions

This chapter illustrates how, in Korean archaeology, the emergence, change and demise of early Korean pottery in a hunting-gathering context has generally been understood in association with subsistence strategy change. However, the presence of elaborate decoration on the majority of these hunter-gatherer ceramics, to which any functional role associated with economic activities is not easily attached, must also be noted. Indeed, the fact that these decorations disappear with the demise of a hunting-gathering lifestyle indicates that the making, decorating and use of these ceramics may have been an integral part of the social practices that facilitated the reproduction of a hunting-gathering way of life in the Korean Neolithic. It can therefore be suggested that one of the more fruitful avenues of research in the future may be to consider the role that these ceramics played in the social life of hunter-gatherers.

Notes

1. We would like to thank Sally Smith and Eunyoung Kim for their invaluable feedback.
2. The term 'archaic' has been used here to differentiate this Neolithic plain ware from the plain ware (*Mumun* pottery) of the following Bronze Age.
3. It should be noted that comb-pattern ware is generally believed to have appeared before 4,000 bc in the central-western region, at a time when impressed ware was being used in the eastern and southern coastal regions of the peninsula (see Ahn 2002).
4. The Seopohang site contains five cultural layers, beginning from around 6,000 bc: layers 1, 2 can be dated to the Initial to Early Neolithic, layer 3 to the Middle Neolithic, and layers 4, 5 to the Late Neolithic (Kim and Seo 1972)
5. 'Seopohang' also referred to as 'Sopkhohan' (e.g., Popov et al. 1997).

References

Ahn, C B (1988) 'Geochang Imbulri excavation report', in The Korean Archaeological Society (ed), *Conference Proceedings of the 12th Annual Korean Archaeology Society Conference*, Seoul: KAS Publications, pp 135–148 (in Korean)

——— (1998) *Prehistoric Farming and Subsistence Production in East Asia*, Seoul: Hakyoun (in Korean)

——— (2002) 'The Neolithic period', in The National Academy of Sciences, Republic of Korea (ed), *A Study of Korean Archaeology (The Prehistory of South Korea)*, Seoul: NAS Press, pp 71–173 (in Korean)

——— (2005) 'Current findings and future work on the study of farming in the Neolithic of Southern Korea', 10 *Journal of the Korean Neolithic Society*, 7–25 (in Korean)

Aikens, C M (1995) 'First in the world: the Jomon pottery of early Japan', in Barnett, W K and Hoopes, J (eds), *The Emergence of Pottery: Technology and Innovation in Ancient Socities*, Washington DC: Smithsonian Institution Press, pp 11–21

Barnes, G L (1993) *The Rise of Civilization in East Asia: The Archaeology of China, Korea and Japan*, London: Thames and Hudson

Chungnam University Museum (CNUM) (2001) *Gado Shell Midden*, Daejeon: CNUM Publications (in Korean)

Choe, C P and Bale, M T (2002) 'Current perspectives on settlement, subsistence, and cultivation in prehistoric Korea', 39(1–2) *Arctic Anthropology*, 95–121

Crawford, G W and Lee, G-A (2003) 'Agricultural origins in the Korean Peninsula', 77(295) *Antiquity*, 87–95

Do, Y H and Hwang, K D (1957a) *Site Excavation Reports (Yoojuk Balgul Bogo) 2: Excavation Report of the Ancient Site of Gungsan*, Pyongyang: Academy of Sciences Press (in Korean)

Do, Y H and Hwang, K D (1957b) 'Interim report of the Jitabri excavation 1·2', 57–5·57–6 *Cultural Properties (Munhwa yoosan)*, 20–24 (in Korean)

Donga University Museum (DUM) (1984) *Sangnodaedo*, Pusan: DUM Publications (in Korean)

Ha, I S (1993) 'Early Chulmun pottery types', 2 *Museum Research Findings (Bakmulkwan Yeongoo Nomunjip)*, 1–38 (in Korean)

——— (2002) 'The establishment and development of raised-design ware', in Dongkuk University Cultural Heritage Institute (ed), *The Environment and Subsistence of the Korean Neolithic*, Kyungju: DUM Publications, pp 153–165 (in Korean)

Han, Y H (1983) 'A comparative regional analysis of the Neolithic', 12 *Essays on Korean History (Hanguksaron)*, 479–521 (in Korean)

——— (1995) 'The Neolithic period', in The Korean Archaeological Society (ed), *A Half Century of Korean Archaeology: Conference Proceedings of the 19th Annual Korean Archaeology Society Conference*, Seoul: KAS Publications, pp 29–74 (in Korean)

Honam Cultural Properties Institute (HCPI) (2003) *Galmuri: Excavation Report of Cultural Properties in the Submerged Area of the Youngdam Dam, Jinan XIII*, Iksan: HCPI Publications (in Korean)

Im, H J (1995) 'The new archaeological data concerned with the cultural relationship between Korea and Japan in the Neolithic age', 35(3) *Korea Journal*, 31–40

Jeju National University Museum (JNUM) (1998) *The Jeju Gosanni* Site, Jeju: JNUM Publications (in Korean)

Kang, C H (2002) 'The establishment and development of the early Neolithic culture of Jeju Island', in The Korean Archaeological Society (ed), *The Archaeology of Marine Exchange: Conference Proceedings of the 26th Annual Korean Archaeology Society Conference*, Seoul: KAS Publications, pp 9–34 (in Korean)

Kangnung University Museum (KNUM) (2002) *The Yangyang Jikyungri Settlement*, Kangnung: KNUM Publications (in Korean)

Kim, E Y (2006) 'Substance-settlement system in the Neolithic period of Korea: the analysis of faunal remains from shell middens of Yeonpyoung Islands', unpublished MA dissertation, Department of Archaeology and Art History, Seoul National University (in Korean)

Kim, J S (2002) 'The Late Neolithic-Early Bronze Age transition and cessation of island exploitation in Central-Western Korea: the spread of territoriality into a mobile hunter-gatherer context', unpublished PhD dissertation, Department of Anthropology, Arizona State University, Tuscon

——— (2003) 'A critical examination of the Neolithic chronology for the central-western region', 5 *Journal of the Korean Neolithic Society*, 5–19

——— (2004) 'The spatial distribution of ceramic decorations in the inland regions of southern Korea during the Middle Neolithic', 54 *Journal of the Korean Archaeological Society*, 5–34 (in Korean)

Kim, J S and Yang, S H (2001) 'A re-examination of the chronology and shell midden exploitation strategy of the Neolithic of the central-western region', 45 *Journal of the Korean Archaeological Society*, 5–44 (in Korean)

Kim, W Y (1986) *Introduction to Korean archaeology*, Seoul: Iljisa (in Korean)

Kim, Y H (2001) 'An examination of the relationship between the Liaoning peninsula and the central-western region of Korea', 8 *Journal of Korean Prehistory (Hanguk Sunsa Gogo Hakbo)*, 55–69 (in Korean)

Kim, Y K (1963) 'Excavation report of the Misongri cave site', 3 *Archaeological Resources (Gogohak Jaryojib)*, 1–19 (in Korean)

Kim, Y K and Seo, G T (1972) 'Excavation report of the prehistoric site of Seopohang', 4 *Archaeology and Folk Essays (Gogo Minsok Nonmunjib)*, 31–145 (in Korean)

Kuz'min, Y V (2006) 'Chronology of the earliest pottery in East Asia: progress and pitfalls', 80 *Antiquity*, 362–371

Lee, D J (2001) 'The comb-pattern ware culture of the central-eastern coastal regions—its relationship with the southern coastal region', 1 *Journal of the Korean Neolithic Society*, 23–55 (in Korean)

——— (2002) 'The origins and characteristics of Korean early Neolithic culture', in The Korean Ancient Historical Society (ed), *Archaeology and the Period of Transition I*, Seoul: Hakyoun, pp 63–111 (in Korean)

Lee, J J (2001) 'From shellfish gathering to agriculture in Prehistorc Korea: the Chulmun to Mumun Transition', unpublished PhD dissertation, Department of Anthropology, University of Wisconsin, Madison

——— (2002a) 'Transition of subsistence from the Neolithic to the Bronze Age at the Gado shell midden', 3 *Journal of the Korean Neolithic Society*, 5–16 (in Korean)

——— (2002b) 'Changes in subsistence strategies in the southern coastal Neolithic—a new interpretation of faunal remains from shell middens', 48 *Journal of the Korean Archaeological Society*, 59–92 (in Korean)

Lim, H J (1983) 'Diachronic change in ceramics', 12 *Essays on Korean History (Hanguksaron)*, 615–654 (in Korean)

Lim, H J and Park, S B (1988) 'The distribution of comb-pattern ware in the central-western coastal region and the Han River area', in Seoul National University Museum (ed), *Oido Shell Midden*, Seoul: SNUM Publications, pp 92–110 (in Korean)

Lim, S T (1993) 'A study of raised-design ware', unpublished MA dissertation, Department of Archaeology and Art History, Seoul National University (in Korean)

——— (1998) 'The characteristics of shell middens—with focus on sites associated with adaptation strategies', 3 *Scientific Archaeology Research (Gwagi Gogo Yeongu)*, 7–64 (in Korean)

——— (2001) 'The regionality of comb-pattern ware culture—with focus on the central-western and Youngdong regions', 1 *Journal of the Korean Neolithic Society*, 1–22 (in Korean)

——— (2003) 'Problems surrounding the relative chronology of the Neolithic in the central region', 5 *Journal of the Korean Neolithic Society*, 21–30 (in Korean)

——— (2004) 'The comb-pattern ware culture of Kangwon province', in The Kangwon Archaeological Society (ed), *Conference Proceedings of the 6th Kangwon Archaeological Society Conference*, Chuncheon: KWAC Publications, pp 23–38 (in Korean)

——— (2006) 'Socieoeconomic changes in Neolithic central-western Korea', unpublished PhD dissertation, Department of Archaeology and Art History, Seoul National University (in Korean)

Misari Prehistoric Site Excavation Team (MPSET) and Kyunggi Province Public Development Branch (KPPDB) (1994) *Misari*, Seoul: Hakyoun

National Museum of Korea (NMK) (1975) *Shinamri I*, Seoul: NMK Press (in Korean)

——— (1989) *Shinamri* II, Seoul: NMK Press (in Korean)

National Museum of Korea (NMK) (1994) *Amsadong*, Seoul: NMK Press (in Korean)

——— (1995) *Amsadong II*, Seoul: NMK Press (in Korean)

National Museum of Korea Kimhae (NMKKH) (2005) *Prehistoric Ceramics in Transition*, Pusan: NMKKH Press (in Korean)

National Museum of Korea Kwangju (NMKKJ) (1989) *Dolsan Songdo I*, Kwangju: NMKKJ Press (in Korean)

——— (1990) *Dolsan Songdo II*, Kwangju: NMKKJ Press (in Korean)

National Research Institute of Cultural Properties (NRICP) (2004) *Gosung Munamri*, Seoul: NRICP Press (in Korean)

Nelson, S M (1993) *The Archaeology of Korea*, Cambridge: Cambridge University Press

——— (2000) 'The questions of agricultural impact on socio-political development in prehistoric Korea', in Lim, H J (ed), *The Origins of Rice Farming in Ancient Korea*, Seoul: Hakyoun, pp 269–300 (in Korean)

Popov, A N, Chikisheva, T A and Shpakova, E G (1997) *Boismanskaia Arkheologicheskaia Kul'tura Yuzhnogo Primor'a [The Boisman Archaeological Culture of Southern Primor'e]*, Novosibirsk: Nauka

Pusan City Museum (PCM) (1993) *Bumbang Shell Midden I*, Pusan: PCM Publications (in Korean)

——— (1997) *Bumbang Shell Midden II*, Pusan: PCM Publications (in Korean)

Pusan National University Museum (PNUM) (1980) *Kumgokdong Yulri Shell Midden*, Pusan: PNUM Publications (in Korean)

——— (1981) *Kimhae Sugari Shell Midden I*, Pusan: PNUM Publications (in Korean)

——— (1994) *Ojinri Rockshelter at Chungdo*, Pusan: PNUM Publications (in Korean)

Seoul National University Museum (SNUM) (1984) *Osanni*, Seoul: SNUM Publications, (in Korean)

——— (1985a) *Amsadong*, Seoul: SNUM Publications (in Korean)

——— (1985b) *Osanni II*, Seoul: SNUM Publications (in Korean)

——— (1988a) *Osanni III*, Seoul: SNUM Publications (in Korean)

——— (1988b) *Oido Shell Midden*, Seoul: SNUM Publications (in Korean)

Song, E S (2002) 'A study on the diffusion of the comb-pattern pottery culture in Neolithic Korea', unpublished PhD dissertation, Department of Archaeology and Art History, Seoul National University (in Korean)

Wonkwang Univerity Mahan-Paekche Research Institute (WUMPRI) (2002) *Noraesum*, Iksan: WUMPI Publications (in Korean)

Yonsei Univerity Museum (YUM) (1984) *Damyang Sangsi I Rockshelter*, Seoul: YUM Publications (in Korean)

Zhang, C (2002) 'The discovery of early pottery in China', XXIX *Documenta Praehistorica*, 29–35

Chapter 5

Review of Early Hunter-Gatherer Pottery in Eastern Siberia

Hugh G McKenzie

Introduction

Eastern Siberia can be defined as that territory delimited by the Yenisei River in the west, the Lena River in the east, the Arctic Ocean in the north, and by a chain of mountains in the south and southeast that extends north from Central Asia to separate the Arctic and Pacific watersheds (Figure 5.1). Less than 10 years ago, this enormous region featured in few discussions on early pottery. Prudence Rice (1999), for example, in her extensive article, 'On the origins of pottery', reviewed existing evidence for very early pottery in Japan, China, western Asia, Africa, South America and Central and North America. The omission of Siberia, and of Central Asia more generally, from this and other reviews is largely understandable as a product of the political and language barriers that for so long kept Russian and western archaeology apart. While archaeologists in eastern Siberia have examined the region's pottery for well over 100 years, relatively little of this research has ever been translated out of Russian or presented at non-Russian academic venues. In addition, radiocarbon dating of eastern Siberian archaeological material has only been done in a widespread and systematic manner over the last 10–15 years (e.g., Konopatskii 1982; Kuz'min 2002; Kuz'min and Orlova 1998, 2000; Kuz'min et al. 1998; Mamonova and Sulerzhitskii 1989; Orlova 1995; Timofeev and Zaitseva 1996; Weber 1995; Weber and Goriunova 2005; Weber et al. 2004, 2005, 2006, 2008). Intriguingly, this radiocarbon evidence confirms that hunter-gatherers in eastern Siberia adopted pottery prior to 10,000 bp,[1] and consequently the region must be

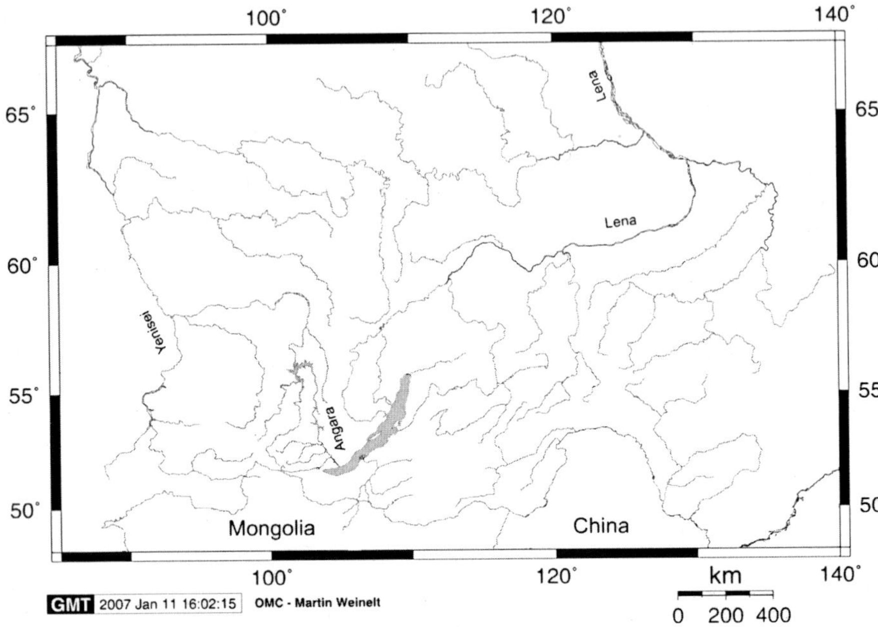

Figure 5.1 Map of discussed regions of sub-Arctic eastern Siberia.

recognized along with southern Japan, southern China, and the Russian Far East as among the world's first centres of pottery use (Kuz'min 2002; Kuz'min and Orlova 1998, 2000; Kuz'min and Vetrov 2007).

Based on a survey of the available literature, the aim of this chapter is to provide a broad regional overview of the evidence for the timing and form of the earliest pottery adopted by hunter-gatherers across eastern Siberia. Unfortunately, from an archaeological perspective, the region has not been evenly researched. Some areas, such as the zone surrounding Lake Baikal, have seen extensive investigation, including intensive excavations and analyses over the past 10 years. Other areas, especially those further north, have received scant attention. This, combined with the massive size of the region, requires that our review be selective in both its geographical and chronological scope. I am not, for example, going to discuss developments north of the Arctic Circle and instead concentrate on describing the evidence from those areas for which there is the most complete information. In particular, I focus on four areas where absolute chronology has best been established through the use of radiocarbon dating: (1) Trans-Baikal, (2) Cis-Baikal, (3) the Central Yenisei Basin, and (4) the Central Siberian Plateau (Yakutia and Evenkia). I will spend proportionately more time describing the material from Trans-Baikal and Cis-Baikal since ceramic typologies in these areas are often used by scholars to construct relative chronologies in other regions. From a chronological standpoint, the discussion is focused on documenting the evidence for the earliest adoption of pottery by hunter-gatherers across eastern Siberia (i.e., the Early Neolithic), and I only briefly consider subsequent developments in each area.

Finally, it is necessary to note that descriptions of eastern Siberian pottery are derived, whenever possible, from original site reports and primary publications; however, I have also relied on synthetic works by Aseev (2003), Khlobystin (1996), Konopatskii (1982) and Okladnikov (1950, 1955), which summarize important information from sources that are either unpublished or are published in venues that are difficult or impossible to access as a result of their small print runs and limited circulation. As elsewhere in northeast Asia, archaeologists in Siberia use the widespread adoption of pottery to define the introduction of Neolithic cultural complexes (Chard 1974: 63–64), and while I retain this convention here, I also point out instances where such terminology has created confusion.

Basic Terminology, Forms and Ornamentation

Although there is discernible variation in pottery forms and ornamentation across eastern Siberia, as well as in the timing of local ceramic sequences, there are also a considerable number of basic elements that recur widely. In order to avoid repetition during the regional summaries, it is useful to have an inventory of some of the basic terminology used to describe the forms of eastern Siberian Neolithic pottery as well as some of the more widespread ceramic styles.

Goriunova and Savel'ev (1981) have designed a system of terminology specifically for describing Neolithic and Early Bronze Age pottery from eastern Siberia. While not universally adopted, aspects of this system are widely used, especially in the areas around Lake Baikal. With respect to basic vessel form, the system first classifies each pot on the basis of its profile into one of the four main geometric forms: sphere, ovaloid (often called egg-shaped), paraboloid (often called mitre-shaped), and ellipsis (Figure 5.2). Each of these geometric forms is then divided into three main categories: simple, complex and ornamental. Simple pots lack a separate and defined rim; complex forms have weakly everted rims, but little in the way of a neck; and ornamental pots feature strongly everted rims that are joined to the body of the pot via an identifiable neck. Finally, each category is then divided into an open form, in which the greatest diameter of the pot is at the rim, and a closed form, in which the largest diameter is found somewhere lower on the body.

In addition to describing the form of pots using this or similar terminology, archaeologists have also identified a number of ceramic types or

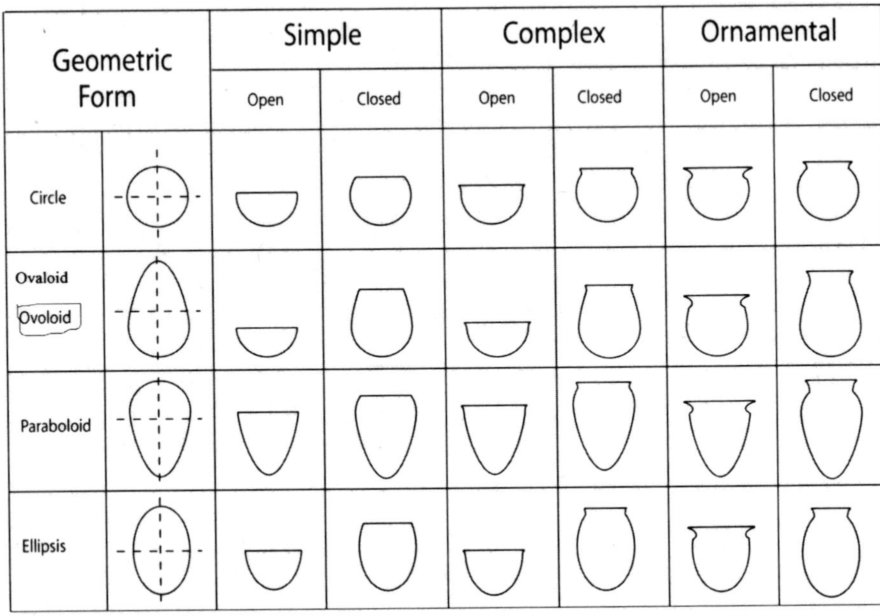

Figure 5.2 Classification of eastern Siberian Neolithic and Early Bronze Age pottery forms (modified from Goriunova and Savel'ev 1981).

styles[2] that are common throughout eastern Siberia. While I will now provide brief descriptions of the five most widespread pottery styles from the eastern Siberian Neolithic (net-impressed, cord-impressed, Khaita, Posol'sk and Ust'-Belaia), specific details regarding the sites from which these pottery types were recovered, their relative stratigraphic positions, associated radiocarbon dates etc., will wait until discussions of the individual regions. Likewise, I will also wait to introduce those pottery types that have only limited geographical or temporal distributions, including the oldest pottery in the region, Ust'-Karenga. Unless otherwise noted, the following descriptions are based on the information provided in Khlobystin (1996), Savel'ev (1989), Sinitsyna (1986) and Weber (1995).

Net-Impressed Pottery

As the name suggests, net-impressed pottery is defined by characteristic imprints of net or grid weaving on the exterior surface of the vessels (Figure 5.3a). Most scholars believe that the nets were used to support the clay when forming the pot, and that the impressions were left behind after the net was removed during the drying or firing process (e.g., Alekseev and Cherosov 1990: 75; Okladnikov 1950: 167–171). Okladnikov (1950:

167–171) suggested that such pots could have been produced by smearing clay onto the walls of a small excavated and net-lined pit. After the clay had dried, the pot could be pulled out of the pit with the net and fired. Vorob'ev (1995), however, considers many of the net-impressed pots to have been paddle-formed, while D'iakonov (2001) sees evidence for the use of woven fabrics derived from flax or nettle (see also Okladnkiov 1955: 84). It seems likely that a number of approaches would have been employed, and there is little doubt that the umbrella phrase 'net-impressed pottery' encompasses significantly more variation than is commonly acknowledged in the literature (D'iakonov 2001). Zubkov (1982), however, does distinguish two general variants: net-impressed I and net-impressed II (also known as Nizhne Seredkino). The second variant differs from the first in that the net impressions were deliberately smoothed out leaving only faint traces of the lines and nodes. In addition, it seems that the net-impressed I vessels rarely have any additional decorations, while net-impressed II pots more often exhibit small pits or stamps around the rims as well as occasional comb decorations on the upper body. The first version is usually considered to be older than the second (e.g., Weber 1995; Zubkov 1982).

In one form or another, net-impressed pottery is known across all of eastern Siberia (Glushkhov and Glushkova 1992). In the Cis-Baikal region, for example, it was recorded, at Gorelyi Les and Ust'-Khaita on the Belaia River (Igumnova et al. 2004; Savel'ev et al. 1974: 164), at Ityrkhei, Tyshkine, Ulan-Khada and Sagan-Zaba on the shores of Lake Baikal (Goriunova 1984a, 1984b, 2001; Goriunova and Khlobystin 1991; Khlobystin 1969), and at Kazachka in the Yenisei region of eastern Siberia (Generalov 1979). It is widespread in Yakutia and Evenkia as well (Alekseev 1996). It is also the only pottery style that can unambiguously be identified from Cis-Baikal Neolithic mortuary sites in addition to campsites. A notable exception to the wide geographical distribution of this type is Trans-Baikal where net-impressed pottery is relatively rare (Aseev 2003; Grishin 1981; Khlobystin 1996: 291).

In most areas of eastern Siberia including Cis-Baikal, Yakutia and Evenkia, net-impressed pottery is found at the bottom of Neolithic deposits, immediately above aceramic Mesolithic layers, and as such is considered the oldest type in these areas (Khlobystin 1996). Net-impressed pottery is also widespread in a chronological sense, however, and, depending on which region you examine and which form of net-impressed pottery you are looking at, dates from at least 7,000 bp all the way through to around 3,000 bp.

Cord-Impressed Pottery

Cord-impressed pottery is defined primarily on the basis of cord impressions that were applied with the use of a string or cord-wrapped paddle

(a)

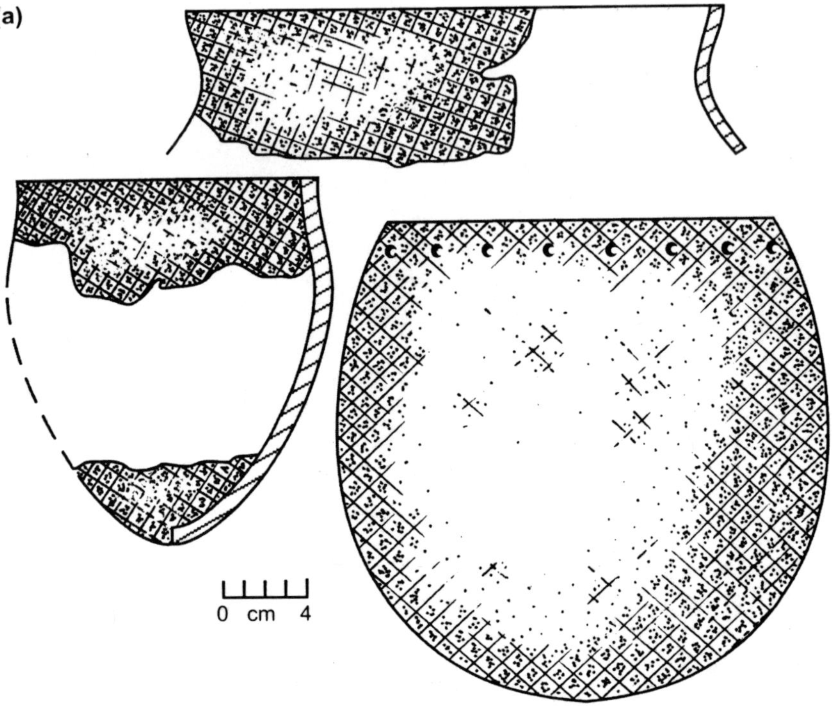

0 cm 4

(b)

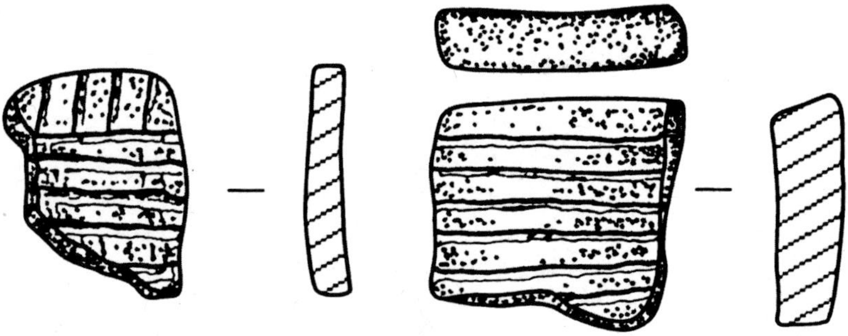

0 cm 4

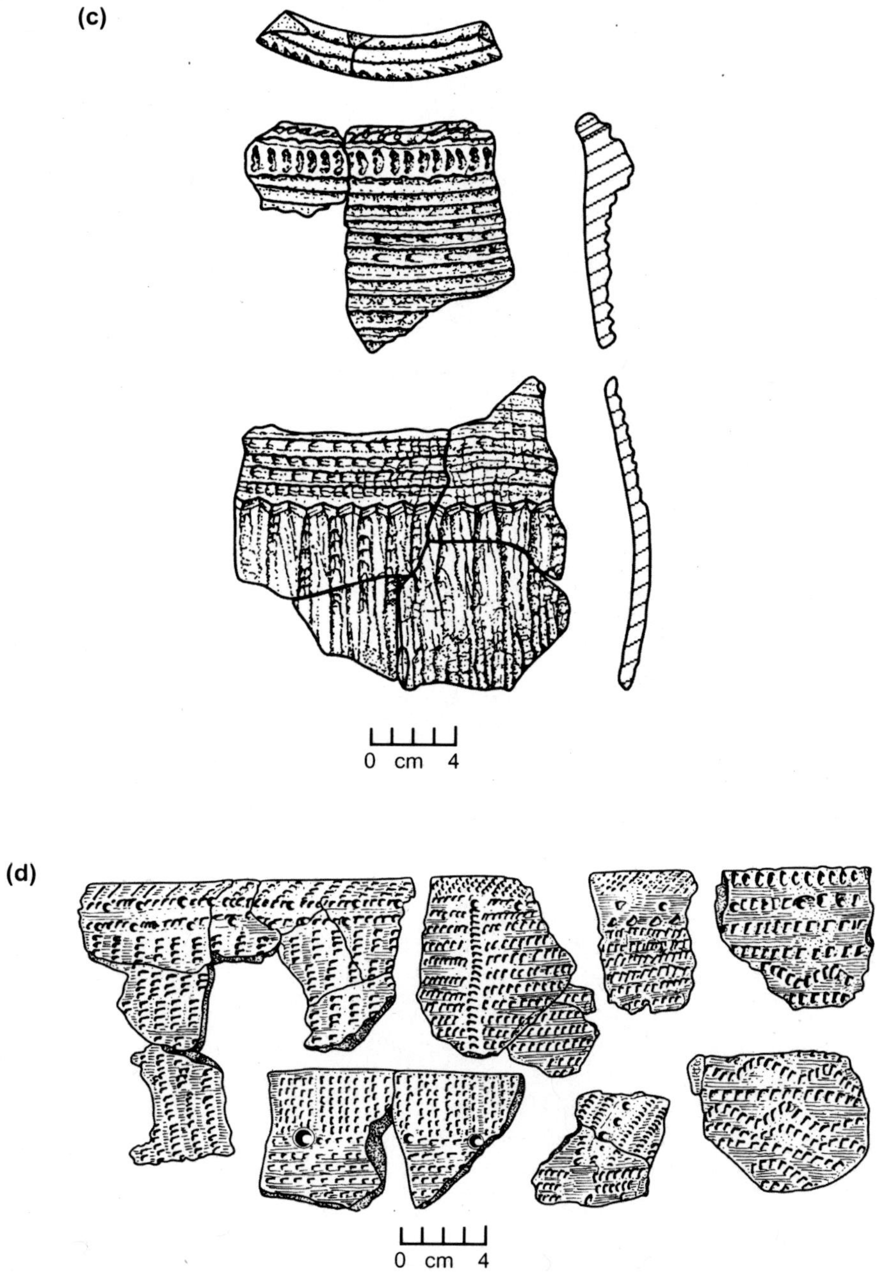

Figure 5.3 Common ceramic styles of eastern Siberia: (a) net impressed (modified from Goriunova nd), (b) Khaita (modified from Nomokonova and Goriunova 2004: 119); (c) Posol'sk (modified from Makarov 2005: 154); (d) Ust'-Belaia (adapted from Khlobystin 1996: 288).

during the final shaping and smoothing phases of the manufacturing process. Like net-impressed pottery, there is considerable variation subsumed under this category, and the number, thickness and orientation of cord impressions varies. The style has the same wide spatial distribution as net-impressed pottery and is found in all areas of eastern Siberia (Khlobystin 1996). In the Central Yenisei Basin (Makarov 2005) and Trans-Baikal (Khlobystin 1996: 291), cord-impressed pottery is considered to be the earliest type (but see discussion of Ust'-Karenga pottery below), while in most other areas it appears to follow net-impressed, although with some temporal overlap. Khlobystin (1996: 291) suggests that this spatial and temporal distribution indicates that cord-impressed pottery originated in southern Trans-Baikal and then spread north through Cis-Baikal to Yakutia.

Khaita Pottery

Khaita pottery was first identified by Savel'ev (1982). The style is characterized by cord impressions akin to those that define cord-impressed pottery, but, in addition, Khaita pots feature decorations on the upper half of the pots consisting of a series of vertical and horizontal incised lines forming herringbone patterns and other geometric motifs (Figure 5.3b; Savel'ev 1982, 1989: 22; Sinitsyna 1986: 10). The similarity in manufacturing techniques utilizing cord-wrapped paddles has led Khlobystin (1996: 291) to suggest that Khaita is a derivative of the basic cord-impressed pottery.

Like net-impressed and cord-impressed pottery, Khaita has a wide spatial distribution that was recovered from sites including: Ust'-Khaita, Ust'-Belaia, and Gorelyi Les in the Angara/Belaia basin (Igumnova et al. 2001; Medvedev et al. 2001; Savel'ev et al. 1974: 164–168), Ulan-Khada, Ityrkhei, Sagan-Zaba, and Kulara III on the shores of Lake Baikal (Goriunova 1984b: 5–7; Goriunova and Khlobystin 1991: 46), Popovskii Lug from the Kachug region on the Upper Lena River (Igumnova et al. 2004), Kazachka in the Yenisei basin (Generalov 1979: 8) and Biriusinsk in the Krasnoiarsk region (Sinitsyna 1986: 89).

Since it is often found in the same stratigraphic layers as net-impressed pottery, Khaita is also considered to be among the oldest pottery types in the region; but like net-impressed pottery, it also persists into later times.

Posol'sk Pottery

The defining feature of Posol'sk ceramics is a pronounced neck and characteristically thickened rim, produced through the addition of a band of clay that is rounded or triangular in cross-section. Such pots are usually decorated with stab-and-drag horizontal lines running around the body of the pot, which are often accompanied by small short lines 'hanging' from the lowest horizontal line such that they resemble a sort of tassel (Figure 5.3c). In addition, a line of small pits often runs below and parallel to the thickened rim, which itself can be ornamented by various incisions or notches.

Like the other three types discussed above, Posol'sk is widely distributed across eastern Siberia, having been recovered from sites including: Ityrkhei, Kulara I, Ulan-Khada, Posol'sk and Tyshkine on the shores of Lake Baikal (Goriunova and Khlobystin 1991: 44; Nomokonova and Goriunova 2004), Popovskii Lug and Makarovo I in the Upper Lena basin (Zubkov 1982: 6), Gorelyi Les and Nizhne Seredkino in the Angara basin (Savel'ev et al. 1974), and Niasha and Eleneva Cave in the Yenisei basin (Makarov 2005; Savel'ev 1989: 23).

Posol'sk pottery is traditionally associated with the Kitoi culture complex (e.g., Okladnikov 1950), which most scholars agree to be the oldest Neolithic tradition in the Cis-Baikal area (Weber 1995; Weber et al. 2005). This association was established prior to the application of radiocarbon dating to the Cis-Baikal material, which resulted in significant changes to the region's culture-history. How these changes affect the relative dating of Posol'sk pottery is still controversial, but it appears that Posol'sk pottery is likely to be c.1,000 years younger than the Kitoi mortuary tradition (Weber 1995). Recently, however, Makarov (2005) has suggested that Posol'sk pottery is the earliest type in the Central Yenisei Basin, with radiocarbon dates that are equivalent to the Kitoi mortuary tradition from Cis-Baikal (see discussion below).

Ust'-Belaia

First described by Savel'ev and Medvedev (1973), Ust'-Belaia pottery can be seen as a younger, more elaborate version of Posol'sk. The thick rims of Posol'sk pottery are gradually replaced by a straight or scalloped rim. Medvedev and colleagues (2001) report that Ust'-Belaia pots were manufactured in parts, with the base being moulded on some sort of block, and the body built up by adding strips and then smoothing them together. Ornamentation on the exterior surface of the pots is quite varied, but is primarily based on 'back-stepped blade' impressions (*otstupaiushchei lopatki*) produced by repeatedly impressing the tip of a small blade or spatula into the clay to form lines and geometric designs across the surface of the vessel (Berdnikova 1986; Figure 5.3d). In addition, stab-and-drag impressions are often found, as are impressions produced by comb and cogged stamps. For the first time, ornamentation covers the entire exterior surface of the pots rather than just the upper portion. Savel'ev (1989) breaks the Ust'-Belaia pottery into an early phase (Kazachinsk) and a late phase (Ust'-Belaia), which he relates to the Neolithic and Bronze Age respectively. Similarly, Sinitsyna (1986) identifies three sequential phases based on stylistic differences. Ust'-Belaia I features slightly inverted rims and horizontal bands of plain or comb stamps covering the entire surface of the vessels. Ust'-Belaia II differs in that the rims are no longer inverted and, while the density of impressions on the pots is lower, the variety of stamps is greater. Such stamps are also often arranged in vertical, rather than horizontal rows. Finally Ust'-Belaia III is characterized by straight rims and an even lower density of stamp impressions and chevron designs.

Notwithstanding the attempts of Savel'ev (1989) and Sinitsyna (1986) to date the material on typological grounds, Medvedev and colleagues (2001) note that Ust'-Belaia is the most difficult to date among the pottery styles discussed so far, and that the problem remains unresolved. Their tentative suggestion, based on the evaluation of evidence from a number of sites, puts the beginning of its use in the Cis-Baikal region at somewhere between 5,800 and 5,500 bp.

As with all the other types discussed so far, Ust'-Belaia pottery has been recovered from sites across a wide geographical area including: Gorelyi Les and Ust'-Belaia along the Belaia river (Sinitsyna 1986; Savel'ev and Medvedev 1973), Niasha and Kazachka from the Yenisei basin (Generalov 1979: 10), Ulan-Khada on the shores of Lake Baikal (Goriunova and Khlobystin 1991), and Popovskii Lug in the Lena valley (Zubkov 1982: 6).

Having described the major widespread ceramic types for Neolithic eastern Siberia, I will now review the evidence for when, where, and in what form ceramics were introduced in Trans-Baikal, Cis-Baikal, the Central Yenisei and the Central Siberian Plateau.

Trans-Baikal

Trans-Baikal (often translated as *Zabaikal'e*) is that area of eastern Siberia located to the south and east of Lake Baikal (Figure 5.1). The region exhibits ecological variation on a latitudinal gradient, with the southern portion sharing affinities with the steppe and steppe-forest ecotones of northern Mongolia, and the northern portion marking the beginning of the dense taiga forests that spread north through Yakutia and the Central Siberian Plateau. The Vitim Plateau bisects Trans-Baikal into a western zone, associated with the Baikal watershed, and an eastern zone that drains into the Arctic Ocean through the Lena and its upper tributaries, the Vitim and Olekma Rivers. In addition, although technically part of the Amur-Pacific watershed, sites located along the Onon, Ingoda, and Shilka Rivers in the

Russian Chita Oblast are usually treated as part of eastern Trans-Baikal rather than as part of the Amur region. Differences in ecology, material culture and history of archaeological research have resulted in parallel culture-historical schemes for eastern and western Trans-Baikal (Ivashina 1979; Khlobystin 1996: 306), and as such it makes sense to discuss each area separately.

Appearance of Pottery in Eastern Trans-Baikal

Ust'-Karenga

Currently, the oldest reliably dated pottery from eastern Siberia comes from the Vitim River basin in northeastern Trans-Baikal and is associated with the Ust'-Karenga cultural complex (Kuz'min 2002; Kuz'min and Orlova 2000; Kuz'min and Vetrov 2007; Vetrov 1981, 1985, 1995, 1997, 2000). First described in the 1970s, the Ust'-Karenga complex has been identified at more than 30 sites at or near the mouth of the Karenga River, a right tributary of the Vitim (Vetrov 1981, 1985, 1995, 1997, 2000).

Initial radiocarbon dates of 11,240±80 bp (GIN-8066) and 10,750±60 bp (GIN-8067) were obtained for charcoal recovered from a hearth in the oldest ceramic-bearing level (Cultural Layer 7) at the eponymous type-site (Vetrov 1995). Subsequent analyses on charcoal from the same layer were commissioned by Vetrov and Kuz'min (2005) and returned dates of 12180±60 bp (AA-60201) and 12170±70 bp (AA-60202). In addition, a series of dates on organic pottery temper confirm the early dating of this complex: O'Malley et al (1999) obtained a date of 10,600±100 bp (AA-21378), Kuz'min and Keally (2001) obtained a date of 11,065±70 bp (AA-38101), and Vetrov and Kuz'min (2005) obtained a date of 10870±70 bp (AA-60667).

The Ust'-Karenga ceramics collection is currently comprised of several hundred fragments and around 20 reconstructed pots (Vetrov 1985, 1995). Vetrov (1985) describes the form of these pots as closed parabolic with slightly rounded rims and pointed bases. He identifies two main sizes: one is 12–14 cm in diameter and 16–18 cm tall, while the other is around 26.5–29.5 cm in diameter and 34.5–38.5 cm tall (Vetrov 1985: 124). The walls of both versions measure approximately 0.5 cm thick, while the bases are slightly thicker and are composed of several layers of clay. All of the pots are primarily plant fibre tempered, but sand, and crushed quartzite were also used (Vetrov 1985).

The pots were paddle-formed,[3] although in the cross-section of some fragments, it is possible to see evidence of joins between separately constructed rims, bodies and bases. Both the inner and outer surfaces of all pots featured smoothed out lines that Vetrov (1985: 124) attributes to the use of a comb to smooth and even the clay during the manufacturing process. Such lines are more pronounced on the inner surface of the pots, where they are oriented horizontally, than on the outer surface where they are oriented vertically.

Surface designs are most commonly located on the upper portion of the pots and are dominated by continuous acute-angled zig-zags composed of either straight or curved lines that were imprinted by walking or rocking a comb across the wet clay (Figure 5.4). Often several rows of these zig-zags encircle the entire pot. Other common decorations include intermittent imprinted lines arranged in a small number of geometric motifs such as zig-zags, diamonds, horizontal and vertical herringbone and individual lines. In some instances, a cogged roulette was run at a diagonal angle on both the inner and outer surfaces of the rim. Conical apertures appear on a number of pots and apparently served for the fastening of lines. At least one entirely undecorated pot was also recovered (Vetrov 1985: 124).

Vetrov (1985, 1995) asserts that the earliest Ust'-Karenga pottery differs significantly from the earliest pottery in adjacent eastern Siberia and the Amur River basin (see also Khlobystin 1996: 270); but he does see similarities with early ceramics in more distant territories. In particular, he points to the acute-angled comb impressions that are well known from the Neolithic of Central Asia (Formozov 1959; Vinogradov 1981), and geometric comb ornamentation documented from western Siberia and the Urals

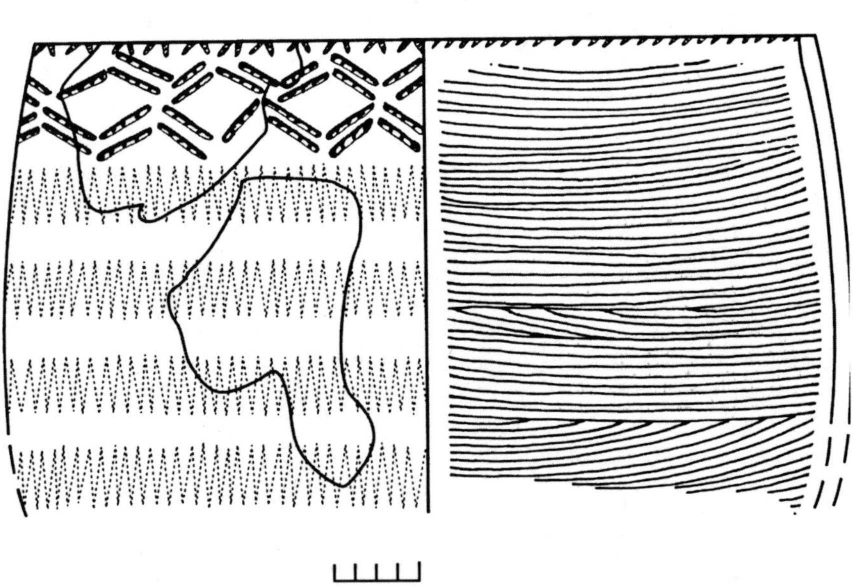

Figure 5.4 Ust'-Karenga pottery from Trans-Baikal (modified from Vetrov 1985).

(Bader 1970: 157–171; Kiparisova 1960; Kosarev 1981; Prokoshev 1940: 21–40). Given the distance between these regions, the lack of examples in intermediate regions and the early dates of the Ust'-Karenga material, however, Vetrov (1985: 127–129) rules out transmission with these regions and explains the similarities as convergent development.

Chindant

Until Vetrov's finds at Ust'-Karenga, the site of Chindant, located on the Onon River in southeastern Trans-Baikal, was often described as the earliest Neolithic site in eastern Siberia (e.g., Khlobystin 1996: 307). Indeed, the Chindant Phase, named after the site, is traditionally associated with eastern Trans-Baikal's Early Neolithic (Aseev 2003; Ivashina 1979; Khlobystin 1996).

Around 20 small pottery fragments were recovered from 11 hearths located within a single 30–50 cm thick cultural layer at Chindant (Aseev 2003: 42). Despite the small quantity, the material was sufficient to determine that the pots were typically thin-walled (0.3–0.5 cm), and featured simple straight walls and a rounded parabolic or mitre-shaped base (Aseev 2003: 42; Khlobystin 1996: 307). As with Ust'-Karenga pottery, paddle forming was the dominant manufacturing technique; however, at Chindant the paddle was wrapped with grass or cords, leaving characteristic imprints on the external surface of the vessels. On some fragments, net or cloth impressions are also reported (Aseev 2003: 42; Grishin 1981). Under the rim, a band of small depressions is often found running in a single line around the pot. Occasionally, the upper portion of the vessels is ornamented by a series of horizontal lines inscribed by a comb or wedge-shaped implement of varying thickness. Vetrov (1985) sees the pottery of Chindant as having closer ties to sites from eastern Mongolia and the Middle Amur than to the Ust'-Karenga material. This opinion is shared by scholars who have examined the lithic material which shows similarities to the Dauria culture complex from the steppe region of southern Siberia and northern Mongolia. The faunal remains recovered from the ceramic-bearing layer at Chindant, including remains of wild ass, wild horses, roe deer, marmots and dogs, also suggest a steppe rather than forest adaptation (Khlobystin 1996: 309).

Unfortunately, radiocarbon data are not available for Chindant, and as a result, there is some controversy over the age of the material (Aseev 2003: 44–45). Traditionally, the Chindant Phase is dated between c. 4,300 and 2,500 uncalibrated BC, based on typological similarities of the lithic assemblage at Chindant and neighbouring regions in the Middle Amur, northern Mongolia, and western Trans-Baikal (e.g., Ivashina 1979; Khlobystin 1996: 307, Kirillov and Rizhskii 1973; Okladnikov and Kirillov 1980; Vasil'evskii and Kirillov 1967). Khlobystin (1996: 307), for example, notes similarities between the large pebble chopping tools at Chindant and those associated with the Gromatukha culture complex in the Middle Amur region. Similarly, as mentioned, Kirillov sees similarities in the microblade technology at Chindant and the Dauria complex of southern Siberia and northern Mongolia (Kirillov and Rizhskii 1973; Okladnikov and Kirillov 1980; Vasil'evskii and Kirillov 1967). Grishin (1981) argues, as opposed to these other scholars, that these dates are too old since net impressions on pottery as well as triangular arrowheads date no older than the 3rd millennium uncalibrated BC. His views are based on comparisions with Cis-Baikal and Yakutia. Most recently, Aseev (2003) cites the chopping instruments, frontal nuclei, and microblade technology at Chindant as evidence that the site represents one of the earliest Neolithic phases, and that it was contemporary with the Mukhino phase in western Trans-Baikal (c. 4,000–2,500 uncalibrated BC), and perhaps even as old as the Ust'-Kiakhta material from western Trans-Baikal which has returned a radiocarbon date of c. 11,000 bp (but see discussion below). Given that radiocarbon dating over the last 15 years has resulted in substantial changes to the chronological frameworks in both the Amur and Cis-Baikal regions with which the Chindant are usually compared, we must view such relative chronologies with some caution (Kuz'min 2006a). Radiocarbon dates from the Gromatukha culture complex in the Amur, for example, suggest that it almost certainly predates 12,000 bp (Kuz'min 2002), and as discussed below, pottery was introduced into Cis-Baikal by at least 6,500 bp and perhaps as early as 8,500 bp. It seems unlikely that the timing of the Chindant material will be resolved without new radiocarbon dates and, perhaps, further excavations to clarify the stratigraphic context of the finds.

Appearance of Pottery in Western Trans-Baikal

Mukhino

As with eastern Trans-Baikal, there is discord between the conventional timing of the Early Neolithic and recent radiocarbon dates. Traditionally, the Early Neolithic in western Trans-Baikal has been associated with the Mukhino phase, originally defined by Okladnikov (1970: 191) on the basis of materials from the type site of Mukhino located along the Ivolga River near the modern settlement of Ulan-Ude (Ivashina 1979). Okladnikov originally assigned the period to c. 4,000–2,500 uncalibrated BC through typological comparisons with material from the neighbouring Cis-Baikal and Yakutia (see also Ivashina 1979). Aseev (2003) suggests that the pottery at Mukhino sites, which are generally paddle-formed, mitre-shaped, and occasionally exhibiting net impressions, confirm Okladnikov's chronology, since net impressions are rare in western Trans-Baikal but common in Cis-Baikal and Yakutia during this time period. In particular, he singles out pots from layer VII at Bel'kachi I in Yakutia, radiocarbon dated between c. 6,000 bp and 4,999 bp, as appropriate analogies.[4] As discussed above, however, net-impressed pottery is known from essentially all over eastern Siberia and for the entire duration of the Neolithic. It is, in fact, the only type of ceramic found in graves belonging to the older Kitoi mortuary tradition (discussed below), which is radiocarbon dated to between c. 7,000 bp and 6,200 bp. In addition, there is now evidence from within western Trans-Baikal that hunter-gatherers may have adopted pottery much earlier than outlined in this conventional model.

Studenoe and Ust'-Menza

The famous Studenoe locality is located on the right bank of the Chikoi River at its confluence with the Studenoe Creek and consists of three excavations (Studenoe 1/1, Studenoe 1/2 and Studenoe 2). Neolithic materials including ceramics, hearth features, ground-stone tools, microblades, flakes and faunal remains are associated with seven cultural horizons (3 through 9) at Studenoe 1/1 and 1/2 (Konstantinov 1994). Besides the ceramic material, the two Early Neolithic layers (cultural layers 8 and 9) contained lithic material indistinguishable from the preceding Mesolithic layers.

Soil humates from the overlying Middle Neolithic Layers 7 and 7b received conventional radiocarbon dates of 9,690±250 bp (GIN-5492)[5] and 10,450±300 bp (GIN-5493), while charcoal from Layers 6 and 5 returned conventional dates of 10,780±150 bp (GIN-4577) and 10,400±155 bp (SOAN 1648) respectively (Konstantinov 1994). Buvit and colleagues (2003: 659), believing that these dates were too old 'since Neolithic cultures did not appear elsewhere in Siberia until approximately 6000 yr [bp]', commissioned new AMS dates in an attempt to clarify the site's chronology. Interestingly, their AMS date on charcoal from Layer 5 returned an even older date of 11,000±90 bp (AA-33037). The authors considered several possibilities to explain the dates, including the option that the dated charcoal could have been contaminated by older carbon given that coal has been found in upper reaches of the Chikoi River, but ruled these out on the basis that the dated charcoal was obtained from a hearth with good stratification and that coal outcrops are not found near the site. The plateau in the atmospheric carbon production occurring between c. 15,000 BP and 13,800 BP also seems inapplicable to the Early Neolithic dates, which are somewhat younger. Instead, given the fact that the Neolithic began earlier than 10,000 bp in parts of eastern Trans-Baikal, and started at least as far back as c. 7,000 bp in Cis-Baikal (see below), it seems reasonable to conclude that the dates for Studenoe might, in fact, be broadly accurate.

At Ust'-Menza 1, located on the Menza River not far from Studenoe, the earliest Neolithic layers (Layers 7 and 8) have not been radiocarbon dated; however, plant remains from the overlying Layer 5 returned a date of 8,715±60 bp (SOAN-3080), which provides a *terminus ante quem* for the appearance of ceramics at the site, and is generally consistent with the Neolithic dates from Studenoe.

Semina (1985) describes the Early Neolithic pottery from both Studenoe and Ust'-Menza as having a simple open parabolic form with sharply pointed conical bases. Little or no neck is distinguishable, and the rounded walls simply terminate at a straight, smooth rim. In only one case was the rim weakly everted, and in this example the imprints of fingertips were also preserved in the clay (Semina 1985: 106).

The pots were paddle-formed and, like at Chindant, the paddle was wrapped with cord, the impressions of which were preserved on numerous fragments. Semina refers to the clay as friable in structure, which she attributes to the inexperience of the potters in selecting and mixing tempering agents, of which the most common were sand, grass and conifer needles. The uneven colouration of vessel surfaces suggests that the pots were likely fired on open hearths. Other than the cord impressions produced during the manufacturing process, decorations are generally limited to occasional simple pits and lines found underneath the rim. In Layer 9 at Studenoe, however, cone-shaped appliqués, approximately 1 cm in diameter and 1–3 cm in length were discovered, and it is believed that they would have been affixed under the rim of the pots (Semina 1985: 106). Such ornamentation is unique to the region.

Although dwelling structures were uncovered from the older Palaeolithic horizons at Studenoe, no such features were found in Mesolithic through Bronze Age layers. Hearths, however, were common in Neolithic layers. Konstantinov (1994) notes that there was a greater proportion of microblades to flakes in Neolithic and Bronze Age horizons compared with Mesolithic layers, suggesting an increased need for composite tools and changing subsistence strategies. How such strategies might be related to the origin of pottery is unclear.

Ust'-Kiakhta

Ust'-Kiakhta (also spelt Ust'-Kjakhta and Ust'-Kyakhta), located on the right banks of the Selenga River just north of the modern Ust'-Kiakhta settlement, was first investigated by the Buriat-Mongolian archaeological expedition headed by A P Okladnikov in 1947. Subsequent excavations in 1976 and 1978 confirmed the presence of two cultural layers, the younger of which (Layer 1) contained around 10 dark brown fragments of pottery located amongst a variety of lithic blades, flakes, nuclei, scrapers and ostrich eggshell beads (Aseev 2003). Judging by the weakly everted rim fragments, the pots were approximately 8–10 cm in diameter, but of unknown height. Ostrich egg shell and crushed rock was used to temper the clay, and the outer surface exhibited a slight gloss, likely a result of burnishing. The inner surface of the fragments exhibited parallel lines; however, no information is provided on manufacturing technique or other aspects of decoration, so it is unclear whether these lines are similar to those seen on the interior of Ust'-Karenga pottery.

Aseev (2003), citing Firsov and colleagues (1972), reports that a radiocarbon date of 11,505±100 bp (SOAN-1552) was obtained for charcoal from Cultural Layer 1, and that charcoal from Layer 2 returned a date of 12,595±50 bp (SOAN-1553). If these dates are accurate, then the earliest pottery from Ust'-Kiakhta was contemporary with, or even slightly older than, that from Ust'-Karenga. It should be noted, however, that in Table 1 of their review of Early Neolithic radiocarbon data, Kuz'min and Orlova (2000: 359) place a question mark beside the provenence information for the Ust'-Kiakhta Layer 1 date. Unfortunately, they discuss neither the date nor the site of Ust'-Kiakhta in the text of their review, so it is unclear as to why they consider the association of the date with the ceramics to be vague (see also Kuz'min 2002: 367).[6] Aseev (2003: 37) accepts the date without comment and points out that the lithic assemblage of Ust'-Kiakhta, which is dominated by blade technology, scrapers as well as chopping pebble tools, shows clear similarities with the Palaeolithic material from the sites of Sannyi Mys and Oshurkovo. The fact that the ceramics were also found in association with ostrich eggshell beads, and that crushed ostrich eggshells were used as temper, appear to support an early date, since ostrich remains are rare in deposits younger than the early Holocene.

Cis-Baikal

Cis-Baikal includes the islands and western shore of Lake Baikal, as well as the drainage basins of the Angara and upper Lena Rivers down to the cities of Ust'-Ilimsk and Kirensk in Eastern Siberia respectively (e.g., Michael 1958: 5; Weber 1995). The region is rich in both Neolithic graves and stratified settlement sites. Unfortunately, despite intense scholarly attention over the last 100 years, cultural materials from Cis-Baikal mortuary and settlement contexts have never been convincingly correlated. In large part, this is because the material culture recovered from Cis-Baikal settlement sites is rarely found at mortuary sites and *vice versa*. A particularly relevant example here is the fact that of the six main ceramic styles identified at Neolithic campsites in the region, only one (net-impressed) has ever been recovered from graves (Bazaliiskii nd; Okladnikov 1950). A second impediment is that while there are currently over 400 radiocarbon dates available from Cis-Baikal mortuary sites, there are fewer than 50 dates available from secure contexts at well stratified settlement sites in the area. Consequently, while we now have strong control over the chronological sequence of Cis-Baikal burial traditions (Weber 1995; Weber et al. 2004, 2005, 2006), the extent to which this sequence is directly applicable to broader cultural developments—including changes in ceramic technology and styles—remains controversial (Bazaliiskii nd; Goriunova 2001, 2003, nd).

Appearance of Pottery in Cis-Baikal Neolithic Graves

Pottery first appears in Cis-Baikal graves just after c. 7,000 bp, and is associated with the Kitoi mortuary tradition. Such examples are extremely rare, however, and of the several hundred known Early Neolithic Kitoi graves, only five have ever produced pottery. In only three of these cases was a complete vessel recovered and all were of the net-impressed style. Vitkovski found a small mitre-shaped vessel with net impressions in Grave Number 13 at the type site Kitoi, located at the mouth of the Kitoi River, plus additional pottery fragments on the surface next to Grave Number 4 at the same site (Okladnikov 1974: 54). A round-bottomed, net-impressed pot was also recovered from Grave 8 at the Lokomotiv (also known as Tsiklodrom) cemetery in the city of Irkutsk (Bazaliiskii 1990). Finally, Okladnikov recovered a mitre-shaped pot with net impressions and a belt of pits below the rim in Grave 3 [1972] at the Shamanskii Mys (also known as Khuzhir) cemetery on Lake Baikal's Ol'khon Island, along with additional fragments from a neighbouring grave (Konopatskii 1982: 42–48). Human bone from the Shamanski Mys grave was radiocarbon dated to 6,310±80 bp (TO-10211), while a date on human bone from the Lokomotiv grave returned a date of 6,670±70 bp (TO-10152). As mentioned, several hundred other dates from Kitoi graves indicate that the Kitoi mortuary tradition lasted from c. 7,000 bp to 6,200 bp (c. 7,800–6,900 BP), which Weber and colleagues (2005, 2006, ndb) use to define the boundaries of the Early Neolithic. It is interesting to note that no pottery has ever been recovered from Kitoi graves east of Lake Baikal (i.e., at the Fofanovo cemetery in the Selenga delta) or at any of the Upper Lena Kitoi cemeteries such as Makrushina,[7] Iushino, Kartukhai, Staryi Kachug, Turuka or Stepnyi Baltai (Bazaliiskii nd).

Appearance of Pottery in Cis-Baikal Neolithic Campsites

As mentioned above, correlation between Cis-Baikal Neolithic mortuary traditions and campsites remains controversial. This is especially the case for the Early Neolithic Kitoi. Traditionally, the Kitoi have been associated with the Posol'sk pottery style (e.g., Khlobystin 1978; Okladnikov 1950). As Weber (1995) notes, however, this association was established prior to the application of radiocarbon dating to the Cis-Baikal material, which resulted in significant changes to the region's culture-history. Specifically, both Okladnikov and Khlobystin considered the Kitoi to be an intermediate stage between the Neolithic Serovo and the Bronze Age Glazkovo traditions and consequently assumed that it should be associated with intermediate Posol'sk-bearing layers at stratified campsites such as Ulan-Khada on Lake Baikal. Radiocarbon evidence now suggests that the Kitoi mortuary tradition should probably correspond to the Late Mesolithic and earliest Neolithic layers at stratified habitation sites (Weber 1995; Weber et al. 2005, 2006, ndb). Posol'sk pottery in Cis-Baikal is c. 1,000 years younger than this and should likely be associated with the Late Neolithic Serovo burial traditions (Weber 1995: 151–152), although it has never been found in graves (Goriunova nd).

However, Makarov (2005) claims that Posol'sk pottery is an Early Neolithic type in the Central Yenisei Basin, with associated radiocarbon dates equivalent to those of the Kitoi mortuary tradition (see discussion below). In addition, diagnostic Kitoi-style fishhooks have been recovered along with Posol'sk pottery in Layer IV at Okunevo IV and Layers VII and VI at Katun I, both located in the Chivyrkuisk Inlet on the east coast of Lake Baikal (Nomokonova 2004; Nomokonova and Goriunova 2004). At Okunevo IV, a variety of other smooth-walled, cord-impressed and hatched pottery was also recovered in the same layers (Nomokonova 2004). Similarly, at Katun I, Posol'sk pottery was also associated with Khaita pottery (Nomokonova and Goriunova 2004). This diversity of pottery styles, combined with perceived similarities of the lithic and bone technology with that found in Serovo graves, has prompted Goriunova to propose that Kitoi and Serovo co-existed in Cis-Baikal between c. 5,500 bp and 4,300 bp (see also Khlobystin 1978). This is at odds with the extensive radiocarbon data from mortuary sites indicating that the Serovo mortuary tradition is at least 1,000 years younger than the Kitoi. It is also unlike the record at other stratified sites where Kitoi fishhooks are found associated with Khaita and net-impressed I pottery only (e.g., Layer 6 at Gorelyi Les; Layers 5 and 5a at Ust'-Khaita; Layer 4 at Sagan-Zaba). At all three of these other sites, Posol'sk pottery is found in younger strata (Layer 5 at Gorelyi Les, Layer 4 at Ust'-Khaita, Layers 3 and 3 [lower] at Sagan-Zaba). Unfortunately, no radiocarbon dates are available from either Okunevo IV or Katun I. Hence, it is difficult to evaluate whether Kitoi fishing technology persisted long after the disappearance of Kitoi mortuary traditions, a situation in which Posol'sk pottery is found much earlier on the eastern side of Baikal than elsewhere in Cis-Baikal, or perhaps taphonomic processes went unrecognized at some of these habitation sites. Given the large diversity of pottery found at these sites, which is a characteristic of Late Neolithic times, the first scenario seems more likely.

At other sites in Cis-Baikal, the earliest Neolithic layers are clearly associated with one or both of net-impressed I and Khaita pottery. The timing of such pottery, however, varies across Cis-Baikal. Currently, the earliest evidence comes from the stratified sites of Gorelyi Les and Ust'-Khaita, both located along the Belaia River.

Gorelyi Les

The initial site report from the stratified campsite Gorelyi Les cited the presence of both net-impressed and Khaita pottery in Cultural Layer 7 (Savel'ev et al. 1974: 164). Four dates ranging between c. 9,000 bp and 6,400 bp have been estimated for this layer. It could reasonably be considered a settlement correlate of the Kitoi mortuary tradition. However, Berdnikova and colleagues (1991: 29–30) report that Layer 7 was aceramic, and Weber (1995) cites personal communication from the original excavators (Savel'ev and Goriunova) that the pottery recovered from Layer 7 was likely a result of post-depositional disturbances. Despite this, the radiocarbon dates continue to be cited without comment by other local researchers as evidence for the timing of the start of the Early Neolithic (e.g., Aseev 2003; Medvedev et al. 2001). Most recently, Igumnova and colleagues (2004),[8] based on the results of excavations conducted in 2002, report that in some areas of the site it is possible to divide Layer 7 into an upper (Layer 7) and lower (Layer 7a) horizon. From the older Layer 7a, 16 dark grey fragments of a single pot were recovered. The context of the finds is apparently secure since they were located together on top of a 75×54 cm paving consisting of 15 dolomite slabs and one quartzite pebble. Based on dates of 8,444±124 bp (Ri-51) and 8,830±300 bp (KRIL-234) for Layer 7, they conclude that pottery was introduced to this part of Cis-Baikal around c. 8,500 bp. If accurate, this is by far the oldest example of pottery in the region and would actually precede the Kitoi mortuary tradition by c. 1,500 years. This proposal is not widely accepted by other researchers in the area (e.g., Goriunova nd), and in order to confirm this dating, Igumnova and colleagues (2004) report that they have collected charcoal from a hearth in the same layer to obtain a new date.

The shape of the vessel was likely open parabolic with a weakly expressed neck and slightly everted rim. The original height could not be estimated since the base was missing, but the reconstructed pot had a height of 10 cm and a diameter around the body of 26 cm. The walls ranged from 0.5 to 0.7 cm in thickness and were tempered with sand and small quantities of quartz. Ornamentation included blade and stamp impressions as well as incisions in the form of herring-bone. Overall, Igumnova and colleagues (2004) point out the resemblance in manufacturing technique and ornamentation between these and the Khaita pots found in the younger Layers 6 and 5a at the same site.

Less controversial is Layer 6 from the same site, in which were discovered pottery fragments of three types: (1) mitre-shaped net-impressed pottery with an everted rim; (2) smooth-walled pottery with an everted rim and no ornamentation; and (3) Khaita. It is particularly significant that a Kitoi-style fishhook was also recovered from this layer. Five radiocarbon dates ranging from 7,890±80 bp to 6,510±100 bp (TO-4839) have been estimated for this layer, which appears to confirm previous proposals (e.g., Weber 1995: 151) that both net-impressed I and Khaita should be considered settlement correlates of the Kitoi mortuary tradition (Savel'ev et al. 2001). Layer 5a, which also contained Khaita pottery, returned dates of 5,430±120 bp (Ri-52) and 5,895±110 bp (SOAN 5601) suggesting that Khaita pottery persisted in the region even beyond the range of the Kitoi mortuary tradition, which ends c. 6,200 bp. Charcoal from the younger Layer 5, which contained fragments of Posol'sk pottery, returned a date of 4,880±180 bp (GIN-4366), which is contemporary with the Late Neolithic Serovo mortuary tradition.

Ust'-Khaita

Ust'-Khaita, located at the confluence of the Khaita and Belaia rivers, has recently been the subject of extensive excavations (Savel'ev et al. 2001). The stratigraphically oldest ceramic-bearing layers (Layers 5 and 5a) contained both net-impressed and Khaita pottery. Like at Gorelyi Les, overlying layers contained a greater diversity of pottery types including Posol'sk and smooth-walled vessels in addition to net-impressed and Khaita (Layer 4).

A large series of radiocarbon dates has been obtained for Ust'-Khaita (Savel'ev et al. 2001). However, as Goriunova (nd) points out, these dates are not entirely consistent with each other. For example, the two dates from Layer 5 are nearly 600 radiocarbon years apart and are statistically non-overlapping: 7,245±150 bp (SOAN 4431) and 6,625±150 bp (SOAN 4647). Likewise, the three dates from Layer 5a returned two overlapping dates and one clear outlier: 6,415±205 bp (SOAN 4648), 6,450±70 bp (SOAN 4439), and 7,190±90 bp (SOAN 4428). Finally, the date from the stratigraphically younger Layer 4 (7,040±40 bp SOAN 4438) is much older than those obtained from Layer 5a. Goriunova believes that rodent disturbances are responsible; but such disturbances are not mentioned in the original report (e.g., Savel'ev et al. 2001). As a result, it is unclear how this site should be interpreted. On the one hand, the dates are internally inconsistent. On the other hand, none of the dates from any of the three oldest Neolithic layers are younger than c. 6,400 bp, which places them all as contemporary with the Kitoi mortuary tradition and in broad agreement with the Early Neolithic Layer 6 at Gorelyi Les.

In other areas of Cis-Baikal, net-impressed and Khaita are also the earliest styles; but radiocarbon data indicate that they appear somewhat later. In the Ol'khon region, for example, net-impressed I pottery is the only type recovered from the earliest ceramic-bearing layer (Layer X) at Ulan-Khada, and soil from a hearth in this strata was dated to 5,495±125 bp (SOAN 3336). Net-impressed pottery is also the only type found in the oldest Neolithic layers at Ityrkhei (Layers V–VI) in the same area, which received dates of 5,700±200 bp (GIN 4881) and 5,680±60 bp (SOAN 3341) for animal bone and hearth soil respectively. Finally, both Khaita and Posol'sk pottery are found in Layers 3 and 3 (upper) at Sagan-Zaba on the western shore of Baikal just south of the Ol'khon region, which returned dates of 5,160±55 bp (SOAN 1376) and 6,000±40 bp (SOAN 1572) for charcoal and animal bone respectively. Overall, these dates suggest that pottery was not widely used at habitation sites in the Ol'khon region until after the Kitoi mortuary tradition had already disappeared around c. 6,200 bp.

It must be acknowledged, however, that older dates do exist in this area, although they are controversial. Timofeev and Zaitseva (1996) report dates of 7,560±80 bp (LE 2277) and 6,310±70 bp (LE 2783) from Ulan-Khada Layer X, which would place them in the same range as the Kitoi tradition. Other scholars, however do not cite these dates at all (e.g., Goriunova 2001, 2003, nd; Konopatskii 1982) implying that they are unacceptable. Similarly, although the original site reports from Sagan-Zaba described net-impressed pottery from Cultural Layer 4, which was subsequently radiocarbon dated to 7,630±45 bp (SOAN 1573), Weber (1995: 137) again cites personal communication with local scholars (Goriunova) suggesting that mechanical disturbances may have been responsible for the presence of pottery in this layer. Also, Konopatskii (1982) eliminates this date. Other Siberian archaeologists, however, accept the Sagan-Zaba date without comment as evidence of Kitoi occupation in the area, and note that a Kitoi fishhook was found in the same layer (e.g., Aseev 2003: 61). Aseev (2003: 61–66) also sees parallels between Layer 4 at Sagan-Zaba and the material recovered from the site of El'gen located on the shores of Lake Baikal, north of the Ol'khon region. In particular, Aseev (2003) points to similarities in the lithic and bone assemblages and the presence of a Kitoi-style fishhook. Pottery fragments at El'gen were not numerous, but exhibited a variety of decorative forms including dashed lines, comb-incisions, stepped-back trowel impressions, cogged stamps, net and cord impressions, incised and thickened rims and apertures. On the basis of two radiocarbon dates of 6,130±115 bp (SOAN 5121) and 6,790±85 bp (SOAN 5122) obtained for charcoal from hearth-like features, Aseev (2003) suggests that at least some of the material is contemporary with

Kitoi times and with Layer 4 at Sagan-Zaba. Unfortunately, the highly compressed nature of the site makes it difficult to separate materials well enough to be confident in the association of the ceramic material with the radiocarbon dates.

Overall, then, if Kitoi graves containing pottery are rare in Cis-Baikal, then ceramic-bearing layers of the same age from the region's campsites must be considered even more elusive. Currently, of the dozens of stratified campsites documented from Cis-Baikal, only Gorelyi Les, Ust'-Khaita from the Belaia valley and Ulan-Khada and Sagan-Zaba from the west coast of Lake Baikal have produced radiocarbon dates older than c.6,000 bp for ceramic-bearing strata, and in each case the dates are controversial. Part of the problem, of course, is the assumption that the Kitoi mortuary tradition ought to have a ceramic component at settlement sites at all. After all, only three vessels are known from the hundreds of Kitoi graves, and evaluation of bone and lithic technology suggest that the Kitoi are direct outgrowths of the local Mesolithic groups. Both Weber (1995) and Savel'ev and colleagues (2001: 346) consider the Kitoi to have both Mesolithic and Neolithic phases. Given this, we should not be at all surprised if pottery was rare or even non-existent at most Kitoi campsites. Furthermore, it seems quite likely, as Bazaliiskii (nd) suggests, that the pottery in Kitoi graves was of a ritual nature and that complementary material might not be found in more domestic settings. Rather than attempting to force the Kitoi into one of the ceramic layers, it seems more prudent to simply accept that many aceramic layers—conventionally defined as 'Mesolithic'—are, in fact, related to the 'Neolithic' Kitoi mortuary tradition (Weber 1995). In the Ol'khon microregion on the west coast of Lake Baikal, for example, the aceramic Layer 7 at Berloga returned a date of 6,525±100 bp (SOAN 3169) and Layer 7 at Ityrkhei returned a date of 7,300±290 bp (SOAN 0402) (Goriunova 2001). Likewise, Layer 7 from Gorelyi Les returned dates spanning from 8,850±300 bp to 6,490±100 bp (TO 6485) (KRIL 0234).

Later Developments in Neolithic Pottery in Cis-Baikal

Between c. 5,700 and 4,500 bp, variously referred to as Developed, Middle or Late Neolithic, there was an explosion in both the quantity and diversity of pottery recovered from habitation and mortuary sites in Cis-Baikal.

Following a c.1,000 calendar-year hiatus during which mortuary remains are entirely absent from Cis-Baikal, the Late Neolithic Isakovo-Serovo mortuary tradition appears at c. 5,300 bp. Some scholars assert that the Isakovo and Serovo graves exhibit enough similarity that they should be considered as a single unit (e.g., Gerasimov 1955; Weber 1995), while others see both temporal and spatial variability. Bazaliiskii (nd), for example, believes that Isakovo and Serovo are sequential phases limited to the Angara valley. In the Upper Lena valley, he instead refers to an 'Archaic' mortuary tradition that resembles the Isakovo, and for the Ol'khon region, he follows Goriunova (1997) in distinguishing a 'Late Serovo' that resembles the classic Serovo in the Angara valley. Regardless of whether or not we treat Isakovo-Serovo traditions as a single unit, it is clear that unlike the previous Kitoi tradition, pottery is ubiquitous in Late Neolithic graves and exhibits a much wider variety of forms and styles.

Bazaliiskii (nd) notes, for example, that 70% of the 58 known Isakovo graves and 50% of the 35 known archaic graves contained pots. Typically, such vessels were of a simple open-mouthed, mitre-shaped form in the net-impressed I tradition. Ornamentation is rare, although occasionally a narrow band of small pits or holes is found below the rim. Pottery is even more common in Serovo graves, almost all of which contain at least one pot located near the head or at the legs. The vessels differ from those in the Isakovo and archaic graves by tending to feature a closed mouth and a wide round bottom. Undecorated net-impressed I style is still common in Serovo graves; but net-impressed II is also identified with occasional comb impressions as well as 'lugs' for suspending the pots (Okladnikov 1950). In the Ol'khon region, there appears to be even greater variation, where Late Serovo pots exhibit a variety of smooth-walled, net and hatched finishes, as well as diverse ornamentation including comb-like stamps, incised lines, spatula impressions and round depressions along the rim (Goriunova 1997: 89). Goriunova (1997) reports that a total of 31 pots were recovered from 19 Late Serovo graves.

At Developed and Late Neolithic habitation sites, pottery is even more diverse. In addition to the net-impressed I and Khaita styles that dominate the Early Neolithic layers, net-impressed II, Posol'sk and comb-impressed types are all introduced and, towards the end of the Neolithic period, Ust'-Belaia (e.g., Layer 5 Gorelyi Les, Ulan-Khada Layer 9, Tyshkine II Layers 8 and 9, Kulara III Layers 1 and 2, Ityrkhei Layers 2–4, Layers 3 and 4 at Sagan-Zaba, Layers V–3 at Berloga, etc). In addition, there is a much wider diversity in ornamentation, largely but not exclusively based on comb, and stab-and-drag techniques. This proliferation of pottery styles seems to correspond with the introduction of the Late Neolithic Isakovo-Serovo culture complex. Both osteological and DNA analyses indicate that Isakovo-Serovo individuals are genetically distinct from the preceding Kitoi, suggesting that they were an immigrant population. In addition, Isakovo-Serovo relied less on fishing and more on a broader spectrum of hunting and fishing with, perhaps, greater mobility throughout the region (Weber and McKenzie 2003; Weber et al. 2002, nda). Details regarding the origin of the migrants or how this relates to the widespread adoption of pottery are currently unknown.

Central Yenisei Basin

The Neolithic cultures residing in the forest-steppe ecotone of the Central Yenisei Basin, centered around Krasnoiarsk on the border between eastern and western Siberia, are considered to have strong links to the Baikal region via the Angara River, as well as to western Siberia (Khlobystin 1996). As such, the region has a complex history exhibiting influences from both areas. In general, it appears that links were strongest with eastern Siberia during Early Neolithic times, and with western Siberian groups during the Late Neolithic and Early Bronze Age (Khlobystin 1996; Makarov 2005).

Although much better investigated than the lower or upper portions of the Yenisei basin, Makarov (2005) notes that much of the archaeological research over the last century into the Central Yenisei's Neolithic was confined to a very small number of sites, was of poor quality, and remains largely unpublished. Makarov (2005) argues that even today there is insufficient evidence to define any culture complexes in the area. He suggests that the periodization of Siberian funeral complexes is inapplicable to the region. Nevertheless, research at a number of stratified sites does make it possible to outline a broad culture-history for the area, including the development of pottery styles (Makarov 2005).

Appearance of Pottery in the Central Yenisei Basin

Currently, the best evidence for the introduction of pottery in the Central Yenisei region comes from the stratified sites of Eleneva Cave and Ust' Kazachka (often just called Kazachka).

Eleneva Cave

Excavations at Eleneva cave revealed 30 cultural layers spanning the Palaeolithic to the Middle Ages (Makarov et al. 1992: 111–120; 1995: 143–145). Pottery was first encountered in Layer 11G, which returned two radiocarbon dates of 6,530±60 bp (SOAN 2907) and 6,900±115 bp (SOAN 3998). The underlying aceramic Layer 11D returned a slightly older date of 7,330±35 bp (SOAN 2908) (Kuz'min and Orlova 2000; Orlova 1995). Such pottery was cord-impressed, and fragments indicate that the pots were typically thin walled (0.4–0.6 cm) and tempered with sand and gravel. Unfortunately, the remains were too fragmentary to provide information on the size or shape of the vessels. Rim fragments were thickened and roughly triangular in cross-section with cogged or serrated impressions. On the upper side of the rim is often found a belt of small pinholes, while encircling the pot below the rim are frequently a series of horizontal lines consisting of back-stepped blade impressions. Perpendicular lines, formed in the same way, 'hang' off the lowest horizontal lines to resemble tassels. On the basis of the combination of these cord impressions, thick end rims, and distinctive ornamentation, Makarov (2005) describes such pottery as Posol'sk. He notes that Posol'sk ceramics are commonly found in the earliest Neolithic layers at other sites in the area. For example, at both Ust'-Karabula (Layer 5) on Lower Angara and Shilka 2 in the Kazachinsk zone of the Yenisei, Posol'sk pottery is found in the first Neolithic layers, immediately above the aceramic Mesolithic strata. If correct, then the radiocarbon dates for Eleneva Cave would place Posol'sk pottery between one and two thousand years earlier than in Cis-Baikal, where, as discussed above, such pottery is currently considered to be a Late Neolithic (c. 5,300–3,000 bp) correlate. Other scholars, while accepting that cord-impressed pottery is found in Early Neolithic layers in the Central Yenisei region, do not refer to these vessels as Posol'sk. They claim that Posol'sk pottery was introduced to the area later (e.g., Khlobystin 1996; Savel'ev 1989). Radiocarbon dates from the Posol'sk-bearing Layer IIIB at Niasha, for example, returned a date of 4,080±60 bp (LE-1415).

The overlying Layer 11 V at Eleneva Cave, also considered by Makarov to represent the Early Neolithic, returned a date of 6,475±190 bp (SOAN 2906) (Makarov 2005: 151; Makarov and Orlova 1992: 171–175). Pottery fragments from this layer are more diverse and include horizontal rows of small oval serrated depressions, crescent impressions, zig-zag stamps, and punctate-comb zig-zag decorations. According to Makarov (2005), such ornamentation is well known from Early Neolithic layers at other sites in the area including the well known Ust'-Kazachka, located on the banks of the Kazachka River.

Ust'-Kazachka

Over 300 sq m of excavation at Kazachka have revealed 20 cultural layers, the top 7 of which (Layers 1 through 7) contained ceramic materials (Generalov 1979). Although Makarov (2005) describes the pottery from the oldest ceramic-bearing layer at the site (Layer 7) as only cord-impressed, others describe the material as both net- and cord-impressed (Generalov 1979; Khlobystin 1996). The fragmentary nature of the finds means that little can be said with respect to vessel form or size. Radiocarbon dates are not available from this layer; but Makarov (2005) considers it to have been contemporary with the Early Neolithic layers at Eleneva Cave on the basis of dates from the underlying Mesolithic Layer 8 (6,850±200 bp) and the overlying Neolithic Layer 6 (6,660±190 bp, LE 1231) (Vorob'eva and Savel'ev 1984: 149).

In Layer 6 at Ust'-Kazachka, fragments exhibiting serrated oval stamps and crescent impressions analogous to those from Layer 11 V from Eleneva Cave were also recovered. The original excavators considered Layer 6 to represent the Developed Neolithic; they also noted that pottery was more diverse in the upper portion of the layer than in the lower portion. On this basis, they dated the lower part of the layer to the second half of the 4th millennium uncalibrated BC and the upper portion to the start of the 3rd millennium uncalibrated BC.

Overall, then, Makarov (2005) dates the introduction of plain and decorated cord-impressed pottery to the Central Yenisei somewhere between 7,000 bp and 6,500 bp. At all other sites in the region, there is considerable compression of cultural materials making it difficult to isolate ceramic complexes of the Early Neolithic. Even sites such as Ust'-Sobakino that, in the past, were described as well stratified and used to outline regional developments (e.g., Kartsov 1929; Khlobystin 1996) are now viewed as considerably mixed (Makarov 2005; Maksimenkov 1966). Based on this material, Makarov (2005) suggests that both net-impressed and Khaita pottery should be considered Early Neolithic forms given that they are found together at the bottom of Neolithic deposits at sites such as Ust'-Biriusa located to the south of Krasnoiarsk.

Later Developments in Neolithic Pottery in the Central Yenisei Basin

Subsequent layers at both Eleneva Cave and Ust'-Kazachka indicate that pottery styles are more numerous and diverse than in older layers. Makarov (2005) defines both Developed and Late Neolithic stages.

The Developed Neolithic is represented by Layers 11B and 11A at Eleneva Cave, the upper portion of Layer 6 at Ust'-Kazachka (Savel'ev et al. 1974), Layers 3A and 3B at Niasha (Pogudin 1989), and Layer 2 at Krasnoiarskaia. Only one radiocarbon date is available: 5,150±95 bp from 11B Eleneva Cave; but Makarov (2005) extrapolates the period dates to roughly the entire 4th millennium uncalibrated BC.

Developed Neolithic pottery characteristically featured simple, open, round-bottomed vessels with straight rims. Makarov (2005) identifies two groups of decorative forms: those with impressions of oval and oval-serrated stamps, and those with diverse variants of stepped-back trowel decorations. This seems to correspond to Savel'ev's (1989) Kazachinsk phase of the Ust'-Belaia pottery type, which was defined largely on the material from the upper portion of Layer 6 at Ust'-Kazachka. Towards the end of this stage, both flat-bottomed (e.g., Layer 4 at Zimoveinaia) and corrugated vessels are also found for the first time. Such forms are taken as evidence of increasing influence or even migration from western Siberia during this period (Makarov 2005).

Late Neolithic material is identified from Layers 10 and 9 at Eleneva Cave and Layer 2 at Zimoveinaia (Makarov 2005). Layer 10 at Eleneva Cave returned three radiocarbon dates: 4,655±30 bp (SOAN 2905), 4,930±55 bp (SOAN 3295) and 5,365±95 bp (SOAN 2921); but Makarov (2005) suggests that the final date should be disregarded as it is the only one that fails to match the stratigraphic column, noting that the underlying Layer 11B received a date of 5,150±95 bp. In general terms, then, he dates the Late Neolithic between the end of the 4th millennium and the first half of the 3rd millennium uncalibrated BC. Pottery from these layers is typically round-bottomed and was manufactured using strips of clay that were subsequently drawn together with the use of a smooth paddle, much as Medvedev and colleagues (2001) described for Ust'-Belaia pottery in Cis-Baikal. The primary form of decoration is that of horizontal and vertical rows of intermittent dashes often referred to as punctate comb impressions (*punktirnaia grebenka*). Savel'ev (1989: 23) first identified punctate comb pottery as a ceramic layer (*plast*) or style for southern Siberia, and dated it to c. 4,500 bp. Okladnikov (1957: 31–36) also noted this decoration at Iadeiki and Bazaikha, and related it to the late Serovo tradition from Cis-Biakal. Makarov (2005) also considers net-impressed pottery to have continued into the Late Neolithic, given that such pottery was found together with punctated comb at Zimoveinaia and at a number of other sites in the area including Ust'-Karabula on the lower Angara (Makarov 2005).

Central Siberian Plateau (Yakutia and Evenkia)[9]

The Central Siberian Plateau is characterized by a taiga ecology, cut with numerous rivers that belong to the Lena watershed in the east (Yakutia) and the Yenisei watershed in the west (Evenkia).

Appearance of Pottery on the Central Siberian Plateau

In the eastern portion of the Central Siberian Plateau that drains into the Central Lena and its major tributaries the Aldan, Olekma and Vilui Rivers, the appearance of ceramics has long been associated with the Early Neolithic Syalakh culture complex (Alekseev 1996; Mochanov 1969; Mochanov et al. 1983). Alekseev (1996) indicates that the Syalakh occupied the entire region of Yakutia, and even extended into the Taimyr Peninsula in the northwest and as far south as the Malaia Ludarskaia cave on the northwestern coast of Baikal. Among the best investigated sites in the region are

Bel'kachi I and Sumnagin I in the Aldan River basin, Tensik 1, Kurung 2 and Bol'shoi Kiusk in the Olekma River basin, and Ust'-Chirkuo in the Upper Vilui River basin (Alekseev 1996; Mochanov 1969; Mochanov et al. 1983). A large series of radiocarbon dates from Bel'kachi I (Layers 7–6) and Sumnagin I (Layers 14–11) demonstrate that the Syalakh spanned the period between c. 6,000 bp and 4,800 bp (Alekseev 1996; Mochanov 1969), which is 500–1,500 years later than in the Cis-Baikal and Central Yenisei regions, depending on which dates one accepts.

As with most of Siberia, the earliest ceramics in Yakutia are net impressed, although D'iakonov (2001) suggests that many of these impressions might be textile rather than net. The vessels are characteristically small and mitre-shaped with open mouths and straight flat rims. Like the early ceramics in both Cis-Baikal and Trans-Baikal, decorations are rare, the few examples being a horizontal belt of round or oval depressions or holes beneath the rim. Occassionally, the rim is thickened by a collar of clay that is, itself, incised at regular intervals. Towards the end of the period, small aperatures are also found as well as slanting lattice decorations formed with some sort of wedge-shaped implement. Comb or comb-like impressions are found under the rim on a small number of pots.

Alekseev (1996) observes similarities between Syalakh vessels and the early net-impressed pottery from Cis-Baikial, and contends that the Syalakh is a product of migration from both Cis- and Trans-Baikal combining with local Yakutian Mesolithic traditions. Kuz'min and Orlova (2000) also suggest that net-impressed pottery might have originated in the Angara river headwaters and spread northward into Yakutia with migrants or other forms of cultural exchange.

In the western portion of the Central Siberian Plateau (often referred to as Evenkia), which drains into the Yenisei River via the Podkamennaia and Nizhniaia Tunguska Rivers and associated tributaries, the Early Neolithic also features net-impressed pottery (Khlobystin 1996: 301–302). However, in archaeological terms, this region is relatively poorly investigated and much of the research that has been conducted remains unpublished (Khlobystin 1996: 301).

In the earliest layers at the region's best researched stratified site, Tura A (also called Tura I), located at the confluence of the Kochechumo and Nizhniaia Tunguska Rivers, the earliest ceramics were egg-shaped vessels with net impressions that seem to share affinities with the Isakovo and Serovo pots from Cis-Baikal (Khlobystin 1996: 301). As elsewhere in eastern Siberia, decorations were rare and typically consisted of only a belt of small depressions or holes beneath the rim. Occasionally, vessels with smooth walls or smoothed out net impressions are found, as are punctuate comb impressions. In addition, imprints of woven textiles are reported from some fragments, but as noted earlier, it is often difficult to distinguish net from textile impressions. Khlobystin (1996: 301) believes that the Early Neolithic sites in this region combine features of both the Syalakh and Serovo. Unfortunately, no radiocarbon data are available to determine when such pottery may have entered the area.

Later Developments in Neolithic Pottery on the Central Siberian Plateau

Radiocarbon dates from Bel'kachi I (Layers 4 through 1) and Sumnagin I (Layers 10 and 9) demonstrate that cord-impressed pottery was introduced in Yakutia somewhere after c. 4,900 bp with the Middle Neolithic Belkachinsk Culture, which spans c. 4,800–3,700 bp (Alekseev 1996). In addition to cord-impressions, vessel forms were more diverse than Syalakh pottery, and included both simple and complex round-bottomed and mitre-shaped pots that occasionally featured lugs for suspension. The everted rims were often thickened as in Posol'sk pottery from Cis-Baikal dating to around the same period. Decorations are common and include a variety of zig-zags, incised lines, triangles, comb impressions and stamps.

Like the Syalakh, the Belkachinsk has a wide geographical distribution across Yakutia, and Mochanov (1969) believes that they were a local development from the Early Neolithic Syalakh with contributions from neighbouring regions.

Faunal remains suggest that the economy of both the Syalakh and Belkachinsk was based on fishing (sturgeon, pike, perch, lenok) and hunting (reindeer, roe deer). Houses were semi-subterranean and likely covered with reindeer skins. Okladnikov (1970) identified storage pits associated with Syalakh dwellings at sites along the Kullaty River. These pits were apparently lined and covered with birch bark and held the remains of meat and fish. Whether pottery was used in such storage pits is unclear.

In the upper layers at Tura A in Evenkia, round-bottomed, cord-impressed pots dominate the assemblages, and such vessels are decorated with the same variety of incised lines, comb-impressions and punctuate stamps, that characterize Belkachinsk pottery from further east.

Discussion and Conclusions

Overall, then, radiocarbon dates can trace the introduction of pottery across eastern Siberia starting in eastern Trans-Baikal sometime between c. 12,000 bp and 11,000 bp. The Ust'-Karenga material from eastern Trans-Baikal is the most securely dated, and such pottery appears to be fairly localized in the area around the Upper Vitim River basin. The origins of these ceramics are not well understood. Although there is currently no direct evidence, it seems possible that there was influence from the neighbouring Amur region, in which pottery was already well developed. In other areas of Trans-Baikal, the evidence is much less secure in chronological terms, but there are hints that pottery may be just as old (Ust'-Kiakhta) or only slightly younger (Studenoe and Ust'-Menza). The cord-impressed pottery from these latter sites seems to have a wider geographical distribution than Ust'-Karenga material, and is found throughout Trans-Baikal. Unfortunately, a lack of radiocarbon dates from other sites makes it difficult to confirm the antiquity of such material. On the basis of stratigraphic evidence, however, it is clear that cord-impressed pottery is the earliest type in most areas of Trans-Baikal, and is found with lithic material that is described as indistinguishable from the preceding Mesolithic period.

In other areas of eastern Siberia, pottery appears to have been adopted much later. In Cis-Baikal, the earliest uncontroversial evidence consists of three net-impressed pots from Kitoi graves dating around 6,500 bp. Evidence for the timing of ceramics at habitation sites in the area is less definite; but the dates from Layer 6 at Gorelyi Les and Layers 5 and 5a at Ust'-Khaita seem to indicate that pottery was also used in a domestic context, albeit rarely, around the same time and perhaps even as far back as 8,500 bp (Gorelyi Les, Layer 7). Unlike Trans-Baikal, net-impressed and Khaita pottery rather than cord-impressed are found in the earliest Neolithic layers.

In the west, radiocarbon data from Kazachka and Eleneva Cave in the Central Yenisei basin suggest that both cord- and net-impressed pottery were introduced around 6,500 bp, and Makarov (2005) also considers Posol'sk and Khaita pottery to have been Early Neolithic forms in the region. Finally, in more northern areas of the Central Siberian Plateau, net-impressed pottery dominates the early assemblages and is introduced around 6,000 bp, slightly later than in the neighbouring southern areas. Cord-impressed pottery is introduced to Yakutia around 1,000 years after this.

I would like to conclude by offering brief thoughts on some areas for future research. To date, the vast majority of research into eastern Siberian pottery has been concerned with the definition of ceramic styles and their chronological and geographical distributions. Obviously, more work on these topics is still required, and the continued application of radiocarbon dating must be seen as a priority. However, in addition, it must be recognized that very little is known currently about the reasons why pottery was adopted in the region and for what purposes it was used. In order to gain a more meaningful understanding of the significance ceramics held for Neolithic hunter-gatherers in eastern Siberia, it is now necessary to start focusing our research on interpreting the social context of pottery production, use and distribution. With respect to the adoption of pottery by hunter-gatherers, Rice has noted that pottery

> can be considered as one end product of a series of cultural transformations beginning with resource intensification (of particular kinds of foods), through socio-economic intensification (emergence of social ranking and aggrandizing individuals or families), to technological intensification (pottery as prestige technology and a durable good) (1999: 45).

All three elements of this equation require greater investigation in the context of eastern Siberia.

First, with respect to resource intensification, it would be useful to determine the sort of resources that might have been stored, served or cooked in the pots. For this we should be looking to residue analysis, pollen and phytolith studies; but such research also requires that we take better stock of the zooarchaeological and palaeobotanical remains at archaeological sites, as well as the potential range of perishable resources that would have existed in the region. Most areas of eastern Siberia have seen little work on this front beyond compiling species lists and rough element counts at archaeological sites. Given that widespread fishing technology seems to coincide with the appearance of pottery in most areas of eastern Siberia, the recovery and analysis of fish remains should be emphasized.

There is also great potential for various social/symbolic models. Hayden (1995, 1998) proposes that pottery can act as a medium for acquiring, displaying and communicating messages of social identity, especially within a social context of competition and differentiation. In Cis-Baikal, for example, it was noted that the earliest securely dated pottery is found in mortuary contexts. Bazaliiskii suggests that such pots would have been ritual in nature. The fact that few pots are found at settlement sites of the same period suggests that pottery may not have played a large role in domestic activities. In later Neolithic times, the greater abundance of both mortuary and settlement contexts suggests that pottery was used for increasingly diverse purposes. The wide diversity of decorative motifs and surface treatments might, for example, relate to their role in distribution and exchage, social identification, etc. In order to investigate the potential of such interpretations, we must take better account of the context in which such vessles were used.

For this, and for much of the research described here, it would be immensely helpful for scholars in the region to start reporting quantitative measures of abundance and distribution—both of the pottery and of the associated lithic and bone assemblages. Although unpublished site reports generally do contain artefact counts and basic statistical analyses, little of this is published widely, and very little statistical work is published comparing the abundance and distribution of pottery between sites or assemblages.

Research into the technological aspects of eastern Siberian pottery would also benefit from more detailed examinations of raw materials and manufacturing techniques. Ceramic petrology would be particularly useful in this regard, as it would lead to a better understanding of ceramic production sites and patterns of trade and exchange.

Finally, it is worth noting that the recent introduction of widespread radiocarbon dating to eastern Siberia has created a crisis of terminology for some scholars. Vetrov (2000), for example, notes that the discovery of ceramic production in Pleistocene–Holocene deposits in Trans-Baikal presents a problem of how to term the Mesolithic and Neolithic. He offers three solutions: (1) retain the term Mesolithic in relation to pre-ceramic cultures, (2) relocate the Mesolithic age, and (3) discard the concept of Mesolithic and adopt the new perspective of the transformation of Palaeolithic industries directly into Neolithic ones. Such superficial terminological debates seem fruitless. Clearly, it is more important to evaluate what kinds of cultural developments occurred and why, than it is to argue over the various labels we might assign to these developments. In fact, in some cases it appears that scholars might have become prisoners of such labelling. As described above, the fact that the Kitoi mortuary tradition was determined to be 'Neolithic' was almost certainly part of the reason that previous scholars were so reluctant to consider the proposition that Kitoi settlement sites might have been largely aceramic. Indeed, it is precisely these aceramic 'Mesolithic' deposits that must hold the key to how, when and why pottery was subsequently adopted by hunter-gatherers in the region. In an area as large and diverse as eastern Siberia, the particular historical processes leading to the adoption and subsequent development of pottery by hunter-gatherers are likely to be equally diverse. As such, the region has much to offer for improving our understanding of more general processes of ceramic innovation and dispersal.

Notes

1. Uncalibrated radiocarbon dates are presented as bp, while calibrated dates are presented as BP. Occasionally BC notation is reported when cited by Russian scholars; however, it should be noted that in many instances, Russian scholars continue to report converted BC dates, calculated by simply subtracting c. 2,000 years from the radiocarbon date (bp), rather than calibrated BC dates. So, for example, it is often noted that the Syalakh culture complex has radiocarbon dates spanning 6,000±100 bp to 4,900±100 bp, which is then converted to span the 4th millennium BC (e.g., Alekseev 1996; Aseev 2003; Ivashina 1979) rather than c. 5th millennium calibrated BC.

2. In Russian archaeological terminology, these pottery styles are referred to as layers [*plasti*], and they are assumed to have chronological significance.

3. Russian scholars distinguish between ceramics formed by drawing the clay with a paddle alone (*metodom vykolachivaniia*) and pottery formed with both a paddle and anvil (*tekhnika nakoval'ni i molotochka*).
4. Again, note that Aseev (2003) converts, rather than calibrates, these dates to the 4th millennium BC, which is how he can consider them to agree with Okladnikov's scheme.
5. Kuz'min and Orlova 2000, citing Konstantinov 1994, report this date as 9,620±250 bp.
6. It is important to note that Orlova was the third author in the original Firsov et al. (1972) article cited by Aseev (2003).
7. Kuz'min and Orlova (2000) cite Vetrov and colleagues (1995) in noting that pottery was found in Kitoi graves at Makrushina; however, our review of this source reveals no mention of pottery at this site.
8. N A Savel'ev is third author on this paper, suggesting that he has revised his views, expressed in personal communication to Weber that the pottery from Layer 7 was not *in situ*.
9. Prior to AD 2,000, there was no official boundary between eastern Siberia and the Russian Far East; but by convention, the Arctic-Pacific watershed and the Lena and Aldan River basins were most often used. In AD 2000, however, the 'Far Eastern Federal District' was officially defined to include the entire 'Sakha Republic (Yakutia)', including that portion west of the Lena Rivers. In this chapter, I retain the geographical rather than political boundary and so only discuss that portion of 'Yakutia' that encompases the Lena watershed. Similarly, I use the term 'Evenkia' to describe the central and western portion of the Central Siberian Plateau, which is not entirely subsumed within the now obsolete Evenk Autonomous Okrug.

Acknowledgements

I am grateful to Professor Yaroslav V. Kuz'min (Vladivostok, Russia) for his useful suggestions and comments relating to an earlier version of this paper. Thanks also to Andrea Hiob for the artefact drawings.

References

Alekseev, A N (1996) *Drevniaia Iakutiia: Neolit i Epokh Bronzi [Ancient Iakutia: Neolithic and Bronze Age]*, Novosibirsk: Izd-vo IFiE SO RAN

Alekseev, A N and Cherosov, N M (1990) *Arkheologiia Iakutii (Kamennyi Vek) Uchebnoe Posobie Slia Studentov [Archaeology of Iakutia (Stone Age): Educational Aid for Students]*, Yakutsk: Yakutskii Gosudarstvennyi Universitet

Aseev, I V (2003) *Iugo-Vostochnaia Sibir' v Epokhu Kamnia i Metalla [Southeast Siberia in the Age of Stone and Metal]*, Novosibirsk: Izd-vo In-ta Arkheologii i Etnografii SO RAN

Bader, O N (1970) 'Ural'skii Neolit [Uralic Neolithic]', in *Kamennyi Vek na Territorii SSSR*, Moscow: Nauka

Bazaliiskii, V I (1990) 'Otchet o nauchnoissledovatel'skoi rabote: mogil'nik "Lokomotiv": istoriia issledovanii, sovremennoe sostoianie issledovanii. Predlozheniia po okhrannym meropriiatiiam i muzeefikatsii [Report on the scientific research work: 'Lokomotiv' cemetery: history of research, current state of research. Propositions for protection measures and muzeification]', unpublished report, Irkutsk State University

Bazaliiskii, V I (nd) 'Mesolithic and Neolithic mortuary complexes in the Baikal region of Siberia', in Weber, A W, Katzenberg, M A, and Schurr, T (eds), *Prehistoric Hunter-Gatherers of the Baikal Region, Siberia: Bioarchaeological Studies of Past Lifeways*, Philadelphia: University of Pennsylvania Press

Berdnikova, N L (1986) 'Ust'-Bel'skii keramicheskii plast (k postanovke problemy) [The Ust'-Bel'skii pottery layer (on formulating the problem)]' in *Chetvertichnaia Geologiia i Pervobytnaia Arkheologiia Iuzhnoi Sibiri Tezisy Dokladov*, Ulan Ude: Nauka

Berdnikova, N L, Lezhnenko, I L, Savel'ev, N A, Medvedev, G I and Georgievskaia, G M (1991) *Ukazatel' Arkheologicheskikh Pamiatnikov Irkutskoi Oblasti. Usol'skii Raion* [Index of archaeological sites in Irkutsk Oblast'], Irkutsk: Irkutskii Gosudarstvennyi Unversitet

Buvit, I, Waters, M R, Konstantinov, M V and Konstantinov, A V (2003) 'Geoarchaeological investigations at Studenoe, an Upper Paleolithic site in the Transbaikal Region, Russia', 18(6) *Geoarchaeology: An International Journal*, 649–673

Chard, C S (1974) *Northeast Asia in Prehistory,* Madison: University of Wisconsin Press

Derevianko, A P (ed) (1998) *Palaeoecology of the Pleistocene and the Stone Age Cultures of Northern Asia and Adjacent Regions (Proceedings of international conference),* Vol. 2, Novosibirsk: Institute of Archaeology and Ethnography Press (in Russian)

D'iakonov, V M (2001) 'K voprosu o tekstil'noi keramike v Neolite Iakutii [On the question of textile pottery in the Neolithic of Iakutia]', *Materialy XII Regional'noi Arkheologo-Etnograficheskoi Studencheskoi Konferentsii,* Barnaul Marta, 25–30

Firsov, L V, Panychev, V A and Orlova, L A (1972) 'Radiouglerodnye daty laboratorii geokhronologii Instituta Geologii i Geofiziki [Radiocarbon dates from the Laboratory of Geochronology in the Institute of Geology and Geophysics]', *Izvestiia SOAN* 38

Formozov, A A (1959) 'Mikroplitícheskie Pamiatniki Aziatskoi Chasti SSSR [Microlith sites in the Asiatic parts of the USSR]', 2 *Sovetskaia Arkheologiia*

Generalov, A G (1979) 'Neoliticheskie kompleksy mnogosloinogo poseleniia Kazachka i ikh znachenie dlia izucheniia neolita Krasnoiarskoi Kanskoi lesostepi [Neolithic complxes in the multi-layered settlement of Kazachka and their significance for investigating the Neolithic of the Krasnoiarsk Kansk forest steppe]', unpublished PhD dissertation summary, AA Zhdanov State University, Leningrad

Gerasimov, M M (1955) 'Osnovy vosstanovleniia litsa po cherepu [Fundamentals of restoring the face according to the skull]', 27 *Trudy Instituta Etnografii: Novaia Seriia*

Glushkov, I G and Glushkova, T N (1992) *Tekstil'naia Keramika kak Istoricheskii Istochnik [Textile Pottery as a Historical Source]*, Tobol'sk: Izdatel'stvo Tobol'skogo Pedinstituta

Goriunova, O I (1984a) 'Kompleksy rannego bronzovogo veka mnogosloinogo poseleniia Tyshkine III [Early Bronze Age complexes in the multi-layred settlement of Tyshkine III]', in Medvedev, V E (ed), *Arkheologiia iuga Sibiri i Dal'nego Vostoka*, Novosibirsk: Nauka, pp 57–69

Goriunova, O I (1984b) 'Mnogosloinye pamiatniki Malogo Moria i o Ol'khon [Multi-layered sites of the Little Sea and Ol'khon]', unpublished PhD dissertation summary, USSR Academy of Sciences, Novosibirsk

——— (1997) *Serovskie Pogrebeniia [Serovo Burials of Priol'khon]*, Priol'khon'ia, Novosibirsk: SO RAN

——— (2001) 'Neolit Priol'khon'ia (oz Baikal) [Neolithic of Priol'khon (Lake Baikal]', in Derevianko, A P and Medvedev, G I (eds), *Sovremennye Problemy Evraziiskogo Paleolitovedeniia*, Novosibirsk: SO RAN, pp 369–373

——— (2003) 'The Neolithic of the Ol'khon region (Lake Baikal)', in Weber, A W and McKenzie, H G (eds), *Prehistoric Foragers of the Cis-Baikal, Siberia,* Edmonton: Canadian Circumpolar Institute Press, pp 15–36

——— (nd) 'The Neolithic of the Ol'khon region (Lake Baikal)' in Weber, A W, Katzenberg, M A and Schurr, T (eds), *Prehistoric Hunter-Gatherers of the Baikal Region, Siberia: Bioarchaeological Studies of Past Lifeways*, Philadelphia: University of Pennsylvania Press, (in review)

Goriunova, O I and Khlobystin, L P (1991) 'Datirovka kompleksov poselenii i pogrebenii bukhty UlanKhada [Dating of settlement and burial complexes in Ulan Khada Bay]', in Masson, V M (ed), *Drevnosti Baikala*, Irkutsk: Irkutskii Gosudarstvennyi Unversitet, pp 41–56

Goriunova, O I and Savel'ev, N A (1981) 'Opyt razrabotki poniatii dlia opisaniia form sosudov Neoliticheskoi i rannebronzovoi keramiki Vostochnoi Sibiri [An attempt to work out concepts for the description of the form of Neolithic and Early Bronze ceramics from Eastern Siberia]', *Problemy Terminologii i Analiza Arkheologicheskikh Istochnikov*, Irkutsk: Irkutskii Gosudarstvennyi Unversitet, pp 115–126

Grishin, I C (1981) *Pamiatniki Neolita, Bronzovogo i Rannego Zheleznogo Vekov Lesostepnogo Zabaikal'ia [Neolithic, Bronze Age and Early Iron Age Sites of Zabaikal'e],* Moscow: Nauka

Hayden, B (1995) 'The emergence of prestige technologies and pottery', in Barnett, W K and Hoopes, J W (eds), *The Emergence of Pottery: Technology and Innovation in Ancient Societies,* Washington DC: Smithsonian Institution Press, pp 257–266

——— (1998) 'Practical and prestige technologies: the evolution of material systems', 5 *Journal of Archaeological Method and Theory*, 1–55

Igumnova, E S, Savel'ev, A N, Teten'kin, A V, Batrakova, N A and Samarina, I V (2001) 'Novoe mnogosloinoe mestonakhozhdenie v doline reki Beloi—Ust'-Khaita', *Istoriko-Kul'turnoe Nasledie Severnoi Azii: Itogi i Perspektivy Izucheniia na Rubezhe Tysiacheletii [A New Multi-Layered Location in the Valley of River Belaia—Ust-Khaita, Historical-Cultural Heritage of Northern Asia: Results and Research Perspectives at the Turn of the Millennium]*, Barnaul: Izdatel'stvo Al'taiskovo Instituta

Igumnova, E C, Morozova, O V and Savel'ev, N A (2004) 'Novye dannye o geoarkheologicheskom ob'ekte "Gorelyi Les"', Traditsionnye Kul'tury i Obshchestva Severnoi Azii (s drevneishikh vremen do sovremennosti): materialy XLIV regional'noi (s mezhdunarodnym uchastiem) arkheologo-etnograficheskoi konferentsii studentov i molodykh uchenykh, Kemerovo, 31 marta–3 aprelia

Igumnova, E C (2004) 'Sosudy—Bliznetsy (po materialam polevykh arkheologicheskikh rabot 2000–2001 gg) [Vessels-Twins (according to archaeological fieldwork materials from 2000–2001)],', Traditsionnye Kul'tury i Obshchestva Severnoi Azii (s drevneishikh vremen do sovremennosti): materialy XLIV regional'noi (s mezhdunarodnym uchastiem) arkheologo-etnograficheskoi konferentsii studentov i molodykh uchenykh, Kemerovo, 31 marta–3 aprelia

Ivashina, L G (1979) *Neolit i Eneolit Lesostepnoi Zony Buriatii [Neolithic and Eneolithic of the forest-steppe zone of Buriatia]*, Novosibirsk: Nauka

Jull, A J T, O'Malley, J M, Biddulpii, D L, Derevianko, A P, Kuz'min, Y V, Medvedev, V E, Tabarkv, A V, Zenin, V N, Vetrov, V M, Lapsiina, Z S, Garkovik, A V and Zhushchikhovskaya, I S (1998) 'Radiocarbon chronology of the earliest Neolithic cultures of the southern Russian Far East and Transbaikal based on direct pottery dating by accelerator mass spectrometry', in Derevianko, A P (ed), *Paleoeconomy of the Pleistocene and the Stone Age Cultures of Northern Asia and Adjacent Regions (Proceedings of International Conference)*, Vol. 2, Novosibirsk: Institute of Archaeology and Ethnography Press [in Russian], pp 63–66

Jull, A J T, Burr, G S, Derevianko, A P, Kuz'min, Y V and Shevkamud, I Y (2001) 'Radiocarbon chronology of the Paleolithic–Neolithic transition in the Amur River Basin (Russian Far East)', in Derevianko, A P (ed), *Current Problems of the Eurasian Paleolithic Studies*, Novosibirsk: Nauka Proceedings of International Conference, pp 140–142

Kartsov, V G (1929) *Materialy k Arkheologii Krasnoiarskogo Raiona [Materials on the Archaeology of Krasnoiarsk Raion]*, Gosudarstvennyi muzei Prieniseiskogo Kraia: Opisanie kollektsii i materialov muzeia, Krasnoiarsk

Khlobystin, L P (1969) 'The stratified settlement of Ulan-Khada on Lake Baikal', 6(1) *Arctic Anthropology*, 88–94

——— (1978) 'Vozrast i sootnoshenie neoliticheskikh kul'tur Vostochnoi Sibiri [Age and correlation of Neolithic cultures of Eastern Siberia]', 153 *Kratkie Soobshcheniia Instituta Arkheologii*, 93–99

——— (1996) 'Vostochnaia Sibir' i Dal'nii Vostok [Eastern Siberia and the Far East]', in Oshibkina, S V (ed), *Neolit Severnoi Evrazii*, Moscow: Nauka, pp 270–329

Kiparisova, N P (1960) *O Kul'turakh Lesnogo Zaural'ia [On the Cultures of Forested Zaural'e]*, Moscow: SA

Kirillov, I I and Rizhskii, M M (1973) *Ocherki Drevnei Istorii Zabaikal'ia [Essays on the Ancient History of Zabaikal'e]*, Chita: Nauka

Konopatskii, A K (1982) *Drevnie Kul'tury Baikala (o Ol'khon) [Ancient Cultures of Baikal (Ol'khon Island)]*, Novosibirsk: Nauka

Konstantinov, M V (1994) *Kamennyi Vek Vostochnogo Regiona Baikal'skoi Azii [Stone Age of Eastern Regions of Baikal Asia]*, Ulan Ude: SO RAN

Kosarev, M F (1981) *Bronzovyi Vek Zapadnoi Sibiri [Bronze Age of Western Siberia]*, Moscow: SA

Kuz'min, Y V (2002) 'The earliest centers of pottery origin in the Russian Far East and Siberia: review of chronology for the oldest Neolithic cultures', 29 *Documenta Praehistorica*, 37–46

——— (2006a) 'Chronology of the earliest pottery in East Asia: progress and pitfalls', 80 *Antiquity*, 362–371

——— (2006b) 'Palaeoenvironment and Chronology', in Nelson, S M, Derevianko, A P, Kuz'min, Y V and Bland, R L (eds), *Archaeology of the Russian Far East: Essays in Stone Age Prehistory*, BAR International Series 1540, Oxford: Archaeopress, pp 13–40

Kuz'min, Y V and Jull, A J T (1997) 'AMS radiocarbon dating of the Paleolithic–Neolithic transition in the Russian Far East', 14 *Current Research in the Pleistocene*, 46–48

Kuz'min, Y V and Keally, C T (2001) 'Radiocarbon chronology of the Earliest Neolithic Sites in East Asia', 43 *Radiocarbon*, 1121–1128

Kuz'min, Y V and Orlova, L A (1998) 'Radiocarbon chronology of the Siberian Paleolithic', 12(1) *Journal of World Prehistory*, 1–53

——— (2000) 'The Neolithization of Siberia and the Russian Far East: radiocarbon evidence', 74 *Antiquity*, 356–364

Kuz'min, Y V and Shevkomud, I Y (2003) 'The Palaeolithic–Neolithic transition in the Russian Far East', 24(2) *The Review of Archaeology*, 37–45

Kuz'min, Y V, Jull, A J T, Orlova, L A and Sulerzhitskii, L D (1998) 'Radiocarbon chronology of the Stone Age cultures, Russian Far East', 40(2) *Radiocarbon*, 675–686

Kuz'min, Y V and Vetrov, V M (2007) 'The earliest Neolithic complex in Siberia: the Ust-Karenga 12 site and its significance for the Neolithisation process in Europe', 34 *Documenta Praehistorica*, 9–20

Maksimenkov, G A (1966) 'Ust'-Sobakinskaia stoianka i ee znachenie dlia izucheniia drevnei istorii raiona Krasnoiarska [Ust'-Sobakinskaia site and its significance for research into the ancient history of Krasnoiarsk Raion]', in *Drevniaia Sibir'*, Novosibirsk: Sibirskii Arkheologicheskii Sbornik, pp 156–179

Makarov, N P (2005) 'Khronologiia i periodizatsiia epokhi Neolita i Bronzy Krasnoiarskoi lesostepi [Chronology and periodization of the Neolithic and Bronze Age of Krasnoiarsk forest-steppe]', 3 *Izvestiia Laboratorii Drevnikh Tekhnologii IrGTU*, 149–171

Makarov, N P and Orlova, L A (1992) 'Mezolit peshchery Eleneva [Mesolithic of Eleneva cave]', in *Paleoekologiia i Rasselenie Drevnego Cheloveka v Severnoi Azii i Amerike. Tezisy Dokladov Mezhdunarodnogo Simposziuma*, Krasnoiarsk, pp 171–175

Makarov, N P, Martynovich, N D, Ovodov, N D, Iamskikh, A F, Iamskikh, F I U, Orlova, L A, Nazarov, V I, Tsipkin, E A, Chikhikvadze, V M (1992) 'Peshchera Eleneva—mnogosloinyi golotsenovyi arkheologicheskii pamiatnik na Srednem Enisee [Elenaeva cave—multi-layered Holocene archaeological site on the Middle Enisei]', in *Problemy Arkheologii, Etnografii, Istorii i Kraevedeniia Prieniseiskogo Kraia*, Krasnoiarsk, Vol. 1, pp 112–121

Makarov, N P, Martynovich, N V, Orlova, N D and Orlova, L A (1995) 'Kompleksnye issledovaniia peshchery Eleneva [Research into the complexes of Eleneva Cave]', in *Obozrenie Resul'tatov Polevykh i Laboratornykh Issledovanii Arkheologov, Etnografov i Antropologov Sibiri i Dal'nego Vostoka v 1993 godu*, Novosibirsk: SO RAN, pp 143–145

Mamonova, N N and Sulerzhitskii, L D (1989) 'Opyt datirovaniia po C14 pogrebenii Pribaikal'ia epokhi golotsena [An attempt to C14 date the Holocene burials of Pribaikal'e]', 29(1) *Sovetskaia Arkheologiia*, 15–20

Medvedev, G I, Slagoda, E A, Lipnina, E A, Berdnikova, N E, Generalov, A G, Rogovskoi, E O, Oshepkova, E B, Vorob'eva, G A and Shmygun, P E (2001) 'Kamennyi vek Iuzhnogo Priangar'ia Bel'skii geoarkheologicheskii raion Tom II [Stone Age of southern Priangar'ia, Bel'skii geoarchaeological district, Volume II]', Irkutsk: Izd-vo Irkutsk Gosudarstvennyi Universitet

Michael, H N (1958) *The Neolithic Age in Eastern Siberia*, Transactions of the American Philosopical Society. New Series, v. 48, pt. 2, Philadelphia: American Philosophical Society

Michael, H N (1992) 'The Neolithic cultures of Siberia and the Soviet Far East', in Ehrich, R W (ed), *Chronologies in Old World Archaeology 1*, Chicago: University of Chicago Press, pp 416–429

Mochanov, I A (1969) *Mnogosloinaiia Stoianka Bel'kachi i Periodizatsiia Kamennogo veka Iakutii [Multi-Layered Site of Bel'kachi and the Periodization of the Stone Age of Iakutia]*, Yakutsk

Mochanov, I A, Fedoseeva, S A and Alekseev, A N (1983) *Arkheologicheskie Pamiatniki Iakutii: Basseiny Aldana i Olekmy [Archaeological Sites of Iakutia: Basins of the Aldan and Olekma]*, Novosibirsk: Nauka

Nomokonova, T I (2004) 'Okunevaia IV—mnogosloinoe poselenie Chivyrkuiskogo zaliva oz. Baikal [Okunevaia IV—Multi-layered settlement of Chivyrkuiskii Bay]', Traditsionnye Kul'tury i Obshchestva Severnoi Azii (s drevneishikh vremen do sovremennosti): materialy XLIV regional'noi (s mezhdunarodnym uchastiem) arkheologo-etnograficheskoi konferentsii studentov i molodykh uchenykh, Kemerovo, 31 marta—3 aprelia

Nomokonova, T I and Goriunova, O I (2004) 'Neoliticheskie kompleksy mnogosloinogo poseleniia Katun [Neolithic complexes of the multi-layered settlement of katun I (Chivyrkuiskii Bay, lake Baikal)]' I (Chivyrkuiskii zaliv oz. Baikal)', 2 *Izvestiia Laboratorii Drevnikh Tekhnologii IrGTU*, 117–123

Okladnikov, A P (1950) 'Neolit i bronzovyi vek Pribaikal'ia (chast' I i II) [Neolithic and Bronze Age of Pribaikal'e (Parts I and II)]' 18 *Materialy i Issledovaniia po Arkheologii SSSR*, Moscow: Izdatel'stvo Akademii Nauk SSSR

——— (1955) 'Neolit i bronzovyi vek Pribaikal'ia (chast' III) [Neolithic and Bronze Age of Pribaikal'e (Part III)]', 43 *Materialy i Issledovaniia po Arkheologii SSSR*, Moscow: Izdatel'stvo Akademii nauk SSSR

——— (1957) *Iz Istorii Etnicheskikh i Kul'turnykh Sviazei Neoliticheskikh Plemen Srednego Eniseia (K Voprosu o Proiskhozhdenii Samodiiskikh Plemen* [From the History of Ethnic and Cultural Connections of Neolithic Tribes of Central Enisei (On the Question of the Origins of Samoyedic Tribes)], Moscow: SA

——— (1970) 'Neolit Sibiri i Dal'engo Vostoka [Neolithic of Siberia and the Far East]', 166 *Materialy i Issledovaniia po Arkheologii SSSR*, 172–193

——— (1974) *Neoliticheskie Pamiatniki Angary (Ot Shchukino do Bureti) [Neolithic Sites of the Angara (from Shchukino to Bureti)]*, Novosibirsk: Nauka

——— (1977) *Istoriia i Kul'tura Buriatii [History and Culture of Buriatia]*, Ulan-Ude: Nauka

Okladnikov, A P and Kirillov, I I (1980) *Iugo-Vostochnoe Zabaikal'e v Epokhu Kamnia i Rannei Bronzy* [Southeast Zabaikal'e in the Stone Age and Early Bronze Age], Novosibirsk: Nauka

O'Malley, J M, Kuz'min, Y V, Burr, G S, Donahue, D J, Jull, A J T (1999) 'Direct Radiocarbon Accelerator Mass Spectometry dating of the earliest pottery from the Russian Far East and Transbaikal', 26 *Mémoires de la Société Préhistorique Française*, 19–24

Orlova, L A (1995) 'Radiouglerodnoe datirovanie arkheologicheskikh pamiatnikov Sibiri i Dal'nego Vostoka. Soobshchenie 1 [Radiocarbon dating of archaeological sites of Siberia and the Far East. Report 1]', in Derevianko, A P and Kholushkin, Y P (eds), *Metody Estestvennykh Nauk v Arkheologicheskikh Rekonstruktsiiakh,* Novosibirsk: Institute of Archaeology and Ethnography Press, pp 207–232

Orlova, L A (1998) 'Radiocarbon dating of archaeological sites in Siberia and the Russian Far East Part 2', in Derevianko, A P, Kholushkin, Y P, Rostovtsev, P S and Voronin, V T (eds), *Statistical Analysis of the Late Paleolithic Complexes in Northern Asia*, Novosibirsk: Novosibirsk State University Press, pp 74–99

Pogudin, V A (1989) 'Mnogosloinaia stoianka Niasha—opornyi pamiatnik epokhi Neolita na Srednem Enisee [Multi-layered site of Niashia—a key Neolithic site on the Middle Enisei]', in *Problemy Izucheniia Sibiri v Nauchno-Issledovatel'skoi Rabote Muzeev. Tezisy Dokladov Nauchno-Prakticheskoi Konferentsii*, Krasnoiarsk, pp 235–261

Prokoshev, N A (1940) 'K voprosu o neoliticheskikh pamiatnikakh kamskogo Priural'ia [On the question of Neolithic sites of Kamskii Priural'e]', in *Arkheologicheskie Pamiatniki Urala i Prikam'ia*, Moscow: Nauka

Rice, P M (1999) 'On the origins of pottery', 6(1) *Journal of Archaeological Method and Theory*, 1–54

Savel'ev, N A (1982) 'Problemy issledovaniia neoliticheskikh poselenii na iuge Srednei Sibiri [Research problems of Neolithic settlement in the south of Central Siberia]', *Problemy Arkheologii i Etnografii Sibiri*, Irkutsk: Irkutskii Gosudarstvennyi Universitet

——— (1989) 'Neolit iuga srednei Sibiri: istoriia osnovnykh idei i sovremennoe sostoianie problemy [Neolithic of the south of Central Siberia: history of basic ideas and the contermporary state of problems]', unpublished PhD dissertation summary, Novosibirsk: USSR Academy of Sciences

Savel'ev, N A, Goriunova, O I and Generalov, A G (1974) 'Raskopki mnogosloinoi stoianki Gorelyi Les (predvaritel'noe soobshchenie) [Excavations of the multi-layered site of Gorelyi Les (Preliminary Report)]', *Drevniaia Istoriia Narodov Iuga Sibiri, (Ch 1),* Irkutsk: Nauka

Savel'ev, N A and Medvedev, G I (1973) 'Rannii keramicheskii kompleks mnogosloinogo poseleniia Ust'-Belaia [Early ceramic complex of the multi-layered settlement of Ust'-Belaia]', *Problemy Arkheologii Urala i Sibiri*, Moscow: Nauka

Savel'ev, N A, Temen'kin, A V, Igymnova, E C, Abdulov, T A, Ineshin, E M, Osadchii, S S, Vetrov, V M, Klement'ev, A M, Mamonov, M P, Orlova, L A and Shibanova, I V (2001) 'Mnogosloinyi geoarkheologicheskii ob'ekt Ust'-Khaita—predvaritel'nye dannye [Multi-layered geoarchaeological object Ust'-Khaita—preliminary data]', in Derevianko, A P and Medvedev, G I (eds), *Sovremennye Problemy Evraziiskogo Paleolitovedeniia*, Novosibirsk: SO RAN, pp 338–351

Semina, L V (1985) 'Keramika epokhi Neolita i bronzy iugo-zapadnogo Zabaikal'ia [Ceramics of the Neolithic and Bronze Age of southwestern Zabaikal'e]', in *Drevnee Zabaikal'e i Ego Kul'turnye Sviazi*, Novosibirsk: SO RAN

Sinitsyna, G V (1986) 'Neoliticheskie pamiatniki Verkhnei Angary (na materialakh poselenii) [Neolithic sites of the Upper Angara (on the basis of settlement materials)]', unpublished PhD dissertation summary, Leningrad: USSR Academy of Sciences

Timofeev, V I and Zaitseva, G I (1996) 'Spisok radiouglerodnykh datirovok Neolita [List of radiocarbon dates of the Neolithic]', in Oshibkina, S V (ed), *Neolit Severnoi Evrazii*, Moscow: Nauka

Vasil'evskii, R S and Kirillov, N N (1967) 'Arkheologicheskie issledovaniia v doline r Onon [Archaeological research in the valley of the Onon River]', *Arkheologicheskie Otkrytiia*, Moscow: Nauka

Vetrov, V M (1981) 'Ust'-Karengskii kompleks stoianok na Vitime [Ust'-Karenga complex of sites on the Vitim]', *Novoe v Arkheologii Zabaikal'ia*, Novosibirsk: Nauka

Vetrov, V M (1985) 'Keramika Ust'-Karengskoi kul'tury na Vitime [Ceramics of Ust'-Karenga culture on the Vitim]', in Konovalov, P B (ed), *Drevnee Zabaikal'e i Ego Kul'turnye Sviazi*, Novosibirsk: Nauka pp 123–130

——— (1995) 'Early ceramics of the Stone Age cultures of Upper Vitim', in Kajiwara, I I (ed), *The Origin of Ceramics in East Asia and the Far East,* Sendai: Tohoku Fukushi University Press [in Russian], pp 31–35

——— (1997) 'Ust-Karenga culture and its place in the system of archaeological sites of adjacent regions', in Medvedev, G I (ed), *Inter-Relations of the Nations of Russia, Siberia, and Oriental Countries: History and Modernity* 2, Novosibirsk: Nauka [in Russian], pp 176–180

——— (2000) 'Archaeology of the Stone Age of the Vitim Plateau (cultures and chronology)', in Medvedev, G I (ed), *Archaic and Traditional Cultures of Northeast Asia: Problems of Origins and Transcontinental Contacts,* Novosibirsk: Nauka [in Russian], pp 28–36

Vetrov, V M, Berdnikova, N E, Altukhov, V V, and Frolov, A V (1995) 'Makrushinskii mogil'nik. Ranneneoliticheskii kompleks [The Makrushino burial ground: Early Neolithic complex]', in V M Vetrov (ed), *Baikal'skaia Sibir' v Drevnosti*, Irkutsk: Irkutsk State University Press, pp 112–132

Vetrov, V M and Kuz'min, Y V (2005) 'K Istorii Izucheniya Drevnei Keramiki na Verkhnem Vitime', in Kharinsky, A V (ed), *Sotsiogenez v Severnoi Azii, Chast 1*, Irkutsk State Technical University, Irkutsk, pp 59–63

Vinogradov, A V (1981) *Drevnie Okhotniki i Rybolovy Sredneaziatskogo Mezhdurech'ia* [Ancient hunters and fishers of the Central Asian inter-riverine zone], Moscow: Nauka

Vorob'ev, S A (1995) 'Ob opyte modelirovaniia neoliticheskoi keramiki Iakutii [About attempts at modeling the Neolithic ceramics of Iakutia]', *Integratsiia Arkheologicheskikh i Etnograficheskikh Issledovanii Chast 1,* Omsk, pp 25–29

Vorob'eva, G A and Savel'ev, N A (1984) 'Problemy issledovaniia kamennogo veka Evrazii [Research problems of Stone Age Eurasia]', *Tezisy dokladov Kraevoi konferentsii,* Krasnoiarsk: Krasnoiarsk State University Press

Weber, A W (1995) 'The Neolithic and Early Bronze Age of the Lake Baikal region, Siberia: a review of recent research', 9(1) *Journal of World Prehistory,* 99–165

Weber, A W, Beukens, R, Bazaliiskii, V I, Goriunova, O I and Savel'ev, N A (2006) 'Radiocarbon dates from Neolithic and Bronze Age hunter-gatherer cemeteries in the Cis-Baikal region of Siberia', 48(3) *Radiocarbo,* 1–40

Weber, A W, Beukens, R and Goriunova, O I (2004) 'Radiocarbon dating of the Bronze Age cemetery Khuzhir Nuge XIV on Lake Baikal', 4 *Archaeology, Ethnography and Anthropology of Eurasia*, 64–72

Weber, A W and Goriunova, O I (2005) 'Khronologiia mogil'nika Kurma XI (ozero Baikal) po arkheologicheskim in radiouglerodnym dannym [Chronology of the Kurma XI cemetery (Lake Baikal) according to archaeological and radiocarbon data]', 3 *Izvestiia Laboratorii Drevnikh Tekhnologii: Irkutskogo Gosudarstvennogo Tekhnicheskogo Universiteta*, 186–190

Weber, A W, Link, D W and Katzenberg, M A (2002) 'Hunter-gatherer culture change and continuity in the Middle Holocene of the Cis-Baikal, Siberia', 21(2) *Journal of Anthropological Archaeology*, 230–299

Weber, A W, Katzenberg, M A and Schurr, T (eds) (nda) *Prehistoric Hunter-Gatherers of the Baikal Region, Siberia: Bioarchaeological Studies of Past Lifeways*, Philadelphia: University of Pennsylvania Press

Weber, A W and McKenzie, H G (eds) (2003) *Prehistoric Foragers of the Cis-Baikal, Siberia,* Edmonton: Canadian Circumpolar Institute Press

Weber, A W, McKenzie, H G, and Beukens, R (2008) 'Relative and radiocarbon dating: cemetery use and regional patterns', in Weber, A W, Goriunova, O I, and McKenzie, H G (eds), *Khuzhir-Nuge XIV, a Middle Holocene Hunter-Gatherer Cemetery on Lake Baikal, Siberia: Archaeological Materials,* Northern Hunter-Gatherers Research Series 4, Edmonton: Canadian Circumpolar Institute Press, pp 185–212

Weber, A W, McKenzie, H G, and Beukens, R (ndb) 'Radiocarbon dating of middle Holocene culture history in Cis-Baikal', in Weber, A W, Katzenberg, M A and Schurr, T (eds), *Prehistoric Hunter-Gatherers of the Baikal Region, Siberia: Bioarchaeological Studies of Past Lifeways*, Philadelphia: University of Pennsylvania Press

Weber, A W, McKenzie, H G, Beukens, R P and Goriunova, O I (2005) 'Evaluation of radiocarbon dates from the Middle Holocene hunter-gatherer cemetery Khuzhir-Nuge XIV, Lake Baikal, Siberia', 32 *Journal of Archaeological Science*, 1481–1500

Zubkov, V S (1982) 'Neolit i rannii bronzovyi vek Verkhnei Leny [Neolithic and Early Bronze Age of the Upper Lena]', unpublished PhD dissertation summary, A A Zhdanov State University, Leningrad

Chapter 6

Early Hunter-Gatherer Ceramics in the Urals and Western Siberia

Natal'ia M Chairkina and Lubov' L Kosinskaia

Introduction

This chapter considers the emergence of the earliest ceramic technology amongst mid-Holocene hunter-fisher-gatherer cultures who occupied the Ural Mountains and the adjacent West Siberia Plain, a low-lying basin drained by the Ob' River, which flows northwards to the Arctic Ocean. The western edge of this drainage area runs up into the Ural Mountains, which stretch from the south to the Kara Sea coast in the north, dividing the East European Plain from the West Siberian lowlands, and forming a line of broken relief some 2,000 km long and around 40–150 km wide. The Urals consist of several ridges, which are interspaced with wide depressions. The region of western Siberia forms an integral part of northern Eurasia, and stretches eastwards from the peaks of the Urals, covering a territory of over 3 million sq km. The area is bound by the Enisei River to the east, the Turgai plateau and Kazakh foothills in the south, and the Altai, Salair and Kuznetsk Alatau mountains in the southeast.

Western Siberia and the eastern flank of the Urals are drained by the Ob' and its main tributaries the Tobol, Ishim and Irtysh, which flow northwards. Their waters pass through several ecological zones—dry grassy steppe, forest-steppe, boreal forest belt of taiga and finally through the bleak windswept tundra. In contrast, the majority of smaller tributary rivers are of latitudinal orientation, many of them originating on the eastern slopes of the Urals. On the other side of these mountain ranges, the rivers flow down from the western slopes into the Volga and Pechora Rivers which belong to the rivers of the East European Plain (Figure 6.1).

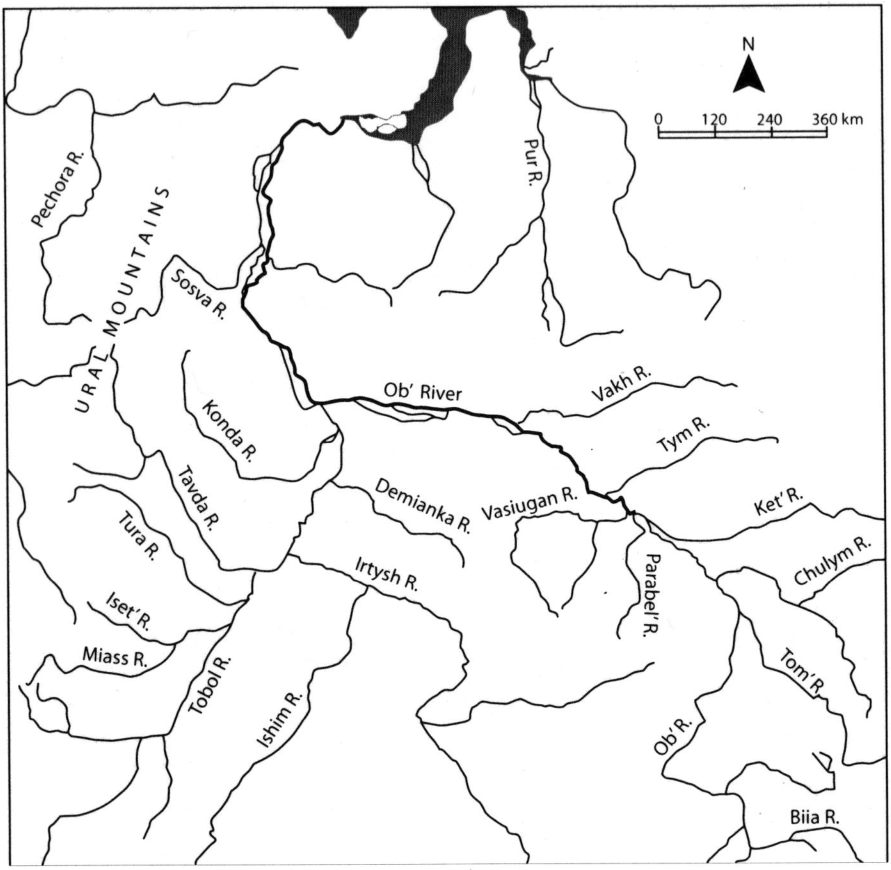

Figure 6.1 The Urals and western Siberia.

The Neolithic in this area is defined by the appearance of the first pottery amongst hunter-fisher-gatherer populations who occupied these areas of the Urals and western Siberia during the middle and second half of the Atlantic period (about 7,000–5,000/4,600 bp), a period characterized by warmer and moister climatic conditions. As conditions of the Atlantic changed into those of the colder, more humid and unstable climate of the early Subboreal period (about 4,600–4,100 bp), we start to see some early use of copper artefacts, and the initial emergence of animal husbandry in the southern margins of the region, in the period defined as the Eneolithic, just prior to the start of the Bronze Age (2,100–1,900 bp) (Chairkina 2005).

In the Atlantic period, local resources were sufficiently plentiful within the varied ecosystems of the steppe, forest-steppe and forest zones (including the southern and middle taiga belts) to support the emergence of an economy based on hunting, fishing and gathering. The specific details of the economy and the relative importance of its different branches depended on local variations in the ecology. In the boreal forest zone, we see the emergence of the 'classic' taiga hunting and fishing economy that persisted with some minor changes well into the historical period, and in places, survives even to the present day. The characteristics of this taiga economy are well described in the ethno-historic literature on the region's local hunter-fisher-gatherer groups—the Khanty, Mansi and Sel'kup—of western Siberia (Dunin-Gorkavitch 1996 [1903]; Fedorova 2000; Golovnev 1993), which involved movements between winter settlements and pit houses, and summer camps with less substantial above-ground structures. In the economy, we see the combination of winter and summer fishing, involving use of various weirs, traps, fish spears and hooked tackle, and the combination of winter and summer hunting of ungulates, game and waterfowl.

In the prehistory of the Urals and western Siberia, the rapid and widespread emergence of earthenware vessels during the Neolithic of the forest and forest-steppe zone appears to coincide with the move away from the more mobile forms of life that had characterized earlier periods (the Mesolithic). This shift began with the transition to a more sedentary mode of life, when groups spent extended periods at riverside and lakeside settlements characterized by pit houses and other substantial structures. This change appears to be linked to the development of fishing and more general improvements in hunting techniques, which resulted in the intensification of food production, and possibly population growth. By far, the most important branch of the economy was fishing which involved a range of specialist technologies. During excavations at settlements, barbed arrow points, fish spears, harpoons, net weights and the remains of wicker fish traps have been found. Hunting was widespread, as evidenced by large volumes of arrow points and faunal remains of ungulates (moose/elk, northern reindeer, roe deer), beaver, forest and waterfowl. On the Pur basin, pit hunting traps were also used in the Neolithic.

Within these wider conditions, pottery vessels came into general use around 8,000–5,000 BP in the Urals and western Siberia, as part of the wider transformations across the northern part of Eurasia. The early Neolithic (8,000–7,000 BP) is represented by two groups of cultures. The first group is characterized by both round- and flat-based containers with incised and 'pricked' ornamentation. It is thought that these cultures reflect a southern cultural 'impulse', possibly linked to various processes including migration, cultural connections through various intermediate groups, and so on, from Kazakhstan, Central Asia and possibly the Caspian. The second group of cultures (8,000–5,000 BP) contains only round-based vessels, with comb stamps, and 'pricked' and incised ornamentation. This ceramic tradition is regarded by archaeologists as being local, but questions about its specific origins, either as a result of cultural influences or through processes of local invention, remain.

Analysis of these early pottery assemblages has played a major role in the archaeology of western Siberia and the Urals, with variations in technological and morphological traits, decorative techniques and ornamental compositions generating important insights into the culture history of local hunter-fisher-gatherer communities. Pottery vessels, ceramic figurines and stone tools have been recovered in abundance, and ceramics have been excavated not only in seasonal and permanent settlements, but also from burial complexes, individual inhumations and from cultic assemblages deposited at what appear to have been ritual sites. However, burials and cultic sites have only been excavated in the southern part of the region (in the forest-steppe and southern forest edge).

Despite our understanding of these general dynamics, the archaeology of this vast region has been unevenly studied, with most research conducted along the major rivers, where recent settlement and other related developments have been concentrated. In contrast, the least investigated areas are found in the northern part of the West Siberian plain, to the north of the Siberian Urals, where only isolated mid-Holocene sites have been found. This uneven coverage, plus the widespread absence of accurate dating for many sites, has meant that the identification and synchronization of many archaeological cultures, and the development of a regional cultural-chronological scheme, has been hindered. As a result, the origins and inter-relations between various archaeological cultures remain issues of considerable academic debate.

Nevertheless, at a local scale, it has been possible to demonstrate that archaeological cultures in both western Siberia and the Urals are characterized by numerous dissimilarities, which have arisen partly from variations in the local ecology and environment, factors which have demanded or encouraged different kinds of subsistence practice, but are also linked to the historical genesis and particularities of local worldviews and traditions. Nevertheless, at a broader analytical scale, it is also clear that these Ural and West Siberian archaeological cultures shared many common features. Throughout the Neolithic and Eneolithic periods, the overall regional coherence distinguished them from contemporary hunter-fisher-gatherer archaeological cultures in both eastern Siberia and the East European Plain.

Early Pottery in the Urals and Western Siberia

Neolithic Earthenware Ceramics—Morphology and Technology

In Russian archaeology, the Neolithic Age in the Urals and western Siberia (8,000–5,000 BP) is a period distinguished by the use of pottery by hunter-fisher-gatherers who had yet to adopt metallurgy or elements of the productive economy. Over 20 distinct types of earthenware have been classified, based on diagnostic features of morphology, technology and ornamentation. Some of these ceramic types are known to be characteristic of specific archaeological cultures, others are of types that cannot be linked to specific cultures due to the absence of wider data from the region. Assigning reliable chronologies to these ceramic types also remains controversial, and the restricted number of radiocarbon dates available so far have been only of limited assistance in determining the origin and sequence of particular local archaeological cultures, and in synchronizing them with others in more distant territories (Kosinskaia 2002: 172–179). These poor regional-scale chronologies continue to hinder more coherent understandings of the Neolithic (defined in this chapter as the emergence of new cultural formations which produce pottery and new kinds of stone tools; these changes are also associated with the increasingly widespread distribution of pit houses and coeval changes in the economy and worldview of local communities), including the emergence of pottery and general patterns of cultural interaction and exchange.

Despite these challenges, in general, the earliest ceramics (of Early Neolithic cultures [about AT-2]) recovered from West Siberia and the Urals seem to have rather similar characteristics; this may be an indication of a closely related Ural-Siberian cultural province. Subsequently, during the later Neolithic (AT-3), the ceramics became more differentiated into smaller and more distinct traditions, located in the west Urals, the east Urals, and the Ob' region (Kosinskaia 2001: 266–275).

All Neolithic ceramics from the Urals and West Siberian region are characterized by the use of tempered clay pastes and bonfire firing. The composition of admixtures in the paste depended on the quality of the natural clay as well as local cultural traditions. Grog and sand were widely used as tempers; talc was also used but only in the forested mountains of the eastern Urals. Finds of talc-tempered ceramics have been made in other areas, for example, in the Kama region of the Volga basin, which suggests that there were social ties between the peoples inhabiting the eastern and western side of these mountains. Organic materials were also added to the clay mixture to improve plasticity and the use of different mixing proportions is notable; there were several distinct paste recipes involving grog, sand and organic matter amongst the ceramics from the Koshkino-Boborykino cultures of the Ishim River basin eastwards of Tobol (Poshekhonova 2004: 23–24).

The pots in this whole region also have similar construction methods, usually involving coil or strip building techniques and often leading to subsequent fracture lines along the clay joints. Pottery of the Bystrino culture appears to have been coil-built and made from the rim downwards, with the body formed from bands 5–7 mm wide. Section of clay was then added to the coils to complete the pot, and the joints were smeared over. The flat-bottomed pots of the Koshkino and Boborykino Ishim cultures were built from the base upwards, starting with a clay flat cake, to which walls of 3–5 mm thickness were then added using the coiling method (Poshekhonova 2004: 25). A similar technique was used to produce the flat-bottom vessels of the Amnya type in the Lower Ob' region (Morozov and Stefanov 1993: 155). Also characteristic of the early ceramics of the Urals and adjoining territories of western Siberia are vessels that have a bulge inwards below a narrower rim. The bulge form was created by adding a final inward-sloping band or coil to the rim (Kalinina and Ustinova (Gadzhieva) 1988: 5–7). In the forest-steppe zone of the upper Ob' and Irtysh Rivers, this bulging form is rare and occasional finds of this pottery suggest cultural influences from the Ural region. Neolithic ceramics are also distinguished by highly smoothed surfaces, both on the inner and outer sides of the vessels. In some rare cases, treatment of the inner surface with a toothed spatula resulted in parallel grooving. In contrast, the outer surface was often burnished and occasionally painted with ochre.

The Boborykino culture (and related cultures) of the eastern Ural slopes and the left bank of the Ob' basin are characterized by diverse ceramic assemblages containing several different forms of pottery, including straight-sided flat-bottomed jars, vertical-ellipsoid vessels and small round-bottomed bowls (Figure 6.2). So far as the non-local characteristics of this pottery are concerned, discussions continue about its origins, genesis and chronology. Early and later Neolithic pottery of this type has been dated in the taiga forest along the River Ob' at 7,220–6,440 bp (Kosinskaia 2004), and has also been dated in the south forest Tobol region at 7,700–7,440 bp (Timofeev 1996; Zakh 1995). In both these regions, the earliest sites with this flat-bottomed pottery are dated at 9,025–8,630 bp (Kosinskaia 2004; Poshekhonova 2004), and later finds at 5,490 bp (Kovaleva 1989), after which this technological tradition appears to die out. Other typical cultures of the Urals and West Siberia are vertical-ellipsoid vessels with a rounded or convex bottom. These are widely distributed and differ only in proportions. Boat-shaped round bottom vessels (Figure 6.2) are rarely found.

Eneolithic Pottery and the Emergence of the Productive Economy

In Russian archaeology, the Eneolithic period can be broadly defined in southern regions by the very early use of pure copper goods, with a partial transition to farming and cattle-breeding. These innovations spread from the primary centres of metallurgical development out to the peripheral areas of steppe, forest-steppe and finally to the hunting and fishing cultures of the northern forest and tundra. This process also involved considerable

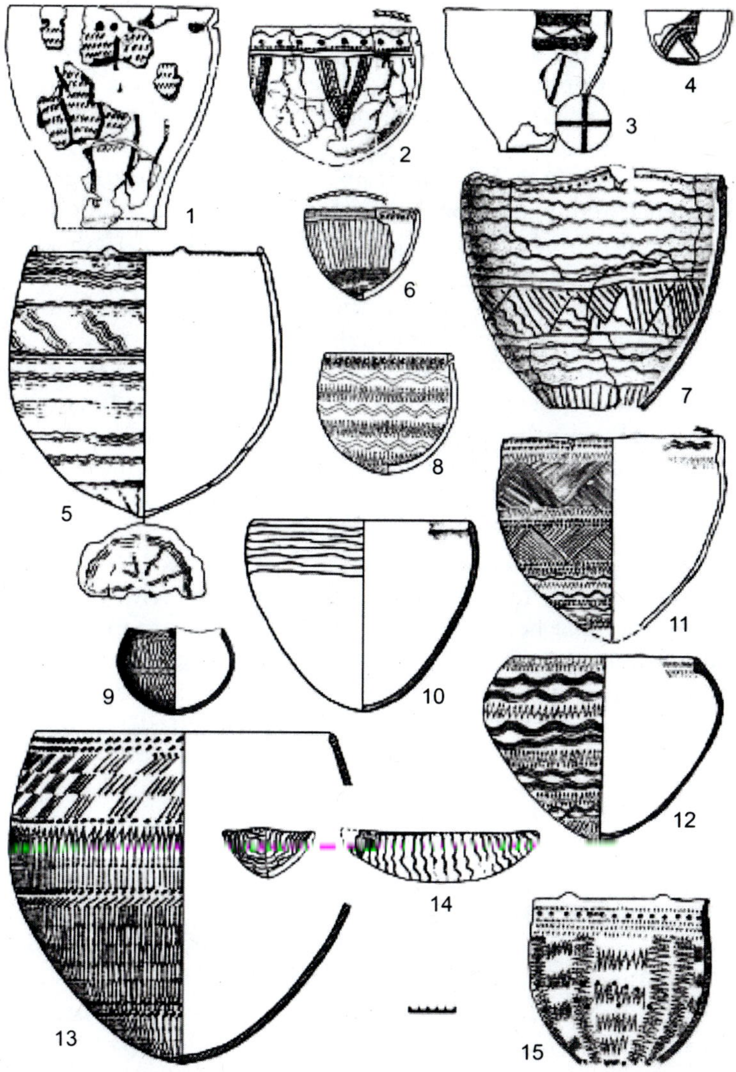

Figure 6.2 Neolithic ceramics from Urals and western Siberia. (1, 2) Amnya I promontory fort; (3) Yurtobor 3 settlement; (4) Mergen 3 settlement; (5) Bystry Kulyogan 66 settlement; (6) Etto I settlement; (7) Sumpania IV settlement; (8) Duvan 5 settlement; (9, 10) Koksharovo Mound; (11, 12) Isetskoe Pravoberezhnoe settlement; (13) Borovoe Ozero I settlement; (14) Barsova Gora II/17 settlement; (15) Kokui I settlement. [1,2, 5, 7 after Kosinskaia, L L (2001, figures 3, 5, 9); 3 after Zakh, V A (1995, figure 2.8); 4 after Zakh, V A and Skochina, S N (2002, figure 5.7); 6 after Kosinskaia, L L (2003, figures 1.3); 8 after Stefanov, V I (1991, figure 5.12); 9, 10 after Shorin, A F (2002, figures 9, 10); 11, 12 after Kerner, V F (1991, figures 3.4, 4.2); 13 after Bader, O N (1951, figure 11); 14 after Chemiakin, Y P (1991, figure 3.6); 15 after Gening, V F and Goldina, R D (1969, table 12.10)].

transformation in all spheres of life, in the economy, social relations, culture and worldview, such that the mastering of copper working was only one element in a wider process of historical change. In short, the Eneolithic can be defined by sites where there is evidence for the use of pure copper objects, although in the taiga zone, the economy remained focused on hunting and fishing, albeit with intensification of production, a move towards a more settled existence and higher population densities. Finally, in the forest zone of the Urals and western Siberia, the introduction of metal coincided with on-going processes of transformation in the older Neolithic cultures, with increasing cultural and ethnic differentiation, re-orientation in social relationships and the formation of new cultural-historical areas, all of which took place in a period of increasing climatic instability. In these changing contexts, the development of new networks of contact, interaction and exchange, linking the northern forests to the centres of metallurgy, were likely to have had far more influence than the low-intensity circulation of limited numbers of copper objects.

In the Eneolithic of the Urals and western Siberia, it is possible to identify several archaeological cultures with distinct earthenware production traditions, all of which show strong continuities with earlier periods. There is considerable heterogeneity in vessel shape, ornamental composition and in the images plotted on vessels, all of which mark out separate areas of ceramic tradition that may have represented distinct ethnocultural communities. For example, during this period the Urals region becomes marked out by the formation of two distinct comb-ware ceramic traditions: the Volosovo-Garino cultural province, situated to the west of the Urals Ridge, and the East Urals-Kazakhstan province, occupying the eastern slope of the Urals and bounded in the east by the upper reaches of the Irtysh, the western spurs of the Altai, and in the north, the Lower Ob' region (Shorin 1999: 23–24).

Eneolithic pottery in the Urals and western Siberia is represented by coil-built hand-moulded jar-shaped vessels with rounded bases, occasionally by vessels with flattened or even flat bottoms. Vessel sides are straight, with rims folded slightly inward or outward, without bulges. The outer surface of vessels is well smoothed, with the inner surfaces smoothed by a toothed scraping tool. Capacity and wall thickness show considerable variation.

Comb-ware traditions prevailed in the Urals and western Siberia. Additionally, stroke-receding pottery decoration technique (rolled design) in the East Urals, and the pit-comb technique in the Tyumen Tobol region were also practised. Comb-pit decorative traditions dominated in the forest-steppe and south taiga areas of western Siberia, the sub-taiga Ishim-Irtysh area, and the Tomsk-Chulym, Vasiugan and Vakh River basin.

Eneolithic pottery was made with a variety of different pastes in the different regions. In the Middle Urals, pots were made from ferruginous clay of mid to low plasticity, usually tempered with sand, some natural mineral inclusions, ground talc, grog of various sizes, and with admixtures of organic material. In the south taiga and forest-steppe zones, the coiled and band-built comb-pit pottery was characterized by the use of monotonous paste recipes, made up of kaolin clays with grog inclusions, the only variation being in the quantity and granulometric composition of the admixtures (Glushkov 1996: 92–93).

Neolithic and Eneolithic Transformations in Pottery Ornamentation

Throughout the Neolithic and Eneolithic periods, pottery vessels were decorated while the clay was still relatively wet. In general, ornamentation was added to the whole outer surface, from the rim to the base, sometimes also around the inner rim, but only rarely to the inner base of the vessel. The only exceptions to this rule are the Neolithic flat-bottomed vessels, which were only decorated in the top and bottom sections, although decoration is entirely absent in some cases.

There appears to have been a common set of ornamentation methods in the Urals-western Siberia Neolithic and Eneolithic pottery, which consisted of using similar tools, employing common designs and following standard ways of filling in the decorative fields with similar design elements, motifs and compositions. Ornamentation tools have occasionally been excavated, and include both pieces of flat stone with teeth cut out to produce comb traces in wet clay, as well as sharpened and toothed bone dies which were used in order to produce similar effects. Other insights into decorative tools have been generated by the study of the impressions themselves. In the Early Neolithic, there was limited use of specially made design tools and the more widespread use of 'natural' objects to produce decoration, including the teeth and jaws of mammals, and the bones of birds and fish. By the end of the Neolithic, the use of natural ornamentation tools decreased, but continued well into the Eneolithic. Pottery of certain regions appears to have been characterized by the preference for some of these tools. For example, sable and marten jaws were used in the Middle East Urals Neolithic; beaver jaws in the Northeast Urals (Kalinina and Ustinova (Gadzhieva) 1995: 78–80); pike jaws and fish spinal columns in the Middle Ob' Region (Bystrino Neolithic culture). Use of these natural tools for the ornamentation of early ceramics in the Urals and western Siberia was part of a much wider phenomenon that extended as far as the pottery-using hunter gatherers of eastern Europe (Kalinina and Ustinova (Gadzhieva) 1993: 87). The wide distribution of this phenomenon (use of natural ornamentation tools) in eastern Europe, the Urals and western Siberia, can probably be related to the economy, and reflects the peculiarities of the worldview of ancient hunter-fishers of the forest zone. There are no grounds to suspect that the use of natural ornamentation in each of these three zones had been adopted from elsewhere.

The region's ceramics are notable for their decoration. A typical technique involved a continuous movement of the ornamentation tool across the clay without taking it off the surface, which is known as the drawn-line technique. Other methods were used, including drawing with periodical pressure ('receding-drawing'), and also a technique with alternating disengagement of one, then the other tool tip known as comb-stepping. Both line drawing and receding drawing were with single-toothed ornamentation tools (called sticks or trowels), and many-toothed tools (for the comb-drawn line, and receding comb techniques). These line drawing techniques prevailed in the pottery of Early Neolithic cultures across the whole region east of the Urals to the Ob' River basin as well, with pricking and stamping techniques occupying an inferior position. In contrast, the pottery of the forest zone in the western Urals was decorated predominantly by comb-stamping, a tradition which extended as far west as the Kama and Volga Rivers. By the Late Neolithic, comb-stamping had also become dominant in areas east of the Urals, as far as the left bank of the Lower Ob'. In the eastern areas of the study region—the right bank of Ob' (the southern forest zone, forest steppe)—the receding, drawn-line and stamping techniques were maintained well into the Eneolithic and Early Bronze age.

In addition, we can track the spread of another design feature, made up of a belt of deep pits running around the outer or inner side of the vessel, just under the rim, and often above another decorative pattern. Initially, this decorative element was more typical of the taiga cultures and is hardly observed in the East Central or western Urals. However, by the Late Neolithic, these pitted designs had also become an integral part of the wider ornamental pattern, and were plotted in parallel rows across the vessel body.

These modes of ceramic techniques persisted from the early pottery of the Neolithic, through to Eneolithic and Bronze Age vessels, prompting Kosarev (1996: 268–269) to distinguish three main ornamentation techniques in the Urals-western Siberia region: the recessive-stroke, comb and comb-and-pit techniques. All of these underwent significant transformations through the course of several millennia, yet remain important diagnostic features marking specific traditions and associated cultural provinces. For example, Eneolithic ceramics decorated with comb-shaped stamping are found in the East Urals and Tyumen area of the Tobol basin, and fall into several cultural types—the Ayat, Shuvakish and Shapkul among others. Differences between these cultures are marked by the shape of the toothed stamp (sub-rectangular, oval, rounded, with one or two rows, straight or inclined tooth cutting, etc.) used to make the impressions, as well as the mode of shaping the form (by stamping, rolling, drawing, stroking, alternating,

alternating-drawing) (Figure 6.3). Debates about the formation, chronology and precise relationships between these cultures remain unresolved.

In addition, there are two other notable ceramic traditions, including vessels of the Lipchinka culture, decorated with recessive-stroke (false corded) ornamentation, ornamentation with a sharp-edged tool (usually animal rib edges, or elbow bones) by means of rolling and recessive-rolling (Kalinina and Ustinova (Gadzhieva) 1988: 15). Finally, the Andreyev culture ceramics had pit-comb ornamentation and were localized in the Tyumen Tobol region. They were distinctive due to belts of comb stamps separated from one another by rows of pressed-in pits (Figure 6.3).

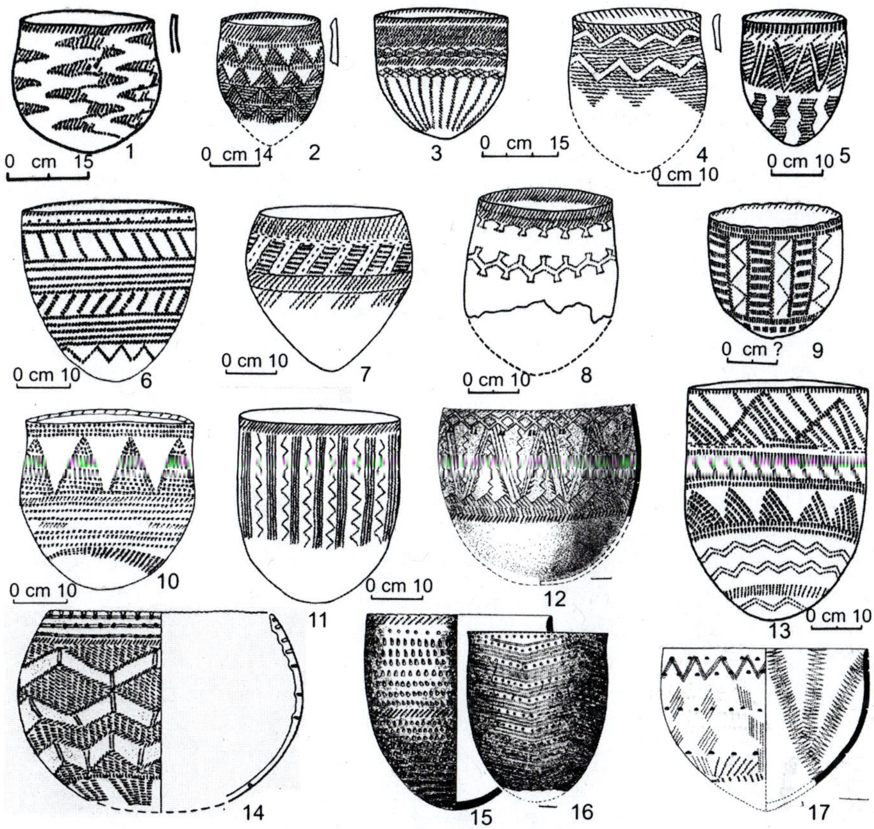

Figure 6.3 Eneolithic ceramics from Urals and western Siberia. (1–11, 13) Ayat settlement; (12) Makusha III settlement; (14) Amnya I fortress; (15) YuAO XII settlement; (16) Vengerovo III settlement; (17) Tukh-Sigat IV settlement [1–13 after Chairkina, N M (2005); 14 after Morozov, V M and Stefanov, V I (1993); 15–16 after Kosarev, M F (1987, figures 90.8, 90.9); 17 after Kiriushin, Y F (2004, figure 58.3)].

Early Ceramics: Decorative Motifs and Compositions

The Neolithic ceramics of the Urals and western Siberia are characterized by a common stock of simple ornamental motifs, which are deployed in different ways according to local traditions. These motifs are straight, wavy and zig-zag lines drawn on vessels either horizontally, vertically or diagonally. Different archaeological cultures are characterized by different stylistic preferences. For example, Kozlov ceramics (Middle East Urals) abound in straight and wavy line motifs; the Sumpania ceramics (North East Urals) have straight and zig-zag lines; and Bystrino pottery (Middle Ob') has wavy lines. In most areas, one finds 'parquet' or 'wattled' patterns made up of inter-penetrating cross-hatched triangles, quadrangles and parallelograms. Research into the comb ware ornaments of the Prikam'e led Kalinina (1974: 170–179) to the conclusion that they originated from different forms of wickerwork.

In addition to these general elements, certain motifs are characteristic of particular cultures. For example, Boborykino (and similar cultures) have triangles with the apexes pointing downwards and vertical zig-zags; Bystrino and Chestiiagskaia cultures have rhombic grids of wavy lines. More figurative images—anthropomorphic, ornithomorphic and tree-like figures—are also found, albeit rarely.

The most widespread type of ornamental compositions on Neolithic vessels is either the use of alternating belts with varying motifs in order to form horizontal decorative zones, or the use of monotonous patterns of horizontal rows of the same repeated motif. A third kind of composition is the vertical division of the ornamental field and the filling of each segment with either repeated motifs or varied ones. There may be combined variants when vertical dividers are located only in the upper part of the vessel body. The pattern on the vessel base may differ in composition from the decoration of the walls, and be designed on the basis of a grid, rosette or radial beam figure. In sum, infinite variations in the decoration of Neolithic pottery were produced through novel combinations of a limited number of motifs produced by different techniques (Figure 6.2).

There are other decorative features on Neolithic ceramics, including vessels with 'ear-like', triangular or semi-oval projections along the rim (Figure 6.4). Others have eyes and a mouth marked by strokes. Less frequently, one can observe zoomorphic heads which are attached to the vessel and are turned inwards or outwards (Figure 6.4). In addition, some Boborykino and Satyga vessels had stuck-on ornamental rollers surrounding the vessel top (Figure 6.4) (Khlobystin 1993: 30–32), and Bystrino pottery is characterized by thorn-like (nipple-like) finishing of the rounded vessel bottoms, which makes them look like female breasts (Figure 6.4).

Pottery from the later Eneolithic of the Urals and western Siberia becomes more varied. The Urals populations practised a broad range of

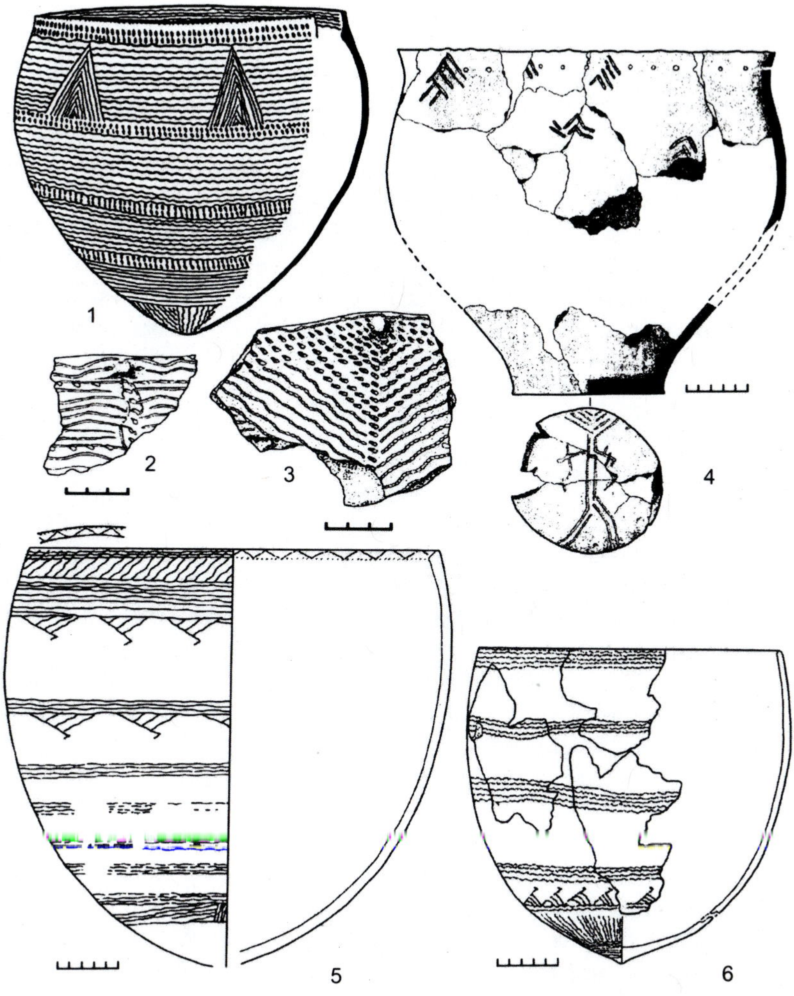

Figure 6.4 Anthropomorphic, zoomorphic and other images on Neolithic ceramics from Urals and western Siberia. (1) Evstyunikha I settlement; (2) Makhtyli Hill; (3) Koksharovo Hill; (4) Pikushka I settlement; (5, 6) Bystry Kulyogan 66 settlement [1–3 after Kovaleva, V T (2002, figures 1.1, 4.4, 4.6); 4 after Kovaleva, V T and Shilov, S N (2002, figure 1.1); 5, 6 after Kosinskaia, L L (2001, figures 1, 4)].

decorative traditions while the pottery of some archaeological cultures (e.g., the Shuvakish and Shapkul) was covered with monotonous belts of horizontal decorative zones. In other areas, the ornamentation (e.g., of Ayat, Atymya, Surtanda and other cultures) was made up of horizontal belts filled with geometric figures (triangles, rhomboids, quadrangles, honeycombs and other complex figures) (Figure 6.3). Lipchinka type pottery was characterized by a combination of vertically and horizontally zoned elements and the usage of specific elements and pattern motifs (Figure 6.3). Less diverse are the decorative compositions on comb-pit and pit-comb pottery of western Siberia. Generally, this decoration consists only of horizontal belts separated by rows of pressed-in pits. Finally, amongst the pottery recovered from the Eneolithic settlement of Tukh-Sigat IV (Vasiugan River), vessels have stuck-on images resembling animal heads (or possibly a beak) (Kiriushin 2004: 49). On the outer surface of a vessel recovered from the Volvoncha I site in the northern part of the East Urals, there are images of snakes made with decorated clay bands that produce a characteristic wavy shape (Koksharov 1990: 20).

Ornamentation Semantics

Ornamentation of early vessels from the Urals and western Siberia also provides some general insights into the spiritual worlds of the region's prehistoric hunter-fisher-gatherer communities. Semantic investigation of Neolithic and Eneolithic pottery has proceeded in two directions. According to the first, ornamentation traditions were related to mythological images that formed part of a totemic worldview. As such, the most ancient notion of the animal-totem was determined by the protective function of its teeth, and by bones which served simultaneously as amulet-protectors, but also as tools for inscribing pottery decoration. The semantic meaning of the ceramic ornamentation is linked to traces of the totem-beast made over the vessel, these marks symbolically covering the vessel with skin and providing protection (Kalinina 1998a: 96–105; 1998b: 116–123; 2000: 55–64). Other marks like 'ears', animal heads on rims and nipple-like images at the base provide further evidence of totemic belief, for example, in making conceptual links between the breast and nurturing (Kalinina 2000: 56). It is suggested that the semantic ideas of the ancient masters consisted of the following: the impressions of teeth, jaw or bones of the totem animal corresponding to an element, or token (for example, the tracks, skin, etc), of that animal. For this reason, in carrying these patterns of natural ornamentation, the ancient master 'clothed' the vessel or marked it with the signs of the totem animal, and in that way the decoration acquired a protective function.

According to the second, individual elements of decoration are likened to elements in a symbolic structure, or encoded text, which together form illustrations of cosmological myths or expressions of other fundamental beliefs about the universe. The picture of the world is represented on vessels as individual mythologems. One distinctive composition found on Evstyunikha-type pottery (from the forest zone of the East Ural mountains) has been interpreted as the World Mountain surrounded by primary waters (Figure 6.2). On Koshkino vessels, ornaments have been translated as symbols representing the World Tree, with wavy lines representing falling waters (Figure 6.2) (Kovaleva 2002: 6–14). Similar symbols have been observed on Boborykino (Middle East Urals) and Amnya (the lower Ob' region) pottery. Koksharovo-Yuryino pottery (Middle East Urals) is also characterized by vertical divisions, perhaps representing individual images of a tree/ladder/pole connecting different world levels, and often accompanied by a zoomorphic image attached to the vessel rim (Figure 6.4). The Boborykino ceramics also contain other symbols, including lens-shaped figures which may represent a seed/bud/embryo, and vertical zig-zags which are reminiscent of either snakes or lightning bolts (Viktorova 2002a: 86–90).

During the Eneolithic there is an increase in the kinds of decorations which may be linked to the expression of cosmological beliefs. Vessels with tree-like decorations are present in all large assemblages from the Middle and south western part of Siberia. The designs vary from high triangles with their apexes turned upward, or quadrangles (trunks) with a fringe (branches). Sometimes the pictures are more realistic, with trunks represented as straight or slightly rippling lines, and the crown as a number of straight short lines, with angles directed downwards or upwards. Slightly different tree styles are depicted on Volvoncha pottery from the Konda basin (Koksharov 1990: 19). In addition, a number of images on East Urals Eneolithic pottery can be interpreted as mountains, and these are especially common on Lipchinka-type ceramics. In contrast, however, there are few pictures of reptiles or snakes on the Eneolithic pottery of East Urals. Lipchinka pottery probably have snake representations in the form of individual triangular zig-zags and ripples placed vertically on the ornamental field of vessels. They often form impressive compositions in combination with high triangles (mountains?) and tree-like figures which serve as either decorative zone dividers or independent images.

Amongst Neolithic vessels we find anthropomorphic figures only on Boborykino-type vessels (Figure 6.4). Of special interest is a fully preserved outline with a tree-like symbol instead of a head (Kovaleva and Shilov 2002: 119–126). More examples of anthropomorphic images are found in the Eneolithic, for example, in the Middle Urals (Figure 6.7) (Baraniy Mys site) and northern East Urals (Volvoncha I settlement), but these are stylistically different from the Boborykino type—front-facing human figures with their legs bent at the knee (Koksharov 1990: 4–28)

The earliest ornithomorphic plots—and so far the only ones recovered from the Urals and West Siberian Neolithic—have been recovered from Bystrino-type pottery from the Surgut Ob' region. Profiled figures of the waterfowl, turned up and down, make up friezes along the vessel walls (Figure 6.4). Possibly, they are depictions of an ancient Proto Uralic world origin myth about the diving bird bringing up soil from the bottom of the primary ocean (Kosinskaia 2001: 57–60).

The waterfowl image enjoys much wider usage on Eneolithic pottery in the Urals, although mainly on Shuvakish-type pottery (Middle East Urals). Here, birds are made with a combed stamp and make up a separate frieze, and are generally placed in the upper part of a vessel's body. Each figure is given a sense of individuality and movement by the careful depiction of its head and neck in a different position (Figure 6.5). Limited numbers of very distinct waterfowl images are found on Lipchinka-type stroke-recession pottery from the Razboinichii Ostrov site in the Middle East Urals (Figure 6.5), and ornithomorphic images have also been observed on the pottery from the Chebarkul site of the southern Urals, and from the Leushi XIX and Leva XII settlement sites in the Konda basin in the northern East Urals. These are highly specific and very different from the Middle East Urals images (Chairkina 1998: 81–104).

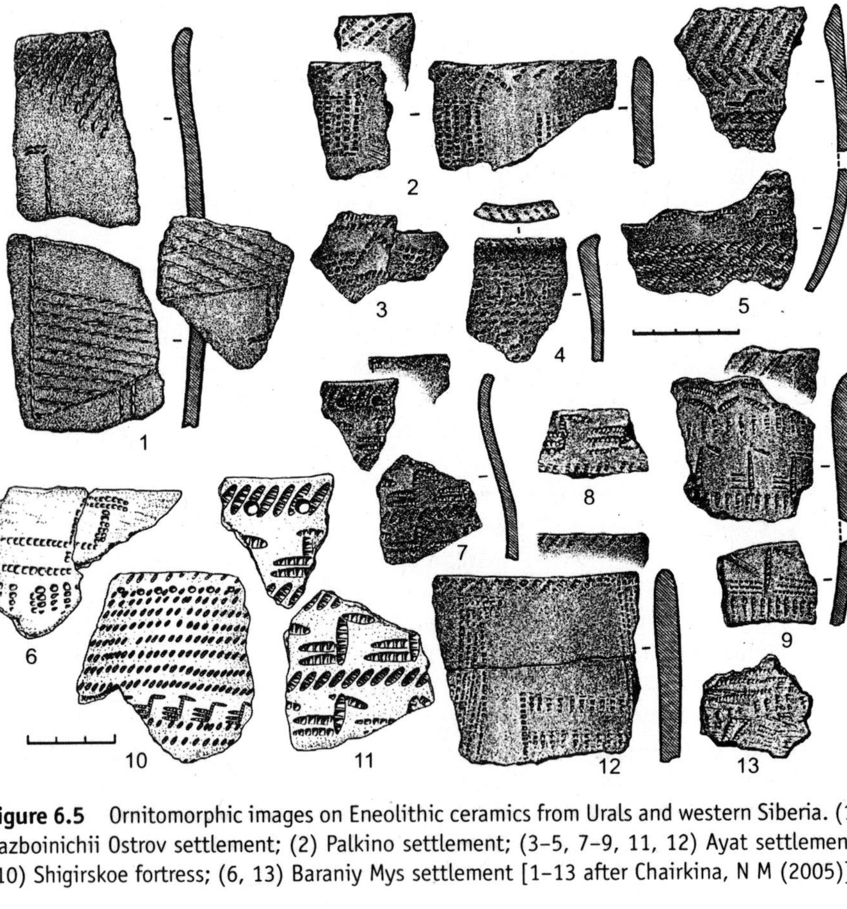

Figure 6.5 Ornitomorphic images on Eneolithic ceramics from Urals and western Siberia. (1) Razboinichii Ostrov settlement; (2) Palkino settlement; (3–5, 7–9, 11, 12) Ayat settlement; (10) Shigirskoe fortress; (6, 13) Baraniy Mys settlement [1–13 after Chairkina, N M (2005)].

Planar depictions of hoofed animals appear only in the Eneolithic age, and they have been observed in the Middle East Urals territory on three Ayat-type Eneolithic vessels from the Karas Lake, Koptyaki 9 and Baraniy Mys sites. The pictures of roe deer are made with a comb stamp, and their stylistic manner is distinguished by contour modelling of figures with four feet, a skeletal interior to the body, and sometimes with line gaps on the animal's head between horns. In front of one of the deer, the frieze is interrupted with triple vertically located zig-zags that also fall into the next zone, with the alternating hollow and cross-hatched rhomboids probably symbolizing traps. There is also the above-mentioned anthropomorphic figure on a pot of the Baraniy Mys site, plus a solar sign in the form of radiating rays whose positioning in relation to the roe deer figures and pictures of traps is not clear (Figure 6.7). The animal on the rim of the pot from Vuryino IV site is portrayed in a somewhat different manner with a long body, long neck, raised head and tail. Serikov (1978: 245–248) considers this to be a running deer or elk. Graphics from the Karasye Lake pots are stylistically closer to the hoofed animals (reindeer) portrayed on a pot from Amnya I in the Lower Ob' region (Morozov and Stefanov 1993: 143–170) (Figure 6.7).

Neolithic and Eneolithic Ceramic Figurines

In addition to pottery vessels, other types of earthenware figurines, dating to the Neolithic and Eneolithic Ages, have been excavated in the Urals and western Siberia. However, these items are found only rarely and are not typical of all the region's archaeological cultures. For example, in the Neolithic dwelling complexes of the Middle East Urals and Tyumen Ob' regions, several kinds of ceramic sculpture have been recovered, including bird figurines, sculptures of bear heads (Evstyunikha-type ceramics assemblages) (Usacheva 1998a: 106–107) and a small anthropomorphic head (Koshkino culture) (Kovaleva and Zyrianova 1998: 177–178) (Figure 6.6). In addition, small-sized ceramic balls and a bird's head made of stone have been excavated from Neolithic contexts at what appears to have been a cultic site at Koksharovsky Hill in the Middle East Urals (Serikov 2002: 131; Shorin 2002: 268). A clay sculpture of a bear's head was also found in the Eneolithic layer of the Shuvakish I settlement (Middle East Urals) (Figure 6.7). Finally, the Shaitanskoye Lake I site (Middle East Urals) also contains several unusual items carved out of Neolithic and Eneolithic pot sherds, which include flat profiled figurines of fish, bird and elk. Despite the mixed ages of the ceramic material, Serikov has suggested that these were part of an Eneolithic cultic assemblage which was ritually deposited at the site (Serikov 2002: 133–134).

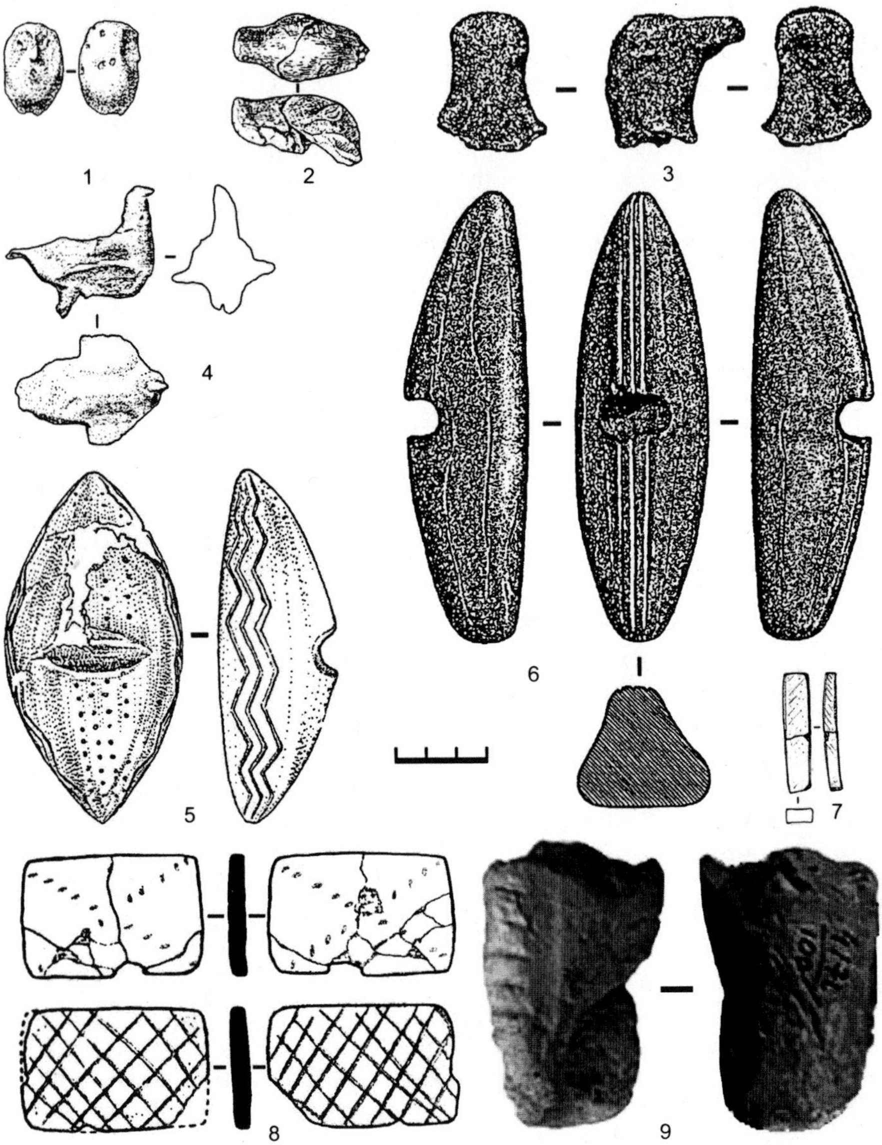

Figure 6.6 Clay plastic art and other artefacts of Neolithic sites from Urals and western Siberia. (1) YuAO XIIIa settlement; (2) Evstyunikha I settlement; (3, 6) Koksharovo Hill; (4) YuAO XV settlement; (5) YuAO XII settlement; (7) Bystry Kulyogan 66 settlement; (8) Serebryanka I settlement; (9) Chernaya 3 settlement [1,5 after Kovaleva, V T and Zyrianova, S Y (1998, figures 4.2, 6.2); 2, 4 after Usacheva, I V (1998a, figures 1.3, 5); 3, 6 after Shorin, A F (2002, figures 5.2, 6); 7, 9 after Kosinskaia, L L (unpublished); 8 after Panfilov, A N (1993, figure 3.10)].

In the broader region, there are wide distributions of so-called 'shaft straighteners' which are oval- or rhombus-shaped with a transverse groove (Usacheva 2005). These are made of stone but also include ceramic versions which are usually decorated. These have originated from dwelling contexts in the Koshkino, Boborykino and other archaeological cultures, and have also been found within the cultic assemblages deposited at Boborykino sanctuary at the Palatki I site (Viktorova 2002b: 46–66) and at the multi-cultural site of Koksharovsky Hill (Figure 6.6). The clay 'shaft straighteners' are not found in Eneolithic contexts.

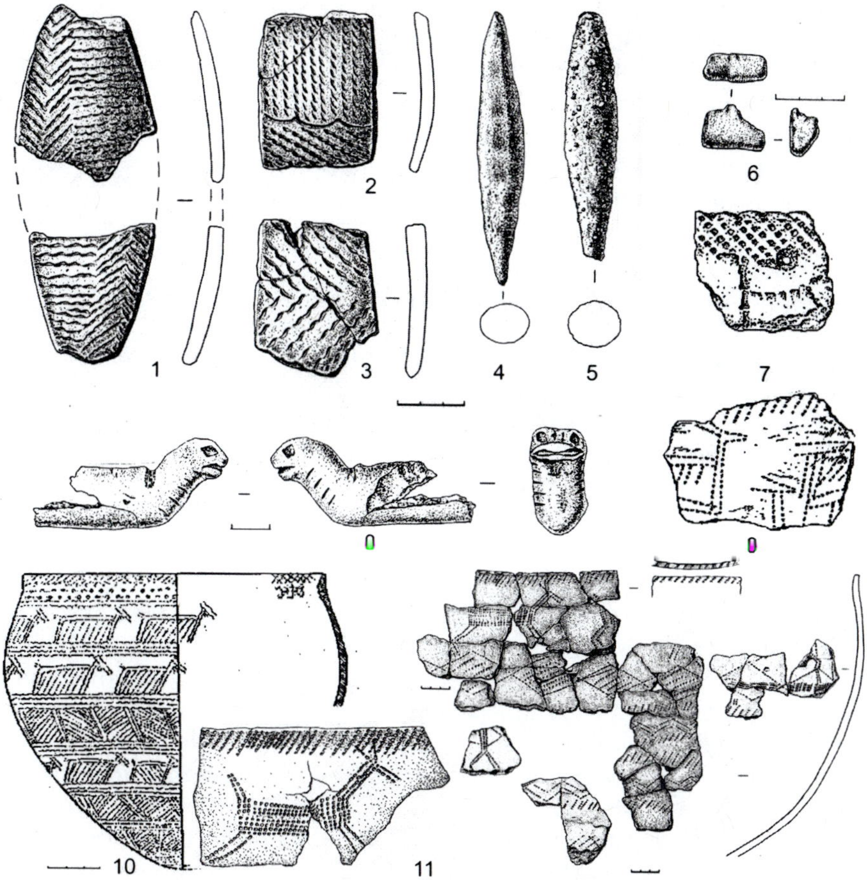

Figure 6.7 Clay plastic art and other artefacts of Eneolithic sites, anthropomorphic and zoomorphic images on Eneolithic ceramics from Urals and western Siberia. (1–3) Razboinichii Ostrov settlement; (4–6) Shuvakish I settlement; (7) Karas lake settlement; (8) YuAO VIII settlement; (9) Koptyaki 9 settlement; (10) Amnya I fortress; (11) Baraniy Mys settlement [1–7, 9, 11 after Chairkina, N M (2005); 8 after Usacheva, I V (1998a, figure 1.1); 10 after Morozov, V M and Stefanov, V I (1993)].

Other kinds of ceramic artefacts have also been found across the region. In the forest-steppe of the Irtysh region, two rectangular clay tablets of unknown purpose have been excavated. These have Neolithic Kokuy type pottery decorations on both sides (Panfilov 1993: 9). Clay bars have also been found. These abrasive tools have strong use-wear traces on their sides (Figure 6.6) and have been found in association with flat-bottomed pottery in Neolithic dwellings of the Middle and Lower Ob' regions, and differently shaped, sometimes decorated, bars have been found with Bystrino type pottery in the Surgut Ob' region. Occasionally, similar bar-like items have been found at Eneolithic sites.

Clay was also used to make fishing net sinkers. In numerous Eneolithic settlements of the Urals and western Siberia, various forms of fishing net sinkers have been recovered which include (Figure 6.7) both 'biconic' and 'cigar-shaped' versions, some of the latter with forked and 'cap-like' ends. The size and weight of the sinkers varied significantly from region to region, and some versions were decorated.

Discussion: Pottery Use in the Urals and Western Siberia

Pottery from the Neolithic and Eneolithic periods is usually found within domestic contexts of the dwelling sites which suggests that its primary role was a form of household ware. The wide range of pot sizes and capacities, and both the presence and absence of cooked food residues on different pots suggests that ceramic vessels were being used in a range of different roles including storage, cooking and serving food and drink. Sherds from broken vessels were recycled in the form of grog, the most widely used inclusion in the clay pastes of all Neolithic and Eneolithic cultures of the region. In the boggy northern parts of western Siberia, there are few good local sources of stone, and so broken sherds of various shapes were reused as scraping and abrasive tools, probably in skin-working tasks (Surgut and the Lower Ob' areas). In other areas, for example, in the Bystrino Culture of the Surgut and Lower Ob' Region, sherds were later shaped into bars which possibly served as whetstones (Figure 6.6). Two small discs, each with narrow holes bored into the centre and the edges chipped and polished, have been interpreted as spindle whorls by Serikov. These were recovered from sites belonging to the Evstyunikha culture (Middle East Urals) and similar items have been recovered from a range of Urals and western Siberian Eneolithic sites.

In the Eneolithic layers of the Razboinichii Ostrov site, 36 sub-rectangular and oval-elongated articles were excavated. These were manufactured from the potsherds of large vessels and were subjected to cutting and edge-smoothing to produce what were probably spatulas. The original ornamentation was retained (Figure 6.7).

Pots were also used in early metallurgical practices in the Eneolithic Period. Small vessels, and sometimes potsherds from larger vessels, have been recovered with traces of slag and also drops of copper adhering to the sides. These finds are particularly frequent at Razboinichii Ostrov and Shuvakish I sites in the Middle East Urals, but also occur in other archaeological sites of the Urals.

Despite the frequent reuse of broken pottery, diverse lines of evidence suggest that the vessels themselves appear to have had an intrinsic value and were probably handled carefully and cautiously. For example, there are frequent indications that attempts were made to repair damaged vessels. Vessels recovered from Gorbunovo peat-bog in the Middle East Urals had rows of parallel holes picked or drilled along either side of cracks, and cords have been passed through these holes to 'sew' up the damage. In one Bystrino-type dwelling of the Surgut Lower Ob', smashed pots had been placed along the walls so that they would not be crushed by the treading of feet. These insights may even suggest that the handling of pottery was guided by both sacral meanings as well as more pragmatic concerns. The special status of some pots is evidenced by distinct forms of decoration whose use may have differed from those in more functional roles. Certain vessels may also have been used as a form of material currency in inter-group practices of gift exchange. In one dwelling of the Bystry Kulyogan 66 site which forms part of the Bystrino culture of the Surgut Ob' area, pottery ornamented by the use of fish bones had been concentrated around only one of the house's two fireplaces, and sherds from large vessels had been scattered over the floor just prior to final abandonment. Usacheva (1998b: 115–116) has argued that isolated ceramic fragments left near the porch and along the walls of a seasonally occupied Eneolithic Sazyk IX campsite of the Tyumen Tobol River region, are the remains marking one's 'own' dwelling space, as opposed to the 'alien' forest one. A similar picture is observed in isolated burials of the forests in East Urals, either within sites or in cave contexts. Serikov (1998: 34, 42, 141) has suggested that small pottery fragments from the Neolithic burials of the Kamen Dozhdevoy grotto, located on the Chusovaya River, Middle Urals, originally functioned as symbols of tribal affiliation, and perhaps even formed elements of a shamanic costume. One potsherd was found near the shoulder of a man buried on Leushi XIV site (Eneolithic, northern East Urals) (Skvortsovskaia and Starostina 1986: 35) and a similar picture has been documented in a child's burial on the Skvortsovskaia mountain Y site (Eneolithic, Middle East Urals). However, placing unbroken pots into a grave is more typical of the burials of the Late Neolithic-Eneolithic Ages of the forest-steppe belt of western Siberia.

Ceramic vessels have been found alongside clay 'shaft straighteners' and small clay items on what appear to have been Neolithic ritual sites. One such assemblage has been recovered from the Palatki I settlement of the

Middle East Urals, where the remains of 13 Boborykino type pots were excavated (Viktorova 2002b: 47–53). In the forests of the East Urals, there are specific cultic sites marked by multi-layered mounds. In Koksharovo mound, inverted Boborykino and Koshkino type vessels have been placed either inside or beside small wooden structures which were then burnt and covered in soil. Repetitions of this action led to the gradual building of the mound (Shorin 2002: 266–275). Many Koksharovo-Yuryino type vessels were also found in the mound's lower layers and had zoomorphic images upon the rims. These figures were moulded and attached during the process of making the pot.

Clay 'shaft straigteners' from cult assemblages as well as ones from dwellings are often broken. In Viktorova and Kerner's (1998: 63–80) opinion, they were, as opposed to stone analogues, of sacral meaning and were subjected to fragmentation in the course of ritual actions, although the exact nature of the ceremonies during which they were broken is not clear.

Specialized cult and sacrificial sites in the East Urals during the Eneolithic age were characterized by a complicated structure and, apparently, some hierarchy: cave sacred places, sacred places (observatories?), sets of petroglyphs. However, the potsherds are, as a rule, not different from pottery at domestic sites. The mere fact of the location of all clay articles, even votive ones, not only at specialized sacrificial and cult places, but also at sites and settlements (at their cult places?) testifies, probably, to the integrity and syncretism of material and spiritual spheres of ancient culture.

Conclusion

In the Urals and western Siberia, ceramics appear in the middle of the Atlantic period, not later than the forest zone of eastern Europe according to the radiocarbon dates. It is supposed that some elements of these early ceramics may have originated in the Urals and West Siberian territory as a result of migrations or cultural impulses from the south, southern Europe or the Near East, but probably not from eastern Siberia. Additional aspects of ceramics arose through more local processes, possibly as a result of the adoption of ceramic technology, into which the local population introduced their own technological and decorative innovations. However, during the earlier stages of the Neolithic, complex vessel ornamentation was characteristic only of ceramics that may have been introduced into the region from other non-local cultures.

Through time, the interactions between these local and non-local traditions led to the constitution of new archaeological cultures during the Eneolithic, which were characterized by pottery with complex ornamentation and figurative images. Notwithstanding external influences, the cultural development of the region proceeded gradually, and Neolithic and Eneolithic cultures were connected by continuity in core traditions. These formed a powerful aboriginal stratum which provided the basis for cultures at the start of the Bronze Age, and, seemingly, have a direct relationship to the ethnogenesis of the Samoyedic and Ugrian populations that are recorded in the Urals and western Siberia from the ethno-historic period onwards.

References

Bader, O N (1951) 'Stoianki Nizhneadischevskaia i Borovoe ozero I na r. Tchusovoi [The Nizhneadischevskaya and Borovoe Lake I sites on the Tchusovaia River]', in Tchernetsov, V N (ed), *Materialy i Issledovaniia po Arkheologii Urala i Priuralia, Volume 3 (MIA, 22),* Moscow: Akademiya Nauk SSSR, pp 7–22

Chairkina, N M (1998) 'Antropo- i zoomorfnye obrazy eneoliticheskikh kompleksov Srednego Zaural'ia [Anthropo- and zoomorphic images of the Eneolithic complexes of the Middle East Urals]', in Kovaleva, V T (ed), *Voprosy Arkheologii Urala 23*, Ekaterinburg: Ural'skii Universitet, pp 81–104

——— (2005) 'Eneolit Srednegó Zaural'ia [Eneolit the Middle East Urals]', in Shorin, A F (ed), Ekaterinburg: UrO RAN

Chemiakin, Iu P (1991) 'Neoliticheskaia stoianka Barsova Gora II/17 [The Barsova Gora II/17 site]', in Krizhevskaia, L Ia (ed), *Neoliticheskie Pamyatniki Urala*, Sverdlovsk: Uralskoe Otdelenie Akademii Nauk SSSR, 184–196

Dunin-Gorkavitch, A A (1996) [1903] *Tobol'skii Sever [The Tobol'sl North]*, Moscow: Liberia Publishing House

Fedorova, E G (2000) *Rybolovy i Okhotniki Basseina Obi: Problemy Formirovaniia Kul'tury Khantov i Mansi [Fishermen and Hunters of the Ob' Basin: Problems in the Formation of the Culture of the Khanty and Mansi]*, Saint Petersburg: Evropeiskii Dom

Gening, V F and Goldina, R D (1969) 'Poselenie Kokui I [The Kokui I settlement]', in Gening, V F (ed), *Voprosy Arkheologii Urala 8*, Sverdlovsk: Uralskii Universitet, pp 30–47

Glushkov, I G (1996) 'Keramika kak arkheologicheskii istochnik [Ceramics as an archaeological source]', in Molodin, V I (ed), Novosibirsk: Institute of Archaeology and Ethnography, Siberian Branch of Russian Academy of Sciences

Golovnev, A V (1993) *Istoricheskaia Tipologiia Khoziaistva Narodov Severo-Zapadnoi Sibiri* [Historical Typology of Subsistence among the Peoples of Northwest Siberia], Novosibirsk: Novosibirsk University Press

Kalinina, I V (1974) 'Ornamentatsiia keramiki Volgo-Kamskogo neolita [Pottery ornamentation of the neolithic in the Volga-Kama region]', 4 *Sovetskaia Arkheologiia*, 170–179

——— (1998a) 'Totem v strukture obraznogo myshleniia [The totem in the structure of figurative thinking]', in *Teoriia i Metodologiia Arkhaiki*, Saint Petersburg: Sankt-Peterburgskii Universitet, pp 96–105

Kalinina, I V (1998b) 'Semantika i tekhnologiia drevnikh ornamentov [Semantics and technology of ancient ornaments]', in Chernykh, E N (ed), *Tverskoi Arkheologicheskii Sbornik* 3, Tver': Tverskoi Gosudarstvennyi Oblastnoi Muzei, pp 116–124

——— (2000) 'Zver'-totem v tekhnologii obrazov [The beast-totem in the technology of images]', in *Teoriia i Metodologiia Arkhaiki* 2, Saint Petersburg: Sankt-Peterburgskii Universitet, pp 55–64

Kalinina, I V and Ustinova (Gadzhieva), E A (1988) 'Opyt ispol'zovaniia metoda eksperimental'nogo modelirovaniia dlia tekhnologicheskogo analiza keramiki (po materialam neoliticheskikh pamiatnikov Polydenka I i Khutorskaia) [Use of the experimental modeling method for the technological analysis of ceramics (on the materials of the Neolithic sites Poludenka I and Khutorskaia)]', 29 *Arkheologicheskii Sbornik Gosudarstvennogo Ermitazha*, 5–17

Kalinina, I V and Ustinova (Gadzhieva), E A (1993) 'Arkhaicheskie ornamentiry dlia keramiki [Archaic ornamentation tools for ceramics]', in Gracheva, G N (ed), *Arkheologicheskie Izyskaniia* 10, Saint Petersburg: 'Farn', pp 83–94

Kalinina, I V and Ustinova (Gadzhieva), E A (1995) 'Ispol'zovanie cheliustei zhivotnykh dlia ornamentatsii drevnei keramiki [Use of animal jaws for ornamentation of ancient ceramics]', 2 *Rossiiskaia Arkheologiia*, 69–83

Kerner, V F (1991) 'Poselenie Isetskoe Pravoberezhnoe [The Isetskoe Pravoberezhnoe settlement]', in Krizhevskaia, L Ia (ed), *Neoliticheskie Pamyatniki Urala*, Sverdlovsk: Uralskoe Otdelenie Akademii Nauk SSSR, pp 46–67

Khlobystin, L P (1993) 'Satyginskii tip keramiki Zapadnoi Sibiri [Ceramics of Satyginskii type of western Siberia]' in Gracheva, G N (ed), *Arkheologicheskie Izyskaniia* 10, Saint Petersburg: 'Farn', pp 29–35

Kiriushin, Y F (2004) *Eneolit i Bronzovyi Vek Iuzhno-Taezhnoi Zony Zapadnoi Sibiri [The Eneolithic and Bronze Ages of the south taiga western Siberia]*, Derevianko, A P (ed), Barnaul: Altai Universitet

Koksharov, S F (1990) 'Opyt rekonstruktsii nekotorykh mifologicheskikh predstavlenii kondinskogo naseleniia epokhi eneolita [On reconstruction of some mythological ideas of the Konda people of the Eneolithic Age]', in Viktorova, V D (ed), *Materialy po Izobrazitel'noi Deiatel'nosti Drevnego Naseleniia Urala,* Sverdlovsk: Ural'skoe Otdelenie Akademii Nauk SSSR, pp 4–28

Kosarev, M F (1987) 'Zapadnaia Sibir v perekhodnoe vremia ot neolita k bronzovomu veku [Western Siberia during the transitional period from the Neolithic to the Bronze Age]', in Bader, O N, Krainov, D A and Kosarev, M F (eds), *Epokha Bronzy Lesnoi Polosy SSSR,* Moscow: Nauka, pp 252–267

——— (1996) 'Neolit Vostochnogo Zaural'ia i Zapadnoi Sibiri [The Neolithic of Eastern East Urals and western Siberia]', in Oshibkina, S V (ed), *Arkheologiia. Neolit Severnoi Evrazii*, Moscow: Nauka, pp 253–269

Kosinskaia, L L (2001) 'Obraz vodoplavaiushchei ptitsy na keramike bystrinskogo tipa [Waterfowl image on the Bystrino-type ceramics]', in Zakh, V A (ed), *Problemy Izucheniia Neolita Zapadnoi Sibiri*, Tyumen: IPOS SO RAN, pp 57–60

——— (2002) 'O nekotorykh problemakh neolitovedeniia Urala i Zapadnoi Sibiri [On some problems of the Neolithic studies in the Urals and western Siberia]', in Golovnev, A V (ed), *Northern Archaeological Congress Papers September 9–14, 2002, Khanty-Mansiisk,* Ekaterinburg: 'Akademkniga', pp 172–179

Kosinskaia, L L (2003) 'Keramika ettovskogo tipa v neolite severa Zapadnoi Sibiri [The Ettovski type of ceramics in the Neolithic of Northwestern Siberia]', in Neskorov, A V (ed), *Ugry Materialy VI Sibirskogo Simpoziuma Kulturnoe Nasledie Narodov Zapadnoi Sibiri (9–11 Dekabrya 2003 g., Tobolsk),* Tobolsk: Tobolski gosydarstvennyi istoriko-arkhitekturnyi muzei-zapovednik, pp 95–101

——— (2004) 'Problemy khronologii zapadnosibirskogo neolita (k voprosu o roli radiouglerodnogo datirovaniia) [Problems of chronology of West-Siberian Neolithic (on the rule of the radiocarbon dating)]', in Timofeev, V I and Zaitseva, G I (eds), *Problemy Khronologii i Etnokulturnykh Vzaimodeistvii v Neolite Evrazii,* Saint Petersburg: IIMK RAN, pp 301–307

Kovaleva, V T (1989) 'Neolit Srednego Zaural'ia [The Neolithic of Middle Urals]', Serikov, Y B (ed), Sverdlovsk: Ural'skii Universitet

——— (2002) 'Arkhaicheskie modeli mirozdaniia zaural'skogo naseleniia (po arkheologicheskim istochnikam [Archaic world creation models of the East-Ural peoples]', in Kovaleva, V T (ed), *Voprosy Arkheologii Urala 24*, Ekaterinburg: Ural'skii Universitet, pp 4–45

Kovaleva, V T and Shilov, S N (2002) 'Proobraz Indry: ob' interpretatsii antropomorfnogo izobrazheniia na sosude [The prototype of Indra: on the interpretation of an anthropomorphic image on a vessel]', in Kovaleva, V T (ed), *Voprosy Arkheologii Urala 24*, Ekaterinburg: Ural'skii Universitet, pp 119–126

Kovaleva, V T and Zyrianova, S Y (1998) 'Istoriografiia i diskussionnye problemy boborykinskoi kultury [Historiography and the discussions on the Boborykino culture]', in Kovaleva, V T (ed), *Voprosy Arkheologii Urala 23*, Ekaterinburg: Ural'skii Universitet, pp 162–183

Molodin, V I (1992) 'Drevnee iskusstvo Zapadnoi Sibiri (Ob'–Irtyshskaia lesostep') [Ancient art of western Siberia (the Ob'–Irtysh forest-steppe)]', Derevianko, A P (ed), Novosibirsk: Nauka

Morozov, V M and Stefanov, V I (1993) 'Amnya I–drevneishee gorodishche Severnoi Evrazii? [Amnya I–the most ancient fortress in the northern Eurasia?]', in Kovaleva, V T (ed), *Voprosy Arkheologii Urala 21*, Ekaterinburg: Ural'skii Universitet, pp 143–170

Panfilov, A N (1993) *Mnogosloinoe Poselenie Serebryanka I v Nizhnem Priirtysh'e. Itogi Polevykh Issledovanii [The Multi-Layer Settlement of Serebryanka I in the Low Irtysh Region. Results of the Field Research]*, Tyumen: IPOS SO RAN

Poshekhonova, O E (2004) 'Osobennosti keramicheskogo proizvodstva u neoliticheskogo naseleniia Tobolo-Ishim'ia [Peculiarities of the ceramics production among the Neolithic peoples of the Tobol-Ishim region]', in Bagashev, A N (ed), *Vestnik arkheologii, antropologii i etnografii 5*, Tyumen: IPOS SO RAN, pp 20–26

Serikov, Y B (1978) 'O nakhodkakh proizvedenii iskusstva na stoianke Yurino IV (Zaural'e) [On finds of pieces of art on the Yurino IV site (East Urals)]', 4 *Sovetskaia arkheologiia*, 245–248

——— (1991) 'Ural'skie Zori II—odnosloinyi neoliticheskii pamiatnik novogo tipa [Ural'skie Zori II—a new type one-layer Neolithic site]', in Krizhevskaia, L Y (ed), *Neoliticheskie pamiatniki Urala*, Sverdlovsk: Ural'skoe Otdelenie Akademii Nauk SSSR, pp 32–45

——— (1993) 'Issledovanie grota na Kamne Dozhdevom (r. Chusovaya) [Investigations of a grotto on the Kamen Dozhdevoy site (the Chusovaya river)]', in Kovaleva, V T, *Voprosy arkheologii Urala 21*, Ekaterinburg: Ural'skii Universitet, pp 120–143

Serikov, Y B (1998) 'Shamanskie pogrebeniia Zaural'ia [The shaman burials of the East Urals]', in Kovaleva, V T (ed), *Voprosy Arkheologii Urala 23*, Ekaterinburg: Ural'skii Universitet, pp 29–47

——— (2002) 'Proizvedeniia pervobytnogo iskusstva s vostochnogo sklona Urala [Pieces of prehistoric art from the eastern slopes of the Urals]', in Kovaleva, V T (ed), *Voprosy Arkheologii Urala 24*, Ekaterinburg: Ural'skii Universitet, pp 127–150

Shorin, A F (1999) 'Eneolit Urala i sopredel'nykh territorii: problemy kul'turogeneza [The Eneolithic Age of the Urals and bordering territories: problems of cultural genesis]', Ekaterinburg: UrO RAN

——— (2002) 'Koksharovskii Holm: osnovaniia i vozmozhnosti istoricheskoi rekonstruktsii [The Koksharovo Hill: justification and possibilities for the archaeological-historical reconstruction]', in Golovnev, A V (ed), *Northern Archaeological Congress Papers, September 9–14, 2002, Khanty-Mansiisk*, Ekaterinburg: 'Akademkniga', pp 266–274

Skvortsovskaia, E M and Starostina, E E (1986) 'Pogrebenie v Nizhnem Priirtysh'e [A burial ground in the Lower Irtysh region]', in Kovaleva, V T (ed), *Problemy Uralo-Sibirskoi Arkheologii*, Sverdlovsk: Ural'skii Universitet, pp 33–38

Stefanov, V I (1991) 'Neoliticheskoe poselenie Duvan V [The Duvan V Neolithic settlement]', in Krizhevskaia, L Ia (ed), *Neoliticheskie Pamyatniki Urala*, Sverdlovsk: Uralskoe Otdelenie Akademii Nauk SSSR, pp 144–160

Timofeev, V I (1996) 'Problemy absoliutnoi khronologii [Problems of absolute chronology]', in Oshibkina, S V (ed), *Arkheologiia. Neolit Severnoi Evrazii*, Moscow: Nauka, pp 330–347

Usacheva, I V (1998a) 'K istokam mirovozzreniia drevnikh ural'tsev (po materialam melkoi glinianoi i kremnevoi plastiki epokhi neolita-bronzy oz. Andreevskogo Tyumenskoi oblasti) [On the sources of the world outlook of ancient Ural peoples (on finds of small-size clay and flint plastics of the Neolithic-Bronze Ages at Andreyevskoye lake in the Tyumen region]', in Kovaleva, V T (ed), *Voprosy Arkheologii Urala 23*, Ekaterinburg: Ural'skii Universitet, pp 105–134

——— (1998b) 'Elementy sakral'nogo otnosheniia k domu u naseleniia Tiumenskogo Priob'ia v epokhu eneolita-bronzy [Elements of sacral attitude to Home among the peoples of the Tyumen Ob' region in the Eneolithic-Bronze Ages]', in Tomilov N A (ed), *Integratsiia arkheologicheskikh i etnograficheskikh issledovanii*, Omsk-Saint Petersburg: Omskii Universitet, pp 115–116

——— (2005) 'Utiuzhki v kulturah Evrazii [Shaft straighteners in cultures of Eurasia]', in Bagashev, A N (ed), *Vestnik Arkheologii, Antropologii i Etnografii 6*, Tyumen: IPOS SO RAN, pp 12–23

Viktorova, V D (2002a) 'Mify na sosydakh [Myths of the vessels]', in Golovnev, A V (ed), *Northern Archaeological Congress Papers, September 9–14, 2002, Khanty-Mansiisk*, Ekaterinburg: 'Akademkniga', pp 80–84

——— (2002b) 'Sviatilishche boborykinskoi kul'tury na pamiatnike Palatki I [Sanctuary of the Boborykino culture on the site Palatki I]', in Kovaleva, V T (ed), *Voprosy Arkheologii Urala 24*, Ekaterinburg: Ural'skii Universitet, pp 46–66

Viktorova, V D and Kerner, V F (1998) ' "Utiuzhki" s neoliticheskikh i eneoliticheskikh pamiatnikov Zaural'ia ["Small irons" from the Neolithic and Eneolithic sites of the East Urals]', in Kovaleva, V T (ed), *Voprosy Arkheologii Urala 23*, Ekaterinburg: Ural'skii Universitet, pp 63–80

Zakh, V.A. (1995) 'Boborykinskii kompleks Yurtobor 3 v Nizhnem Pritobol'e [Complex (association) of Boborykino type Yurtobor 3 in the Lower Tobol region]', in Zenko, A P (ed), *Drevniaia i Sovremennaia Kul'tura Narodov Zapadnoi Sibiri,* Tyumen: Tyumenskii Universitet, pp 12–28

Zakh, V A and Skochina, S N (2002) 'poselenie Mergen 3 [The Mergen 3 settlement]', in Bagashev, A N (ed), *Vestnik Arkheologii, Antropologii i Etnografii 5*, Tumen: IPOS SO RAN, pp 37–56

CHAPTER 7

EARLY POTTERY MAKERS IN EASTERN EUROPE: CENTRES OF ORIGINS, SUBSISTENCE AND DISPERSAL

Pavel M Dolukhanov, Andrei M Mazurkevich and Anvar M Shukurov

INTRODUCTION

The emergence and early development of pottery holds an important place in the debate about the European Neolithic. This debate includes the localization of its initial centres and the detailed chronology of the subsequent spread of pottery and related technological innovations. Also important are social mechanisms involved in the development of pottery making traditions and their consequences and finally, the role of pottery making in the process of Europe's neolithization.

In his initial definition of the Neolithic, Sir John Lubbock (1865) highlighted the use of polished tools and pottery, together with the growing of crops and the taming of animals as the main criteria of the Neolithic. Since then, western European interpretations of the Neolithic have become focused on the subsistence economy, more precisely, on the shift from foraging to the cultivation of domesticated cereals and animal rearing. This has been viewed as a material base for the development of pottery production, together with the proliferation of permanent villages, monumental structures and social hierarchy (Case 1969).

This view has been contested by several writers, particularly Julian Thomas (1996, 2003), who argues against the existence of a fixed and universal 'Neolithic package'. Thomas views the Neolithic as a range of different processes, generating a considerable variety of subsistence practices. Archaeobotanical studies (Hather and Mason 2002: 4–5) show that clear distinctions between agriculture and hunting-gathering are often impossible, as 'hunter-gatherers may undertake agricultural practices and vice versa'. Evidence shows that wild plant species were extensively gathered in most areas of Neolithic Britain (Robinson 2000).

To date, an influential archaeological school in the former USSR identifies the Neolithic with the large-scale manufacture of ceramic ware and polished stone and bone tools by predominantly hunter-gatherer communities (Oshibkina 1996a: 7). In many cases, the pottery-making hunter-gatherer groups in the boreal forests of Eurasia display attributes of complex societies, such as sedentism, high population density, intensive food procurement, technological elaboration, exchange networks (that may include their agricultural neighbours), social differentiation and territorial control (Oshibkina 1996a: 8; Zvelebil 1996: 331).

It is conceivable that the distinction between the two forms of the Neolithic could arise from a single wave of advance that spread across a strongly varying habitat thereby developing distinct traditions. For example, the subsistence strategy could have changed from farming to foraging in the boreal zone of northeastern Europe. However, careful analysis of radiocarbon dates from that region does not support the idea of spread from a single centre (Davison et al. 2007a, 2007b); in many cases, pottery making in the east pre-dates the farming cultures in the west and south of Europe. This leads us to suggest that the Neolithic as a pan-European phenomenon resulted from at least two processes, one of which involved primarily farming, and another, pottery making. The two processes had apparently different centres or origins and were not simultaneous. To test this hypothesis, we returned to the analysis of old and newly obtained evidence from various areas of the East European Plain.

Overview

Over the past few decades an area of early pottery making has been identified in the southeast of the east European Plain. This region, characterized by steppe and semi-desert landscapes, includes the Lower Volga, the Volga-Ural interfluve and, partly, the north Caspian Lowland. Early types of pottery found there were decorated by notches and incised lines and, specifically 'triangular notches'. Based on stylistic peculiarities, several cultural varieties were distinguished: the Tenteksor and Djangar-Varfolomeyevka groups in the western part of this region, and Yelshanian culture further east. It was presumed that at a later stage these groups merged together and moved to the west, forming the Middle Don culture (Siniuk 1986, 2000).

The sites of the Yelshanian culture were studied in more detail (Mamonov 2000a, 2000b). These were small, apparently seasonal occupations lying close to water channels, scattered in a vast area of the steppe between the Lower Volga and Ural Rivers (Figure 7.1). The pottery is made from silty clay tempered with organic matter, fish scales and bone. The early vessels are small, with straight or S-shaped rims, and flat or conic bases. They are decorated with imprints of pits, notches, incised lines and lines forming rows, rhombi, triangles and zig-zags (Figure 7.2). Eight radiocarbon dates have been obtained for *Unio* molluscs collected at these sites. Five of these

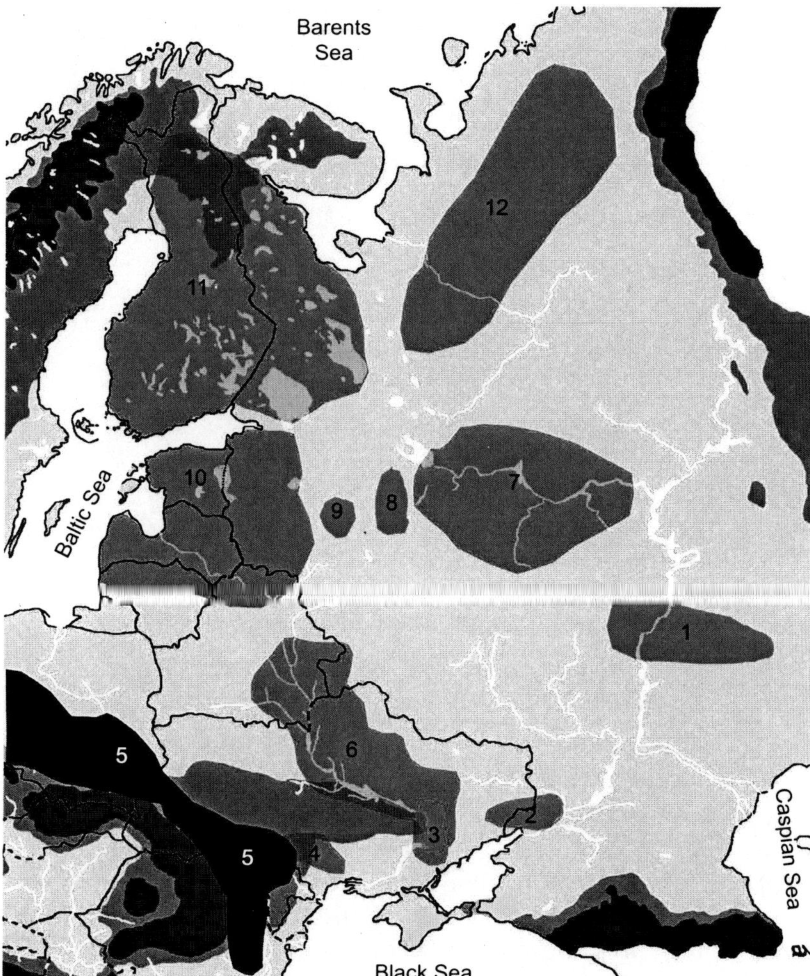

Figure 7.1 Early pottery cultures in Europe: (1) Ryazan-Dolgov; (2) Southern-Don; (3) Sursk-Dnieper; (4) Bug-Dniester; (5) LBK (Linear Pottery Ware); (6) Dneper-Donetz (Russian) / Dniepro-Donets (Ukrainian); (7) Volga-Kama; (8) Upper Volga; (9) Valdai; (10) Narva; (11) Sperrings Comb Ware; (12) Pechora/Dvina/Kargopol culture group.

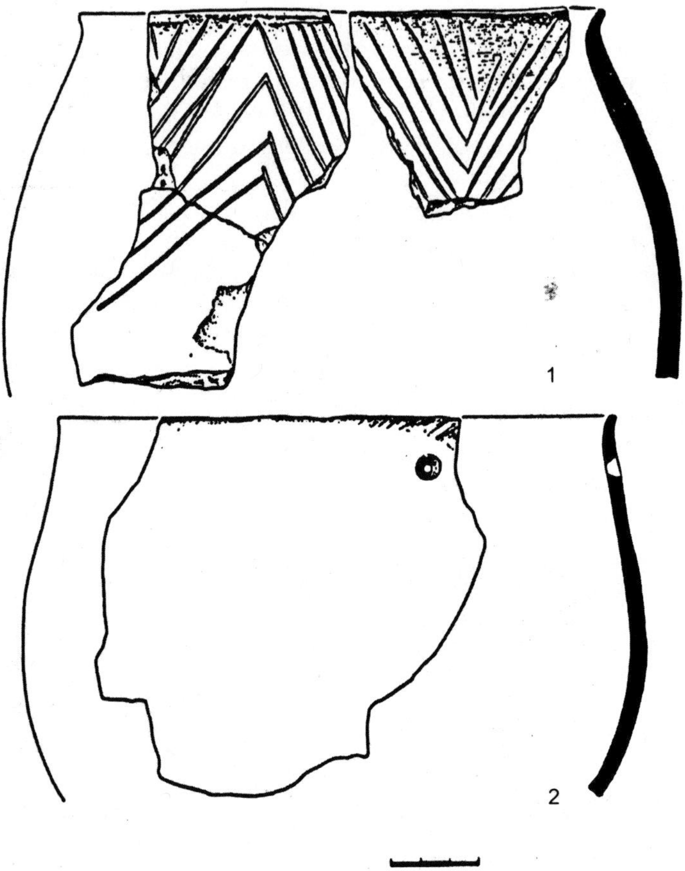

Figure 7.2 Examples of Yelshanian pottery.

dates form a cluster with the mean age of 6,910±58 BC (Dolukhanov et al. 2005). The remaining dates show an older age of 8,025–7,475 BC.

Early pottery making was also found in the Lower Don area and is exemplified by the Rakushechnyi Iar culture (Figure 7.2, *1*,*2*) (Belanovskaia 1995). Ceramic ware from the Rakushechnyi Iar's lower strata are similar to Yelshanian pottery, notably in the use of triangular notches, the 'zoned-geometric' decoration pattern as well as the specific coiling technique of manufacturing both conical and flat-bottomed vessels. Notable are the S-profiled vessels, with thorn-like bottoms, 'pearls' or pits below the rims, networked patterns, rows of tooth-stamp impressions, or complete absence of decoration (Figure 7.3). The dates from the lower layers have been processed statistically (Dolukhanov et al. 2005). Six dates satisfy the criterion for contemporaneity, yielding the age of 5,863±130 BC. The remaining

Figure 7.3 Examples of Rakushechnyi Iar pottery.

dates include one younger date (5,000 BC) and three older ones (6,850–6,550 BC).

Both Yelshanian and Rakushechnyi Iar sites feature hunting-gathering subsistence, which included the hunting of red deer, roe deer, fox, hare and birds, and collecting of edible molluscs. Rare bones of domesticates (sheep, goat, cattle, dog and horse) apparently result from the penetration of more recent deposits. There are no indications of any crops, yet there is evidence of intensive use of plant food in the steppe area since the Mesolithic (Jacobs 1994; Lillie and Richards 2000).

The earliest evidence for the use of crops comes from the Bug-Dniester sites in the forest-steppe area. About 40 sites of seasonal character are located on the lower terraces of the Dniester (Nistru) and Pyvdenyi Buh Rivers and their tributaries (Figure 7.1) (Danilenko 1969; Markevich 1974). Impressions of three varieties of wheat (emmer, einkorn and spelt) were found on the pottery. At earlier sites, about 80% of animal remains belong to wild species, mostly roe deer and red deer. The occurrence rate of domestic animals (pig, cattle and sheep/goat) increases at sites of more recent age. Archaeological deposits contain large amounts of *Unio* molluscs and tortoise shells, as well as remains of fish (roach, pike, sturgeon and catfish) and birds (sparrow hawk, honey buzzard and wood pigeon). The flint industry is of Mesolithic character, based on the prismatic core technique with the common occurrence of retouched blades, backed blades, and small-sized circular end-scrapers, and numerous geometrics (trapezes and triangles). Several blades at Soroki 1 show a sickle gloss. The bone and antler industry includes points, awls, mattocks, chisels and 'hoe-like' tools.

The Bug-Dniesterian pottery includes deep bowls with an S-like profile and hemispherical flat-bottomed beakers made of clay tempered with organic matter and crushed shells. Ornamental patterns consist of rows of shell-rim impressions, finger impressions, notched ornament and incised lines forming zig-zags and volutes (Figure 7.4). Kotova (2002) notes that conical vessels with notched ornaments have direct analogies in the Rakushechnyi Iar culture. Danilenko (1969) argued that the earlier varieties of the Bug-Dniesterian pottery reflect influences stemming from the steppe area in the east and, particularly, the site of Djeitun in western Turkmenistan. Several writers (Danilenko 1969; Dolukhanov 1979; Kotova 2002) note the western influences, certain categories of ware and, particularly, the barbotine ornament having analogies in the early Starčevo-Körös-Criş pottery. Imported potsherds of Linear Pottery (with 'music-note' patterns) were found at several sites on the Pyvdenyi Buh River belonging to later stages of the Bug-Dniesterian culture. It has been suggested (Dolukhanov 1979) that both domesticated animals and

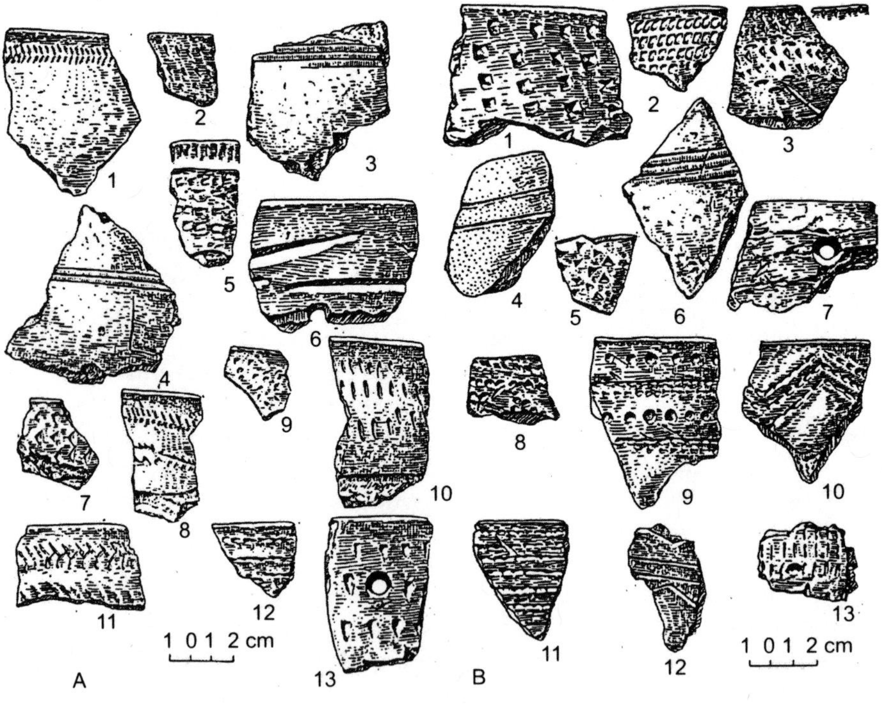

Figure 7.4 Bug-Dniester Pottery.

crops resulted from trade links between predominantly foraging Bug-Dniesterian groups and neighbouring farming communities. Contrary to earlier suggestions, the radiocarbon dates suggest a limited duration of the Bug-Dniesterian culture. All recently available measurements from the sites on the Pyvdenyi Buh form a coherent cluster with the mean age of 6,121±143 BC.

The early Neolithic in the central part of the East European Plain features the 'notch-and-comb decorated pottery', which exhibits several stylistic varieties, notably the Upper Volga (Figure 7.1) and Valdai (Figure 7.1) cultures. The sites of the former consist of small-sized habitations clustering along the rivers of the Upper Volga basin, the larger sites are found in wetland landscapes, in bogs and mires. The subsistence was based on hunting the forest game (elk, red deer, roe deer, aurochs, wild boar and fur animals) supplemented by fishing and food-collecting. The pottery corpus consists of small vessels (15–30 cm in diameter), conical or flat-bottomed, made of chamotte-tempered clay and ornamented by impressions of notches, combs, cords and incised lines forming simple geometric patterns. At a later stage, small round-bottomed cups appear, and the mineral tempering becomes more frequent. The radiocarbon dates of earlier Upper Volga sites lie in the range of 6,200–6,000 cal BC.

As follows from its title, the sites of the Valdai culture are located in the Valdai Hills, where several middle Russian rivers, including the Volga, Lovat and Western Dvina, take their sources. The sites are usually found on the dunes along the banks of these rivers and on the lake shores. The age, material culture and subsistence are basically similar to those of the Upper Volga culture (Gurina 1996). The main peculiarity consists in the richness and diversity of the flint industry. This area is rich in outcrops of high-quality flint, which was exported as far as Finland.

Narva culture sites (Figure 7.1) are focused in the eastern Baltic area clustering in Latvia and Estonia. Early Narva sites (as exemplified by Zvidze in eastern Latvia) are dated around 5,300–5,200 cal BC.

The Sperrings culture (or the I:1 Style of the Finnish writers) (Figure 7.1) are located on ancient sea and lake shorelines in a vast territory encompassing southern and central Finland and Ladoga and Onega Lake basins in Russian Karelia (Oshibkina 1996b). The pottery corpus consists of large conical vessels with straight rims decorated with impressions of cord, incised lines and pits forming a simple zoned ornament. The lithic industry manufactured from quartz, schist and rarely, flint (presumably imported from Upper Volga) retains a Mesolithic character. Earlier age assessments based on the gradients of the shoreline displacements (Siiriäinen 1982) suggest an age of 5,300–5,000 BC. New radiocarbon dates obtained for early Sperrings sites near Saint Petersburg suggest an older age of 5,600–5,200 cal BC (Vereshchagina 2003).

The subsistence of both these cultures was based on seasonally varying hunting-gathering strategies with the prominence of seal-hunting in the coastal area. Over the past years, several writers reported proxy evidence suggestive of primitive agriculture in Lithuania, Latvia and Estonia allegedly related to early Neolithic settlements (Polska and Saarse 2002; Rimantiene 1999; Seglinš et al. 1999).

The mean age of the entire dataset of early pottery-bearing sites in the boreal East European Plain is 5,417±30 BC, with the spread (standard deviation) of σ_c = 160 years. The remaining dates include those which are older (6,200–5,800 BC) and younger (5,200–4,200 BC) than the coeval sample.

This general overview based on the available radiocarbon dates indicates an unexpectedly early appearance of pottery making in the eastern region of the steppe zone, somewhere between 8,000 BC and 7,000 BC. Broadly around the same time (8,600–5,800 BC) this technology was able to spread further westwards, until it had eventually encompassed the entire boreal zone by 6,200–5,400 BC.

The Western Dvina–Lovat Case Study

The area which lies immediately south of the terminal moraines of the last ice age is often referred to as the Russia's Lake District. It includes numerous sites of various ages that have been systematically excavated with the use of multidisciplinary techniques since the 1960s (Mikliaev 2000). The evidence obtained offers a unique opportunity to follow up the early stages of pottery making in the boreal zone with a high temporal resolution (Dolukhanov et al. 2004).

The oldest pottery in that area, referred to as Serteyan, has been identified at several sites within the Serteya Valley (Figure 7.1), a small tributary of the Western Dvina River, 20 km east of the town Velizh. Based on the stratigraphy of the sites, three phases of the early Serteyan pottery can be distinguished (denoted A, B and C).

The oldest Phase A is radiocarbon dated to 6,300–6,100 cal BC. The vessels were manufactured with the use of coiling technique, coils added either spirally or as rings. Straight or sharpened rims correspond to the shape of the upper coil. The ceramic paste consists of local clay tempered with organic matter, crushed shells and coarse sand. The exterior surface was slightly polished and decorated. The ornament consists of notches and incised lines forming simple geometric patterns and covering the entire surface. The mitre-like conical vessels are normally 30–40 cm tall, with a neck diameter of 16–22 cm and 6–9 mm thick walls (Figure 7.6). The analogies to this pottery can be found in the areas close to the centre of initial

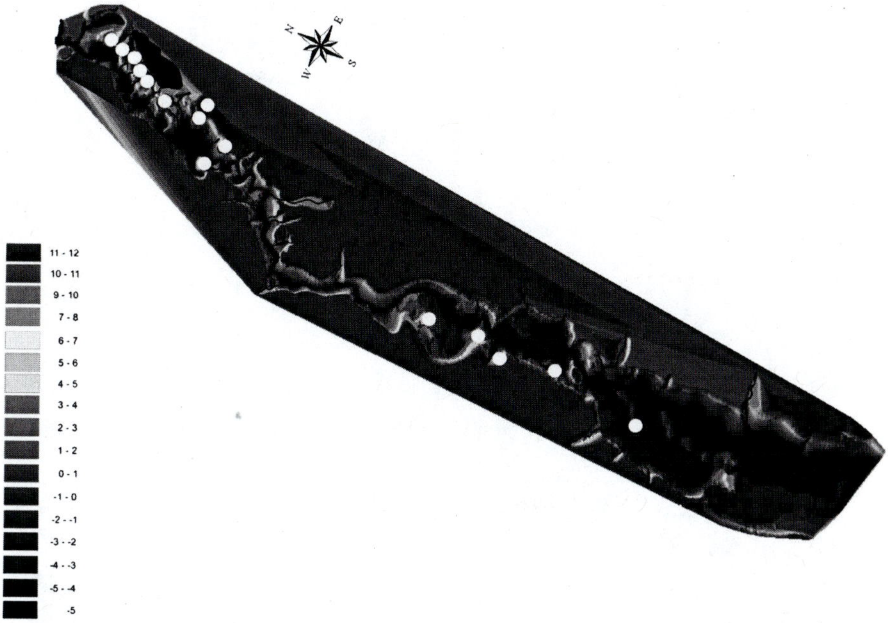

Figure 7.5 Serteya Valley and its Neolithic sites (three-dimensional projection).

pottery making in the south-east of the East European Plain encompassing the Lower Volga and Middle Don basin.

Phase B dates to 6,000–5,900 cal BC. The technology remains basically the same as in Phase A, the only innovation being the paddle and anvil technique used for smoothing over the joints. The vessels are conical and mitre-shaped and the walls are 6–9 cm thick. Several flat-bottomed varieties were found. The decoration is restricted to the vessels' upper part, and consists of notches forming either triangles or more complicated geometric patterns (Figure 7.7). Rims are straight with either flat or sharpened tips. Analogies are found in the Upper Don area (Karamyshevo 5 sites) and also in the area of Upper Volga and Valdai cultures. At one of the sites (Serteya X), remains of an oval-shaped dwelling have been found, 2.5 m × 3 m in size, 50–70 cm deep into the ground, with postholes along the perimeter and a hearth in the middle.

The age of Phase C pottery is estimated as 5,300–5,000 cal BC. The pottery consists of thick vessels (7–9 mm) manufactured with a slightly different technique—docked coils, with the addition of clay rings. Both external and internal surfaces were smoothed by toothed stamps (Figure 7.8). Undecorated cauldrons are either conical or flat-bottomed. This pottery finds analogies in the Upper Don area (Karamyshevo 5 site), and is particularly common in the materials of Rakushechnyi Iar.

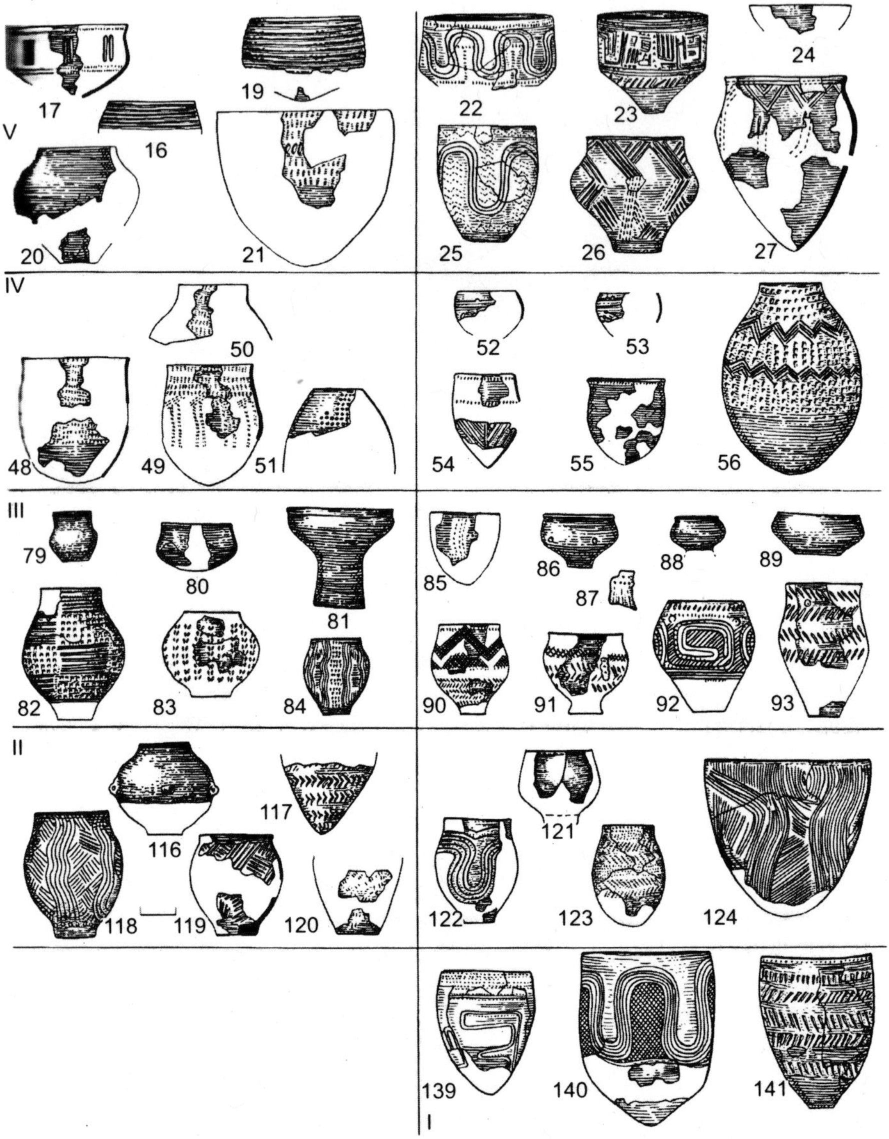

Figure 7.6 Serteyan pottery, phase A.

At Serteya XIV, the corresponding layer includes an oval-shaped ground dwelling, 5.1 m × 3 m, with postholes along the periphery and the central axis, and a hearth in its western part. A more complex dwelling appeared at this site at a later stage: a sub-rectangular structure, 8.8 m × 6.3 m in size, with two rows of postholes along the central axis, and a large oval

Figure 7.7 Serteyan pottery, phase B.

Figure 7.8 Serteyan pottery, phase C.

stone-laid hearth it its northern part. The postholes surrounding the hearths suggest the occurrence of a wind protecting device.

A new tradition of pottery making appeared in that area between 5,300 and 4,900 cal BC. Although the technology remained basically the same, both the shapes and ornamentation of the vessels became clearly distinct. Three types of vessels are distinguished: those with 'S' profiles, 'CS' profiles and conical beakers. Rims are either rounded or bent inside; the bases are conical, with thorn-like tips. Vessels with 'CS' profiles are decorated in their upper parts with either horizontal or diagonal rows of pits, nail notches and comb impressions. Conical beakers bear similar ornaments beneath the rims. Vessels with 'S' profiles are usually not decorated. Both the pottery and the accompanying lithic, bone and antler inventories have direct analogies in the Narva culture in the eastern Baltic area, and particularly in the area of Lubanas Lake in eastern Latvia. Hence the suggestion that, distinct from the earlier stages, this type of pottery penetrated from the west, moving upstream along the Western Dvina River.

The faunal and botanical records indicate that the livelihood of early pottery-making communities in that area firmly relied on wildlife resources, with a round-the-year procurement of meat and fur animals and fishing. At least 30 edible wild plants were utilized for sustenance. The earliest signals of agriculture (apparently of swidden type) came from the levels of Usvyatian culture dated to 5,500 and 4,000 cal BP.

Underwater investigations of the submerged site Dubokrai V, on the Sennitsa Lake in the same area, yielded around 1,000 potsherds of vessels manufactured from clay tempered by coarse lake sand, crushed shells and organic matter. The decoration consists of incised lines, pits and notches which form geometric curvilinear patterns. Rim tips and adjusting interior surface are decorated with pits and notches (Figure 7.9). Analogies to this pottery may be found in some earlier varieties of Linear Pottery in the areas of Lower Rhine, Maas, Elbe and Saale. Unfortunately, the stratigraphic

Figure 7.9 'LBK'-type pottery.

position and age of this pottery in Russia's Lake District remain uncertain, but the possibility of an influx of population from the LBK area to the east cannot be ruled out.

Discussion

As follows from the above review of archaeological materials, the early pottery of the East European Plain consistently shows influences from sources located in the east of the steppe zone, approximately in the lower Volga-Ural interfluves. There is evidence suggesting that the early wave of pottery making spread far to the west, reaching the western confines of Europe. This wave includes the 'groups' La Hoguette and Limbourg in the western area of the LBK, as well as 'Epicardial' and Roucadour groups in southern France. Sites belonging to these groups have archaic-looking pottery: oval and cauldron-like vessels, usually with narrow rim manufactured from clay tempered with crushed bones and shells. Decoration consists of incised lines and the impressions of three-toothed stamps, forming single or multiple bands (Figure 7.10) (van Berg 1990). Radiocarbon dates of earlier 'Epicardial' and Roucadour sites suggest the age of 7,500–6,500 cal BC (Roussot-Laroque 1990). At the same time, numerous elements of the pottery-bearing cultures of East European Plain show clear relationships to early farming cultures in southeast and central Europe, the origins of which lie in the Fertile Crescent of the Near East. Thus, the archaeological evidence supports the hypothesis of two sources of European Neolithic suggested above.

We further tested this hypothesis using mathematical modelling based on a comprehensive set of radiocarbon dates of Neolithic sites encompassing

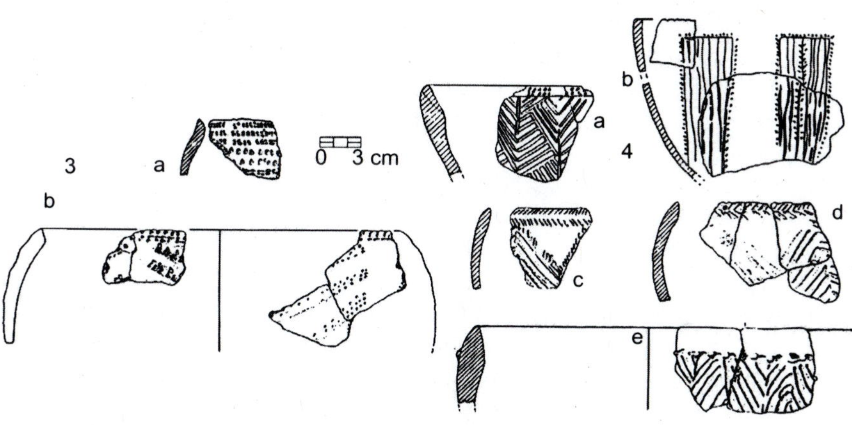

Figure 7.10 La Hoguette pottery.

the whole of Europe (492 dates). The quality of the models, discussed in detail by Davison and colleagues (2007a, 2007b), was controlled by comparing their results with radiocarbon dates using various statistical criteria. The modelling involved numerical solution of the reaction-diffusion equation of population dynamics in two dimensions with allowance for coastlines, major rivers and altitude. The simulations made it possible to model the arrival time of the wave of advance to a given site and then to compare it with the relevant radiocarbon dates. Two basic models have been studied.

The first model explored the assumption that all Neolithic sites in Europe could have originated from a single source located in the Near East. Comparison of the ages predicted by the model with the radiocarbon dates shows satisfactory agreement for central and western Europe: the model prediction agrees within 500 years with the radiocarbon age for 68% of early Neolithic sites in that area. However, only 38% of sites in eastern Europe exhibit similar degree of agreement. Accordingly, another model was developed, which included one more source. The location and activation time of the additional source were chosen as to optimize the agreement of the results with the radiocarbon dates. This places it in the southern Urals, and the spread from this source starts about 1,500 years before that in the Near East. The introduction of this source improved very significantly the agreement between the modelled and radiocarbon ages of pan-European Neolithic sites, especially in the eastern area. About 70% of the total number of radiocarbon dates agree with the model to within 500 years (as compared to 55% for the single-source model). More importantly, the degree of agreement between the modelled and radiocarbon ages is now quite uniform across the whole of Europe. The model strongly suggests that the Neolithic sites in the East European Plain could have resulted from the combined impact of the eastern and near eastern centres. The model further provides the average speed of the front propagation: 1.1 km/year for the wave originating in the near east, and 1.7 km/year for the source in the east; the difference in the spread rates can, presumably, be attributed to the difference between the behaviours of farmers and hunter-gatherers, respectively.

Conclusions

Both archaeological evidence and mathematical modelling support the suggestion that the spread of the Neolithic in Europe involved at least two waves propagating from distinct centres. We argue that an earlier wave spread from the east via the 'steppe corridor'. This wave resulted in the adoption of pottery making, the establishment of permanent and semipermanent dwelling sites, advanced forms of hunting and food gathering and, in some cases, elements of stock-breeding. This was supplemented by a later wave which originated in the near east, bringing to Europe advanced techniques of cereal husbandry, architecture, agriculture-based symbolism and, possibly, Indo-European speech.

REFERENCES

Belanovskaia, T D (1995) *Iz Drevneishego Proshlogo Nizhnego Podon'ia. Poselenie Vremeni Neolita i Eneolota Rakushechnyi Iar (The Distant Past of the Lower Don Area. The Neolithic-Eneolithic Settlement of Rakushechnyi Iar),* Saint Petersburg: Saint Petersburg University Press (in Russian)

Case, H J (1969) 'Neolithic explanations', 43 *Antiquity*, 176–186

Danilenko, V N (1969) *Neolit Ukrainy* (The Neolithic in the Ukraine), Kiev: Naukova Dumka (in Russian)

Davison, K, Dolukhanov, P M, Shukurov, A, Sarson, G R and Zaitseva, G I (2007a) 'Multiple sources of the European Neolithic: mathematical modelling constrained by radiocarbon dates', 203 *International Quaternary*, 10–18

——— (2007b) 'Mnozhestvennye istochniki evropeiskogo neolita' [Multiple sources of the European Neolithic] in Zaitseva, G (ed), *Radiouglerod v Arkheologicheskikh i Paleoecologicheskikh Issledovaniiakh (Radiocarbon in Archaeological and Palaeoecological Studies),* Saint Petersburg: Teza, pp 172–184 (Russian, English summary)

Dolukhanov, P (1979) *Ecology and Economy in Neolithic Eastern Europe,* London: Duckworth

Dolukhanov, P, Shukurov, A, Arslanov, K, Mazurkevich, A N, Savel'eva, L A, Dzinoridze, E N, Kulkova, M A and Zaitseva, G I (2004) 'The Holocene environment and transition to agriculture in Boreal Russia (Serteya Valley Case Study)', 17 *Internet Archaeology*, (http://intarch.ac.uk/journal/issue17)

Dolukhanov, P, Shukurov, A, Gronenborn, D, Timofeev, V and Zaitseva, G (2005) 'The chronology of Neolithic dispersal in Central and Eastern Europe', 32 *Journal of Archaeological Science*, 1442–1458

Gurina, N N (1996) 'Valdaiskaia kul'tura [The Valdai Culture]', in Oshibkina, S V (ed), *Neolit Severnoi Evrazii [The Neolithic of Northern Eurasia]*, Moscow: Nauka, pp 184–193

Hather, J G and Mason, S L (2002) 'Introduction: some issues in the archaeobotany of hunter-gatherers', in Mason, S L R and Hather, J G (eds), *Hunter-Gatherer Archaeobotany*, London: Institute of Archaeology, UCL, p 4

Jacobs, K (1994) 'Human dento-gnathic metric variation in Mesolithic/Neolithic Ukraine: possible evidence of demic diffusion in the Dnieper Rapids region', 95 *American Journal of Physical Anthropology*, 1–26

Kotova, N S (2002) *Neolitizatsiia Ukrainy [The Neolithization of the Ukraine]*, Lugansk: Shliakh

Lillie, M C and Richards, M (2000) 'Stable isotope analysis and dental evidence of diet at the Mesolithic-Neolithic transition in Ukraine', 27 *Journal of Archaeological Science*, 965–972

Lubbock, J (1865) *Pre-historic Times, as Illustrated by Ancient Remains, and the Manners and Customs of Modern Savages*, London: Williams and Norgate

Mamonov, A E (2000a) 'Khronologicheskii aspekt izucheniia yelshanskoi kul'tury [Chronological aspects in the study of Yelshanian Culture]', in Timofeev, V I (ed), *Khronologiia Neolita Lesnoi i Stepnoi Zon (Chronology of the Neolithic in the Boreal and Steppe Zones)*, Saint Petersburg: Institute for History of Material Culture, pp 50–52

——— (2000b) 'Rannii neolit. Yelshanskaya kul'tura [Early Neolithic. Yelshanian Culture]', *Voprosy Arkheologii Povolzh'ia (Problems in Archaeology of the Volga Basin),* 1. Samara: Samara Pedagogical University Press, pp 1–12

Markevich, V I (1974) *Bugo-Dnestrovskaia Kul'tura na Territorii Moldavii (The Bug-Dniester Culture on the Territory of Moldavia).* Kishinev: Stiinca

Mikliaev, A M (2000) 'Kamennyi–zheleznyi vek v mezhdurech'e Zapadnoi Dviny i Lovati [Stone–Iron Ages in the Western Dvina-Lovat Interfluve]', 9 *Peterburgskii Arkheologicheskii Vestnik*, 5–39

Oshibkina, S V (1996a) 'Introduction', in Oshibkina, S V (ed), *Neolit Severnoi Evrazii [The Neolithic of Northern Eurasia]*, Moscow: Nauka, pp 6–9 (in Russian)

——— (1996b) 'Sever vostochnoi Evropy [The north of Eastern Europe]', in Oshibkina, S V (ed), *Neolit Severnoi Evrazii [The Neolithic of Northern Eurasia]*, Moscow: Nauka, pp 210–242

Polska, A and Saarse, L (2002) 'Vegetation development and introduction of agriculture to Saarema Island, Estonia: the human response to shore displacement', 12/5, *The Holocene,* pp 555–568

Rimantiene, R (1999) 'Traces of agricultural activities in the Stone Age sites of Lithuania,' in Miller, U, Hackens, T, Lang, V, Raukas, A and Hicks, S (eds), *Environmental and Cultural History of the Eastern Baltic Region*, PACT 57, Rixensart: European Network of Scientific and Technical Cooperation for the Cultural Heritage, pp 275–290

Robinson, M A (2000) 'Further consideration of Neolithic charred cereals, fruit and nuts', in Fairburn, A S (ed), *Plants in Neolithic Britain and Beyond*, Oxford: Oxbow, pp 85–90

Roussot-Laroque, J (1990) 'Rubané et Cardial: le poid de l'ouest', in Cahen, D and Otte, M (eds), *Rubané et Cardial*, ERAUL 39, Liège: Universite de Liège, pp 315–360

Seglinš, V, Kalnina, L and Lacis, A (1999) 'The Lubanas Plain, Latvia, as a reference area for long-term studies of human impact on the environment', in Miller, U, Hackens, T, Lang, V, Raukas, A and Hicks, S (eds), *Environmental and Cultural History of the Eastern Baltic Region*, PACT 57, Rixensart: European Network of Scientific and Technical Cooperation for the Cultural Heritage, pp 105–129

Siiriäinen, A (1982) 'Shore displacement and Archaeology in Finland', 134 *Annales Academiae Scienc. Fennica*, Ser. A III, Geol.-Geograph., 173–184

Siniuk, A T (1986) *Naselenie Basseina Dona v Epokhu Neolita (The Population of the Don Basin during the Neolithic Epoch)*, Voronezh: Voronezh University (in Russian)

——— (2000) 'Problemy khronologii neolita Srednego Dona [The problems of chronology of the Middle Don Neolithic]', in Timofeev, V I (ed), *Khronologiia Neolita Lesnoi i Stepnoi Zon (Chronology of the Neolithic in the Boreal and Steppe zones)*, Saint Petersburg: Institute for History of Material Culture, pp 71–72

Thomas, J (1996) 'The cultural context of the first use of domesticates in Continental and Northwest Europe', in Harris, D R (ed), *The Origins and Spread of Agriculture and Pastoralism in Eurasia*, London: UCL Press, pp 310–322

——— (2003) 'Thoughts on the "repacked" Neolithic revolution', 77 *Antiquity*, 75–86

van Berg, P-L (1990) 'Céramique de Limbourg et la néolithisation en Europe du nord-ouest', in Cahen, D and Otte, M (eds), *Rubané et Cardial*, ERAUL 39, Liège: Universite de Liège, pp 161–208

Vereshchagina, I V (2003) 'Poselenie Khepojarvi v iuzhnoi chasti Karel'skogo peresheika [Khepojarvi settlement in the southern part of the Karelian Isthmus]', in Timofeev, V I (ed), *Neolit-Eneolit Iuga i Neolit Severa Vostochnoi Evropy [The Neolithic Chalcolithic in the South and the Neolithic in the North of Eastern Europe]*, Saint Petersburg: Institute for History of Material Culture, pp 140–151

Zvelebil, M (1996) 'The agricultural frontier and the transition to farming in the circum-Baltic region', in Harris, D R (ed), *The Origins and Spread of Agriculture and Pastoralism in Eurasia*, London: UCL Press, pp 323–345

CHAPTER 8

EARLY HUNTER-GATHERER CERAMICS IN KARELIA

Konstantin German

INTRODUCTION

This chapter focuses on the characteristics and potential origins of early hunter-gatherer 'Sperrings' and 'Säräisniemi I' ceramics in Karelia. 'Sperrings' ceramics were first defined in the 1930s by a famous Finnish archaeologist Eurapaeus-Äyräpää (1930), who named the pottery after a suburb of Helsinki—Espoo Sperrings. Later, Gurina (1961) classified Sperrings as an independently invented local type of pottery found firstly on the territory of Karelia. Currently, there are some 250 archaeological sites in Karelia from which over 1,500 Sperrings-type vessels have been recovered.

Säräisniemi I ceramics are another form of early pottery, which are distinguished as a separate type on the basis of ornamentation and the area of usage, and named after a settlement in northern Finland. It was first identified as a distinct pottery type in 1950 in Norway by the researcher Simonsen (1957, 1961), while Pesonen (1991) was the first to define it as an independent ceramic type in Karelia. More than 100 vessels have been recovered from over 50 archaeological sites in Karelia. Sperrings ceramics are found in sites in the southern and central part of Karelia, and along the shorelines of Lake Onega, while sites with Säräisniemi I pottery are typical of northern Karelia and the White Sea basin (Figure 8.1). This chapter discusses the origins, characteristics, ornamentation and potential uses of this early hunter-gatherer pottery in Karelia, and explores the wider ecological, social and economic contexts in which these first ceramics appear.

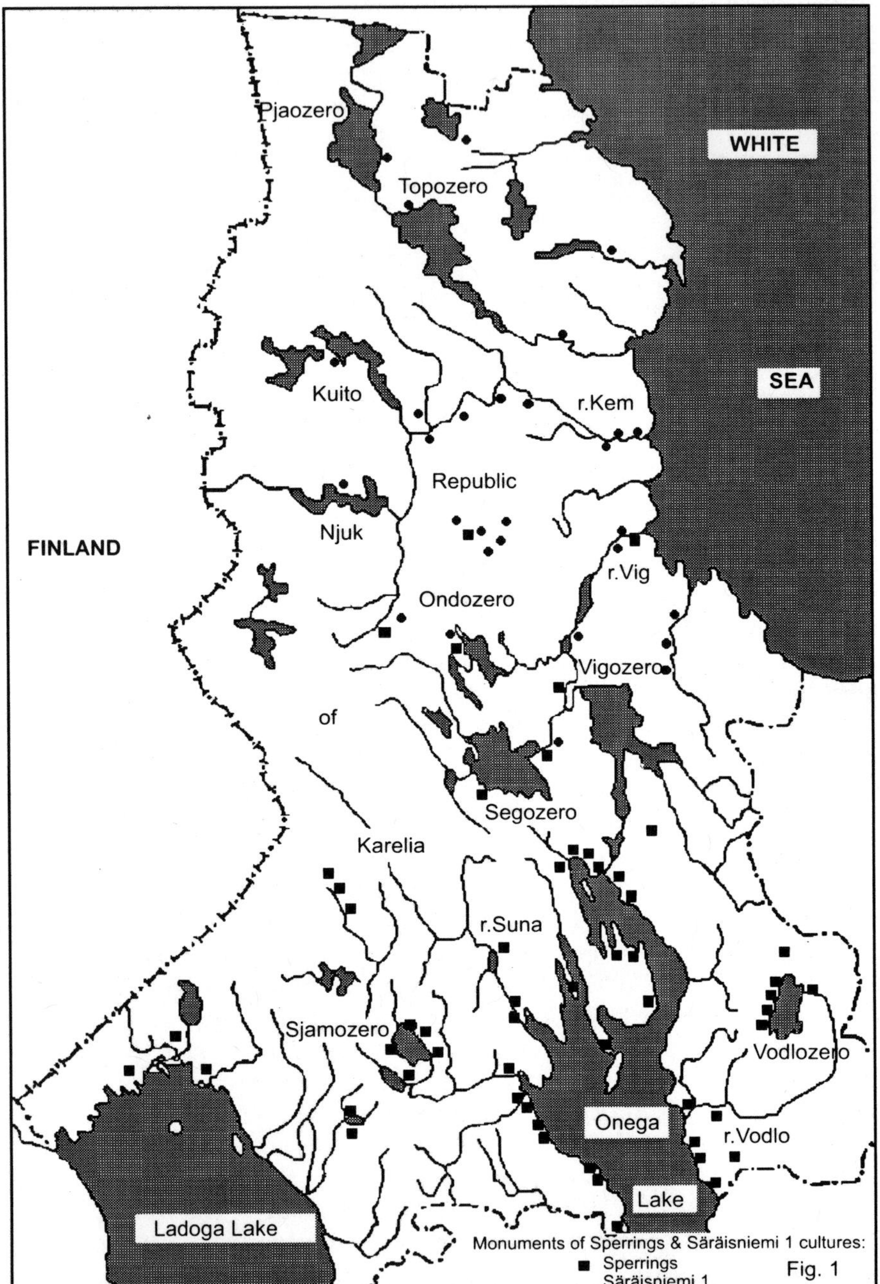

Figure 8.1 Map of monuments of Sperrings and Säräisniemi I ceramics in Karelia.

Environmental Conditions in Karelia (6,500–5,500 bp)

The majority of archaeological sites with Sperrings ceramics are located in the Lake Onega basin, with others recorded in the basins of Lake Ladoga and the White Sea. In contrast, the Säräisniemi I ceramic settlements are found along the ancient coastline terraces of the White Sea, as well as around inland lakes and rivers. Current environmental conditions are considerably different from those in prehistory. Shoreline displacements and other natural processes which occurred during the Early and Middle Holocene are well documented in the literature (Kvasov 1976, 1990; Markov et al. 1934; Zemliakov 1936, 1940). Changes in the local flora during the Atlantic period have also been studied (Deliushina 1989; Elina 1980, 1981; Ivanter 1986; Khomutova 1976; Klimanov 1978; Savvateev and Vereshchagin 1978).

Following the melting of the Fennoscandian glaciers around 13,000–11,000 bp, there was a significant shoreline displacement, although this was uneven around Onega, Ladoga and other big lakes in Karelia. The northern and northwestern shores rose faster than the southern ones. This was reflected in the fact that prehistoric settlements located on the northern and northwestern shores of these lakes are found around 7 to 20 m above the present water levels, while the southern and southeastern shores do not exceed 1.5–2 m above lake levels, with many parts already eroded or heavily undermined. Studies have revealed impressive altitude indexes at Säräisniemi I settlements along the White Sea coast ranging from 30 to 40 m above the present sea level. Other investigations by geologists and palaeontologists have demonstrated that Sperrings settlements along the shores of Lake Onega and the White Sea coast are located on the terraces formed in the second part of the Atlantic period around 6,500–4,700 bp (Deviatova 1976: 9, 1988). In terms of location, the majority of the Sperrings and Säräisniemi I settlements are located on the shores of palaeobays and the arms of the channels which were running deep inland. These are similar to earlier site locations. In contrast, sites where hunter-gatherer pottery has been found in Finland are typically located on the ancient coast terraces of the Baltic Sea, or along the shores of lakes whose waters were connected to the sea.

Climate, Vegetation, Landscape and Fauna

The Atlantic period represented a climatic optimum for all of northern Eurasia, and led to the northerly spread of broadleaved forests (Dolukhanov and Khotinskii 1974: 212; Khotinskii 1977, 1996: 10; Khotinskii and Savina 1985: 18). That period in Karelia can be characterized by the following average annual temperatures: +2°C in the north of Karelia and

+3°C in the south. July temperatures reached +18°C on the White Sea coast, and +19°C on the shores of Lake Ladoga. Annual precipitation was also 50–60 mm above the present level and reached 500–600 mm (Klimanov 1978: 182–183). Shoreline regression of Lake Onega led to an overgrowth of old Lake shores and formation of extensively waterlogged plains (Ekman et al. 1988: 215). The vegetation of those new landscapes differed, both qualitatively and quantitatively, from that of the boreal period. It took on more warmth-loving southern plants.

During the Atlantic period, seal, white hare, beaver, water rat, sable, marten, otter, wolf, brown bear, reindeer and elk inhabited Karelia. A large number of bird species including wood-grouse, blackcock, white partridge, goose and different kinds of ducks were also native to Karelia. Most of them are typically taiga species. However, as a result of warming, forest-steppe species including bison, roe deer, wild boar, wildcat, grey heron, tern and others also inhabited the north (Ivanter 1986: 12–13). Faunal analyses of Sperrings settlements located on the White Sea coast have shown that seal, beaver and reindeer were the main sources of subsistence. Bones of reindeer, elk, beaver, marten, fox, bear, lynx, seal and perch have been found in Lake Onega settlements; bones of beaver, wolf, bear, elk, reindeer, blackcock and fish were recovered at Lake Sjamozero settlements (Ivanichev 1997: 26; Savvateev and Vereshchagin 1978: 181–215). Thus, the analyses of faunal assemblages found at Sperrings sites indicate that early people on the White Sea coast were involved in marine mammal hunting, and those who lived around inland water bodies were engaged in hunting and fishing.

At some lakeshore sites, fish appears to have been a mainstay in the local diets of the early people. High levels of phosphorus have been recorded in the cultural layers at these sites suggesting that fishing was an important household activity. On the other hand, pollen core evidence suggests that elderberry grew abundantly around Sperrings settlements located on the southern shores of Onega. Elderberry is known to thrive in soils rich in nitrogen and grows close to human habitations. High saturation of nitrogen in the soil residues may be found in settlements where hunting dominates over fishing (Deviatova 1984: 50).

Considering the climatic conditions in the second part of the Atlantic period, it is possible to reconstruct the seasonality pattern of Sperrings sites over the territory of Karelia. Early Neolithic[1] people approached the White Sea coast in spring/autumn when optimal climatic conditions favoured marine mammal hunting after snow and ice had melted. The sites are small in size (not exceeding 200–300 sq m), with few ceramic vessels (no more than 2–3) and specialized stone tools for hunting, butchering and processing of animals (drills, chisels, scrapers, sagae, spear points, burins). During winters, people of the Sperrings culture moved further inland and hunted reindeer and elk, and fished. These sites measure approximately 1,000–1,200 sq m and have yielded 10–12 ceramic vessels and stone tools used for hunting, fishing, wood processing and household activities. Few Sperrings sites have been recorded in the inner parts of the coastal White Sea region. This is probably because the region remains relatively unexplored to date by archaeologists.

Sperrings sites on the White Sea coast can be dated to 5,990±100–5,460±80 bp (German 2002b: 264), that is, late Mesolithic culture. The percentage of vessels ornamented with fish vertebrae lowers to 20–30% of the total number of vessels recovered. This evidence may support the idea that marine mammal hunting in the late Sperrings period was more popular than terrestrial animal hunting and fishing.

Seasonality of sites along Lake Onega seems to be much clearer. The spring/autumn period sites are found on islands and promontories, when fishing and fish diet dominated. This is reflected in pottery ornamentation (see below). Winter sites were located along water bodies. This is supposedly due to the fact that, unlike the White Sea coast, the climatic conditions were more favourable and the people did not need to go further inland in search of warmer places. According to Deviatova (1984: 50), hunting was becoming an important, though not dominant, occupation.

No permanent semi-underground dwellings have been found at Sperrings sites. Apparently these early people built light above-ground structures like the Sami 'chum', or a Native American wigwam. On the one hand, climatic conditions may have become favourable enough so that early populations stopped building permanent warm dwellings. On the other hand, natural resources may have increased in abundance and the people could have become less dependent on certain territories.

In Finland, where the majority of Sperrings sites are found on ancient terraces of the Baltic Sea, the situation was slightly different. Specialized seal hunting came to replace terrestrial hunting at around the same time as ceramics came into usage amongst these hunter-fisher-gatherer communities. This may have been caused by a sharp reduction, even extinction, in elk and other populations hunted for meat, such as beaver, hare and bear. The decline in elk populations may be attributed to changes in climate. Warming led to a fall off in elk populations but also to increases in the population sizes of marine fauna (Siiriainen 1982: 18). With this switch in focus from a terrestrial to a marine-based economy in which seal hunting played an important role, the key to the emergence of early ceramics may lie in the need to render and store seal fat in ceramic vessels (Siiriainen 1980: 19, 1982: 18). Other researchers also support this suggestion (Dolukhanov 1986: 5; Núñez 1990: 38).

However, this hypothesis has its opponents. A Finnish zoologist, M Hiekkanen (Hiekkanen 1990: 315–316), has argued that the seal was dominant in hunter-gatherer economies in Finland only in coastal areas prior to the emergence of pottery during the Mesolithic in Finland, where seal bone accounts for 80% of all the osteological material of the Early Neolithic period. Thus, seal hunting was likely to occur only in marine areas, but hunting for forest animals (deer, beaver, etc.) became widely distributed.

Studies from early hunter-gatherer settlements in Estonia where pottery was being used have indicated that there are equal quantities of seal bones at seaside and inland settlements (Paaver 1965: 357). As the quantity of elk bones decreases, bones of hoofed mammals of southern origin increase in number. That is probably connected with vegetation and climatic displacements to the north (Paaver 1965: 3).

A Finnish zoologist, P Ukkonen (1993: 261) is probably correct in arguing that the reduction of elk bones in Finnish sites is not linked to elk population depletion induced by hunting, but to a diversification in the economy and a greater focus on aquatic resources. Researchers also suggest that Finland can hardly have earned a reputation as 'a land of reindeer hunters', because bones of these animals only appear for the first time in the Atlantic period, and even then are extremely rare finds in settlement contexts (Ukkonen 1993: 259).

Thus, Sperrings sites in the territories of Karelia and Finland are considered to have great differences in their location and economies. If marine animal hunting dominates in Finland, then fishing and elk and reindeer hunting are typical of Karelia. Taking into consideration the practically simultaneous appearance of early pottery in Finland and Karelia, it seems wrong to presume that it was used only for fat storage. In general, it appears that the earliest pottery was probably used for a variety of purposes, though saw only minor usage during the initial phases of adoption. Much more widespread usage of pottery occurred somewhat later, achieving more general currency only during the Late Neolithic–Eneolithic.

Although clay used for Sperrings pottery has not been analyzed, it is still possible to identify such pottery as locally produced. In Karelia and Finland, there exist a large number of high-quality band clay deposits, for example, on the southern shore of Lake Onega at Tudozero V settlement. This proves that earthenware was not produced elsewhere and subsequently exchanged or transported into these local communities.

Ceramics of the Sperrings Culture

More than 8,000 fragments of Sperrings pottery have been recovered from prehistoric settlements in southern and central Karelia. After individual features in rims, walls and ornamentation have been analyzed, it became clear that these were fragments of 1,473 vessels. Four groups of vessels have been classified on the basis of their form:

1. Very large vessels: neck diameter is 30–50 cm; average thickness of walls 0.8–1.3 cm; half egg-shaped form with a conical or convex base; also included are half egg-shaped vessels with a footring.
2. Large vessels: neck diameter is 20–30 cm; average thickness of walls 0.4–0.7 cm; straight or slightly profiled rims (bent outside or bent inside) with a conical or convex base.
3. Small vessels (cup-like, with plain bottom): neck diameter is 20 cm; average wall thickness 0.4–0.6 cm.
4. Very small vessels (small saucer-like, cup- and half-egg-shaped pots): average neck diameter is 12 cm; average thickness of walls 0.3–0.5 cm, rare thick-walled pots are 0.7 cm thick. Only 1–5 vessels per settlement have been found.

Vessels of Group 2 are most abundant and make up 50–60% of the total. Although Sperrings ceramics from the Lake Onega sites yielded only finished pottery it does not mean that the ceramics were not produced locally. Small pieces of burnt clay, about 2 × 1.3 × 1.5 cm in size, are often found during excavations. Some researchers interpret them as patterns to be used on clay bands (Kriiska 2002: 175). Sperrings ceramics were produced from 4–6 cm-wide clay bands by coiling or moulding (Andreeva and Gusentsova 1996: 226; Vitenkova 1996: 70). The authors consider both methods to be widely used.

Early Sperrings ceramics were made from clay with the addition of coarse and fine sand and sometimes quartz. Sperrings potters started to include bird droppings into the clay at a later period, a characteristic feature of southern Lake Onega settlement ceramics (Adel 1995: 61; Vitenkova 1996: 70). Larger vessels were made from clay with more admixtures and poorly fired; probably the open-fire method was used, which resulted in low quality and uneven surface. As a result, the colour of large vessels was dark or greyish-brown. Some recovered sherds are crumbly and brittle. In smaller vessels, the proportion of inclusions decreases, the quality of firing improves and vessels are generally red or light brown in colour. Experimental production of replica Sperrings pots was carried out in field conditions during the summer season in 1998. It demonstrated that vessels made from a mixture of clay and coarse and fine sand could withstand the test of boiling water over an open fire for 10–12 minutes (Sizova 2001: 5). The experiment has shown that Sperrings vessels may have been used not only for storage, but for cooking. Earthenware might also have been used for boiling elderberries in the Upper Volga settlements, which are adjacent to the Sperrings settlements (Engovatova 2000a: 211). It should be noted, though, that Sperrings ceramics lack visible signs of organic residues, and it is difficult to draw conclusions about their usage as cooking pots.

The ornamentation of Sperrings ceramics includes several features. The first element is fish vertebrae impressions (Figure 8.2; Figure 8.3: 1, 2, 6, 7; Figure 8.4: 4–8, 10; Figure 8.5: 5). These are found on more than 50% of all vessels and include the vertebrae of pike, bream, perch, roach and whitefish (Andreeva and Gusentsova 1996: 226; Titov 1972: 36). The pot surface is drawn or punctated with one or two sides of fish vertebrae. Vertebrae ends are used for making pits. The branchiate cartilage of humpback salmon and pike and perch bones are used for making comb-like impressions (Ivanichev and Ivanicheva 2000: 292).

The second and third elements of ornamentation are lines either drawn, or punctated on raw clay (Figure 8.3: 3–5; Figure 8.4: 1–3, 9; Figure 8.5: 1–4, 6, 7). About 35% of the vessels are decorated with these impressions made either with fish vertebrae, or other tools including slate rods, animal ribs, notched bones or animal teeth (Ivanichev and Ivanicheva 2000: 292). The width of these drawn and punctated lines ranges from 0.1 to 0.5 cm. These impressions are made with the same kind of tool, but by different techniques. Pots are often decorated using both techniques.

The fourth kind of ornamentation, found on around 7% of the vessels, represents 'cord-like' impressions of various lengths (Figure 8.6). Kalinina (2000: 267), an archaeologist from Saint Petersburg, has put forward two

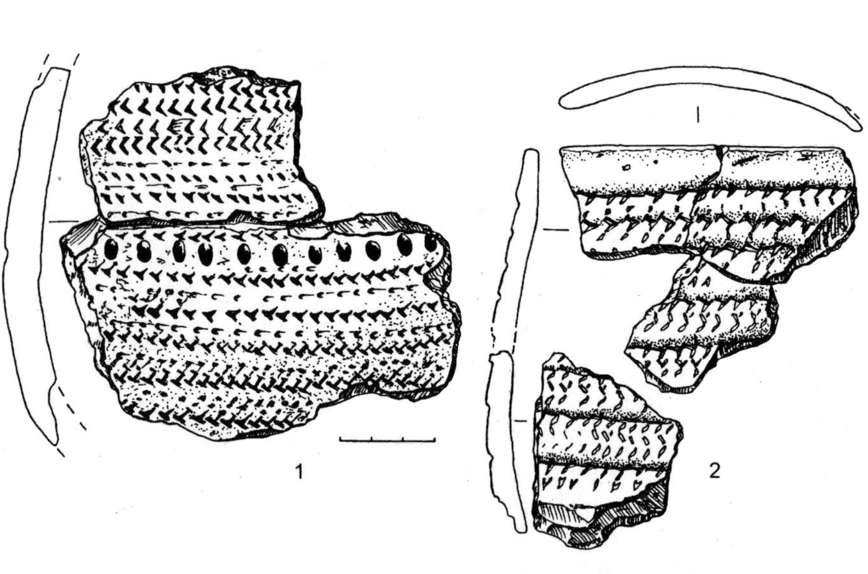

Figure 8.2 Fragments of Sperrings ceramics with vertebrae impressions.

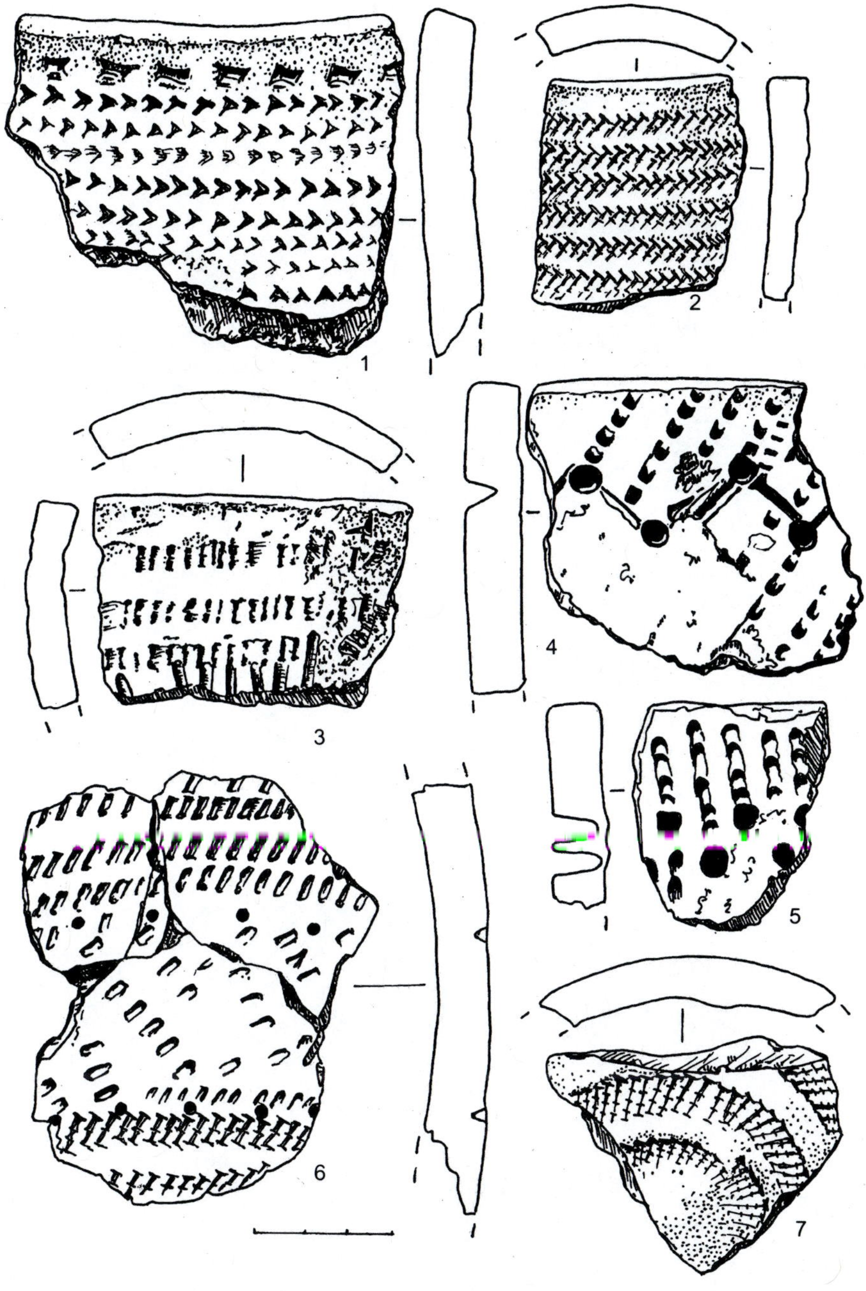

Figure 8.3 Fragments of Sperrings ceramics with vertebrae impressions (1, 2, 6, 7) and lines either drawn or impressed on the raw clay (3–5).

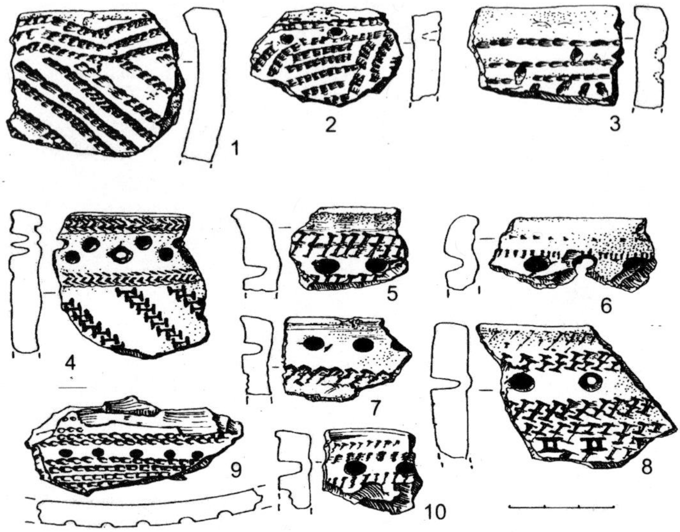

Figure 8.4 Fragments of Sperrings ceramics with vertebrae impressions (4–8,10) and lines either drawn or impressed on the raw clay (1–3, 9).

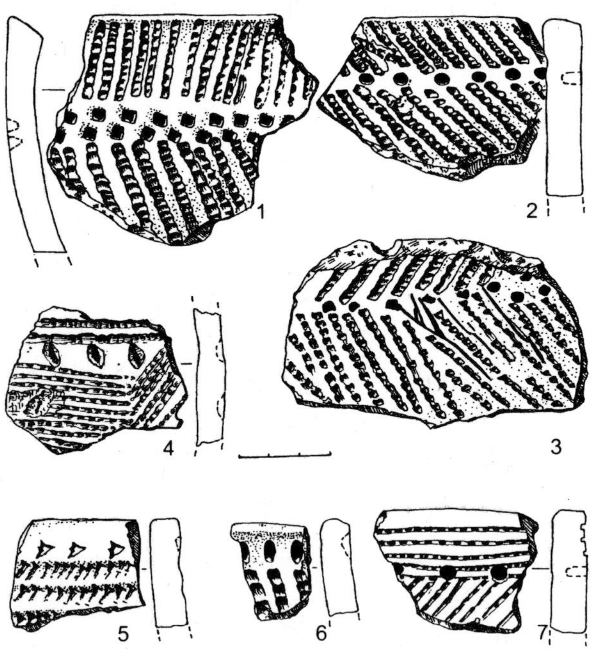

Figure 8.5 Fragments of Sperrings ceramics with vertebrae impressions (5) and lines either drawn or impressed on the raw clay (1–4, 6, 7).

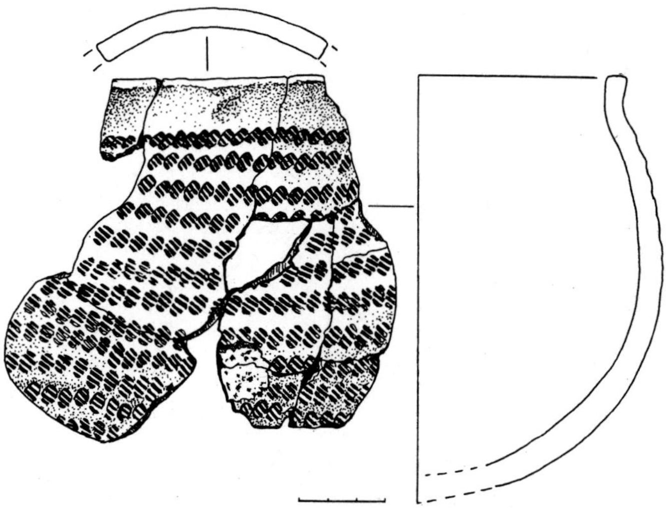

Figure 8.6 Fragments of Sperrings ceramics with 'cord-like' impressions.

hypotheses on the origin of cord-like impressions: either a cord was used for winding around tool handles or a cord itself was applied as a tool for making decorations.

Decorations of Sperrings ceramics are made with one kind of tool, and each individual pot is decorated with a unique, never repeated element. Pits of various shapes (roundish, squarish, right-angled) can be considered as additional design elements. These are made by pinning and found on more than 60% of the vessels. Fish vertebrae, slate rods, animal ribs, notched bones or animal teeth might have been used for making pits. In some rare cases, the entire ornamentation of the vessels was done with this technique (less than 1% of the vessels).

To sum up, ornamentation of Sperrings ceramics includes vertebral impressions (60–80% of the vessels), punctated lines (20–30%), drawn lines (10%), cord-like (5–6%), and comb-like stamps (2–4%), pits (60–70%), notches made with one or two sides of vertebrae (1–5%). Punctating and drawing (70–80%), stamping and pinning (20–30%) are thought to be the predominant Sperrings ornamentation techniques.

Säräisniemi I Ceramics

A total of 797 fragments of Säräisniemi I ceramics have been recovered from archaeological sites in northern Karelia. From these assemblages, 117 separate vessels can be identified on the basis of individual differences in their forms, rims, walls and ornamentation (German 1997: 61). In general, these vessels are badly fired. The clay is tempered with coarse and fine sand, and, in rare cases, organic matter and mica were also added. The walls of the larger pots have an average thickness of 0.8–1.2 cm. Smaller pots have walls 0.5–0.6 cm thick, with their thickness increasing from rim to body.

The outer surface of some vessels is evenly painted with red ochre. One of the red vessels has been partly reconstructed. It represents a bowl with a rounded bottom, thin and straight rim, and walls thickening towards its base to 1.4 cm (Pesonen 1986: 19, figures 10, 11). This pot is similar to Säräisniemi I ceramics recovered from sites in northern Finland. M. Torvinen (2000: 24), a Finnish archaeologist, has divided the Säräisniemi I pottery from these sites into two groups on the basis of rim dimensions and diameter. The first group consists of larger pots whose walls are 0.1–0.5 cm thick, 20–30 cm in diameter, 40 cm high, and with 8–10 litre capacity. The second group includes smaller 'beaker'-type vessels with walls 0.8–0.9 cm thick. These may have been used for drinking (Torvinen 2000: 24). The bowl found in Karelia can be attributed to the second group.

Eight elements of Säräisniemi I ornamentation can be distinguished:

1. Cord-like impressions (Figure 8.7: 4);
2. Comb-like impressions (Figure 8.7: 5–7; Figure 8.8: 2, 3, 7);
3. Lattice-like impressions (Figure 8.6);
4. Impressions of a shell (cardium) (Figure 8.9: 5);
5. Rectangular and u-shaped notches (Figure 8.7: 1; Figure 8.8: 1, 4, 5; Figure 8.9: 1, 4);
6. Punctated lines (Figure 8.7: 2, 3);
7. Drawn lines (Figure 8.9: 3);
8. Pits (Figure 8.9: 2).

The combinations of the first six ornamental elements on the surface of an individual vessel are rarely met. Only two combinations have been found: comb stamps and u-shaped notches, and drawn lines and rectangular notches.

All the Säräisniemi I vessels were entirely ornamented on the outer surface. The pots found did not repeat ornamental elements. According to Torvinen (2000: 24), each vessel was ornamented with a tool specially made for the purpose and the impressed design was unique to every individual pot (Torvinen 2000: 24).

Owing to low firing temperatures, Säräisniemi I sherds found in Karelia were crumbly and badly preserved. This makes a detailed reconstruction of ornamental compositions difficult. It should be noted that the early potters

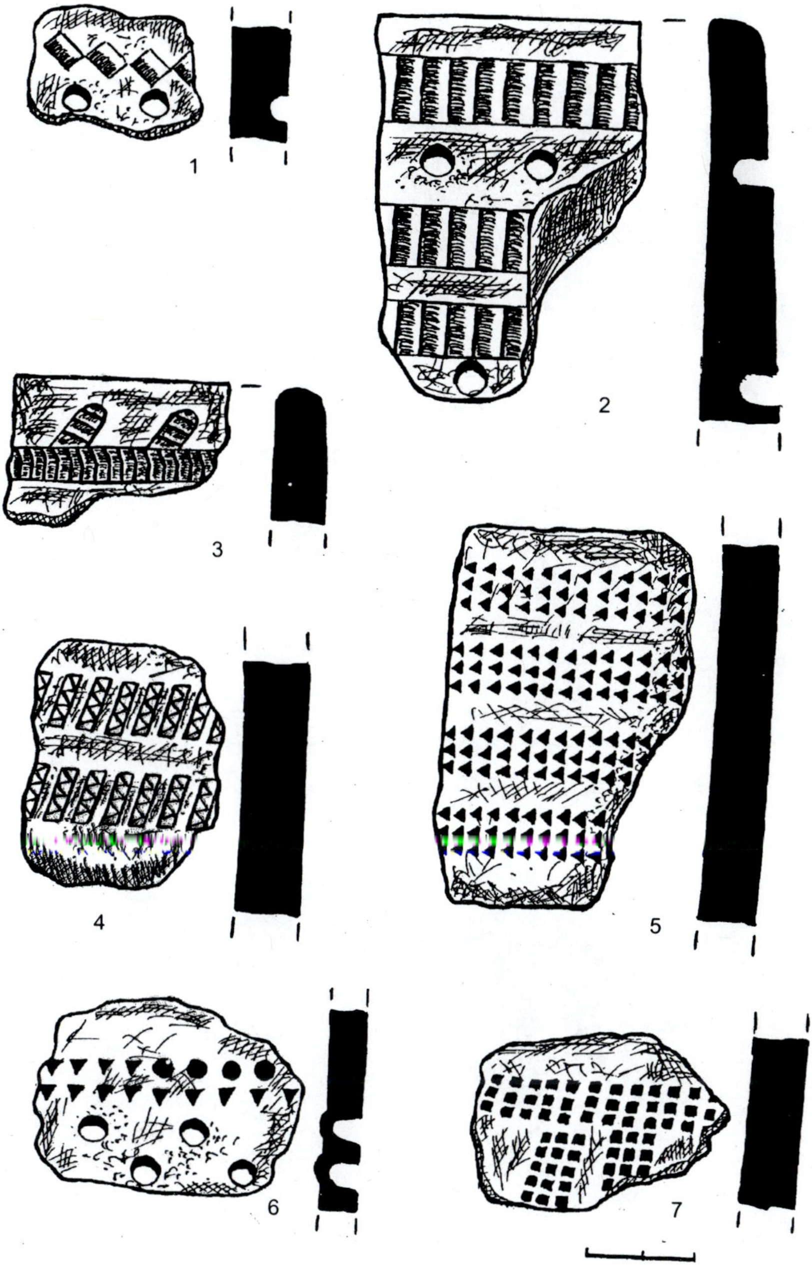

Figure 8.7 Fragments of Säräisniemi I ceramics with rectangular and u-shaped notches (1), impressed lines (2, 3), 'cord-like' impressions (4), 'comb-like' impressions (5–7).

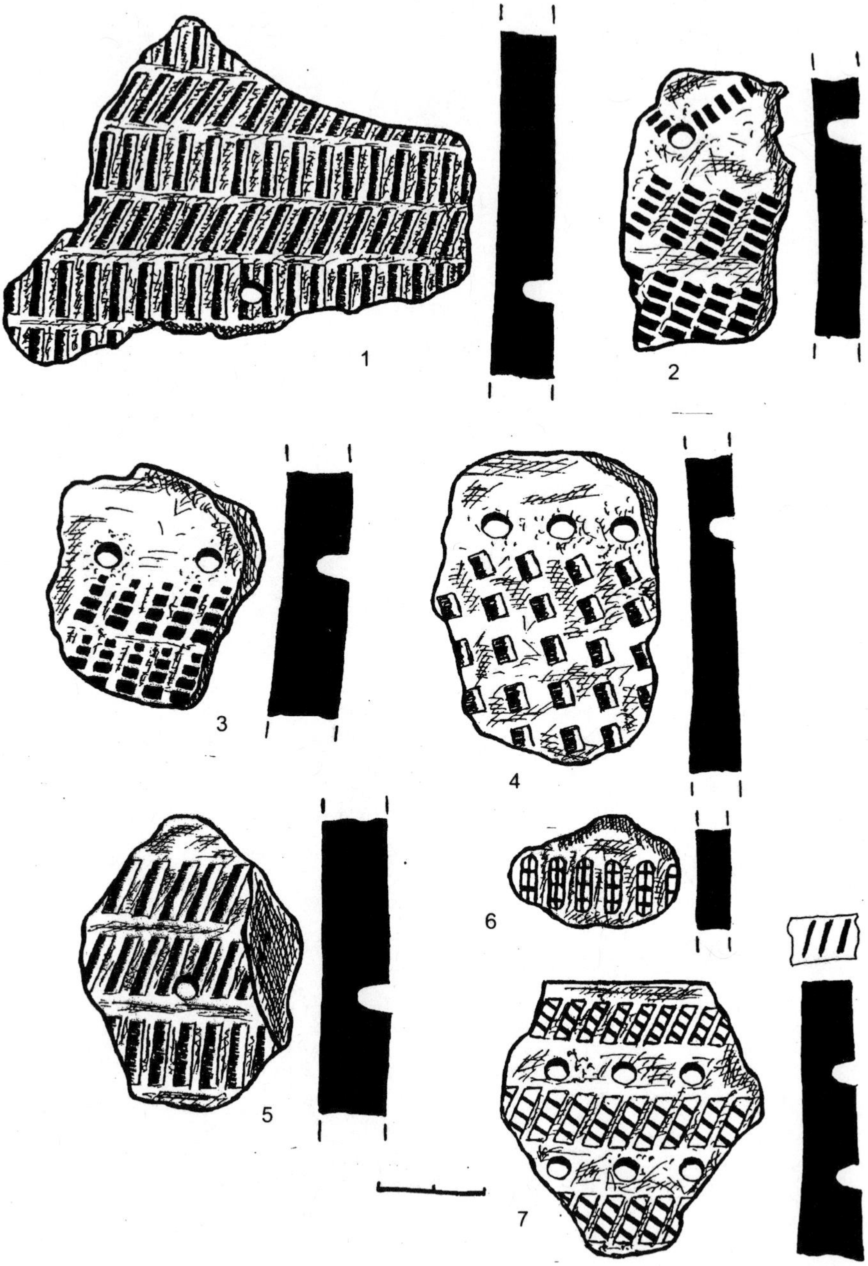

Figure 8.8 Fragments of Säräisniemi I ceramics with rectangular and u-shaped notches (1, 4, 5), 'comb-like' impressions (2, 3, 7), 'lattice-like' impressions (6).

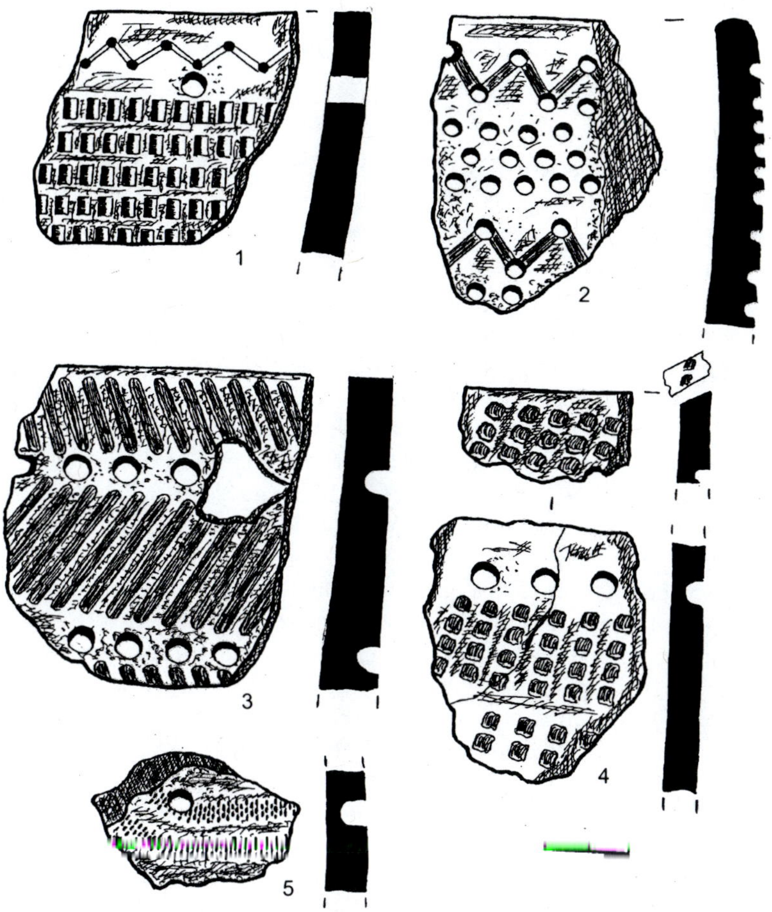

Figure 8.9 Fragments of Säräisniemi I ceramics with rectangular and u-shaped notches (1, 4), pits (2), drawn lines (3), impressions of shell (5).

of the Sperrrings culture did not know complicated designs. Shumkin (1996: 69–70) writes about Sperrings ceramics: 'It is not the variety of elements or complexity of design, but a means of stamping that makes the ceramics look attractive and splendid'.

The Säräisniemi I and Sperrings ceramics have common features in the forms of vessels and their rims. Moreover, vessels of both types are painted with red ochre, especially the early pottery. However, the major differences and distinguishing features are found in the structure of ornamental elements and the decoration composition. Sperrings vessels do not include lattice-like stamps, shell impressions, Z-shaped, zig-zag and comma-shaped impressions which are typical of the Säräisniemi I ceramics of northern

Finland, northern Norway and the Kola Peninsula (Gurina 1997: 47; Simonsen 1957: 240; Torvinen 2000: 4). In contrast, Sperrings ceramics have fish vertebrae impressions of the kind noted above.

Origins of Sperrings and Säräisniemi I Ceramics Amongst Karelian Hunter-Gatherers

In the earlier archaeological literature on Karelian prehistory, it was assumed that local hunting, fishing and gathering communities were responsible for the independent invention of pottery, and that the Sperrings culture represented a purely local phenomenon that had not arisen out of culture contacts with other areas (Filatova 1972: 30; Pankruchev 1978: 43; Pesonen 1991: 84; Titov 1972: 50; Vitenkova 1996: 76). There are strong continuities in several technological traditions including diverse stone tool traditions (axes, adzes, chisels, knives and scrapers), and manufacturing techniques of slate instruments used by local communities both prior to and after the emergence of ceramics.

More recently, both Karelian and Finnish archaeologists have linked Sperrings ceramics with pottery types recovered in the 1970s from the Upper Volga Early Neolithic sites, where hunter-gatherers were also using pottery styles that may have arrived by diffusion from areas occupied by the Volgo-Kama and Dnepr-Don archaeological cultures (Eurapaeus-Äyräpää 1955: 157; Pankruchev 1978: 40; Pesonen 1991: 69; Vitenkova 1996: 76). Núñez (1990: 32) and Timofeev (1997: 20) have argued that the diffusion of ceramic technology into these regions has taken place without major population movements or transformations in economy or settlement. However, the suggestion that knowledge of pottery production reached Karelia from western regions has been strongly rejected by Finnish researchers (Eurapaeus-Äyräpää 1955: 157; Luho 1957: 159; Meinander 1982: 22).

Säräisniemi I ceramics have also been argued to be a local and independent invention according to Finnish researchers. This pottery first appeared in the area of Lake Oulujarvi (Huurre 1983: 142–143; Torvinen 1997: 26, 1998: 40) but later spread northwards to the Arctic Ocean. Some have linked this rapid diffusion of Säräisniemi I ceramics to processes of ethnic differentiation, which later gave rise to the present-day Sami peoples (Huurre 1991: 61). Others suggest that the population of northern Finland had been ethnically isolated as far back as the Early Holocene, during the time of the Suomusjarvi archaeological culture (Torvinen 1998: 44).

According to Gurina (1997: 132–133) and Shumkin (1993: 35), early ceramics first appeared in the Kola Peninsula among local Mesolithic tribes whose material culture has strong similarities with that of northern Karelia.

To understand the origin of the Sperrings culture, it is necessary to consider the evidence recovered from neighbouring Early Neolithic cultures. To the south, Sperrings sites along the Gulf of Finland and the Karelian Isthmus border the areas occupied by sites of the Narva culture (Figure 8.10). However, there is no interrelation between these two archaeological cultures as regards pottery ornamentation and stone tools. Only one fragment of Sperrings ceramics with crushed shells has been found in Berezye settlement on the southern shore of Lake Ladoga.

The timing of the Narva culture (in the territory of what is today Estonia) has been defined by radiocarbon analysis and dated to approximately 6,023±95 bp (Tln-1989)–5,268±95 bp (Tln-1992) (Kriiska 1996: 382). At present, on the basis of geological evidence and radiocarbon dating, Sperrings ceramics in the territory of Finland has been dated as follows: in southern Pochjanmaa, 6,100 bp; in southern Finland, 6,200 bp; in southeastern Finland, 6,350 bp (Salomaa and Matiskainen 1985: 153). Based on these dates, it appears that Narva ceramics in Estonia, and Sperrings pottery in Finland emerged at approximately the same time. This may lead one to assume that the use of organic temper in both traditions was the outcome of cultural exchange between these co-existing and adjacent pottery-making

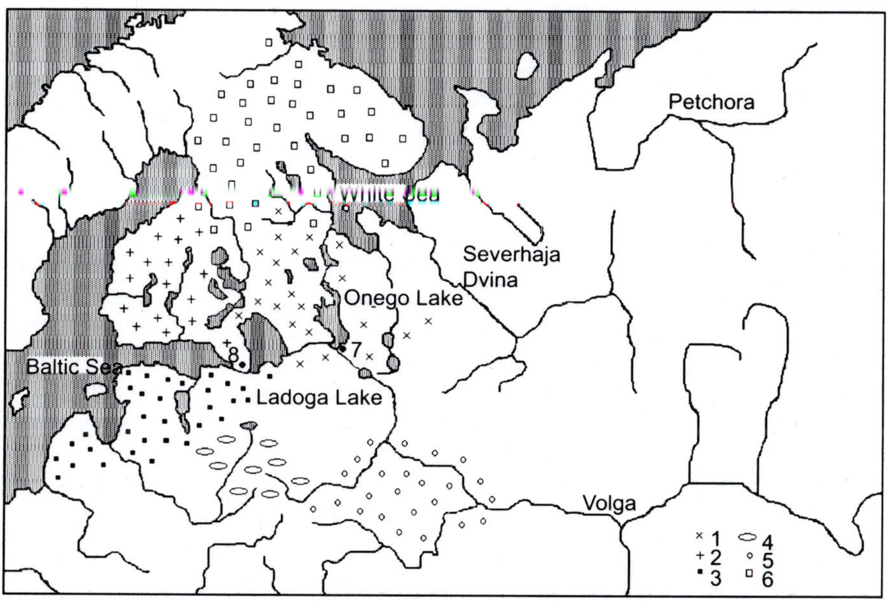

Figure 8.10 Location map showing early pottery cultures and archaeological sites mentioned in the text: (1) Sperrings culture; (2) sites with ceramics Ka I:I (Sperrings); (3) Narva culture; (4) Valdai culture; (5) Verkhniaia Volga culture; (6) Säräisniemi I culture; sites: (7) Tudozero V; (8) Khepojarvi.

traditions. On the other hand, it could be that Narva ceramics were also an unrelated, independent and localized innovation amongst early hunter-gatherer communities in the area (Kriiska 1996: 382).

In the south, the Sperrings culture is adjacent to the Valdai Early Neolithic culture. The earliest type of Valdai ceramics, so-called 'kotchishchensky' type, dates back to 7,000 bp (Sinitsyna and Zaitseva 1995: 36), also suggesting very early origins. This may also account for very few similarities between Sperrings and Valdai ceramics, either in ornamentation or the tempers used. Both may be considered independent local innovations.

There has been greater focus on potential similarities between the ceramics of the Sperrings and the Upper Volga archaeological cultures. These cultures have been subjected to a detailed chronology, and the occurrence of the first pottery has been dated to 7,200–6,800 bp. (Engovatova 2000b: 94). The Sperrings culture is most similar to the Upper Volga pottery of the second stage of the Early Neolithic, which is dated to 6,500–6,000 bp (Engovatova 1998: 242).

Vessels with a convex base appear only in the second stage of the Upper Volga pottery. The outer surface is sometimes painted with red ochre, and clay is sometimes mixed with organic matter, crushed shell, and also coarse and fine sand. Basic ornamentation is represented by cord-like impressions, line drawings and comb stamps.

Given these pottery features, it may be noted that the Upper Volga vessels resemble the Sperrings ceramics. This similarity is reflected in the form of the vessels, painting of the vessels with red ochre, as well as in common elements of both ornamental technique and decorative composition. Some Upper Volga and Sperrings pottery are devoid of decoration (Kostyleva 1994: 55; Krainov 1996: 167, 169).

Despite these broad similarities, we must note that the Upper Volga and Sperrings ceramics have significant differences. For example, the Upper Volga pottery does not have fish vertebra impressions and has few punctated lines. While cord-like and comb stamps are used by both traditions, the forms and sizes of the decorations differ substantially (Engovatova 1998: 240; Kostyleva 1994: 55). Moreover, the decoration designs on Sperrings pottery are also significantly different, although some patterns are similar to the Upper Volga vessels (Kostyleva 1994: 55, figures 9, 11). Timofeev (1997: 20) has suggested that these broad similarities in ornamental compositions are found not only in the Upper Volga and Sperrings ceramics, but also in the ornamentation of other early hunter-gatherer pottery, in the Middle Don and Volgo-Kama archaeological cultures.

Recent discoveries in Karelia have opened new dimensions in the Sperrings and early pottery debates. It now appears that Sperrings vessels may not have been the earliest pottery in the area of Lake Onega. Vessels ornamented with comb- and figure-stamped impressions have been recovered from the shoreline hunter-fisher-gatherer settlement of Tudozero V, located at the southern shore of Lake Onega, in culture layers below those containing Sperrings ceramics. These older pots are decorated with vertebra impressions and receding lines. It has been argued that these earlier ceramics went through three phases of development at Tudozero V. The first stage is characterized by comb-decorated wares, the second by comb decoration and vertebra impressions. Then the punctating technique comes into existence. Finally, reduction in comb stamps and increasing use of the punctating technique by natural artefacts can be observed. Complex decorative designs also appear during this later stage (Ivanichev and Ivanicheva 2000: 292).

The first stage of Tudozero V site has been dated to 7,240±60 bp (TA-2354), which means that the pottery recovered from this site represents the earliest ceramics found anywhere in the territory of European Russia. However, the author believes that this radiocarbon dating may be connected to the Mesolithic complex of Tudozero V settlement. It is only at the second stage of Tudozero V settlement that the oldest Sperrings type pottery has been found, which is associated with the dates of 6,110±100 bp (TA-7662), 6,230±120 bp (TA-7663), 6,250±50 bp (TA-8050) and 6,075±20 bp (TA-6699). More Sperrings ceramics are found at the third stage of Tudozero V settlement, several pots ornamented with punctated lines having close resemblance to Sperrings pots from Obonezh'e. Besides, comb pots also appear at Tudozero V at that time.

Gerasimov (2003: 20) argues that the Sperrings pottery used by hunter-gatherer cultures of the Karelian Isthmus had their origins in earlier local populations, but was also influenced by the Upper Volga culture. The author thinks that this is also supported by two radiocarbon dates from Hepojarvi, the Sperrings ceramics settlement: 6,480±80 bp (TA-1412) and 6,380±60 bp (TA-1411) (Vereshchagina 2003: 149). In addition, the ornamentation of the Sperrings ceramics has some similarities with the pottery of the second stage of the Upper Volga culture (Kostyleva 1994: 55; Krainov 1996: 167, 169).

Sufficient radiocarbon data from the earliest Sperrings sites of the Lake Onega basin has recently been obtained. Pegrema IX, the Sperrings site located on northern shore of Lake Onega, is dated to 6,510±90 bp which coincides with the geo-morphological and pollen data (Deviatova 1986: 55). Other researchers also support this view (Pesonen 1991: 66; Vitenkova 1996: 77–78). All the Sperrings pottery recovered from Pegrema IX site is decorated with fish vertebrae designs, which suggests that this ornament was invented by the early people who inhabited the northern shores of Lake Onega.

Sholtozero XI and Yya III sites located on the southern shores of Lake Onega were radiocarbon dated to 6,480±70 bp (TA-1312) and 6,770±80 bp (TA-2352) respectively, placing them 500 years apart. Their earthenware fragments have been collected from the surface layer. Thus, they cannot serve as the basis for dating the Sperrings ceramics. Moreover, the monument represents a mixed complex where, along with Sperrings pottery, Mesolithic stone tools and some fragments of pit-comb and asbestos pots were found. According to Kosmenko (1992: 52–54), the hearth (6,400±150 bp [TA-1552]) recovered from Shettima I site with a mixed complex can be attributed to the Sperrings complex. This hypothesis is supported by fragments of early pots found at the site.

Conclusion: The History of Early Ceramic Innovations in Karelia

This chapter focuses on the earliest ceramics of the prehistoric hunter-gatherer communities in Karelia, and follows studies presented by the author in recent publications (German 2000, 2002b). On the basis of radiocarbon dating and the location of archaeological sites, and in relation to known shoreline displacements, it can be concluded that Sperrings ceramics were first produced along the northern and northwestern shores of Lake Onega around 6,500–6,400 bp. Settlements with ceramic assemblages during an early stage of Sperrings ceramics culture occupy the top terraces of the former lake shoreline. Many radiocarbon dates have been obtained (German 2002b: 264–266). Early pottery production, along with the skilled use of fish bones for decoration of the vessels, started here and was quickly diffused among hunter-fisher-gatherer communities inhabiting the lake shores. Sperrings ceramics also appear to have diffused rapidly to further settlements along the nearby lake shores of Sjamozero and Vodlozero, as well as, possibly, to the eastern shores of Lake Onega. Knowledge of Sperrings ceramics was carried out of the area, but eventually struggled to diffuse across the watersheds between the Baltic and White Sea basins. The eventual diffusion of Sperrings ceramics to the White Sea coast and into the lower reaches of the River Vyg only took place at the same time as the arrival of Pit-Comb Wares into the area around 5,990±100 bp–5,460±80 bp (German 2002b: 264).

We may safely conclude that Sperrings ceramics are some of the earliest kinds of pottery originating in this area, and that Sperrings sherds are found at lower stratigraphic levels than those containing Pit-Comb Wares.

Understanding the emergence of Säräisniemi I ceramics has proved more difficult, although the dating of organic residues on the vessels has enabled the first occurrence of Säräisniemi I ceramics in northern Finland to be dated to 6,100 bp around Lake Oulujarvi, 6,000 bp in the Kiiminkijoki region and 5,800 bp around Lake Inari (Torvinen 1998: 40). However, in Karelia, the studies of Säräisniemi I sites have not been brought up to date, so the chronology and periodization of Säräisniemi I pottery is poorly understood. By drawing analogies with the Finnish data, we can suggest tentatively that Säräisniemi I ceramics probably appeared in northern Karelia around 6,200–6,000 bp, and in a short time, had been diffused further to the White Sea coast, so that by 5,900–5,800 bp, Säräisniemi I pottery was being produced in hunter-fisher-gatherer communities inhabiting an extensive area from the Barents Sea and to northern Karelia and northern Finland, and from the Gulf of Bothnia to the White Sea.

Note

1. This chapter employs the 'eastern' definition of Neolithic, i.e., pottery-using hunter-gatherers. See introductory chapter (Jordan and Zvelebil) for further discussion of the various definitions of Neolithic.

References

Adel, V V (1995) 'Nekotorye metodologicheskie problemy izucheniia keramiki sperrings' [Some methodological aspects of Sperrings ceramics studies]', 3 *Vestnik Karel'skogo Kraevedcheskogo Muzeia*, Petrozavodsk: Karelian Local History Museum, pp 53–62

Andreeva, N A and Gusentsova, T M (1996) 'Nekotorye osobennosti izgotovleniia ranneneoliticheskoi keramiki Mezozer'ia [Some characteristics of Early Neolithic pottery making in Mezozer'e]', 2 *Tverskoi arkheologicheskii sbornik*, 226–234

Biske, G S, Lak, G Z and Lukashov, A D (1971) *Stroenie i Istoriia Kotloviny Onezhskogo Ozera [Structure and History of Lake Onega Basin]*, Petrozavodsk: Karelian Research Centre, Russian Academy of Sciences (KRC RAS)

Deviatova, E I (1976) *Geologiia i Palinologiia Golotsena i Khronologiia Pamiatnikov Pervobytnoi Epokhi v Iugo-zapadnom Belomor'e [Holocenic Geology and Palinology, and Chronology of Pre-Historic Monuments in Southwestern Coast of the White Sea]*, Leningrad: Nauka

——— (1984) 'Paleogeografiia stoianok Sheltozera [Palaeography of Sheltozero sites]', in *Arkheologicheskie Pamiatniki Basseina Onezhskogo Ozera*, Petrozavodsk: Karelian Research Centre, Russian Academy of Sciences (KRC RAS), pp 25–27

———(1986) *Prirodnaia Sreda i ee Izmeneniia v Golotsene (Poberezh'e Severa Tsentra Onezhskogo Ozera) [The Environment and its Developments in the Holocene (Northern Shoreline of Central Lake Onego)]*, Petrozavodsk: Karelian Research Centre, Russian Academy of Sciences (KRC RAS)

——— (1988) 'Paleogeografiia i osvoenie chelovekom Karelii [Palaeography and human colonization of Karelia]', in *Poseleniia Drevnei Karelii*, Petrozavodsk: Karelian Research Centre, Russian Academy of Sciences (KRC RAS), pp 7–18

Deviatova, E I, Filatova, V F and Lobanova, N V (1987) *Paleogeografiia Arkheologicheskikh Pamiatnikov Pindushi i Muromskoe [Paleography of Archeological Monuments at Pindushi]*, Petrozavodsk: Karelian Research Centre, Russian Academy of Sciences (KRC RAS)

Deliushina, I V (1989) 'Razvitie prirody Priladozh'ia v golotsene [Developments in the Holocene environment in Lake Ladoga basin]', in *Istoriia Ozer,* Leningrad: Nauka, pp 43–44

Dolukhanov, P M (1986) 'Natural environment and the Holocene settlement pattern in the north-western part of the USSR', III *Fennoscandia Archaeologica*, 3–16

Dolukhanov, P M and Khotinskii, N A (1974) 'Paleogeograficheskie rubezhy golotsena i mezo-neoliticheskaia istoriia Evropy [Palaeographic borders of the Holocene and Mesolithic history of Europe]', in *Pervobytnyi Chelovek i Prirodnaia Sreda,* Moskow: Nauka, pp 211–216

Ekman, I M, Lukashov, A D, Kolkanen, A M and Liiva, A A (1988) 'Dinamika razvitiia ozer i ozernosti Karelskoi ASSR po dannym radiokhronologii (za poslednie 13000–12000 let) [Dynamics of lakes and watersheds in the Karelian ASSR (for the period of 13,000–12,000 years)]', in *Izotopno-Geokhimicheskie Issledovaniia*, Leningrad: Nauka, pp 206–217

Elina, G A (1980) 'Dinamika lesov i bolot Karelii v golotsene po palinologicheskim dannym [Dynamics of Karelian forests and mires in the Holocene according to palinological data]', in *Bolotno-Lesnye Sistemy Karelii i ikh Dinamika*, Petrozavodsk: Karelian Research Centre, Russian Academy of Sciences (KRC RAS), pp 5–51

——— (1981) *Printsypy i Metody Rekonstruktsii i Kartirovaniia Rastitel'nosti Golotsena [Principles and Methods of Holocene Vegetation Reconstruction and Mapping*], Leningrad: Nauka

Elina, G A and Lebedeva, R M (1990) 'Paleogeografiia severnogo Pribelomor'ia Karelii v golotsene [Palaeography of the north White Sea coast in Holocene]', 1 *Chetvertichnyi Period: Metody Issledovaniia, Stratigrafiia i Ekologiia,* 194–195

Engovatova, A V (1998) 'Khronologiia epokhi neolita Volgo-Okskogo mezhdurech'ia [Chronology of the Neolithic Volga-Oka watershed]', 3 *Tverskoi arkheologicheskii sbornik*, 238–246

——— (2000a) 'O kharaktere ispol'zovaniia keramicheskikh sosudov naseleniem Volgo-Okskogo mezhdurech'ia v neolite' [Use of ceramic vessels by the humans in the Volga-Oka watershed]', 4 *Tverskoi arkheologicheskii sbornik*, 210–212

——— (2000b) 'Khronologiia epokhi neolita Volgo-Okskogo mezhdurech'ia [Chronology of the Neolithic Volgo-Oka watershed]', in *Khronologiia Neolita Vostochnoi Evropy. Tezisy Dokladov Mezhdunarodnoi Konferentsii Posviashchennoi Pamiati N.N. Gurinoi,* Saint Petersburg: Nauka, p 94

Eurapaeus-Äyräpää, A (1930) 'Die relative Chronologie der steinzeitlischen Wohnplatzfunde in Finland', I *Acta Archaeologica*, 165–220

——— (1955) *Den Yngre Stenalderns Kronologi I Finland Och Sverige,* LXII *Finskt Museum*, 5–52

Filatova, V F (1972) 'K voprosu O sviazi kamennykh orudii pamiatnikov s chistym kompleksom keramiki sperrings i pozdnemezoliticheskikh [On relationship of stone tools with pure sperring complex and late-Neolithic]', in *Arkheologicheskie Issledovaniia v Karelii,* Leningrad: Nauka, pp 10–33

Gerasimov, D V (2003) 'Istoriia izucheniia, khronologiia i periodizatsiia pamiatnikov epokhi neolita iuga Karel'skogo peresheika [History of exploration, chronology and periodization of Neolithic monuments in the south of the Karelian isthmus]', 1 *Arkheologicheskoe Nasledie Sankt-Peterburga,* 12–24

German, K E (1997) 'Neoliticheskaia keramika v Severnoi Karelii (Säräisniemi 1) [Neolithic ceramics in northern Karelia (Säräisniemi I)]', 1 *Arkheologiia Severa*, 63–74

——— (2000) 'Problemy khronologii rannego neolita Severo-vostochnoi Fennoskandii [Problems of chronology of the Early Neolithic in northeastern Fennoscandia]' in *Khronologiia Neolita Vostochnoi Evropy. Tezisy Dokladov Mezhdunarodnoi Konferentsii Posviashchennoi Pamiati N.N. Gurinoi,* Saint Petersburg: Nauka, pp 18–19

——— (2002a) 'Problemy khronologii rannego neolita Severo-Vostochnoi Fennoskandii [Chronology of the Early Neolithic in eastern Europe]', in *Khronologiia Neolita Vostochnoi Evropy: Tezisy Dokladov Mezhdunarodnoi Konferentsii Posviashchennoi Pamiati N.N. Gurinoi,* Saint Petersburg: Nauka, pp 18–19

——— (2002b) 'Khronologiia i periodizatsiia kul'tury sperrings [Chronology and periodization of the Sperrings Culture]', 5 *Tverskoi Arkheologicheskii Sbornik*, 264–273

——— (2004) 'Problemy khronologii nachal'nogo etapa rannego neolita Severo-Vostochnoi Fennoskandii [Chronology of the first stage of the Early Neolithic in north-western Fennoscandia]', in *Problemy Khronologii i Etnokul'turnykh Vzaimodeistvii v Neolite Evrazii,* Saint Petersburg: Nauka, pp 56–59

Gurina, N N (1961) 'Drevniaia istoriia severo-zapada evropeiskoi chasti SSSR [Pre-history of the USSR European north-west]', 87 *Materialy i Issledovaniia po Arkheologii SSSR*, 1–524

——— (1997) 'Istoriia kul'tury drevnego naseleniia Kol'skogo poluostrova [Culture history of ancient population at the Kola Peninsula]', 32 *Arkheologicheskie Izyskaniia*

Hiekkanen, M (1990) 'Popytka istolkovaniia morskoi deiatel'nosti naseleniia mezolita i rannego neolita [Approach to define marine activities of the Mesolithic and Early Neolithic populaton]', 1 *Sovetskaia Arkheologiia*, 315–316

Huurre, M (1983) 'Pohjos-Pohjanmaan ja Lapin esihistoria', I *Pohjos-Pohjonmaan ja Lapin historia,* 2–143

——— (1991) 'Oulujokilaakson esihistoria', *Oulujokilaakson Historia,* 4–32

Ivanichev, A M (1997) 'Drevnosti Vytegorii [Pre-history of Vytegoriya], 1 *Vytegra:* [illegible]

Ivanichev, A M and Ivanicheva, M V (2000) 'Tudozero—poselenie pozdnego mezolita–rannego neolita v Iuzhnom Prionezh'e [Tudozero—Late-Mesolithic–Early-Neolithic site in Southern Prionezh'e]', 4 *Tverskoi Arkheologicheskii Sbornik*, 284–295

——— (2004) 'Khronologiia pamiatnikov rannego neolita Iuzhnogo Prionezh'ia [Chronology of Early Neolithic sites in southern Prionezhye]', in *Problemy Khronologii i Etnokul'turnykh Vzaimodeistvii v Neolite Evrazii*, Saint Petersburg: Nauka, pp 60–69

Ivanter, E I (1986) *Mlekopitaiushchie* ['*Mammals*'], Petrozavodsk: Karelia

Kalinina, I V (2000) 'Verevochnyi ornament v neolite [Rope-like ornament in the Neolithic]', 4 *Tverskoi Arkheologicheskii Sbornik*, 263–267

Kalinina, I V and Gadzhieva (Ustinova), E A (1993) 'Arkhaicheskie ornamenty dlia keramiki [Archaic ornaments for ceramics]', 10 *AD POLUS: Arkheologicheskie Izyskaniia*, 83–94

Khomutova, V I (1976) 'Geokhronologiia donnykh otlozhenii po rezul'tatam palinologicheskogo analiza [Geochronology of bottom sediments according to palinological analysis]' in *Paleolimnology of Lake Onego*, Leningrad: Nauka, pp 45–73

Khotinskii, N A (1977) *Golotsen Severnoi Evrazii [Holocene in Northern Eurasia]*, Moskow: Nauka

Khotinskii, N A (1996) 'Prirodnye usloviia v neoliticheskuiu epokhu [Natural environment in the Neolithic]', in *Neolit Severnoi Evrazii*, Moskow: Nauka, pp 10–17

Khotinskii, N A and Savina, S S (1985) 'Paleoklimaticheskie skhemy territorii SSSR v boreal'nom, atlanticheskom i subboreal'nom periodakh golotsena [Palaeoclimatic zones of the USSR territory in the Borealic, Atlantic and Sub-Borealic periods of the Holocene]', 4 *Izvestiia Akademii Nauk SSSR: Seriia geografiia*, 18–34

Klimanov, V A (1978) 'Klimaticheskie usloviia Karel'skoi ASSR v golotsene [Climatic conditions at the territory of the Karelian ASSR in the Holocene]', in *Bolota Evropeiskoi Chasti SSSR*, Leningrad: Nauka, pp 177–185

Kosmenko, M G (1992) *Mnogosloinye Poseleniia Iuzhnoi Karelii [Multi-layer Settlements of Southern Karelia]*, Petrozavodsk: Karelian Research Centre, Russian Academy of Sciences (KRC RAS)

Kostyleva, E L (1994) 'Ranneneoliticheskaia keramika Verkhnego Povolzh'ia [Early-Neolithic ceramics of the Upper Volga region], 1 *Tverskoi Arkheologicheskii Sbornik*, 53–57

Krainov, D A (1996) 'Verkhnevolzhskaia kul'tura [Culture of the Upper Volga region]', in *Neolit Severnoi Evrazii*, Moskow: Nauka, pp 166–173

Kriiska, A (1996) 'The Neolithic pottery manufacturing technique of the lower course of the Narva River', V(3) *Pact* 51, 373–384

——— (2002) 'Formation and development of the Stone Age settlement at Riigiküla, northeastern Estonia', I(8) *Pact* 57, 173–183

Kvasov, D D (1976) 'Proiskhozhdenie kotloviny Onezhskogo ozera [Origin of the Lake Onega basin]', in *Paleolimnologiia Onezhskogo Ozera*, Petrozavodsk: Karelian Research Centre, Russian Academy of Sciences (KRC RAS), pp 7–40

Kvasov, D D (1990) 'Onezhskoe ozero [Lake Onega]', 2 *Istoriia Ozer SSSR*, 77–115

Luho, V (1957) 'Frühe Kammkeramik', 58 *Suomen Muinaismuistoyholistuksen Aikakauskirja*, 141–159

Markov, K K, Poretskii, V S and Shliapina, E V (1934) 'O kolebaniiakh urovnia Ladozhskogo i Onezhskogo ozer [Fluctuations in Lake Onega and Ladoga levels]', 4(1) *Trudy Komissii po Izucheniiu Chetvertichnogo Perioda*, 71–130

Meinander, K F (1982) 'Finny–chast' naseleniia severo-vostoka Evropy [The Finns as part of the northeastern population of Europe]', in *Finno-Ugorskii Sbornik*, Moskow: Nauka, pp 10–32

Núñez, M (1990) 'On subneolithic pottery and its adoption in late mesolithic', VII *Finland Fennoscandia Archaeologica*, 27–57

Paaver, K L (1965) *Formirovanie Teriofauny i Izmen'chivost' Mlekopitaiushchikh Pribaltiki v Golotsene*, Tallinn: Academia

Pankruchev, G A (1978) *Mezolit i Neolit Karelii II: Neolit [Mesolithic and Neolithic in Karelia Part II: Neolithic]*, Leningrad: Nauka

Pesonen, P E (1986) 'Stoianki v nizov'e reki Keret [Sites at lower course of the Keret River]', in *Novye Dannye ob' Arkheologicheskikh Pamiatnikakh Karelii*, Petrozavodsk: Karelian Research Centre, Russian Academy of Sciences (KRC RAS), pp 5–28

——— (1991) 'Khronologiia i periodizatsiia kul'tury sperrings [Chronology and periodization of the Sperrings culture]', in *Khronologiia i Periodizatsiia Arkheologicheskikh Pamiatnikov Karelii*, Petrozavodsk: Karelian Research Centre, Russian Academy of Sciences (KRC RAS), pp 65–85

Salomaa, R and Matiskainen, H (1985) 'New date on shorelines displacement and archaeological chronology in southern Ostrobothnia and northern Satakunta', 5 *Iskos*, 141–155

Savvateev, U A and Vereshchagin, N K (1978) 'Okhtinskie promyslovye zhivotnye i kamennyi inventar' naseleniia Karelii i iuzhnoi chasti Kol'skogo poluostrova epokhi neolita i rannego metalla [Okhta hunted animals and stone tools used by humans in Karelia and the southern Kola Peninsula in Neolithic and Early Metallic]', in *Mezoliticheskie Stoianki Karelii*, Petrozavodsk: Karelian Research Centre, Russian Academy of Sciences (KRC RAS), pp 181–215

Shumkin, V Y (1991) 'Etnogenez saamov (arkheologicheskii aspekt) [The Sami onthogenesis]', in *Proiskhozhdenie Saamov (Po Dannym Antropologii i Arkheologii)*, Moskow: Nauka, pp 129–149

——— (1993) 'Rannii kamennyi vek zapadnoi chasti Evropeiskoi Arktiki (mezolit Severnoi Skandinavii) [Early Stone Age in western part of the European Arctic Region (Mesolithic at northern Scandinavia)]', in *Drevnosti Severo-Zapada*, Saint Petersburg: Nauka, pp 34–59

——— (1996) 'Neolit Kol'skogo poluostrova [The Neolithic Kola Peninsula]', 1 *Drevnosti Russkogo Severa*, 67–74

Siiriainen, A (1980) 'On the cultural ecology of the Finnish Stone Age', *Suomen Museo-Finskt Museum*, 5–40

——— (1982) 'Recent studies on the Stone Age economy in Finland', 1 *Fennoscandia Archaeologica*, 17–25

Simonsen, P (1957) 'Boplatserne ved Noatun i Pasvikdalen', 58 *Suomen Muinaismuistoyholistuksen Aikakauskirja*, 233–267

——— (1961) 'Varanger-funnene II. Fund og udgravninger på fjordens sydkyst', VII *Tromso Museums Skrifter,* 1–502

Sinitsyna, G V and Zaitseva, G I (1995) 'K voprosu o datirovke valdaiskoi kul'tury [The problem of the Valdai culture dating]', in *Drevnosti Severo-Zapadnoi Rossii,* Saint Petersburg: Nauka, pp 35–39

Sizova, O (2001) 'Opyt èksperimental'nogo izucheniia ranneneoliticheskoi keramiki [Experimental studies on Early Neolithic culture]', 3 *Problemy Arkheologii i Etnografii Karelii*, 3–7

Timofeev, V I (1997) 'K probleme neolitizatsii lesnoi zony Vostochnoi Evropy [The problem of dating of the forest zone in eastern Europe]', in *Slaviane i Finno-Ugry*, Saint Petersburg: Nauka, pp 14–20

Titov, U V (1972) 'O kul'ture Sperrings [On the Sperrings culture]', in *Arkheologicheskie Issledovaniia v Karelii*, Leningrad: Nauka, pp 34–51

Torvinen, M (1997) 'Sär 1—comb ware of the Säräisniemi style', in *Slavyane i Finno-Ugry*, Saint Petersburg: Nauka, pp 21–31

——— (1998) 'Sär 1—keramiikkaa kayttanyt vaesto—etnisia kysymyksia', 4 *Muinaistutkija*, 38–45

——— (2000) 'Säräisniemi 1 Ware', XVI *Fennoscandia Archaeologica*, 3–36

Ukkonen, P (1993) 'The post-glacial history of the Finnish mammalian fauna', 30 *Annales Zoologici Fennici*, 249–264

Vereshchagina, I V (2003) 'Poselenie Khepojarvi v iuzhnoi chasti Karel'skogo peresheika [Khepojarvi site at southern part of the Karelian isthmus]', in *Neolit—Eneolit Iuga i Neolit Severa Vostochnoi Evropy*, Saint Petersburg: Nauka, pp 40–51

Vitenkova, I F (1996) 'Kul'tura Sperrings [Sperrings culture]', in *Arkheologia Karelii*, Petrozavodsk: Petrozavodsk: Karelian Research Centre, Russian Academy of Sciences (KRC RAS), pp 65–81

Zemliakov, B F (1936) 'Neoliticheskie stoianki vostochnogo poberezh'ia Onezhskogo ozera [Neolithic sites at eastern shore of Lake Onega]', 9(1) *Trudy Instituta Antropologii, Arkheologii i Etnografii Akademii Nauk SSSR: Seriia Arkheologiia*, 133–141

——— (1940) 'Geologicheskaia istoriia Karelo-Finskoi respubliki v chetvertichnoe vremia v sviazi s problemoi zaseleniia chelovekom severa Evropy [Geological history of Karelian-Finnish Republic in connection with the problem of human colonization of the north of Europe]', VII *Kratkie Soobshcheniia Instituta Istorii Material'noi Kul'tury*, 21–27

Chapter 9

Ceramic Anthropomorphic Sculptures of the East European Forest Zone

Ekaterina Kashina

Introduction

The tradition of making fired clay objects is extremely ancient and has its roots at least as far back as the Palaeolithic. Two Upper Palaeolithic sites, Dolní Věstonice and Pavlov (Czech Republic) are renowned as a result of unique finds of clay figurines, depicting various mammals and Venus women. Clay anthropomorphic sculptures are also found widely in the early agricultural societies of Europe and Asia, with hundreds, if not thousands of clay figurines, recovered from southern Europe, Asia Minor and Mesopotamia, and subsequently described and studied in detail (Antonova 1977). Most of these are feminine and bovine images, occasionally decorated with ornaments, sometimes made of differently coloured clay pastes, or even supplied with wooden joint-pins (Balabina 1997: 122).

This chapter focuses on a lesser known group of clay anthropomorphic figurines made by the highly developed hunter-gatherer cultures of the eastern forest zone of Europe; these societies had already acquired knowledge of how to produce pottery vessels some 2,000 years earlier. As detailed in Chapter 7, ceramic technology appeared in these areas much before the transition to farming. The 'forager' tradition of making figurines was probably a localized cultural innovation because no similar objects are known in adjacent regions. The whole assemblage, recovered mainly from Russia and the eastern Baltic territories, numbers around 50 objects, with the height of individual pieces varying between 3.5 and 9 cm. The sculptures are assumed to date broadly from the end of the 4th to the beginning of the 2nd millennium BC, that is, during the period usually referred to as the Middle and Late Neolithic[1]. In Russian archaeology, the final stage (the second half of the 3rd to the beginning of the 2nd millennium BC) is usually called Eneolithic, despite the paucity of metalware finds.

Two sculptures have been recovered from burial contexts; the rest all originate from numerous multi-layered sites where Neolithic-Eneolithic levels are usually poorly preserved. Sites are located mostly along riverbanks or lakeshores. Most of them were inhabited during a more or less prolonged period, judging by the numerous vessel fragments from the northern and central Russian sites. It is not clear in most cases whether those sites were visited by Neo-Eneolithic people once or several times, especially if the site was excavated before the 1980s. In addition to hunting and gathering, fishing was also important. The remains of dwellings (rectangular, as a rule) are also sometimes found at these sites. The inventory of finds usually includes flint tool assemblages and ceramics. Waterlogged sites are notable for their well-preserved bone, antler and wooden tools. Burials are situated near or even within dwellings, for example, normally on the site territory. There are not only individual tombs, but also graveyards with anything from a handful to dozens of burials. Sometimes the graves contain several skeletons, buried simultaneously with or without goods. Burial goods may include flint, bone and antler tools, as well as pendants made of animal teeth. Sometimes amber pendants and ceramic vessels are present. These communities made not only anthropomorphic clay figurines, but also a range of other anthropomorphic and zoomorphic figurines of various materials, such as bone, antler, amber, wood and flint (Kashina 2002). For the most part these can be dated to the second half of the 3rd millennium BC.

Ceramic figurines are found over a vast area which includes (Figure 9.1) the following regions: Latvia, Estonia, Karelia, the north and part of the central area of European Russia. Scholars have tended to attribute those clay figurines to various regional archaeological cultures:

1. Comb-Pitted Ware (northern area of the European Russia, Karelia, eastern Baltic, 4th to the beginning of the 3rd millennium BC);
2. Pitted-Comb Ware (central area of the European Russia, dated to the same period);
3. Asbestos Ware (Karelia, second half of the 3rd millennium BC);
4. Porous Ware (eastern Baltic, 3rd millennium BC, mostly the second half).

It is also worth noting a group of similar Finnish finds, belonging to the Comb-Pitted Ware (normally called the Typical Comb-Pitted Ware,

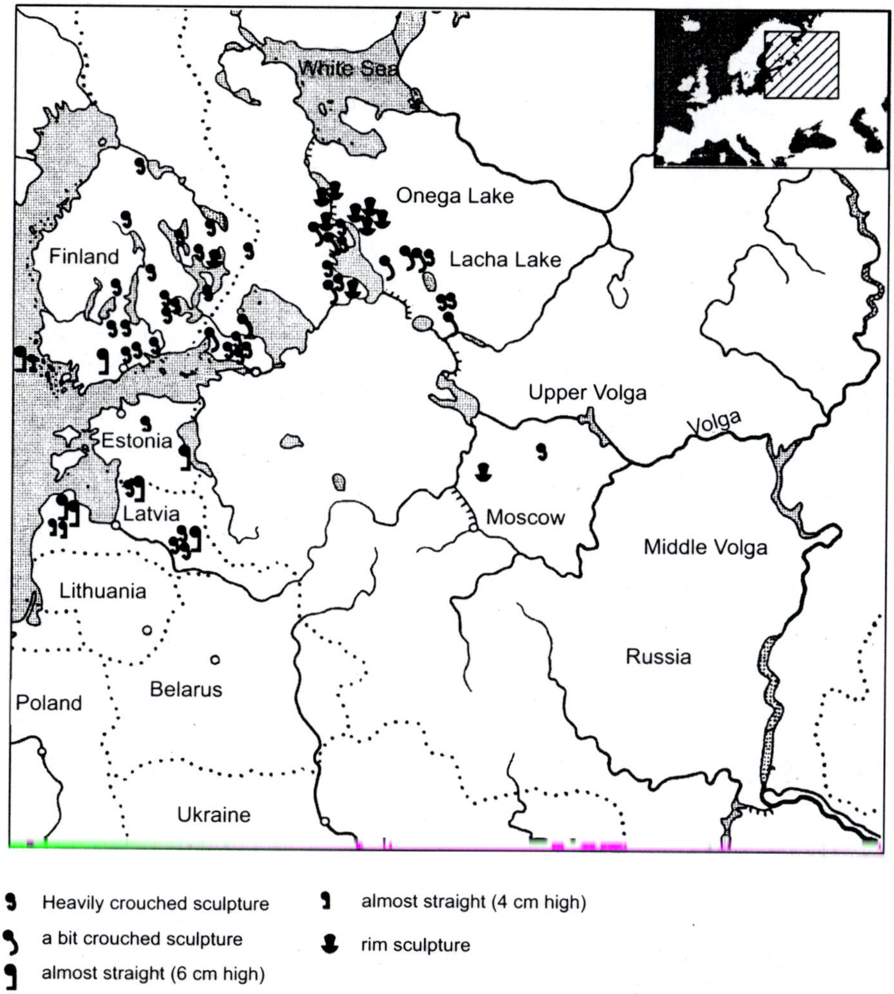

Figure 9.1 Clay sculpture of eastern Europe and Finland.

mainland Finland) and the Swedish Pitted Ware culture, which seems to be related to the Porous Ware (mainly from the Åland Islands). These Finnish figurines are dated to the same time periods and consist of around 100 pieces including fragments (Núñez 1986).

The main aim of this chapter is to consider the clay anthropomorphic sculptures made by the hunter-gatherer cultures in Russia and eastern Baltic as a combined group of artefacts. Similar regional-scale studies of Finnish sculpture have already been made (Núñez 1986; Wyszomirska 1984). The significance of these figurines has been debated in Russia for over 50 years, although research into this theme remains difficult for several reasons.

Most figurines have been recovered from sites which include not only Neolithic (hunter-gatherers with pottery), but also Mesolithic (hunter-gatherers without pottery), Bronze and even Early Iron Age cultural layers. Moreover, for many sites there are no reliable radiocarbon dates. The bulk of finds were made before the 1940s, and detailed information about the circumstances of the finds is usually lacking. The objects are fragile and are often found damaged and in fragments.

Four main topics will be discussed: (a) the figurines' morphological variety in relation to chronology and geographic location; (b) the use of sculptures; (c) the decoration of sculptures and (d) their potential symbolic meaning. Around half of the 50 sculptures were studied using published drawings, photographs and descriptions. The rest (also published) were studied from among the handful of museum collections.

Morphology of the Clay Anthropomorphic Sculptures

There are debates as to the degree to which the appearance of specific figurines is 'anthropomorphic' in nature. Moreover, figurines tend to be significantly different from one another, making classification difficult. In order to generate an overview of variation in the material, two-level classifications were employed. The first was derived as follows (Figure 9.2):

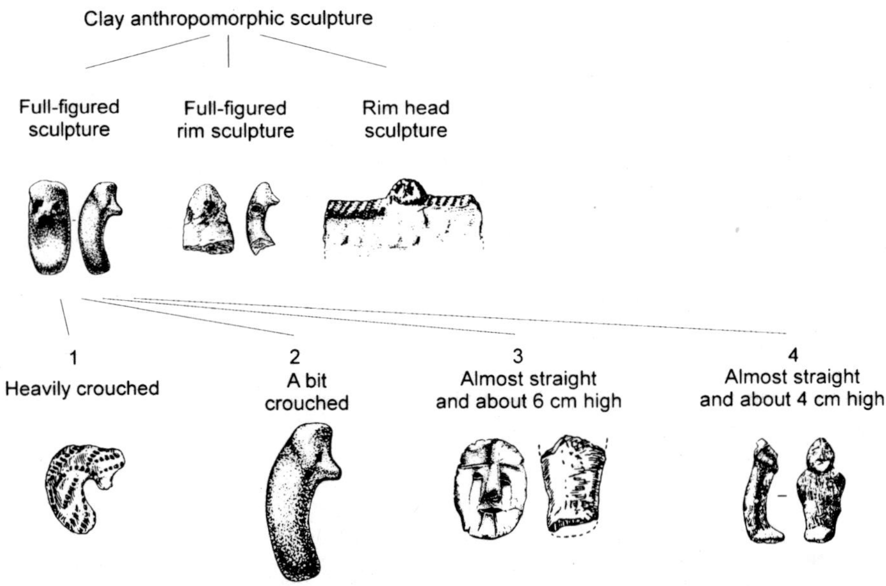

Figure 9.2 Clay sculpture classification.

1. full-figured sculpture (about 42 pieces);
2. full-figured rim sculpture (3 pieces);
3. rim head sculpture (6 pieces).

In the second level of classification, only morphological traits of type 1 (full-figured sculpture) were included. These were grouped according to posture and height (Figure 9.2):

1. heavily crouched and about 4 cm high (about 18 pieces);
2. a bit crouched and higher than 6 cm (about 8 pieces);
3. almost straight and higher than 6 cm (7 pieces);
4. almost straight and about 4 cm high (2 pieces).

Many of the figurines were damaged although height could be estimated on the basis that figurines of each group were of regular proportions. For example, a head fragment measuring 3 cm could only be a part of a 6-cm high figurine and not a shorter one.

Several other traits are not included in these classifications but should also be mentioned. The figurines of group 1 usually have a large protruded nose, a back ridge made by fingertips, a round cross-section torso, and sometimes, two legs. Eyes (if present) are shown by pinholes, and ears by two small protuberances. The group 2 figurines are slightly different: they possess a smaller nose, an oval cross-section torso, eyes and a mouth made by pinholes (sometimes, eyes are shown by fingertip hollows). A ridge along the back is absent. Figurines from groups 3 and 4 have the head and torso flattened, and a regular round, oval or rhombic face. All facial details (small nose, eyes and a mouth made by incisions) look rather realistic.

Finally, it should be mentioned that the difference in heights between sculpture groups (reaching at least 2 cm) was not random, but seems to be traditional, exactly matching the set necessary elements for each figure—a back ridge, or a nose or a round face, etc.

Sculptures in the four groups have several common traits. As for the whole-figured sculpture fastened to a vessel rim, morphologically it is very close to group 2—they are also slightly crouched and nearly 6 cm high. Sculptural heads on vessel rims can be characterized only by their height, which is similar to the height of the slightly crouched sculpture heads. Another similarity between the full-figured sculpture and rim sculpture is the same facial modelling: more or less protruded noses and eyes, made by pinholes or hollows.

The range of traits shared by these ceramic figures suggests that the clay sculpture tradition was a universal phase in the general cultural development of hunter-gatherer communities in eastern Europe.

Figure 9.3 groups the finds according to region and chronology, and demonstrates that there are consistent changes according to chronology across the entire eastern European forest zone (including Finland). The earliest pieces are heavily crouched full-figures, with occasional head sculptures on rims, and belong to the Comb-Pitted Ware, dated approximately to the end of 4th to the beginning of the 3rd millennium BC. Later traditions are characterized by figures that are less crouched, and by full-figure rim sculptures—these are included in the later period of the Comb-Pitted Ware and the Asbestos Ware period, dated mainly to the second half of the 3rd millennium BC. The upright body sculptures of groups 3 and 4 tend to coexist with the Porous Ware and Swedish Pitted Ware cultures (the Åland Islands). In sum, there appear to have been simultaneous changes in sculpture styles across this large region.

The Use of Sculptures

Almost all scholars who have studied these clay anthropomorphic sculptures mention the hurried and careless manner in which the images appear to have been fashioned (Loze 1970; Miettinen 1965; Oshibkina 1978; Zhul'nikov 1999). In their opinion, this may be evidence that the images were meant for short-term use or even of a disposable nature, perhaps even

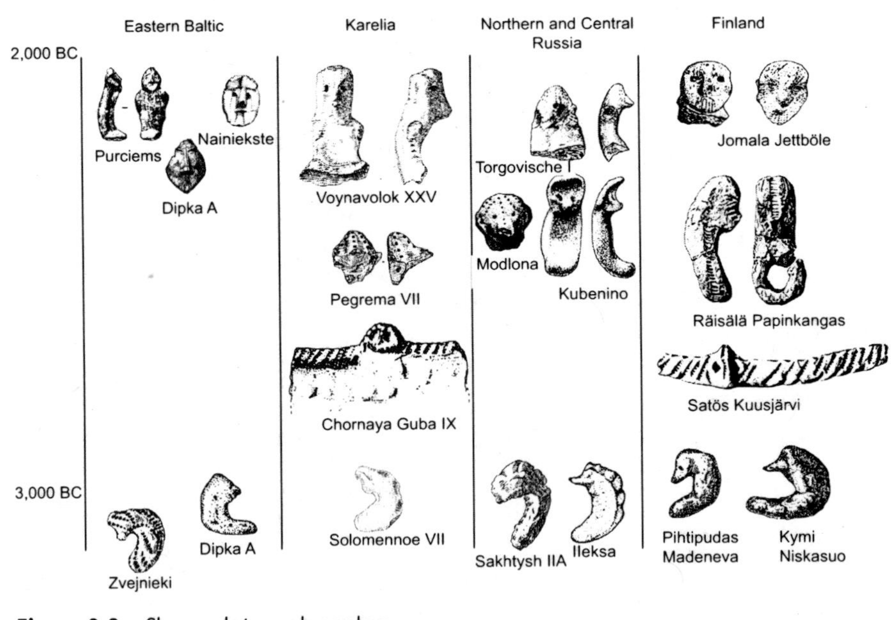

Figure 9.3 Clay sculpture chronology.

deliberately broken and discarded after rituals. Núñez suggests that this was the case for the Åland figurines, where only fragments were found (Núñez 1986: 26). He also assumes that most crouched figures recovered from mainland Finland were more or less intact, that is, they were not broken by people. Regrettably, the suggestion about the figurines' deliberate destruction cannot be proved, for the patterns of breakage could also be the result of post-depositional factors. Some Åland figurines are relatively large and could be prone to breakage; moreover some eastern Baltic and Karelian fragmented sculptures are made from fragile pottery paste, containing organic temper. By contrast, small, compact and crouched figurines were made of more durable pottery paste, perhaps ensuring better preservation.

Several of the figurines have other features which have been used to make other inferences. First, some figurines were covered with red ochre. Two figurines from Karelia (Voynavolok XXV site; see Figure 9.4, 7–8) were wholly painted and one from Latvia (Dipka A site; see Figure 9.3) had the lower part of a face painted with ochre (Loze 1998; Zhul'nikov 1999). According to Finnish studies, many of the Åland figurines also had traces of the red ochre painting (Núñez 1986: 21). Thus, the usage of ochre is recorded on the chronologically later pieces, which are straight and full-figured, as well as those fastened to the rim, and are dated to the second half of the 3rd to the beginning of the 2nd millennium BC.

Second, the sculptures appear to have been kept in dwellings, sometimes near the hearth, judging from the archaeological contexts in which the figurines were recovered. Some finds from Karelia (Voynavolok XXV, Chornaya Guba IX sites; see Figure 9.4, 7–9), northern Russia (Kubenino site; see Figure 9.4, 3–4) and Latvia (Purciems site; see Figure 9.4, 5–6) bear witness to this fact. These pieces belong to both rim sculptures and full-figure upright figurines and date from the middle of the 3rd to the beginning of the 2nd millennium BC, similar to the ochre painted pieces. In Karelia, these painted pieces are also found within dwellings. Interestingly, however, the Åland figurines, found in the Jomala Jettböle site, were concentrated in a small area (Götherström et al. 2002: 45), but no dwelling was mentioned, though certain important contextual features of the site could have been overlooked when the site was investigated about a hundred years ago (Cederhvarf 1912).

Third, a number of sculptures are made from clay pastes not used for making vessels, suggesting that the decision to use different materials for sculptures was deliberate. For example, about 10 sculptures were partly analyzed visually by Dr. Y B Tsetlin, who concluded that they were made of pottery paste without gravel and with low sand content. This is a very unusual recipe for northern Russia, where Comb-Pitted Wares were produced alongside these sculptures. On the other hand, two figurines, probably

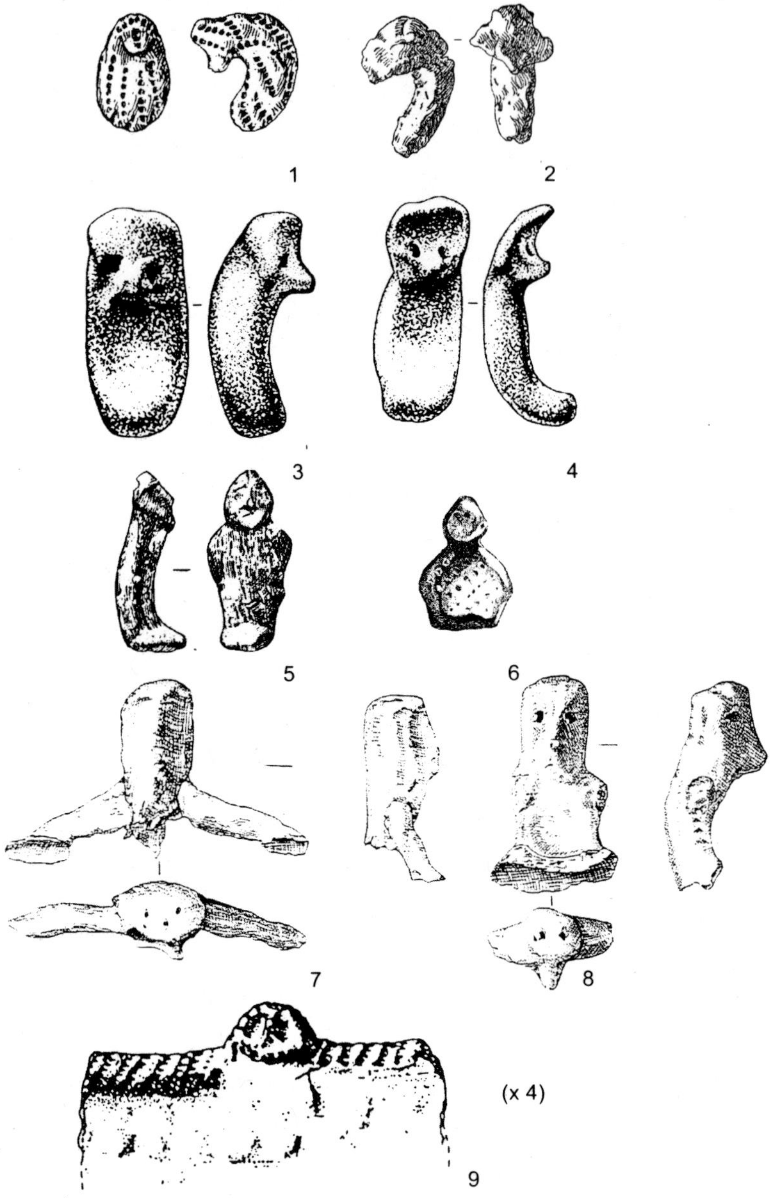

Figure 9.4 Clay sculpture found in particular contexts.

fastened to an Asbestos Ware vessel, were made of the same pottery paste as the vessel (see Figure 9.5). All these facts point to the conclusion that there was immense variation in the manufacture of clay sculptures.

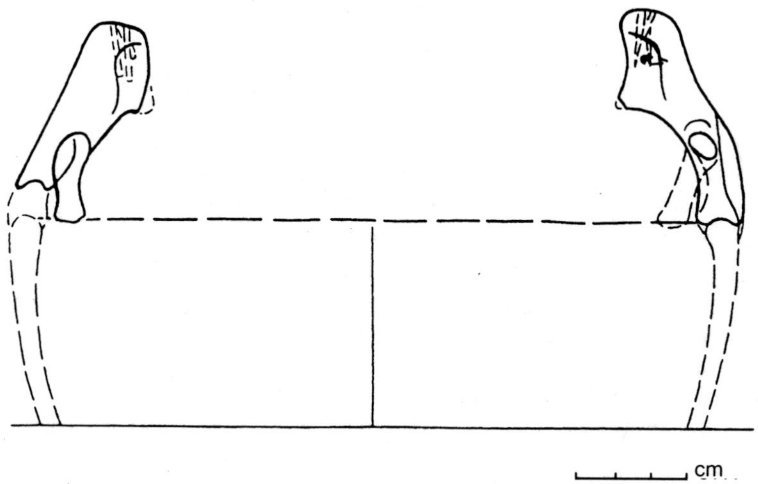

Figure 9.5 Reconstruction of the vessel supplied with two rim sculptures (Voynavolok XXV site, Zhul'nikov 1999, figure 72).

Fourth, some full-figured pieces were found in burials. One of them, from Latvia (Zvejnieki cemetery; Figure 9.4, 1), was found near the left leg of an adult buried in a collective grave. Another, from central Russia (Sakhtysh IIA; Figure 9.4, 2), was under the neck of an adult woman who was buried alone. Both skeletons were extended and lay on their back (Krainov et al. 1993; Zagorskis 1987). Both figurines are heavily crouched, have a back ridge and a protruding nose. There are obvious indications that in these contexts the figurines formed some sort of burial rite, although it is important to note that they are both from the earliest period, approximately dated to the end of the 4th millennium BC. Sculptures from later periods are recovered from different contexts and have no necessary association with burial rites.

Finally, the vessels with rim sculptures must have had some special use or function. Similar vessels, decorated with both zoomorphic and anthropomorphic heads, are well known from a handful of the Late Stone Age and Bronze Age sites across northern Asia from the Urals to the Far East (Khlobystin and Konstantinov 1996; Moshinskaia 1976). It is possible that these images protected the contents of the vessel, or that the vessels were used for serving or preparing food in a wider set of rituals in which the decorations had particular meanings. Similar suggestions could be made about the Karelian vessels with rim sculptures. Interestingly, vessels from Voynavolok XXV and Chornaya Guba IX sites (Karelia, see Figure 9.4, 7–9) have heads facing towards the inside of the bowl, while others are full figures with arms, as if the person is bending over, looking into, or embracing the vessel and contents. In addition, both vessels had traces of ochre painting on their surface and the one from Chornaya Guba IX site had been broken and repaired by drilling holes, sewing and then sealing with resin (Vitenkova 2002; Zhul'nikov 1999). This suggests that such ochre-painted and figurine-decorated vessels were highly valued, and used carefully in only special (ritual) contexts.

Decoration of Sculptures

Both figurines and pottery vessels share modes of ornamentation, perhaps suggesting that they also shared similar symbolic meanings in the hunter-gatherer societies of the day.

Four main motifs were revealed on all ornamented sculpture pieces: comb imprints, a row of pinholes, incisions and the mixture of comb imprints and incisions (Figure 9.6). The comb imprint is one of the basic elements of the Comb-Pitted Ware and Pitted-Comb Ware vessel decoration and, obviously, the same stamps were used for vessel and sculpture decoration. Other motifs (pinholes and incisions) are relatively less used, but could be found on the vessels of all cultures, producing these clay sculptures.

The Russian scholar, M J Foss, who undertook an integrated study of figurines and pottery from Purciems (Latvia) and Jomala Jettböle (the Åland Islands), concluded that they shared a single decorative tradition which may have arisen from what were originally tattoo or body-painting motifs (Foss

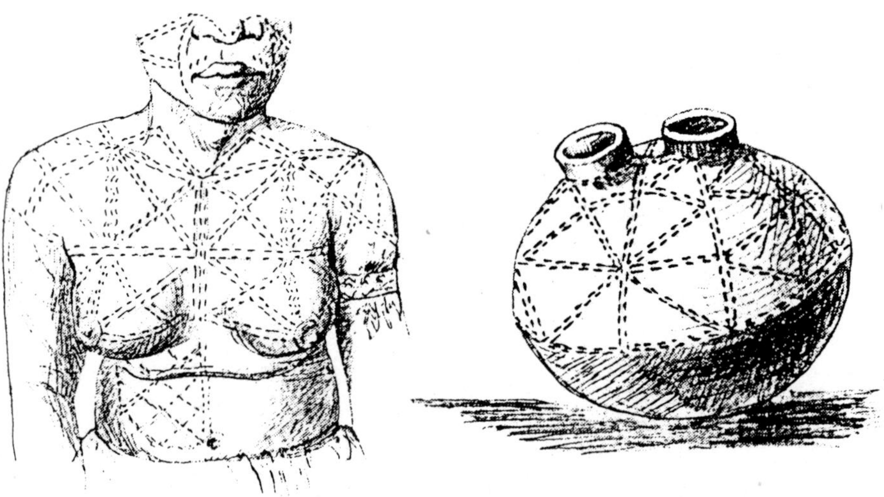

Figure 9.6 Examples of tattoo motifs similar to vessel decorations (Mikloukho-Maklay 1993: 304, 252).

1952). In making these suggestions, she draws broad inspiration from the ethnographic work of Mikloukho-Maklay in Papua New Guinea, where tattoo motifs coincide with local styles of pottery ornamentation (Figure 9.7). Probably, the traces of ochre painting, registered in the lower facial part of the figurine fragment from Dipka A site (Latvia) (Loze 1998), could be a direct copy of locally practised human face-painting techniques.

Another idea proposed by Núñez (1986: 26) is that the figurine face and body decoration could be interpreted as a ritual mask and a costume, like those of Siberian shamans, with feathers, embroidery and other ritual adornments. Thus, according to Núñez, the figurines could represent people, specially dressed for performing some rites. However, nearly half of the sculptures (21 pieces) are not decorated at all. The rest of them have only the head (mainly the face) and sometimes the body decorated. The head decoration was registered on 21 pieces, torso decoration on only five. One ornamented torso is a fragment of a full-sized figurine.

Apparently, the tradition of figurine face decoration was also changing through time. Most undecorated pieces are also heavily crouched and recovered from earlier periods, (as noted above). The distribution of decorated pieces is shown in Figure 9.8.

A row of diagonal comb imprints is characteristic of several of the heads forming rim decorations, which are dated to a later period: the first half of the 3rd millennium BC.

The fork-like motif of three comb imprints located on the face is common for both heavily and less crouched pieces, and is also dated to the same period (mainly to the first half of the 3rd millennium BC). In one case,

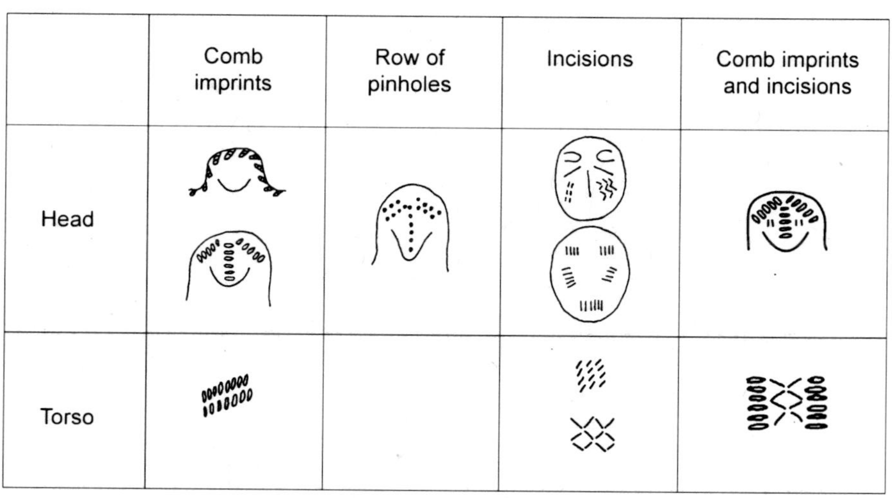

Figure 9.7 Sculpture decoration motifs.

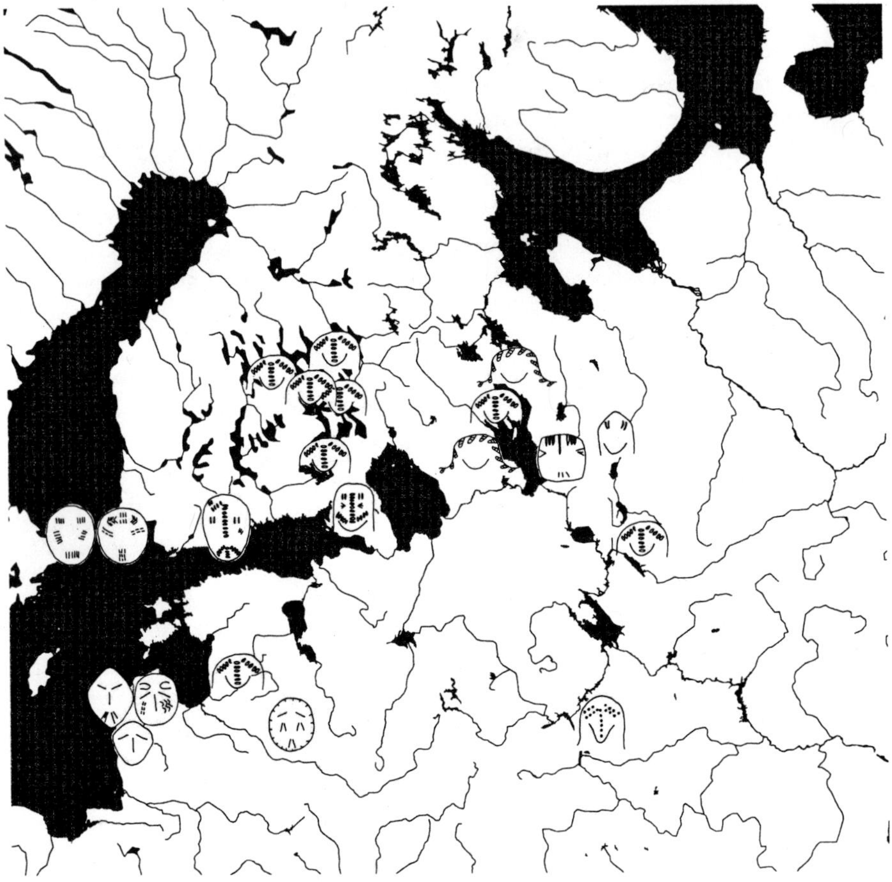

Figure 9.8 Distribution map of decorated sculptures (with facial motifs).

such decoration is made up of a row of pinholes. The torso is sometimes covered with similar comb imprints. Straight figurines of the eastern Baltic region possess different ornamentation with three incised lines marking the nose and the brows, and additional ones on the cheeks and the chin. The torso is commonly covered by similar incisions, grouped in different ways. This decoration brings them close to the famous Jomala Jettböle figurines (the Åland Islands). The motif of paired short incised lines on the brows and the chin, sometimes mixed with the comb imprints and repeated on the torso, is known in southern Finland, Karelia and northern Russia (Chornaya Rechka III and Kubenino sites). All these figurines are believed to date from the second half of the 3rd to the beginning of the 2nd millennium BC.

It should also be mentioned that small holes on the top of some Åland figurines find their parallels in two Karelian Asbestos Ware pieces, fastened to the vessel edge (Figure 9.5). It is believed, that these resemblances and mixings of decoration motifs might reflect regular interactions between different regions of the study area, especially during the 3rd millennium BC. Nevertheless, it is still impossible to say which exact region influenced another in terms of the affiliation of particular decoration motifs. It seems that fork-like comb imprints decorating the face appeared earlier and then turned into three incised lines. The eastern Baltic 'realistic' figure silhouette is similar to that of the Åland Islands. The multiple facial incised lines of the Åland figurines seem to have parallels in mainland Finland and northern Russian paired facial decoration. On the other hand, the Finnish and north Russian figurines possess faces more typical of the crouched sculptures, and sometimes have comb imprints added. In sum, variations in the decoration of figurines hint at different forms of group interaction, although it is difficult to determine the exact forms and influences of these cultural and stylistic exchanges.

Symbolic Meanings

Morphological analysis of the sculptures indicates that there were significant changes in the forms of the figurines through time. In particular, one could suggest that the figurine style evolves from smaller figurines, which grow in size and become straighter, and in some cases, take the form of a separate head which then decorates vessel rims (Figure 9.9). The earliest images (4th millennium to the beginning of the 3rd millennium BC) appear

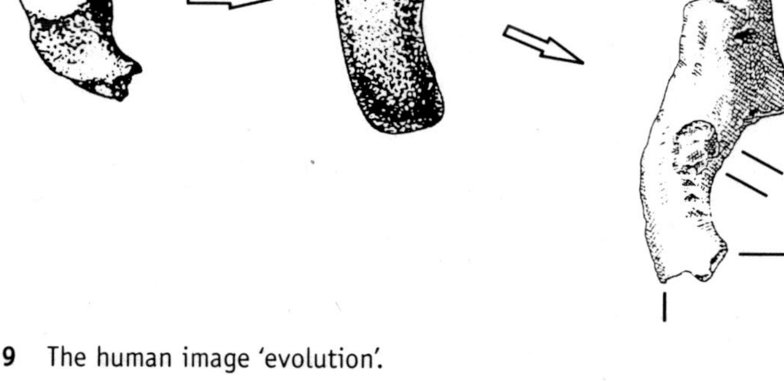

Figure 9.9 The human image 'evolution'.

to represent a fantastic creature, perhaps with mixed animal and human traits. Later figurines (the middle of the 3rd millennium BC) are characterized by the predominance of traits that are more human in form, although it remains clear that they represent continuities with the previous styles. Only the later stages (the end of 3rd and 2nd millennia BC) include images that are unequivocally human. Clearly, these images were highly symbolic and indicate continuities in the traditions of thought despite changes in form through time and in the different areas.

The early figures are undecorated, but originate in societies that decorate their pottery. This suggests that the symbolic meaning of the decorated vessels was different to that of the sculptured images, which appear to be associated with other notions relating to the crossover of animal and human forms, resulting in zoomorphic-fantastic sculptures.

In some cases, it appears that the figurines were used in pairs. For example, in two cases, a pair of images was found not far from one another in the same dwelling (Kubenino site, northern Russia; Purciems site, Latvia, Figure 9.4, 3–4, 5–6). Another pair was attached to the vessel edge opposite each other (Voynavolok XXV site, Karelia, Figure 9.4, 5, 7–8). However, the extraordinary vessel with four sculptural heads facing inwards from the rim, (from Chornaya Guba IX site, Karelia, Figure 9.4, 9), still remains a unique artefact. There are tiny differences within every pair: in height, or in the pattern of eye modelling, or in the number of pinholes on the head top. All of them may symbolize mythical twins or rival heroes.

It remains impossible to identify whether males or females are depicted. Cederhvarf and Núñez have suggested that some Åland figurines have breasts, although there exists a unique figurine with breasts (Cederhvarf 1912; Núñez 1986). Loze reports that a similar torso fragment was found in Latvia, on Dipka A site (Loze 1998). In general, she concludes that most of the upright human sculptures are males, due to their rhombic or oval faces decorated with incisions, reminiscent of a beard. However, to my mind, the rhombic or oval figurine faces decorated with incisions cannot be interpreted only as bearded ones. For example, Núñez preferred to recognize them as masked shamans in ritual garments rather than classify them on the basis of sex (Núñez 1986).

After decades of research into these figurines, we are still struggling to understand their meanings. Interpretations have ranged from the suggestion that they are phallic, to zoomorphic, and finally anthropomorphic, and connected with a shamanistic ritual worldview. However, two perspectives appear to be particularly plausible. First, Loze and Oshibkina proposed that there were close relationships between the figurines and the hearth and home, due to the details of the domestic contexts in which they were found (Loze 1970; Oshibkina 1978). In addition, Miettinen, the Finnish researcher, has suggested that the crouched and ridge-backed figurines could represent human embryos (Miettinen 1965). This second interpretation may be correct in that the very general zoomorphic or even mammalian appearance of crouched figurines could hardly be compared with some specific form of a particular animal. Moreover, even the upright and undoubtedly anthropomorphic figurines with wide shoulders have no arms or carefully made legs, which make them look unusual, as if they are wrapped.

Whatever the specific significance of the figurines, it is clear that they were important cultic objects. Scholars have generally associated pottery, and the transmission of ceramic craft traditions, with female work, matrilineal descent and marriage ties (Zhul'nikov 1999: 23). One might suggest that the figurines were also connected with specifically feminine domains of activity, which included pottery making. This may suggest that the sculptures, especially those on the rims, were made by women as elements of a purely feminine spiritual world. Moreover, the overtly 'domestic' context of some finds—in dwellings, near hearths—adds weight to this interpretation. Closer inspection of the finger imprints along the back ridges of several crouched figurines also reveals that the hands of those making the sculptures were small, and perhaps those of women rather than men.

Conclusion

This chapter has examined the ceramic sculptures made by hunter-gatherer societies of the East European forest zone. Analysis reveals:

1. the common essence of all recognized sculpture, despite the known difference in forms (autonomous sculpture or on the vessel) and details (presence or absence of arms, legs, eyes, etc);
2. the intensive exchange of ritual art traditions between different areas, which are sometimes very far from each other;
3. the image changing in time from the 'zoomorphic-fantastic' to the 'anthropomorphic-realistic' nature and, judging by the context of some finds, probably the changing place and meaning of the sculpture in ritual practice (one can suggest, from funeral to 'domestic').

Whilst it is clear that the images were used in ritual practice, assigning specific meanings and interpretations to the sculptures is challenging. Nevertheless, consideration of a range of facts does point to potential associations between the ceramic images and notions of home, hearth, abdomen and children; the sculptures are found in dwellings, sometimes near the hearth. Moreover, the images encode sensual and material connections with vessels and fire (baked clay), with wrapped figures and embryos, and with other domains of activity, spirituality and identity that have strong feminine associations. These include concepts of home, family and fireplace, and female fertility and birth.

The numerous clay anthropomorphic figurines of agricultural societies appear to have been venerated and worshipped to obtain an abundant harvest. In contrast, the figurines produced in hunter-gatherer societies may have meant something different. Perhaps these 'forager' figurines were needed for the protection of the particular house, of family members and especially children.

Acknowledgements

The author is thankful to Dr. Y B Tsetlin (Institute of Archaeology, Russian Academy of Sciences, Moscow) for valuable advice and friendly support; Dr. N V Lobanova (Institute of Language, Literature and History, Petrozavodsk) for permission to use the data of an unpublished figurine from Chornaya Rechka III site; Professor S V Oshibkina (Institute of Archaeology, Russian Academy of Sciences, Moscow) and Dr. S V Studzitskaia (State Historical Museum, Moscow) for consultation and helpful discussion.

Note

1. This chapter employs the 'eastern' definition of Neolithic, i.e., pottery-using hunter-gatherers. See introductory chapter (Jordan and Zvelebil) for further discussion of the various definitions of Neolithic.

References

Antonova, E (1977) *Antropomorfnaia Skul'ptura Drevnikh Zemledel'tsev Perednei i Srednei Azii [Anthropomorphic Sculptures of the Ancient Farmers of the Near East and Central Asia]*, Moscow: Nauka

Balabina, V (1997) *Figurki Zhivotnykh v Plastike Kukuteni-Tripol'ia [Figurines of Animals in the Plastic Arts of Kukuteni-Tripol'ia]*, Moscow: Staryi Sad

Cederhvarf, B (1912) 'Neolitiska lerfigurer från Åland', XXVI *Suomen Muinaismuistoyhdistyksen Aikakauskirja*, 307–323

Foss, M (1952) 'Znachenie ornamentiki rodovogo obshchestva dlia voprosov etnogeneza [Significance of ornamentation of clan society for questions of ethnogenesis]', 29 *Materialy i Issledovaniia po Arkheologii SSSR*, 64–77

Götherström, A, Stenbäck, A and Storå, J (2002) 'The Jettböle Middle Neolithic site on the Åland Islands—human remains, ancient DNA and pottery', 5(1) *European Journal of Archaeology*, 42–69

Kashina, E (2002) 'Anthropomorphic flint sculpture of the European Russian forest zone', 113 *Anthropologica et Praehistorica*, 51–60

Khlobystin, L and Konstantinov, M (1996) 'Neolit Primor'a i Priamur'a [The Neolithic of Primor'e and Priamur'e]', in Oshibkina, S (ed), *Arkheologiia. Neolit Severnoi Evrazii*, Moscow: Nauka, pp 310–325

Krainov, D, Kostyleva, E and Utkin, A (1993) 'Pogrebeniia i ritual'nye kompleksy na stoianke Sakhtysh IIA [Burials and ritual complexes at the Sakhtysh IIa site]', 2 *Arkheologicheskie Vvesti*, 20–30

Loze, I (1970) 'Seno ticējumu un tradīciju atspoguļojums akmens laikmeta mākslā austrumbaltijā', IX *Archeoloìija un Etnogrāfija*, 9–30

——— (1998) 'Arheoloģiskie un starpdisciplinārie pētījumi Çipkas kāpās un paleoezerā', in *Zinātniskās Atskaites Sesijas. Materiāli par Archeologu 1996. Un 1997 Gada Pētījumu Rezultātiem*, Rīga: Latvijas vēstures institūta apgāds, pp 41–43

Miettinen, T (1965) 'En idol från Hietaniemi i Luopioinen', 1964 *Finskt Museum*, 34–44

Mikloukho-Maklay, N (1993) *Sobranie Sochinenii v Shesti Tomakh. (2) Puteshestviia 1874–1887 gg. Dnevniki, Putevye Zametki, Otchety [Collected Works in Six Volumes (2). Travels 1874–1887, Diaries, Travel Notes, Reports]*, Moscow: Nauka

Moshinskaia, V (1976) *Drevniaia Skul'ptura Urala i Zapadnoi Sibiri [Ancient Sculpture of the Urals and Western Siberia]*, Moscow: Nauka

Núñez, M (1986) 'Clay figurines from the Åland Islands and mainland Finland', III *Fennoscandia Archaeologica*, 17–34

Oshibkina, S (1978) *Neolit Vostochnogo Prionezh'ia [Neolithic of Eastern Pronezh'e]*, Moscow: Nauka

Vitenkova, I (2002) *Pamiatniki Pozdnego Neolita na Territorii Karelii [Late Neolithic Sites on the Terrotiry of Karelia]*, Petrozavodsk: Karel'skii Nauchnyi Tsentr RAN

Wyszomirska, B (1984) *Figurplastik och Gravskick hos Nord-och Nordösteuropas Neolitiska Fångstkulturer, Acta Archaeologica Lundensia*, Bonn-Lund: University of Lund

Zagorskis, F (1987) *Zveinieku Akmens Laikmeta Kapulauks*, Rīga: Zinātne

Zhul'nikov, A (1999) *Eneolit Karelii. Pamiatniki s Poristoi i Asbestovoi Keramikoi [Eneolithic of Karelia. Sites with Porous and Asbestos Ceramics]*, Petrozavodsk: Gosudarstvennyi Mezhmuzeinyi Tsentr RK

Chapter 10

Pottery of the Stone Age Hunter-Gatherers in Finland

Petro Pesonen and Sirpa Leskinen

Introduction

In this chapter, we discuss the arrival of early pottery-making traditions in Finland. Ceramic vessels first came into use around c. 5,300 BC, at a time when the economy was still based on hunting, fishing and gathering. From 5,300 BC to 1,500 BC, these forager societies maintained various pottery-making traditions with strong differences in decorative patterns through time and important regional variations in temper, for example, the use of asbestos in eastern Finland and organic tempers in the coastal areas. These variations may reflect different social interaction spheres between coastal and inland areas. We suggest that Finland's Late Mesolithic hunter-gatherer population adopted ceramics through cultural contacts with their eastern hunter-gatherer neighbours. The spread can be explained through mating networks, where women brought the pottery knowledge rapidly into new areas. Ceramics were adopted as a useful technology in a hunter-gatherer economy in which storage was an essential part, especially during the season for mass hunting of seal and fish. In addition to the storage of animal and plant products, the containers were also employed for food preparation. There is evidence that damaged vessels were often repaired, rather than being discarded as expendable items, which suggests that ceramics held significant value. The substantial time involved in pottery manufacture and difficulties in the transportation of these large and fragile vessels meant that adoption and subsequent use of ceramics by early hunter-gatherer societies took place during a move towards increasing sedentism. With the transformation from a hunter-gatherer to a mixed economy through experiments with animal husbandry and eventually also agriculture (c. 2,500–1,500 BC), the pottery traditions were also changing in western and southern Finland under the influence of the Corded Ware culture. In the inland areas, there was a continuity of the older hunter-gatherer asbestos ceramic traditions.

Chronology

In Finland, the question of the function and social significance of Stone Age ceramics has been rather neglected and many contributions have concentrated upon chronological problems, where ceramics no doubt have a special position because of the easily recognizable stylo-chronological changes. Furthermore, in Finland, pottery was widely adopted in hunter-gatherer communities during the Late Mesolithic in most parts of the country. Only in a few regions, the use of ceramics did not occur until the Early Bronze Age. The wide distribution of ceramics has further increased its potential as a chronological marker.

In Finland, the term 'Neolithic' has been used as a chronological term while its traditional meaning actually describes the economy of the society. As a result, 'sub-Neolithic' means 'pottery-Mesolithic' as defined by, for example, C F Meinander (1961) and Christian Carpelan (1979). For the sake of convenience, the ceramic period of Finnish Stone Age is here called simply sub-Neolithic, which is divided into Early, Middle and Late sub-Neolithic (Table 10.1, see also Carpelan 1979).

Table 10.1 Chronology of Finnish sub-Neolithic ceramics according to Christian Carpelan (1979, 1999, 2002).

Early sub-Neolithic	
Ka 1:1, Early Comb Ware 1	5,200–4,500 BC
Ka 1:2, Early Comb Ware 2	4,500–3,900 BC
Sär 1, Säräisniemi 1 Ware	5,000–4,500 BC
Jäkärlä Ware	4,300–3,800 BC
Kaunissaari Ware	4,300–3,800 BC
Middle sub-Neolithic	
Ka 2, Typical Comb Ware	3,900–3,500 BC
Ka 3, Late Comb Ware	3,600–2,800 BC
Pyheensilta Ware	3,200–2,800 BC
Corded Ware	3,200–2,300 BC
Kierikki Ware	3,800–3,100 BC
Pöljä and Jysmä Ware	3,100–1,500 BC
Late sub-Neolithic	
Kiukais Ware	2,300–1,500 BC

There are two very important factors beyond the traditional stylistic typology which have made ceramics very useful in dating Stone Age settlement sites. One of them is shoreline chronology and the other is AMS dating of the crust and birch bark pitch in ceramics. The former is a method based on the observation that after the last Ice Age, the earth's crust has been lifted up and the water level of the Baltic Sea and its predecessors has correspondingly been lowered. The uplift has similarly caused changes in the water levels of the inland lakes. Almost as a rule, hunter-gatherer sites have been situated on the shores of the sea and lakes and thus the older sites are now at a higher elevation above sea level than the younger ones. The tilting has been more marked in northwestern Finland than in the southeastern part of the country.

It is beyond the scope of this chapter to describe details of the shoreline chronology. In brief, the first archaeologists to use it in the dating of Stone Age sites were Julius Ailio and especially Aarne Äyräpää, who practically created the chronology of Stone Age ceramics in Finland by the 1930s (Europaeus-Äyräpää 1930). Äyräpää's chronology of the Comb Ware has been in use for over 70 years with only some minor corrections made to it. The absolute dates presented by Äyräpää are no longer valid, but they have been corrected by others when the radiocarbon dating method came into general use. Most important studies were conducted by Ari Siiriäinen (e.g., Siiriäinen 1974). Nowadays, the shoreline chronology is a practical tool for dating prehistoric sites on the ancient coast of Finland and at the major inland lakes.

At first, the radiocarbon dating method could be used only for samples containing relatively large amounts of datable material. Current Stone Age chronology in Finland is based mostly on the charcoal dates from fireplaces in settlement sites as well as on shoreline chronology, in which the charcoal datings are in turn used. During the 1990s, it became possible to use the AMS technique for dating smaller samples in Finland. Apart from the hearth charcoal and wooden implements, mortar, etc., minor residues could now be used for dating. Furthermore, charred crust and birch bark pitches on the ceramic vessel walls can be assumed to have a limited age of their own. So far, approximately 200 AMS datings have been made of crust or birch bark pitch in prehistoric ceramics in Finland. A total of 110 datings were made of Stone Age ceramics, most of them of Typical Comb Ware (Figure 10.1).

AMS datings of ceramics form an independent factor in the chronology. So far, the dates have differed surprisingly little from shoreline chronology and charcoal datings. The oldest ceramic styles seem to fit almost perfectly into the traditional scheme. Some discrepancies arise during the later phases, however.

The oldest ceramics in Finland belong to the group of comb ceramics. The earliest representatives of comb ceramics are Early Comb Ware 1 (Ka 1) and Säräisniemi 1 Ware (Sär 1). The latter has a markedly northern distribution—

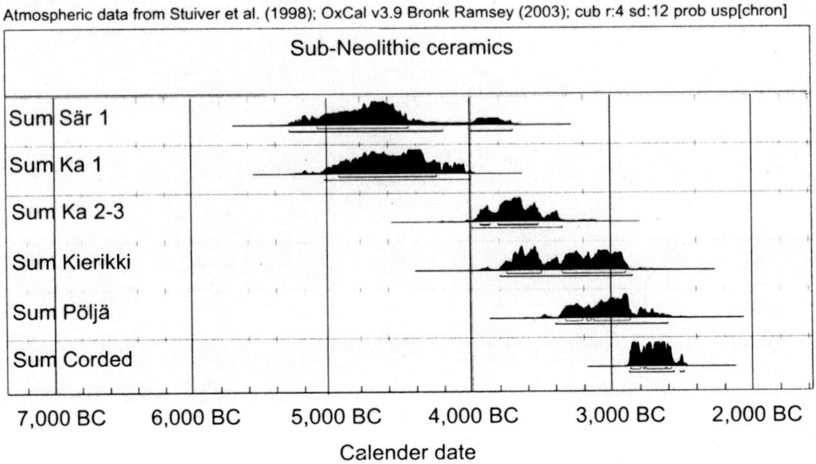

Figure 10.1 Probability distribution for the dates of 110 samples from sub-Neolithic ceramics in Finland. Explanations: Sär 1 = Säräisniemi 1 Ware (9 dates), Ka 1 = Early Comb Ware 1 and 2 (19 dates), Ka 2–3 = Typical and Late Comb Ware (61 dates), Kierikki Ware (7 dates), Pöljä Ware (13 dates, includes 1 Jysmä Ware date), Corded Ware (1 date). OxCal v. 3.9 (Bronk 1995, 2001).

in Finland, most southern finds are in North Karelia, while the northern limit of Ka 1:1 distribution lies in southern Lappland (Torvinen 2000). According to AMS datings from the crust in the Sär 1-style ceramics, this style is dated around c. 5,300–3,700 BC, which is a considerably longer period of time than traditionally suspected. This may indicate continuity in Sär 1 ceramics in the north until the Typical Comb Ware period (Carpelan 2004: 29). The oldest crust date from ceramics in Finland comes from a Sär 1 sherd at the Pyhänniska site in Utajärvi, North Ostrobothnia (Figure 10.2), which is dated around 5,260–4,860 BC (Hela-148, 6,140±105 bp, Torvinen 2000: 16).

The AMS dates of Ka 1:1 and Ka 1:2 are c. 5,200 BC–4,000 BC. Both the oldest and the youngest date are from the Ankkapurha site in Anjalankoski, southeastern Finland (Schulz 2004: 46), which suggests a continuation between styles. Two additional local ceramic traditions existed during the Ka 1:2 period, organic-tempered Jäkärlä Ware in southeastern Finland and asbestos-tempered Kaunissaari Ware in eastern Finland. So far, neither of these ceramic groups has AMS datings (Edgren 1966; Pesonen 1996, 2001).

Most of the AMS datings have been obtained for Typical Comb Ware (Ka 2), and these datings have also been discussed in some articles (Leskinen 2003; Pesonen 1999, 2004). Typical Comb Ware vessels have been dated more often because they were repaired with a method involving birch bark tar as a sort of sealing material or putty. It should also be noted that Ka 2 is

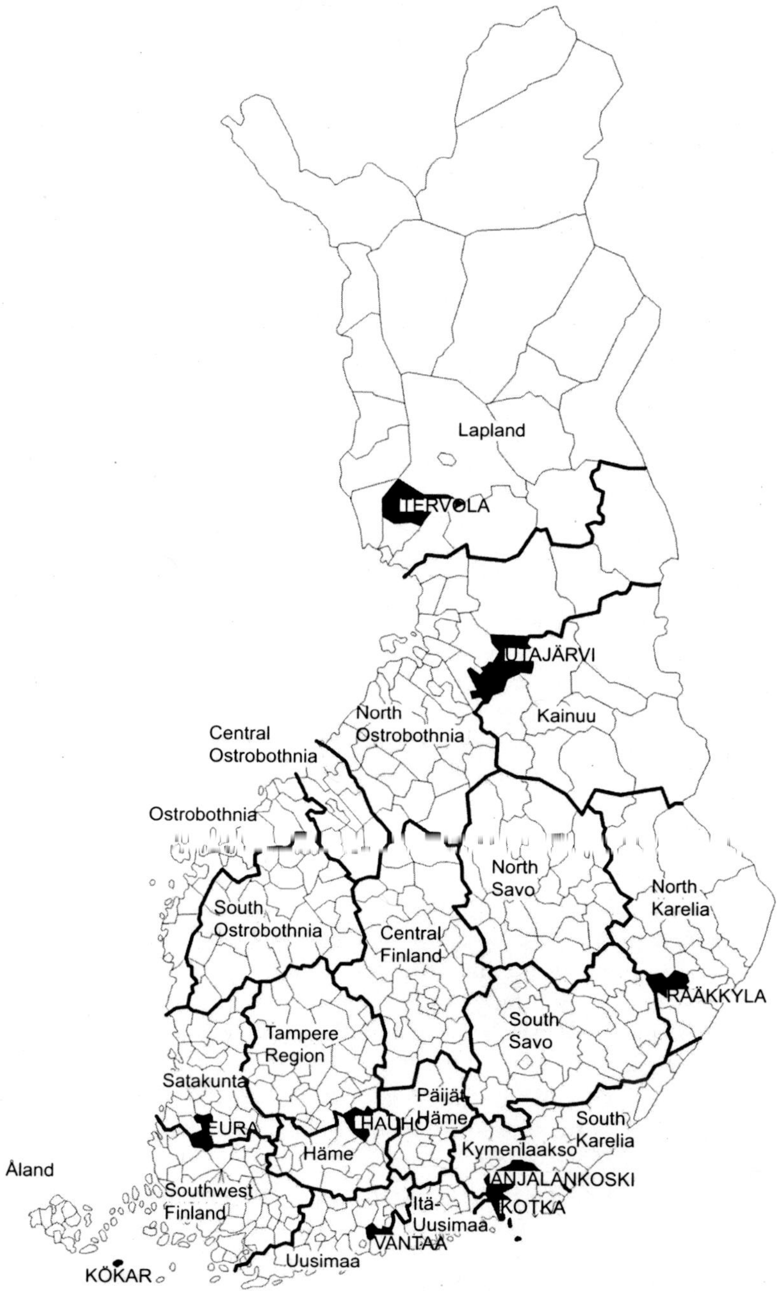

Figure 10.2 The provinces of Finland and some important localities mentioned in the text (drawing by Petro Pesonen).

the most common Stone Age ceramic type in Finland. Late Comb Ware (Ka 3) is traditionally considered a younger ceramic group than Ka 2. In light of the AMS datings, the situation is not so simple. Ka 3 seems to exist simultaneously with Ka 2, at least in the later phase of Ka 2 (c. 3,800–3,400 BC). The division of ceramics into Ka 2 and Ka 3 styles is somewhat arbitrary and chronologically suitable perhaps only for the coastal areas of Finland (see, e.g., Leskinen 2003; Pesonen 2004; Räihälä 1996).

Typical and Late Comb Ware periods form a quite uniform ceramic horizon in southern and eastern Finland. The ceramic styles did not spread north of Rovaniemi and Kemijärvi in Lapland and the most western examples can be found in eastern Sweden around River Kalix (Halén 1994). Traditionally, the spreading of Typical Comb Ware has been seen as a rapid diffusion or even as an immigration from the southeast, from the direction of Karelian Isthmus (e.g., Carpelan 1999: 258; Meinander 1984: 35–37). The AMS datings of the ceramics do not necessarily support this view, however. The oldest dates for Typical Comb Ware are not in eastern Finland but, on the contrary, in southwestern Finland (Kolmhaara site in Eura, Hela-362, 5,155±60 bp) and in North Ostrobothnia (Törmävaara site in Tervola, Hela-78, 5,160±100 bp). Nevertheless, when calibrated, these datings are very close to the ones from the sites near the Karelian Isthmus. Actually, the datings seem to support the view that Typical Comb Ware spread rapidly in its whole distribution area.

While comb ceramic traditions continued in coastal and southern Finland as Ka 3 and Pyheensilta Ware, the inland area had its own ceramic styles after the 'uniform' period of Ka 2. There is a succession of asbestos-tempered ceramic wares, of which Kierikki, Pöljä and Jysmä Wares are counted as sub-Neolithic styles preceding several Early Metal period ceramics in eastern and northern Finland. There seem to be some discrepancies between the datings and the styles, since Kierikki AMS dates stretch from c. 3,800 BC to 2,900 BC and Pöljä dates from c. 3,500 BC to 2,500 BC. The only Jysmä date available is c. 3,400–2,900 BC. Thus, the datings overlap strongly and it is obvious that the whole asbestos ceramic sequence still needs a basic chronological and typological study (cf., Pesonen 2004: 92).

In the southern and coastal areas, comb ceramics were partly substituted by Corded Ware spreading from the south around c. 3,200 BC. The carriers of the Corded Ware culture are thought to have practised early forms of husbandry and even agriculture. However, stronger evidence of agriculture in southern Finland is found only during the next period, the Kiukais culture. This period also had its own ceramic form: the flat-bottomed Kiukais vessels. There is only one AMS date of Corded Ware, which falls well within the sphere of the traditional chronology. Kiukais ceramics have no AMS dates so far.

The Adoption of Ceramics in Late Mesolithic Finland

Finland was probably populated already during the Early Mesolithic, soon after the melting of the Scandinavian ice sheet. The climate was warming up and reached its optimum during the 6th and 5th millennia BC in the Late Mesolithic.

Prehistoric society in Late Mesolithic Finland was already divided into several regional groups with different economies depending on environmental factors. Seal hunting was performed in the coastal regions of the Baltic Sea, and to some extent, also in the inland lake area. The primary livelihood in the lake area was a mixed hunting economy where big game like elk, wild reindeer and also beaver were hunted, but fishing played an important role as a balancing year-round food resource. Northern Finland's economy was based on wild reindeer and also beaver hunting, but fish must have been used where and when available. Due to poor preservation, the salmon bones have not survived in the soil, but no doubt salmon was an important nutrient in the riverside and river estuary sites. The role of gathering during the Late Mesolithic is poorly known, but there is no reason to assume that it was not practised. Gathering should have comprised at least 20% of the subsistence base in the economy (e.g., Nuñez 1990, 1991).

There are no pottery pre-forms in Finnish prehistory. The only examples of using clay in the Late Mesolithic are possibly fish-baking burnt clay pieces found at Gröndal 2 in Vantaa (Figure 10.2), southern Finland (Saunaluoma 2001). The origins of Finnish pottery have therefore been sought elsewhere from the neighbouring areas, usually in the east, where the most probable sources mentioned are the pre-Lialovo pottery of Upper Volga culture as well as pottery types in Desna and Zosh valleys, possibly also in the Upper Duna areas (e.g., Nuñez 1990: 33). Nevertheless, according to new radiocarbon datings, several old pottery traditions appear in the European part of Russia (Timofeev and Zaitseva 1997). The relationship of these potteries to Early Comb Ware has not been discussed yet, but in any case it took some 1,000 years until the ceramics spread from the Upper Volga to Finland (Carpelan 1999: 253). The spreading of ceramics into Finland has been explained as a diffusion process, where the most plausible method would have been the practice of exogamy (e.g., Carpelan 1999: 254; Nuñez 1990: 35). Mating networks across western Russia and Finland would best explain the relatively rapid diffusion of pottery-making craft without other changes in archaeological assemblages. The potters were presumably women, who did not bring other artefactual novelties with them (Nuñez 1990: 35).

The adoption of ceramics among the Mesolithic populations of Finland presupposes some need for ceramic vessels. The most frequently suggested explanation is the rise in the seal and fish catches due to warming climatic conditions. To some extent, there was also a shift from terrestrial game to maritime game. Increased use of maritime resources would have led to surplus production and eventually to the need for more effective and easier storage methods, especially for the seal products during the mass-hunting seasons. In this situation, the ceramic vessels would have been readily adopted (Edgren 1982; Nuñez 1990, 1999; Siiriäinen 1981). This economic change may also be seen as a shift towards a more stable settlement pattern; at least the base camps would have become more permanent because of storage. The use of ceramics has in itself been thought of as a sign of permanency because heavy clay vessels would hardly have been moved around and pottery-making requires long stays at the same site (Kokkonen 1982; Nuñez 1990: 35–38; Nuñez and Okkonen 1999: 106). According to some scholars, one of the primary functions of pottery was the preparation of alcoholic beverages (e.g., Salo 1989: 8). This may also have been one of the reasons for the rapid spreading of ceramics.

It has also been argued that ceramics were readily adopted by hunter-gatherer societies for more social and symbolic reasons. Its manufacture brought an important new task in society and potters may have been valued members of their groups (e.g., Carpelan 1999: 254). Comb ceramic storage vessels are large and quite impressive and they can be interpreted as a material manifestation of a group. Pottery expressed the social status of the family or the group, the skillfulness and artistic capabilities of women, the plenitude of food and thereby the competence of hunters (Salo 1989: 10).

However, it is interesting to note that ceramics did not spread everywhere in Finland. For instance, large areas in Lapland were aceramic during most of its prehistory. In northern Lapland, Säräisniemi 1 Ware was used in the beginning of the sub-Neolithic, but the next ceramics in the area were Lovozero Ware almost 3,000 years later. In southern and eastern Finland, ceramics came to stay though in some areas ceramics were used only sporadically. The issue has not been discussed thoroughly. A number of reasons may lie behind the delay in the introduction of pottery making in some areas. It may be that the population in these areas either had no need for pottery because of a different economy or did not have the connections for acquiring the knowledge of making pottery.

Practical Uses of Pottery—Vantaa Maarinkunnas Site as an Example

Some studies have shown that the main functions of ceramic vessels were storage and food preparation (e.g., Edgren 1982; Salo 1989). The vessels could have also been used for drinking, beverage preparation, water heating and rendering train oil from blubber (e.g., Alakärppä 1999; Hulthén 1982; Ihalainen and Jussila 1998; Pesonen 1996; Salo 1989). A couple of years ago, the function of vessels was studied by new means—lipid analysis. The 1997 excavation at the Maarinkunnas site in Vantaa, southern Finland (Figure 10.2) yielded a large collection of finds, the majority consisting of Typical and Late Comb Ware. Some of the pot sherds contained a black, organic crust on their surface which could possibly be the remains of food. Attempts were made to analyze this crust and some promising results were obtained. The following is based on an article by one of the authors (Leskinen 2003).

The crust in Maarinkunnas pot sherds as well as in other Finnish ceramics is pitch black and has a flaky consistency with a cracked surface (Figure 10.3). It often occurs as a fairly thick layer. In the Maarinkunnas assemblage, the crust occurs most commonly on the inner surface of the vessels, but also on the rim edge and outer surface. There is a similar crust, for

Figure 10.3 Late Comb Ware with crust inside the vessel wall (KM 30464: 14065, Vantaa Maarinkunnas), Typical Comb Ware with birch bark tar repair (KM 28013: 9374, Rääkkylä Pörrinmökki) and asbestos-tempered Kaunissaari Ware with clay paste repair (KM 25817: 136). KM = Collections of the National Museum of Finland (photo Markku Haverinen, National Board of Antiquities).

example, in Swedish Funnel Beaker pottery and Danish Ertebølle culture vessels (Andersen and Malmros 1985: 83; Hallgren and Possnert 1997: 124). The crust in the latter vessels has been suggested to indicate the boiling of a soup-like concoction (Andersen and Malmros 1985: 91). Several studies have concluded that the organic crust deposits on pottery derive from fermented food. This kind of food made by fermentation could have been some kind of blood porridge or cake. The crust samples have had high iron content (cf., blood) together with vegetable lipids that are particularly abundant in cereals (Arrhenius 1985, 1987; Arrhenius and Lidén 1988; Arrhenius and Slytå 1981; Hallgren and Possnert 1997; Østmo et al. 1996).

A total of 11 samples from Maarinkunnas were analyzed for traces of fat, proteins and carbohydrates, but the latter two were not detected. Traces of lipids such as cholesterol and their decomposition products, on the other hand, were found in all of the samples. In addition, some samples contained minor components such as benzoic acid. The majority of the fatty acids consisted of palmitic and stearic acid, which are common among fats in all kinds of food. The certain ratios of stearic acid to palmitic acid as well as palmitoleic acid to oleic acid suggest that the fats derive from marine animals. These, in turn, suggest fish, but especially seals, the oil and meat of which can be the source of cholesterol as well (Hopia et al. 2003). This view is supported by the bone analysis from Maarinkunnas, which is dominated by seal bones, although fish bones, especially pike, are also present.

Eight of the analyzed samples also contained phytanic acid, which derives from chlorophyll. Phytanic acid occurs in animals that have consumed food rich in chlorophyll. It is thus found in the fats of marine and terrestrial herbivores; particularly high concentrations occur in some fish oils. In addition to lipids, fats and oils contain several other components, such as sterols. The only sterol present in the studied samples was cholesterol. It was found in four samples. Cholesterol is an indication of animal tissue, since it is not found in plant matter. The high cholesterol content strongly suggests that food derived from the animal kingdom was stored or prepared in the vessels (Hopia et al. 2003).

Benzoic acid is found naturally in cowberry (*Vaccinium vitis-idaea*) and cranberry (*Vaccinium oxycoccos/microcarpum*), and the modern food industry uses it as a preservative (E210) (Hopia et al. 2003). It is possible that cowberry and cranberry have been part of the diet in Maarinkunnas; they may have been eaten as well as used in food storage to prevent moulding or fermentation. If cowberry and/or cranberry were part of the diet, we are probably dealing with autumn fare, or at least food prepared or stored during the fall. In addition, the blubber layer of seals is at its thickest in the autumn, and late fall is a profitable period for netting seals close to the shoreline (Ylimaunu 1999: 130, see also Ylimaunu et al. 1999: 140, 148). For the sake of comparison, results from the organic crust on a pot sherd from the Otterböte site in Kökar, the Åland Islands (Figure 10.2), suggest that it derives from marine animals: the high lipid content indicates seal products. Train oil from seals works very well in the long-term storage of berries, roots, meat and fish (Gustavsson 1997: 93–95).

Since mashed cowberries keep through the winter, it was also possible to use them as a nutritional supplement for several months. It is obvious that prehistoric diet included a variety of berries, in addition to nuts and grains. It is also common knowledge that some Native Americans habitually prepared pemmican, a high-energy travel fare that consisted of dried and ground bison meat, fat and berries such as wild cherries or mountain ash berries. It is thus not impossible that meat and berries were combined into a single dish during the Comb Ceramic Period as well. Among some Pacific Coast Indians, the meat in pemmican was replaced by fish. Furthermore, Eskimos are known to make a kind of ice cream (*Akutuq*) from seal oil or caribou fat by first warming the oil and then whipping it by hand in the open air; as the oil slowly cools, a rich foam is produced. *Akutuq* is usually flavoured with cowberries (Jones 1983).

Results of the crust analyses from Maarinkunnas differ from those of neighbouring countries. In contrast to the samples analyzed from this region, Swedish and Danish samples often show a total absence, or very small amounts, of cholesterol, whereas proteins and sometimes carbohydrates are present (Arrhenius 1985; Arrhenius and Lidén 1988). The results often suggest a diet derived from vegetable matter, which is natural, since agriculture and animal husbandry were already practised in southern Scandinavia in the 5th century BC, whereas in Finland, tangible evidence of agriculture at this time is absent. During the Comb Ceramic Period, the economy of Finland was still based on hunting, fishing, sealing and gathering. It is also noteworthy that benzoic acid is apparently not present in the Swedish crust samples. Nevertheless, it can be said that Neolithic pottery was used for food preparation in addition to storage, regardless of whether the method was cooking or fermentation.

Results derived from analysis of crusts probably relate only to the last episode of use of the vessels. On the other hand, the crust may contain residues from previous foodstuffs, and the resulting view may represent a kind of melange, an average of the different substances stored or prepared in the vessel. This might explain the curious combinations, such as the presence of benzoic acid and cholesterol in a single sample.

Valuable Pottery—Repairing of the Vessels

In subarctic regions, the manufacture of ceramics was an especially laborious task because of the long drying stage, which requires predictable and extended periods of warm and dry conditions (e.g., Nuñez and Okkonen 1999: 106). The climatic optimum may explain the larger amount of ceramics during the comb ceramic periods in Early and Middle sub-Neolithic. Thinner vessel walls were preferred in the Middle and Late sub-Neolithic and such temper materials which made this possible were sought. There was a growing tendency away from coarse sand and crushed stone tempers towards asbestos and talc during the Middle sub-Neolithic.

Due to challenging production conditions, ceramics were valuable and we see evidence of careful treatment and efforts to repair the pots. Two methods of repair have been observed in Stone Age ceramics. The first one, repairing with clay paste (Figure 10.3), has been used especially in Early Asbestos Ware, though examples of repaired Comb Ware vessels also often appear. Thinned vessel walls could also have been mended by using clay paste (Pesonen 1996: 12). Another method included the use of birch bark pitch or tar (Figure 10.3), which was used as sealing material or putty in the cracks of the vessel walls, especially in Typical Comb Ware (Pesonen 1999). It is not clear whether the cracks formed during the drying stage, firing or use of the vessel. On both sides of the crack, holes were drilled through the wall, a cord made of elk's vein was tied through the holes and both the crack and the holes were sealed with the birch bark pitch. Usually only the large storage vessels were repaired—apparently the repairs were not durable enough in cooking vessels. So far, no archaeological structures related to the manufacture of the birch bark pitch have been found in excavations.

Apparently, pottery was not a disposable product, but a valuable household item, frequently used and worth repairing. Moreover, the repairing of vessels may be seen as further evidence of a permanent lifestyle, where ceramics were not abandoned in the dumps after single use, but seen as a part of the future life of the site.

Regional Pottery—Uniform Style with Temper Variations

There are some periods in the Finnish Stone Age when the style of ceramics was quite uniform in large areas, but pottery manufacturing technique, especially the use of temper materials varied regionally. Though Early Comb Ware 1 may already have had local variants (in addition to the Säräisniemi 1 Ware, which is a northern style contemporary with Early Comb Ware 1, but not an actual variant [Torvinen 2000]); Early Comb Ware 2, in particular, has been interpreted to have been divided into at least two local style variants: southwestern and southeastern ones (Edgren 1966). In addition to the style diversions, there were also two ceramic groups contemporary with Early Comb Ware 2, the southwestern Jäkärlä Ware with organic temper and the eastern Kaunissaari Ware with asbestos temper (Figure 10.3). In spite of the distinct difference in the structure of the ceramic paste, the decoration is essentially the same in these two wares: oblong, broad-combed stamps in horizontal rows without any pits, of which the last trait is characteristic also of the 'common' Early Comb Ware 2 (Edgren 1966; Pesonen 1996).

Temper materials have been regional in other periods too. The most obvious regional temper material is asbestos, which was quarried mainly from the deposits in North Savo (Figure 3.2), exploited during the Stone Age (Lavento and Hornytzkyj 1995, 1996). The use of asbestos began in the Early Comb Ware 2 period in eastern Finland—the earliest examples of asbestos temper are found in the vessels of this style. Soon, a distinct ceramic style, Kaunissaari Ware, came into use in eastern Finland, where its main distribution areas are in the provinces of Karelia and Savo (Pesonen 1996). Asbestos was used to some extent also in Typical and Late Comb Ware, but not as manifest as in Kierikki, Pöljä and Jysmä ceramics, which are the 'proper' asbestos wares in Finnish prehistory. The use of asbestos in the Stone Age was restricted mainly to eastern Finland, though Kierikki and Pöljä ceramics spread to Ostrobothnia and Lapland. Some sites are known also in the provinces of Häme and Uusimaa. Sporadic finds have been discovered even in southern and southwestern coastal regions. Asbestos actually 'broke out' of its regionality only during the Early Metal Period, when Säräisniemi 2 Ware was used from Häme to Lapland. At this time the local asbestos deposits were discovered and used (Lavento and Hornytzkyj 1996).

The use of organic tempers is another regional feature of the wider ceramic technique. During the Comb Ware periods, organic tempers were used in coastal regions in addition to sand and crushed stone tempers (e.g., Pesonen 2004: 93). In time, organic tempers were used in larger quantities in the inland regions as well. In Kierikki and Pöljä ceramics, organic tempers were widely used in addition to asbestos tempers (e.g., Huurre 1984: 46). The variability in the use of different temper materials may depend on the availability of raw materials, pottery function and even on the season when pottery was manufactured (Ikäheimo and Panttila 2002). An interesting observation has been made concerning the porosity of thick- and thin-walled vessels. It seems that thick-walled and often porous vessels are better suited for storage use while thin-walled and compact vessels are suitable for cooking. The choice of temper does not seem to affect the pottery function as much as the porosity does (Ihalainen and Jussila 1998).

The clay material of the pottery has been an object of study more rarely than the temper materials. Some studies exist, however. Most commonly, diatom flora have been studied in order to find out the provenance of the clay. Rarely, Atomic Absorption Spectrometry (AAS) has also been used to trace elements of clay and temper materials. Results of the diatom studies indicate that potters usually used clay available in the proximity of the dwelling sites. However, there are exceptions. In Åland, Ancylus clay was used, which was not available in the archipelago and thus the ceramics or the clay itself must have been transported to the islands during the Stone Age. The AAS analysis together with stylistic studies suggests that a Pit and Comb Ware vessel found in Kotka, southeastern Finland (Figure 10.2), probably originates in East Karelia (Alhonen et al. 1980; Alhonen and Väkeväinen 1981; Matiskainen and Alhonen 1984).

Beginning of Agriculture and Animal Husbandry

The beginning of change in the economies of the Stone Age societies is manifested by the emergence of the Battle Axe culture in Finland. From 3,200 BC, Battle Axe cultures spread in Europe from the Rhein to the Dnieper and from Switzerland to the Baltic and southern Fennoscandia. The characteristic artefact forms of the Battle Axe culture in Finland are the battle axes and cord-decorated pottery. In Finland, this phase is thus usually called the Corded Ware culture. This, more often than any other phase in the Finnish Stone Age, has been postulated as a result of immigration. The distribution of Corded Ware in Finland was restricted to the southern and western provinces, in the areas best suited for agriculture.

Almost everywhere, the Battle Axe culture groups have practised cultivation or animal husbandry, but in Finland, no direct evidence of agriculture has yet been found. The signs have fruitlessly been sought from the artefact forms, pollen and macrofossil data and also from imprints in ceramic vessels (Edgren 1970, 1984; Vuorela 1982, 1999). The research has especially concentrated on the Perkiö site in Hauho (Figure 10.2), which is one of the largest Corded Ware sites in Finland. About 30,000 sherds of ceramics have been found there. No imprints of cereals have been found on the vessel walls, nor were any cereal pollens found in the vicinity of the site. Christian Carpelan has proposed that this site was not an actual settlement site, but a pottery production site (Carpelan 1999: 264). This would explain the missing cereal pollens as well as missing cereal imprints. The research at this site does not refute the possibility of agriculture in Corded Ware culture.

The vessel form and manufacture do differ markedly from the earlier 'comb ceramic' model. The Corded Ware vessels were, in general, smaller than Comb Ware vessels, chamotte was the only temper used and the vessels were flat-bottomed while comb ceramics were round- or spit-bottomed (Figure 10.4). It is speculated that Corded Ware vessels could have been used to store milk and cereal products (e.g., Edgren 1982: 51). In addition to the location of settlement, which now favoured meadows suitable for cattle grazing, this is an indirect proof of animal husbandry (e.g., Siiriäinen 1981; Zvelebil 1978). More clearly, animal husbandry from this time is

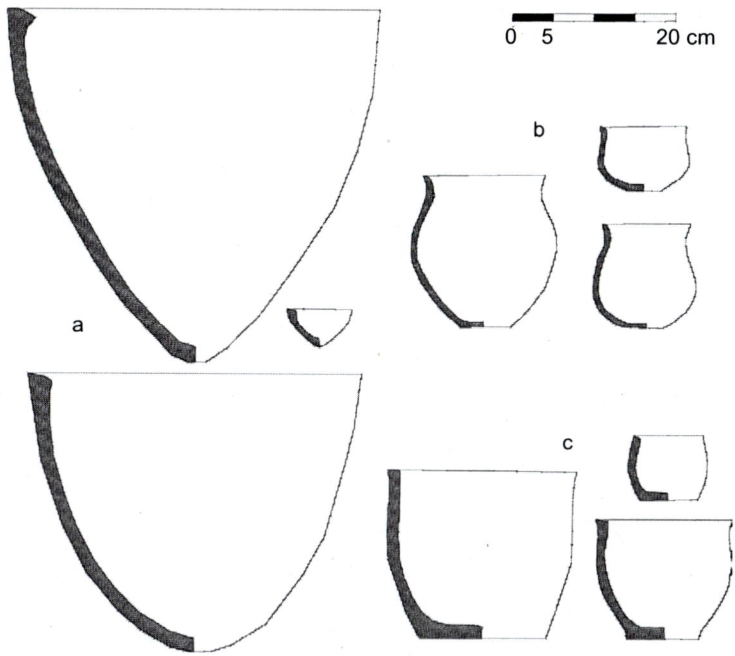

Figure 10.4 Vessel shapes and size of (a) Comb Ware, (b) Corded Ware and (c) Kiukais Ware (drawing by Sirpa Leskinen).

indicated by fragments of unerupted juvenile bovine teeth from the Jokiniemi site in Vantaa (Nuñez 1999: 137).

The oldest carbonized seeds of barley derive from the sites of the Kiukais culture. Pollen records indicate that cultivation was clearly present during and after this period (Lempiäinen 1999; Vuorela 1999). Coastal Kiukais culture has been interpreted as a maritime hunting-based economy, which also practised agriculture and animal husbandry. The vessel form is more reminiscent of the Corded Ware than the Comb Ware tradition (Figure 10.4). While pottery production in the Corded Ware culture seems to have been concentrated in special production sites, the vessels of the Kiukais culture were probably made by local groups. The pottery tradition of the Kiukais culture was continued in the vessel forms of the western and southern Finnish Bronze Age.

In the inland lake area and in Ostrobothnia and Lapland, the ceramic tradition continued among hunter-gatherer communities. Here the Pöljä and Jysmä asbestos ceramic traditions were followed by several new styles which were influenced by textile ceramics from the south and east during the Bronze Age. It is possible that the function of ceramics gained new dimensions beyond cooking and storage (e.g., Hulthén 1991).

Conclusion and Outlook

Pottery saw widespread use amongst the early hunter-gatherer societies of Finland, from 5,300 BC onwards. The ceramics were adopted here as a ready product, since no pottery pre-forms have been found and this is true also for the neighbouring areas. The first pottery in Finland, Early Comb Ware, is found in Late Mesolithic sites with no other artefact novelties. This can be taken as an evidence for exogamic practice as a pottery diffusion mechanism. Pottery was adopted among Finland's Mesolithic populations because there was an apparent need for effective storage methods. This need was most likely caused by the increased use, and eventually surplus production, of maritime resources, especially seal and fish. The lipid analysis from the crust in Typical and Late Comb Ware shows that vessels were probably also used in food preparation, involving either fermentation or cooking. Besides storage and food preparation, the adoption of pottery may also be due to social reasons. Potters may have been valued members in society because they had unique skills. Knowledge of pottery making may have influenced the rapid adoption of pottery and eventually led to long-distance exogamy. Pottery was not only interesting, but also valuable, which is indicated by the fact that the vessels were often repaired.

Though changes took place in the artefactual and socio-economic culture among the hunter-gatherer societies in Finland, it seems that there were no major changes in the pottery traditions prior to the coming of the Battle Axe culture with its Corded Ware ceramics around c. 3,200 BC. Before that, only the decoration motifs developed, but vessel shapes and sizes remained constant. Instead, there were local variations in the use of temper, for example, asbestos in eastern Finland and organic tempers in the coastal areas. These may be seen as technological adaptations for local conditions, while the decoration represents a cultural uniformity. Corded Ware broke the Comb Ware tradition only in southern and western Finland, but hunter-gatherers in eastern and northern Finland continued the old tradition until the Early Metal Period, when the new ceramic traditions arrived there from the east.

The study of hunter-gatherer ceramics in Finland has long concentrated on the typology and chronology. In the future, research interests may shift towards examination of the actual uses of pottery—not only the everyday function, but also the social function of ceramics in hunter-gatherer societies. It must be stressed, though, that new analytical methods have helped a great deal in the study of the everyday function of vessels. Also, AMS datings have helped shed light on the chronology of the ceramics and this is vital: the chronology can never be accurate enough. These analytical methods enable a better picture of the chronology and function of pottery which in turn helps us draw valid conclusions about the social meaning of pottery. We are now able to connect Finnish pottery with the development of pottery traditions in a wider perspective.

Acknowledgements

The authors wish to thank Mr Trevor Doyle and Mr Miikka Tallavaara for reading and commenting on the manuscript.

References

Alakärppä, J (1999) 'Kuoppa-asumusten lämmityskokeilu Kierikissä', 4 *Muinaistutkija*, 50–57

Alhonen, P, Kokkonen, J, Matiskainen, H and Vuorinen, A (1980) 'Applications of AAS and diatom analysis and stylistic studies of Finnish Subneolithic pottery', 52:2 *Bulletin of the Geological Society of Finland*, 193–206

Alhonen, P and Väkeväinen, L (1981) 'Diatom analytical studies of Early Comb Ceramic vessels from Åland', *Suomen Museo*, 67–77

Andersen, S H and Malmros, C (1985) ' "Madskorpe" på Ertebøllekar fra Tybrind Vig', *Aarbøger for Nordisk Oldkyndighed og Historie*, 78–95

Arrhenius, B (1985) 'Chemical analyses of organic remains in archaeological context', in Edgren, T (ed), *Iskos 5*, Helsinki: The Finnish Antiquarian Society, pp 339–344

——— (1987) 'Prehistoric diet and archaeology in the laboratory', *Swedish Archaeology 1981–1985*, Stockholm: The Swedish Archaeological Society, pp 113–122

Arrhenius, B and Lidén, K (1988) 'Fisksoppa eller vegetabilisk gröt? Diskussion kring matresterna från Tybrind Vig', in Arrhenius, B (ed), 3 *Laborativ arkeologi*, 7–17

Arrhenius, B and Slytå, K (1981) 'Fermenterade blodgrötar', in *Rapport från Stockholms Universitets Arkeologiska Forskningslaboratorium nr 1, 1981*, Stockholm: Stockholms universitet, pp 80–108

Bronk, R C (1995) 'Radiocarbon calibration and analysis of stratigraphy: the OxCal program', 37(2) *Radiocarbon*, 425–430

——— (2001) 'Development of the Radiocarbon Program OxCal', 43(2A) *Radiocarbon*, 355–363

Carpelan, C (1979) 'Om asbestkeramikens historia i Fennoskandien', *Finskt Museum*, 5–25

——— (1999) 'Käännekohtia Suomen esihistoriassa aikavälillä 5,100–1,000 eKr.', in Fogelberg, P (ed), *Pohjan Poluilla. Suomalaisten Juuret Nykytutkimuksen Mukaan. Bidrag till Kännedom av Finlands Natur och Folk 153,* Helsinki: Societas Scientiarum Fennica, pp 249–280

——— (2002) 'Esihistorian vuosiluvut, ajoitukset ja kronologia', in Grünthal, R (ed), *Ennen, Muinoin. Miten Menneisyyttämme Tutkitaan,* Helsinki: Suomalaisen Kirjallisuuden Seura, pp 18–27

——— (2004) 'Environment, archaeology and radiocarbon dates. Notes from the Inari region, northern Finnish Lapland', in Lavento, M (ed), *Early in the North, Vol. 5, Iskos 13,* Helsinki: The Archaeological Society of Finland, The Finnish Antiquarian Society, pp 17–45

Edgren, T (1966) *Jäkärlä-gruppen. En västfinsk kulturgrupp under yngre stenålder,* Suomen Muinaismuistoyhdistyksen Vuosikirja 64, Helsinki: The Finnish Antiquarian Society

Edgren, T (1970) *Studier över den snörkeramiska kulturens keramik i Finland,* Suomen Muinaismuistoyhdistyksen Vuosikirja 72, Helsinki: The Finnish Antiquarian Society

——— (1982) *Formgivning och funktion. En kamkeramisk studie*, Iskos 3, Helsinki: The Finnish Antiquarian Society

——— (1984) 'On the economy and subsistence of the Battle-Axe culture in Finland', in Edgren, T (ed), *Fenno-Ugri et Slavi 1983. Iskos 4*, Helsinki: The Finnish Antiquarian Society, pp 9–15

Europaeus-Äyräpää, A (1930) 'Die relative Chronologie der steinzeitlichen Keramik in Finnland', (I) *Acta Archaeologica,* 165–190, 205–220

Gustavsson, K (1997) *Otterböte. New Light on a Bronze Age Site in the Baltic,* Theses and Papers in Archaeology B:4. Archaeological Research Laboratory, Stockholm University

Halén, O (1994) *Sedentariness during the Stone Age of Northern Sweden in the Light of the Alträsket Site, c. 5,000 B.C. and the Comb Ware Site Lillberget, c. 3,900 B.C. Source Critical Problems of Representativity in Archaeology,* Acta Archaelogica Lundensia, Series 4, Number 20, Lund.

Hallgren, F and Possnert, G (1997) 'Pottery design and time. The pottery from the TRB site Skogsmossen, in view of the AMS-datings of organic remains on potsherds', (29) *TOR*, 113–126

Hansson, A-M and Isaksson, S (1994) 'Analyses of charred organic remains', in Arrhenius, B (ed), *Laborativ arkeologi 7*, Stockholm: Arkeologiska forskningslaboratoriet Stockholms universitet, pp 21–29

Hopia, A, Reunanen, M and Pesonen, P (2003) 'GC-MS analysis of organic residues in the potsherd samples from Vantaa Maarinkunnas', *Finskt Museum* (1995), 44–55

Hulthén, B (1982) 'Keramik', Kompendium i arkeologi, unpublished stencil, Department of Archaeology, University of Stockholm

——— (1991) *On Ceramic Ware in Northern Scandinavia During the Neolithic, Bronze and Early Iron Age,* Archaeology and Environment 8. Umeå: Department of Archaeology, University of Umeå

Huurre, M (1984) 'Kainuu from the Stone Age to the Bronze Age. Finds and cultural connections', in Edgren, T (ed), *Fenno-Ugri et Slavi 1983. Iskos 4*, Helsinki: The Finnish Antiquarian Society, pp 42–50

Ihalainen, P and Jussila, T (1998) 'Esihistoriallisen keramiikan huokoisuustutkimuksia', 3 *Muinaistutkija*, 24–34

Ikäheimo, J and Panttila, H (2002) 'Explaining ceramic variability: the case of two tempers', (XIX) *Fennoscandia Archaeologica*, 3–11

Isaksson, S (1997a) 'Soil-lipids in cultural layers', in Edgren, T (ed), *Iskos 11*, Helsinki: The Finnish Antiquarian Society, pp 242–253

——— (1997b) 'Laboratory report concerning the lipid content and fatty acid pattern of the organic material on four sherds from Otterböte, Kökar, Åland', Appendix 6 in Gustavsson, K (ed) *Otterböte. New Light on a Bronze Age Site in the Baltic,* Theses and Papers in Archaeology B:4, pp 169–176

Jones, A (1983) *Nauriat Nigiñaqtua—Plants That We Eat.* Kotzebue, Alaska. Maniilaq Association Traditional Nutrition Program

Kokkonen, J (1982) 'Defining the potter's craft. Problems of systematics in Comb Ceramic studies', in *Studia Minora. Helsingin Yliopiston Arkeologian Laitos, Moniste No. 29*, Helsinki: Department of Archaeology, The University of Helsinki, pp 1–12

Lavento, M and Hornytzkyj, S (1995) 'On asbestos used as temper in Finnish Subneolithic, Neolithic and Early Metal Period pottery', (XII) *Fennoscandia Archaeologica*, 71–76

——— (1996) 'Asbestos types and their distribution in the Neolithic, Early Metal Period and Iron Age pottery in Finland and eastern Karelia', in Kirkinen, T (ed), *Pithouses and Potmakers in Eastern Finland. Reports of the Ancient Lake Saimaa Project. Helsinki Papers in Archaeology, no. 9,* Helsinki: Department of Archaeology, The University of Helsinki, pp 41–70

Lempiäinen, T (1999) 'Hiiltyneet viljanjyvät ja maanviljelyn alku Suomessa', in Fogelberg, P (ed), *Pohjan Poluilla. Suomalaisten Juuret Nykytutkimuksen Mukaan. Bidrag till Kännedom av Finlands Natur och Folk 153,* Helsinki: Societas Scientiarum Fennica, pp 152–54

Leskinen, S (2003) 'On the dating and function of the Comb Ceramics from Maarinkunnas', *Finskt Museum* (1995), 5–43

Matiskainen, H and Alhonen, P (1984) 'Diatoms as indicators of provenance in Finnish Sub-Neolithic pottery', 11 *Journal of Archaeological Science*, 147–157

Meinander, C F (1961) 'De subneolitiska kulturgrupperna i norra Europa', 39 B:4 *Societas Scientiarum Fennica, Årsbok*, 1–23

——— (1984) 'Kivikautemme väestöhistoria', in Åström, S-E (ed), *Suomen Väestön Esihistorialliset Juuret. Bidrag till Kännedom av Finlands Natur och Folk 131,* Helsinki: Societas Scientiarum Fennica, pp 21–48

Nuñez, M (1990) 'On Subneolithic pottery and its adoption in late Mesolithic Finland', (VII) *Fennoscandia Archaeologica*, 27–52

——— (1991) 'On the food resources available to man in Stone Age Finland', *Finskt Museum* (1990), 24–54

——— (1999) 'Role of food production in Stone Age Finland', in Fogelberg, P (ed), *Pohjan Poluilla. Suomalaisten Juuret Nykytutkimuksen Mukaan. Bidrag till Kännedom av Finlands Natur och Folk 153,* Helsinki: Societas Scientiarum Fennica, pp 133–142

Nuñez, M and Okkonen, J (1999) 'Environmental background for the rise and fall of villages and megastructures in North Ostrobothnia 4,000–2,000 cal BC', in Huurre, M (ed), *Dig It All. Papers Dedicated to Ari Siiriäinen*, Helsinki: The Archaeological Society of Finland, The Finnish Antiquarian Society, pp 105–115

Østmo, E, Hulthén, B and Isaksson, S (1996) 'The Middle Neolithic settlement at Auve', in Arrhenius, B (ed), *Laborativ Arkeologi* 9, Stockholm: Arkeologiska forskningslaboratoriet Stockholms universitet, pp 31–40

Pesonen, P (1996) 'Early Asbestos Ware', in *Pithouses and Potmakers in Eastern Finland. Reports of the Ancient Lake Saimaa Project. Helsinki Papers in Archaeology, no. 9,* Helsinki: Department of Archaeology, The University of Helsinki, pp 9–39

——— (1999) 'Radiocarbon dating of birch bark pitches in Typical Comb ware in Finland', in Huurre, M (ed), *Dig It All. Papers Dedicated to Ari Siiriäinen*, Helsinki: The Archaeological Society of Finland, The Finnish Antiquarian Society, pp 191–200

——— (2001) 'Kiteen Sarvisuo—lisää varhaisesta asbestikeramiikasta' in Ranta, H (ed), *Kentältä Poimittua 5, Kirjoitelmia Arkeologian Alalta, Museoviraston Arkeologian Osaston Julkaisuja no* 9, Helsinki: Museovirasto, pp 34–56

——— (2004) 'Neolithic pots and ceramics chronology—AMS-datings of Middle and Late Neolithic ceramics in Finland', in Uino, P (ed), *Fenno-Ugri et Slavi 2002, Dating and Chronology, Museoviraston arkeologian osaston julkaisuja no.* 10, Helsinki: Museovirasto, pp 87–97

Räihälä, O (1996) 'A Comb Ware house in Outokumpu Sätös—some remarks on the application of ceramic typologies', in Kirkinen, T (ed), *Pithouses and Potmakers in Eastern Finland. Reports of the Ancient Lake Saimaa Project. Helsinki Papers in Archaeology, no.* 9, Helsinki: Department of Archaeology, The University of Helsinki, pp 89–117

Salo, U (1989) 'Astian kulttuurihistoriasta Suomessa ja naapurialueilla', *Suomen Museo*, 5–48

Saunaluoma, S (2001) 'Gröndal 2—mesoliittisen kivikauden asutusta Vantaan Kaivokselassa', in Ranta, H (ed), *Kentältä Poimittua 5, Kirjoitelmia Arkeologian Alalta, Museoviraston Arkeologian Osaston Julkaisuja no.* 9, Helsinki: Museovirasto, pp 5–10

Schulz, E-L (2004) 'Ankkapurhan arkeologisen aineiston radiohiiliajoitukset', in Uino, P (ed), *Ammoin Ankkapurhassa. Kymenlaaksossa Kivikaudella*, Helsinki: Museovirasto and Stora Enso Anjalankosken tehtaat, p 46

Siiriäinen, A (1974) 'Studies relating to shore displacement and Stone Age chronology in Finland', *Finskt Museum* (1973), 5–22

——— (1981) 'On the cultural ecology of the Finnish Stone Age', *Suomen Museo* (1980), 5–40

Timofeev, V I and Zaitseva, G I (1997) 'Some aspects on the radiocarbon chronology of the Neolithic cultures in the forest zone of the European part of Russia', in Edgren, T (ed), *Proceedings of the VII Nordic Conference on the Application of Scientific Methods in Archaeology, Iskos 11,* Helsinki: The Finnish Antiquarian Society, pp 15–22

Torvinen, M (2000) 'Säräisniemi 1 Ware', (XVII) *Fennoscandia Archaeologica*, 3–35

Vuorela, I (1982) 'Pollen stratigraphy and chemical analyses of a mineral soil profile at a Corded Ware dwelling site in southern Finland compared with those of local organic sediments', in *Pact 7 Part I. Second Nordic Conference on the Application of Scientific Methods in Archaeology,* Strasbourg, 175–193

——— (1999) 'Viljelytoiminnan alku Suomessa paleoekologisen tutkimuksen kohteena', in Fogelberg, P (ed), *Pohjan Poluilla. Suomalaisten Juuret Nykytutkimuksen Mukaan. Bidrag till Kännedom av Finlands Natur och Folk 153,* Helsinki: Societas Scientiarum Fennica, pp 143–151

Ylimaunu, J, Ylimaunu, T and Okkonen, J (1999) 'Hylkeenpyynnin kehityksestä ja merkityksestä Itämerellä esihistoriallisena aikana', (XXII–XXIII) *Faravid, Pohjois-Suomen Historiallisen Yhdistyksen Vuosikirja,* 131–158

Ylimaunu, T (1999) 'Iin Hangaskankaan keittokuopan rasva-analyysi', (XXII–XXIII) *Faravid, Pohjois-Suomen Historiallisen Yhdistyksen Vuosikirja*, 125–130

Zvelebil, M (1978) 'Subsistence and settlement in the north-eastern Baltic', in Mellars, P (ed), *The Early Postglacial Settlement of Northern Europe*, London: Duckworth, pp 204–241

CHAPTER 11

CERAMICS AS A NOVELTY IN NORTHERN AND SOUTHERN SWEDEN

Ole Stilborg and Lena Holm

This article seeks to investigate the characteristics of early ceramics in hunter-gatherer and early agricultural societies in northern and southern Sweden. The aim of this paper is to investigate how early ceramics were introduced into prehistoric Sweden. We explore different trajectories of introduction and use of ceramics in northern and southern Sweden.

In the coastal region of northern Sweden as indeed in the rest of northern Scandinavia, datings of Comb Ware between 6,500 and 5,800 BP (4,500–3,800 BC) indicate that there was early use of pottery as far north as the Barents Sea. It was introduced as a novelty in social settings far away from agrarian contexts. This is discussed in terms of representativity, technology and design. Later, in the Early Neolithic—after farming had arrived in the south—there was an aceramic period in northern Sweden and subsequently new technologies and traditions of pottery were introduced from farmers living further south. These new pottery traditions indicate greater diversity in function and probably even an increasingly active and different social role.

In Scania, the southernmost province of Sweden, the knowledge of making ceramic vessels was introduced into hunter-gatherer societies at about the same time as in northern Sweden, around 4,500 BC. Studies of the technology, design and use of vessels made and used by the hunter-gatherer groups of the Ertebølle culture produce a picture of an experimental handicraft, a sparse and delineated design range and a limited functionality. They seem not yet to have been directly integrated into the ensemble of everyday objects framing and being manipulated by our actions—the material universe of daily life. As the practice of farming was adopted more widely in the economically and culturally defined Early Neolithic, the changes in the ceramics point to a more accomplished craft, a greater range in functionality and indications that pottery served a more active social and symbolic role in these early farming societies.

Introduction

The social importance of exotic materials and ready-made objects, especially in the shape of adornments, acquired through trade and exchange is a widely discussed subject within Stone Age research (Taffinder 1998: 32 with references). Most often, they can be seen to have been acquired, or their import and distribution controlled, by elite groups and individuals in the society with means to restrict access to the exotics.

In contrast, less attention has been given to constructing the social history of the introduction of new, perhaps more functional or practical techniques or tools, such as ceramic containers. As difficult as they may be to answer, the questions related to the adoption of novel technologies are nevertheless well worth asking and discussing. In the case of pottery, there seems to have been no initial phase, where pottery was introduced as imported foreign products in Scandinavia before the skills to make pottery were acquired.

Based on social anthropological studies (for example in Polynesia), we may assume that ceramics as an innovation was instigated or controlled by informal opinion leaders in society (Bargatzky 1986). Bargatzky defines opinions leaders as individuals esteemed because of their adherence to norms and values of the group, while not necessarily with high social status. They have more important outside contacts than other group members and thereby more choices (Bargatzky 1986: 7–8).

But considering that ceramic containers hardly fulfilled a critical need in the subsistence of society (Brown 1986: 2–3), what would be the incentive for Stone Age opinion leaders to introduce pottery making? Could any prestige or advantage be gained from a handicraft, which produces rather inconspicuous products? In the first phase of the introduction of iron, for example, it was used for jewellery rather than for weapons and tools, although it was for these products that the new material was an innovation. Likewise, the first ceramics could have been figurines, but instead the new material was used for making practical containers from the start.

In the history of archaeology, ceramics have traditionally been seen as a paraphernalia of 'neolithization', but in both the northern and southern parts of Sweden, the ceramic craft was first practised in hunter-gatherer societies. The question is whether potting was introduced as a craft for its own sake or whether the pots were only a part of a more general acquisition of new imported processes or customs. If the latter was the case, could the prestige or advantage be attached to a particular, desirable product made in the ceramic containers? Following on from that, what social connections were established with the pots themselves as a consequence of their function?

In order to approach some possible interpretations, we need an encompassing analysis of the making, design and use of pots as well as of the contexts of their use. In this chapter, we aim to present an overview of the present knowledge of the history of the introduction of pottery into societies primarily with a hunter-gatherer economy in the coastal regions

Table 11.1 Chronological table of the culture groups/pottery styles and the subsistence economy in southern and coastal northern Sweden during the periods discussed in the article. The information for southern Sweden builds on Karsten and Svensson 2004: 18–19.

BC	S. Sweden	N. Sweden
6,000	Kongemose. Semi-sedentary HG	Late Mesolithic Mobile HG
5,500	Early Ertebølle Sedentary HG	
5,000		
4,500	Pottery Late Ertebølle Sedentary HG	Pottery Early Comb Ware Mobile HG in N. Sweden, Norway and Finland
4,000	Early Funnel Beaker Domestication	
3,500	Middle Funnel Beaker Sedentary farmers	
3,000	Late Funnel Beaker. Sedentary farmers	Typical Comb Ware Mobile HG
2,500	Battle Axe Culture Sedentary farmers	Neolithic Mobile HG
2,000	Final Neolithic Sedentary farmers	Pitted Ware, Corded Ware, Kiukais Ware & local technology Semi-sedentary HG
1,500	Bronze Age	Final Neolithic

of northern Sweden and into similar groups in southern Sweden. We will focus on the interplay between technology, representativity, design and use of early ceramics during the introduction of a new technology into existing social and economic contexts. In short, can the archaeology of early ceramics tell us something about how this novelty was received in northern and southern Sweden? We will start in the northern regions of Sweden and then move to the southernmost part of Sweden.

Case Study 1: Early Ceramics in Northern Coastal Regions

The landscapes of northern Scandinavia are heterogeneous in character and it is vital to understand their importance as subsistence grounds to prehistoric societies. The differences concerning topography, vegetation and climate are significant both within this vast area and compared to southern parts of Scandinavia. These conditions created different economic opportunities and constraints for northern and southern Stone Age societies and were of importance for human strategies and for our understanding of prehistoric societies.

Northern landscapes were dominated by moraines with vegetations attracting a rich and diverse fauna. Soon after deglaciation, birch forests covered vast areas, including northernmost areas like Finnmarksvidda and Varanger. But by the Late Mesolithic and Early Neolithic, pine replaced birch and dominated forests in large areas of northern Scandinavia (Skandfer 2003a: 12–13). These forests affected the soils. For agriculturists, this was, and still is, an aggravating circumstance and a limiting factor.

The fact that most of the north Swedish topography laid, and still lies, above the highest coastline means that marine and other fine sediments are lacking in vast areas. The consequences of the latest glaciation are still obvious, seen in land uplift and isostatic movements, conditions which were more accentuated during the Late Mesolithic and Neolithic. When understanding cultural traditions and subsistence in north Scandinavian Stone Age societies, the circumstances described above should be seen as contextual elements. Compared to the south, northern conditions were unsuitable and in a sense obstructing for agriculture.

Stone Age societies were organized as hunter-gatherer groups either with coastal fishing, hunting and gathering or inland hunting, gathering and fishing at well-structured settlements. Agriculture spread more slowly into these northern areas. The dramatic landscape relief is yet another hindrance for farming, but large areas with more fertile soils, like some coastal areas of the Bothnian Sea in the east and the region of Storsjön to the southwest, were more favourable and did provide opportunities for successful arable lands since prehistoric times.

Early Northern Comb Ware

In the Late Mesolithic, certain items in the archaeological data indicate widespread communications with other regions. For example, early ceramics suggest links to the east, to the vast area of Comb Ware distribution (Figure 11.1). The technology and shapes of Early Comb Ware in north Scandinavia are of the same origin as the eastern and southeastern technological traditions. Datings indicate early use of ceramics in this northernmost region of Europe long before the arrival of elements of agro-pastoral farming (Table 11.1). The earliest dated ceramics of northern Fennoscandia represent the earliest phase of the use of pottery in northern Europe. In this chapter, the focus lies principally on Early Comb Ware and to a certain extent on the later development named Typical Comb Ware (Figure 11.2). Comb Ware has an eastern distribution covering northern Norway and Sweden, Finland, and western parts of Russia during the Late Mesolithic and Neolithic (Figure 11.3).

All documentations of Comb Ware sherds from northern Scandinavia indicate a certain kind of homogeneity, regarding form as well as decoration

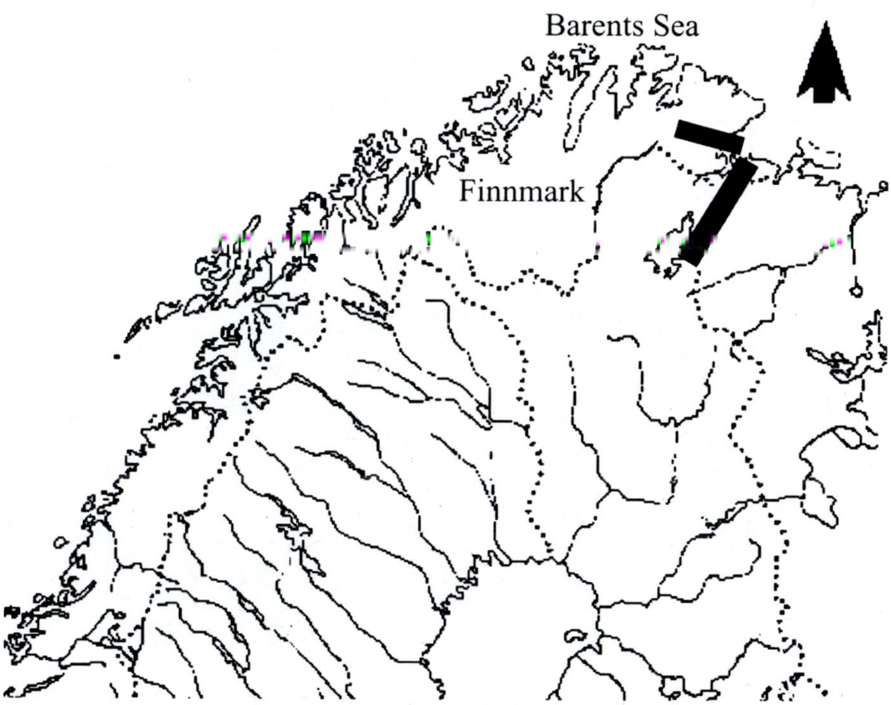

Figure 11.1 Map of Finnmark in north Norway indicating the sites with Early Comb Ware mentioned in the text (from Skandfer 2003a).

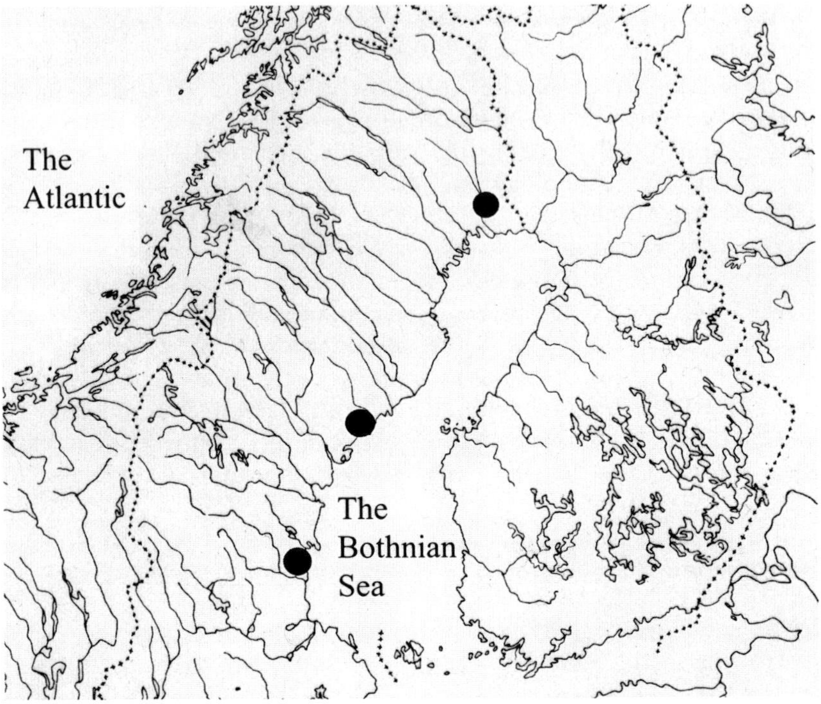

Figure 11.2 Map of northern Sweden and the Bothnian Sea indicating the area with Early and Typical Comb Ware mentioned in the text and the site of Hedningahällan.

Figure 11.3 Example of Early Comb Ware, called Säräisniemi, from Varanger in Finmark, Norway.

Table 11.2 Datings of organic material from Early Comb Ware and pottery at Hedningahällan mentioned in the text.

Lab no.	Site	Character of sample	14C BP	Reference
TUa-3028	Nordli, No.	Organic, so called crust	6,570±60	Skandfer 2003a: 440
Hela-148	Utajärvi, Fi.	Organic, so called crust	6,140±105	Torvinen 2000: 16
TUa-3022	Mennikka, No.	Organic, so called crust	5,795±55	Skandfer 2003a: 440
Hela-40	Turpeenniemi, Fi.	Organic, so called crust	5,520±185	Torvinen 2000: 16
Ua-13333	Hedningahällan, Sw.	Organic, so called crust	4,085±115	Holm 2003: 95
Ua-13331	Hedningahällan Sw.	Organic so called crust	4,075±80	Holm 2003: 95
Ua-13332	Hedningahällan, Sw.	Organic, so called crust	4,045±115	Holm 2003: 95
Ua-5024	Hedningahällan, Sw.	Organic, so called crust	4,020±100	Holm 2003: 95
Ua-25338	Mosjön, Sw.	Carbonised grain	3,790±55	Färjare & Olsson 2000: 34
Ua-5080	Hedningahällan, Sw.	Carbonised grain	3,780±110	Schierbeck 1994: 41

Table 11.3 Calibrated dates of 14C-datings mentioned in the text.

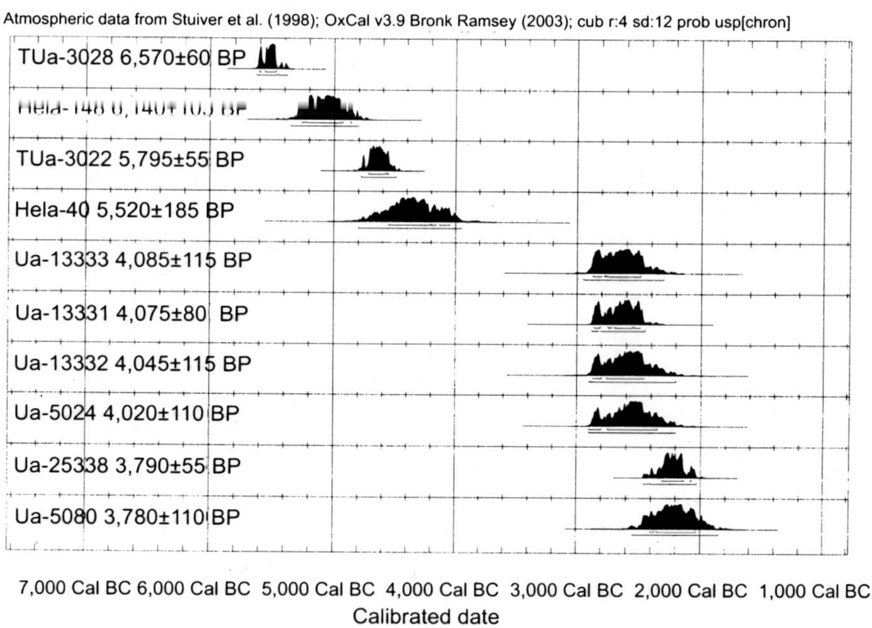

and technique (Figure 11.4). In general, the manufacturing tradition remains relatively unchanged during the long period of manufacture and use while decoration varies in terms of surface coverage and combinations of the basic imprints. Technological traits such as tempering and vessel sizes vary from site to site as well as within sites. Comb Ware is generally characterized by large vessels. The largest Typical Comb Ware vessels, with comparatively straight walls, usually with straight rims, ending in pointed bases, have a capacity of 35–45, even up to 70 litres. Decoration was dominated by various impressions, mostly made by denticulate tools and stamps leaving cord marks and lines; even pits on occasion. All the elements occur in various quantities, more or less covering both walls and rims.

Before proceeding to the Comb Ware sites of northern Sweden, a regional overview is presented. With reference to present datings, the earliest dated Comb Ware in Scandinavia is found in Finnmark in the northeastern part of Norway, in Varanger on the coast of Barents Sea and in the valley of Pasvik at the borders of Finland and Russia, at a small number of dwellings dated to the Late Mesolithic/Early Neolithic (Figure 11.1). On the basis of several radiocarbon datings of organic material from the inside of the vessels, the so-called crust, a more exact chronology of the use of Early Comb Ware can be established. These datings vary between 6,570 ±60 BP and 5,795±55 BP (Skandfer 2003a: 233). With all respect for source criticisms, these datings do indicate early use of pottery in northern Europe (Tables 11.2 and 11.3).

The dating of organic materials of Early Comb Ware in northern Finland vary between 6,140±105 BP and 5,520±185 BP (Skandfer 2003a: 241; Torvinen 2000: 19, 29). This indicates contemporaneousness with the regions of Varanger and Pasvik. Hitherto, the earliest Comb Ware found in northern Sweden is younger. Datings by other methods, principally based on estimations of ancient shorelines, indicate use from c. 4,600 BC to 3,600 BC (Skandfer 2003a: 252). In cases where exact dating methods cannot be used, the ceramics are dated as either Early or Typical Comb Ware based on the location above present-day sea level and comparative studies of decoration in Finnish materials (Halén 1994: 136; Skandfer 2003a: 97).

No ceramic analyses have been conducted on Early Comb Ware, yet the technology is described in various contexts, although not in detail. Early Comb Ware was tempered with crushed rock or sand, resulting in more or less coarse ware. Vessels were built of coils with straight walls and rims, smoothed with clay and water, and finally often painted with red ochre. Compared to Typical Comb Ware, Early Comb Ware vessels are 'small', with dimensions indicating volumes of eight to ten litres (Torvinen 2000: 6–7). In the Varanger–Pasvik region, rim diameters of up to 34 cm are frequent but they can be as large as 48 cm (Skandfer 2003a: 126). The decoration encompasses a variety of numerous stamp impressions and pits made with denticulate tools and cord impressions in manifold geometric, non-figurative formations and lines (Torvinen 2000: 7–8).

In Early Comb Ware contexts in Varanger and Pasvik in Finnmark, sherds are deposited in small amounts, mostly some hundred at each site and at a maximum no more than 3,800 sherds at 13 dwellings in Varanger and in Finnmark. At some dwellings, only a few sherds were recovered (Skandfer 2003a: 120).

With a contextual perspective and by using binary oppositions like interior–exterior, horizontality–verticality, bottom–rim, pit impressions–comb impressions and individuality–generality, Skandfer has analyzed northern Early Comb Ware (2003a: 105–106). With this approach of dichotomizing sherds, she found clear similarities between the sites, but also significant variations in decoration indicating individuality among the vessels in Varanger and Pasvik. The results suggest systematic variations between dwellings (Skandfer 2003a: 139). Furthermore, the analyses indicate that vessels were manufactured at various sites where different technological strategies were used.

In the coastal areas of northern Sweden, larger deposits—more than 8,500 sherds from three dwellings, one of which is excavated—indicate more intensive use. Pottery from the dwelling sites of Stor-Brändberget and Lillberget in Norrbotten (Figure 11.2) has not been analyzed, yet Halén states that the sherds contain quartz temper (1994: 78–79, 132–133). Also, datings are not based on valid methods and explicitly described. As mentioned, no radiocarbon samples have been dated. But by comparison with Finnish materials, the ceramics at Stor-Brändberget have been interpreted as Early Comb Ware (Skandfer 2003a: 97; Torvinen 2000: 20) and the finds at Lillberget as Typical Comb Ware (KaII:1) (Halén 1994: 136). The analyses indicate morphological and technological similarities with Eastern Comb Ware in Finland (Skandfer 2003a: 97). To infer routes of influence on the basis of these similarities is not relevant for the understanding of the roles of the earliest pottery in northern Sweden.

The total weight of ceramics found at Lillberget was approximately 16 kg. The sizes of the pots vary with rim diameters estimated to be at most 49 cm and a volume of up to 70–80 litres (Färjare 1996: 89; Halén 1994: 131). The decorated sherds indicate that the surface of the pots was more or less covered with comb imprints ordered in various geometrical patterns (Halén 1994: 135).

The discussion of whether the earliest dated Comb Ware in northern Fennoscandia should be typologized as Säräisniemi 1 (Sär 1) or Kamkeramik 1:1 (Ka 1:1) has brought up a series of datings, both from samples of organic materials on pottery and from charcoal samples from dwelling contexts (see, for example, Skandfer 2003a, 2003b; Torvinen 2000). This discussion is of methodological interest, of necessity focusing on cultural and social contexts in relation to typology. Anyhow, the earliest dated ceramics of northern Fennoscandia—in this paper designated Early Comb Ware until the subject of the methodological discussion progresses—represents the earliest phase of ceramics in the northernmost part of Europe.

To summarize, a clear line of very early datings of Early Comb Ware may be traced as far north as the Varanger Bay on the Arctic Sea coast. Early as well as Typical Comb Ware has even been documented in the northern parts of the Bothnian Sea. The pottery is of the same type as the ceramics encountered further east and towards the southeast in Finland and Russia. These similarities indicate a vast network of contacts in an east–west direction. It is important to note that the pottery was used by Stone Age hunter-gatherer societies at a time when farming was still an unknown way of life. Cultivation and animal farming were not introduced until several thousand years later in this area. The earliest pots were well made within an evolved craft tradition and often ornamented. They were of a range of different sizes but mostly fairly large and most likely used in connection with various household activities.

Early Ceramics along the Bothnian Coast

The history of early ceramics along the Bothnian coast is interesting due to the insights into cultural contacts and exchange that it provides. Over the years, several excavations within a large area along the Bothnian coast have resulted in comparatively large assemblages of Neolithic ceramics representing different craft traditions. Ceramics were introduced into hunter-gatherer societies with a subsistence economy based on marine fishing along with gathering and inland hunting. However, the idea as well as the technology of farming were known. Results from recent archaeological investigations indicate that barley, and even to some extent wheat, was used by the people living along the Bothnian Coast.

Carbonized grains of barley were found in larger quantities at Hedningahällan, and in one example, at Lill-Mosjön. The radiocarbon datings indicate contemporaneous use of barley—3,790±55 BP (Ua-25338) and 3,780±110 BP (Ua-5080)—at the two sites (Färjare and Olsson 2000: 34; Holm 2003: 95). The most recently excavated settlement at Bjästamon also produced finds of a few grains of barley (Holback et al. 2004).

On the present limited evidence, it is not possible to decide whether these finds of grain should be interpreted as the result of local cultivation or as something brought to the settlements from somewhere else. In any case, these novelties were clearly introduced via far reaching contacts towards the south and southeast. In the inland areas inhabited by hunter-gatherer groups, neither pottery nor other objects indicating contacts with farmers were present at this time. A relatively large number of excavations of their settlements have produced no such objects.

There is no uniform ceramic technology, decoration or chronological settings along this coastal zone that can be of typological guidance. Both Comb Ware, Pitted Ware, Kiukais Ware and Corded Ware as well as local traditions and combinations of manufacturing techniques have been described (Arbman 1945; Baudou 1995; Holm 2003, 2006).

At the Hedningahällan site (Heathen Cliff in English), more than 130 kg of pottery sherds have been recorded through the years (Figure 11.2). The pottery has been described as Pitted Ware, Corded Ware, Comb Ware and Kiukais Ware with chronological differences (Arbman 1945; Schierbeck 1994) (Figure 11.4). The site is unique because of its topographical location and the extraordinary context of finds with only a few dwelling features. During the Middle Neolithic/Late Neolithic, it was situated on a small prominent cliff in the outer archipelago, exposed to the sea. The amount of pottery sherds, the mixture of ceramic manufacture traditions, the presence

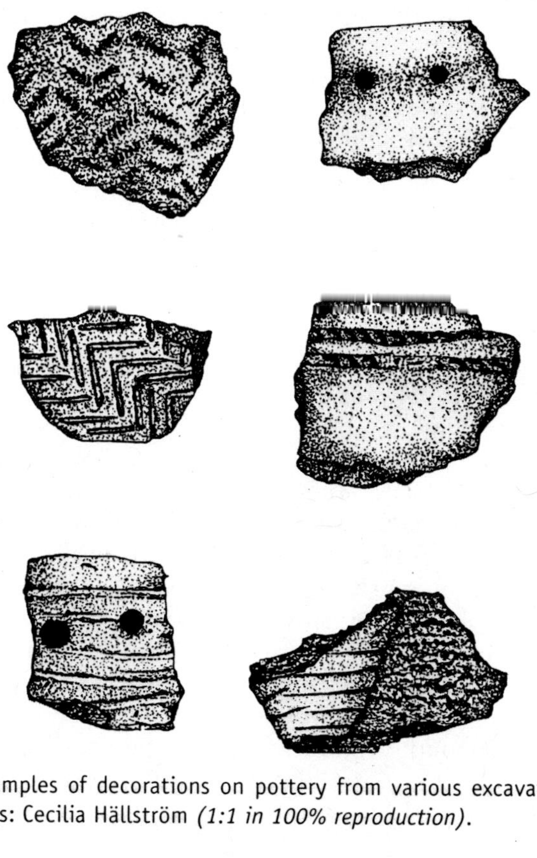

Figure 11.4 Examples of decorations on pottery from various excavations at Hedningahällan. Drawings: Cecilia Hällström *(1:1 in 100% reproduction)*.

of clay figurines and the large amount of seal bones together with grains of barley may imply a site with extraordinary activities. Sherds with cord impressions, pits, grooves and combinations of pits and cord impressions are probably contemporary. The datings of crusts range between 4,085±115 BP, 4,075±80 BP, 4,045±115 BP and 4,020±100 BP (Ua-13331–13333 and Ua-5024, Tables 11.2 and 11.3), which indicate contemporariness (Holm 2003: 95).

Different vessel shapes and decoration elements are visible at Hedningahällan. The rim forms indicate straight walled or slightly convex vessels with both rounded and flat bases. One base also indicates a vessel with a protruding foot. Vessels were coil built in the N-technique. Probably, vessels modelled from a lump are also present in the material (Brorsson 2000: 2). Of the analyzed wares, 50% were made of fine non-calciferous clay tempered with crushed granite. Temper qualities were rather varied with comparatively small maximum sizes, between 2.5 and 4 mm, and temper volumes between 16 and 23%. Some organic material and grog was also observed in the fine clay. A few sherds with fine clay were tempered with burnt bone, in one case in combination with crushed granite. Half of the surfaces were later smoothed (Brorsson 2000: 3–5).

The results indicate a fair degree of homogeneity concerning the choice and use of temper and clay, while the choice of decorative elements varies. However, one of the sherds of the same date as the others differs with a higher share of crushed granite in a mica-poor fine clay (Brorsson 2000: 6) This sherd is of Kiukais-type. Kiukais pottery is recorded mostly in southwestern Finland in Late Neolithic contexts. The ceramic situation at the Hedningahällan is interpreted as a confirmation of the idea of far-reaching communication with other societies (Holm 2000).

At Hedningahällan, a variety of decoration types indicates a complex ceramic situation. The Neolithic people decorated their wares in manifold ways based on various cultural traditions, using all known types of decoration. On the other hand, technological strategies were fairly homogeneous irrespective of decoration tradition. One of the carbonized grains of barley dates to 3,780±110 BP (Ua-5080) (Schierbeck 1994: 41). This dating is obviously younger. With the present datings in mind, the pottery tradition at Hedningahällan is interpreted as older than the arrival of agriculture techniques and ideas. It is interesting that the site was used over several centuries and that new technologies and knowledge were brought there.

The interpretation of this remarkable site is not an easy task. The archaeological descriptions and interpretations over the years have been based on the ceramic variation of the site, the exposed position towards the sea and the large amount of grains found at the site. The available space on the rock surface does not allow the erection of any type of housing. The large amount of seal bones indicate large-scale butchering, while the significant variation in ceramic traditions shows us that different groups of people may have been here at the same time. Later, large amounts of grains were transported to Hedningahällan. So far, archaeologists working with the site have interpreted it as a cult site where people from the surrounding region and beyond met for religious and social purposes. This is supported by other finds from the site consisting of, among other things, clay and bone figurines and smashed battle axes (Holm 2006). It should be added that these people knew and practised both potting and later on farming.

On the basis of the knowledge we have today, it is not possible to decide whether pottery and grain were produced regionally or had been transported here through interregional contacts. At Hedningahällan, people met carrying with them fundamental knowledge and practical experience of pottery making from the south and across the sea from the east. They used their different ceramic traditions incorporating different shapes and ornamentations as a means of communication in order to identify themselves during encounters with others. Even so, pottery as a cultural legacy expressed a mutual way of thinking shared by people in the region, even if the idea itself came from far away. The contacts and the exchange were probably much more complex than what is revealed by the archaeological material (Holm 2006).

Another ceramic tradition is found further north along the Bothnian Coast, about 250 km north of Hedningahällan. Recently, at the sites of Lill-Mosjön and Bjästamon, comparatively small amounts of sherds were recorded (Holback et al. 2004; Hulthén 2000; Figure 11.2). At Lill-Mosjön, only 700 gm of sherds were documented. The vessels were made from a fine clay built of coils in the N-technique with a smoothed surface. The temper consisted of crushed rock of 2 to 4 mm in size. Also, asbestos-tempered ceramics were found in small quantities. The decoration consisted of various impressions, like fish-tail impressions and cord impressions together with lines in angles and pits. Recorded sherds of bases indicate flat bases (Hulthén 2000: 90–91). The ceramic analysis indicates a homogeneous and skilful craft tradition. The vessels are probably products of local manufacture by people familiar with local ideas and preferences, for example, including use of a local rock (Hulthén 2000: 91–92).

Compared with other ceramic traditions, no evident parallels are discernible with other northern sites, like for example Hedningahällan. One detail worth mentioning is that protruding feet are found at both sites (Hulthén 2000: 92). In addition, compared with the ceramics at the nearby site of Mjäla, except for cord impressions, no parallels are found either in vessel shape or in the use of temper, clay and decoration elements (Hulthén 1991: 10–13, 2000: 92).

Far reaching contacts within regions as well as between regions are discernable in the archaeological materials. Communications with others result in a wider knowledge of new technologies such as agriculture. Indications of occasional agricultural activities among Middle Neolithic groups along the Bothnian coast imply the arrival of new technologies and traditions. Nevertheless, the traditional subsistence of hunting, gathering and fishing still dominated.

Three different relations between coastal economy and ceramics may be discerned:

(a) Northernmost Sweden and Norway, 7,000 BP–4,800 BP. Sites with a hunter-gatherer economy, which produced and used Early Comb Ware vessels and later on Typical Comb Ware vessels in their household activities, but did not adopt any aspect of farming;

(b) Bothnian Coast including the site Hedningahällan, 4,100 BP–3,800 BP. Sites with hunter-gatherer economy, but including limited quantities of cultivated (imported?) grains. The people used pots representing a wide variation of different pottery styles like Pitted Ware, Comb Ware, Kiukais Ware and Corded Ware;

(c) Bothnian Coast including the site Lillmosjön, 4,300 BP–4,000 BP. Sites inhabited by hunter-gatherers exploiting marine resources and with a few finds of carbonized grains reported. The people produced and used well developed ceramics of local character.

Case Study II : Ertebølle Pottery—The First Ceramics in Southern Sweden

The story of the introduction of pottery in Scania, the southernmost province of Sweden, is in many ways different from what happened in northern Sweden as described above—except for the appearance of the first pots in hunter-gatherer communities in both areas.

The province of Scania largely consists of a plain in the southwest and a more hilly landscape in the northeast gradually rising towards the highland areas of Småland. This landscape is broken off in the northeast corner by a smaller plain around the town of Kristianstad. Large parts of Scania belong to the most fertile agricultural areas in southern Scandinavia (Figure 11.5).

Ceramics were first introduced to southern Scandinavia between 5,000 BC and 4,500 BC appearing within the Late Mesolithic Ertebølle culture which had a hunter-gatherer economy. The Late Mesolitihic settlement of Scania was dominated by large-scale sites occupied all year round on beach ridges close to the sea (Karsten 2004; Figure 11.8). It is possible that inland sites existed here as well (Andersen 1975)—for example by the lakes in

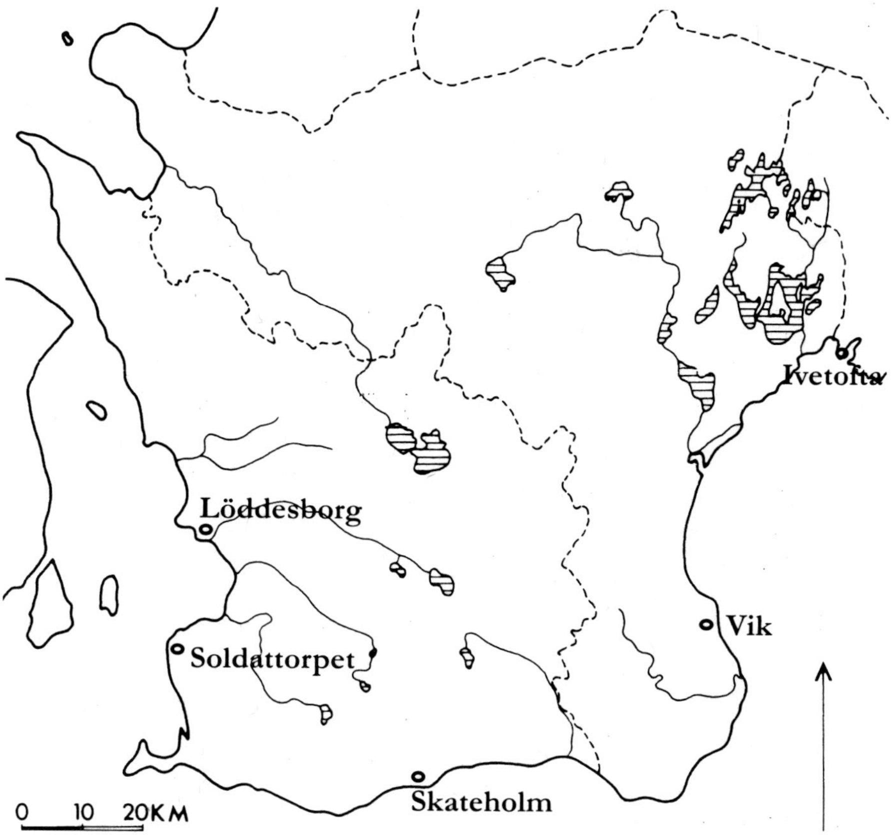

Figure 11.5 Map of Scania indicating the two sites mentioned in the text and including the three other coastal sites with Ertebølle pottery.

central Scania—which were exploited in the Early Mesolithic, but no sites have been positively identified yet. The inland was heavily forested and the easiest communication between settlements was in dugouts paddled along the coast. The economy was to a large extent based on the fishing, hunting and collecting of marine resources, but even the hunt for game on land, and in particular, the collecting of hazelnuts was of major importance.

The nearest agricultural societies to the south at this time were the Linear Band Ceramic (LBK) groups on the north European loess plains. The common opinion regarding the neolithization of southern Scandinavia is that agricultural products, and subsequently, the knowledge of farming gradually spread northwards from the LBK culture. A few finds in Scania indicate that cultivated wheat (Jennbert 1984: 93) and possibly even meat from livestock (Andersson 2004: 144) were imported from areas with an agrarian economy. Traceable objects like the Danubian shaft-hole axes

(*Schuhleistenkeile*) indicate that the other imports most likely have come from the south as well (Andersson 2004: 146; Jennbert 1984: 145–146). The traditional social explanation for these imports is the existence of gift-relations—direct or 'down-the-line'—between the hunter-gatherer groups and the LBK groups. On the basis of the common finds of parts of harpoons for catching eels, Hulthén has suggested that the Ertebølle people were supplying eels as their contribution to the exchange relationship with inland dwellers to the south (Hulthén 1980). The overall picture of the economic development of a subsistence economy in southern Sweden during the Late Mesolithic and the Early Neolithic is one of a long gradual transition (Andersson 2004: 144). During the whole Early Neolithic period, hunting and gathering continued to play a major role for the people of the Funnelbeaker culture living in Scania. Pottery was introduced in the beginning of this long transition.

In this region, in contrast to the eastern influences in the north, the idea of making pottery and the knowledge of basic raw materials most likely reached Scandinavia from the south (Hulthén 1977: 49–50). However, the technology of building the pots, and indeed the skills to do it, were developed by the Ertebølle hunter-gatherer people themselves. This is most evident in the 'Ertebølle ERT' pottery from the eastern part of southern Scandinavia. The simple coiling technique used in the Ertebølle pots found in Scania (and on Sealand) is more primitive than the building techniques evident in the ERT pots from sites on Jutland (Andersen 1975) and clearly inferior to the technique of Swifterband pottery in the Netherlands (Raemaekers 1999: 33; Stilborg and Bergenstråhle 2001: figure 5). Since there were no craft traditions to rely on, the ways pots were made probably reflect not only skill, but also the intended functionality and role of ceramics to a larger extent than in subsequent periods with well-founded ceramic craft traditions. The lack of traditional rules and guidelines for the craft should mean that the choices of the potter in the production of a pot were more dependent on the immediate objectives for this pot. Even so, a consensus on the basic design of the pots seems to have been formed very early on. A possible explanation for this is discussed further on in the chapter.

The find materials of ERT pottery at Scanian sites are both small and very fragmented. The amount varies between two and five kg per site which should be seen in relation to an estimated weight of four to five kg for one whole vessel of the larger beaker-type (Figure 11.6). On the sites of Skateholm I and III on the south coast of Scania (Figure 11.5), around two kilos of pottery was found. Excavations at the turn of the century at another coastal site named Soldattorpet yielded almost five kilos of sherds (Bergenstråhle 1995; Stilborg and Bergenstråhle 2001: 31). The high degree of fragmentation in the Scanian find material severely limits our knowledge about the types and sizes of vessels. However, the pointed bases together

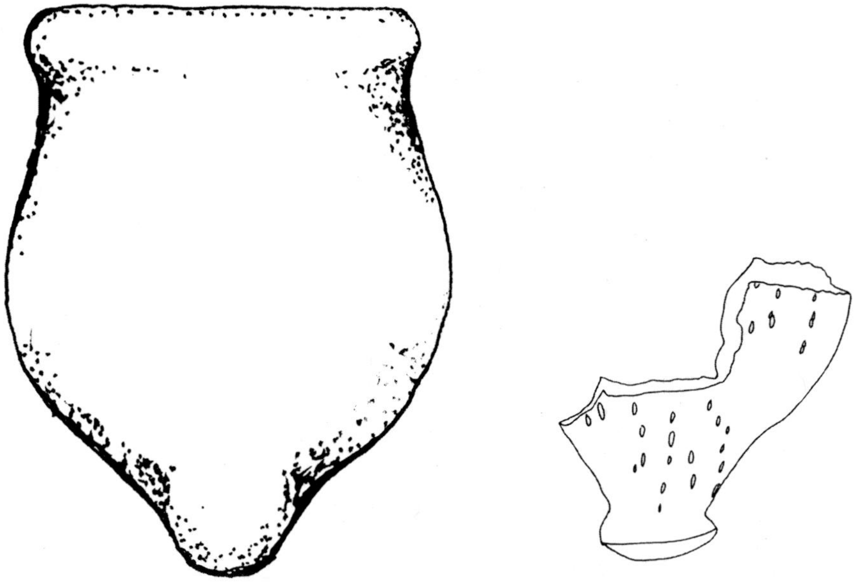

Figure 11.6 The largest preserved piece of a pointed-based Ertebølle vessel from Soldattorpet, Scania (to the right) and an illustration of the general shape of a medium sized pointed-based vessel (to the left, E. Grönberg del.) *(1:2 in 100% reproduction).*

with the outward turned rims are good indications that a number of Scanian vessels had shapes very similar to better preserved Danish finds. The S-shaped profile ends in a pointed base or even a solid tap. The latter is a special east Scandinavian development as Birgitta Hulthén has pointed out (1977: 42, figure 19). The more well preserved material from Jutland indicates three size-groups for the beakers—small (height 8–10 cm, rim diameter 5–6 cm), medium (height 20–30 cm, rim diameter 10–15 cm) and large (height 40–50 cm, rim diameter 15–20 cm) (Andersen 1975: 59, 1998: 39)—of which at least the latter is represented at Skateholm. The medium and the large group with capacities between two and ten litres are represented at Soldattorpet. The other common shape is a small oval 'boat-shaped' bowl (Figure 11.7). This has been interpreted to have been used as a blubber lamp, possibly connected to the catch of eels, attested by the finds of eel spears on coastal sites in southern Scandinavia (Hulthén 1980). From Ivetofta a necked, globular vessel is known (Hulthén 1977: 38). One sherd from Soldattorpet (Layer e) and one sherd from Mölleholmen in central southern Scania (Kelm 1994: 24) may be other examples of globular vessels judged primarily on their similar decoration with rows of finger tip imprints. This type of ornamentation was not found on any sherd clearly deriving from a pointed based beaker.

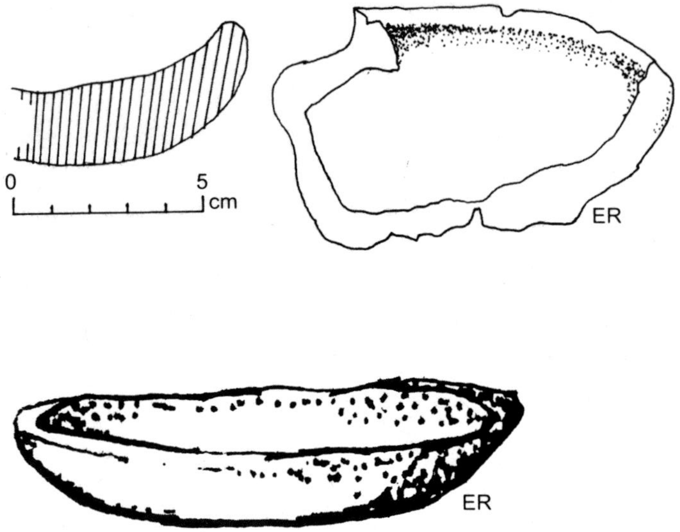

Figure 11.7 Sherd of a boat-shaped Ertebølle lamp from Soldattorpet (upper) and an illustration of the general shape of the lamps (E. Grönberg del.).

Most of the beakers were built of coils in the H-technique pressing a new coil vertically down onto the preceding coil with the fingertips. Some pots were coilbuilt using the U-technique smoothing down the sides of the coil onto the preceding one. Both methods, and especially the H-technique, tend to produce very thick vessel walls (Andersen 1975: 56–57; Hulthén 1977: 26–35; Stilborg and Bergenstråhle 2001: 31). The lamps were hand modelled from a single lump of clay (Hulthén 1977: 27).

The wares of the vessels analyzed from southern Sweden were made from a variety of fine to very coarse non-calciferous clays tempered with crushed rock or sand in very varying qualities. As a contrast, several of the lamps were made from fine calciferous clays tempered with organic material in addition to either crushed stone or grog (Hulthén 1977: 26). Other lamps were made from very coarse, non-calciferous clay without any added temper. As a whole, the ware composition and quality is highly variable in contrast to the limited shape variation.

In southwestern Scania, a large portion of the beaker-shaped pots are characterized by a remarkable, almost stereotypic, decoration. With the exception of the occasional nail and/or finger imprints on the rim, the beakers at Skateholm and Soldattorpet are ornamented with rows of shallow, elliptical to more rounded, oval impressions (Figure 11.8 a-d). According to Kjellmarks' meticulous study of the ornaments at Soldattorpet, the impressions vary in size from 3 × 2 mm to 7.5 × 4.5 mm. They are often aligned

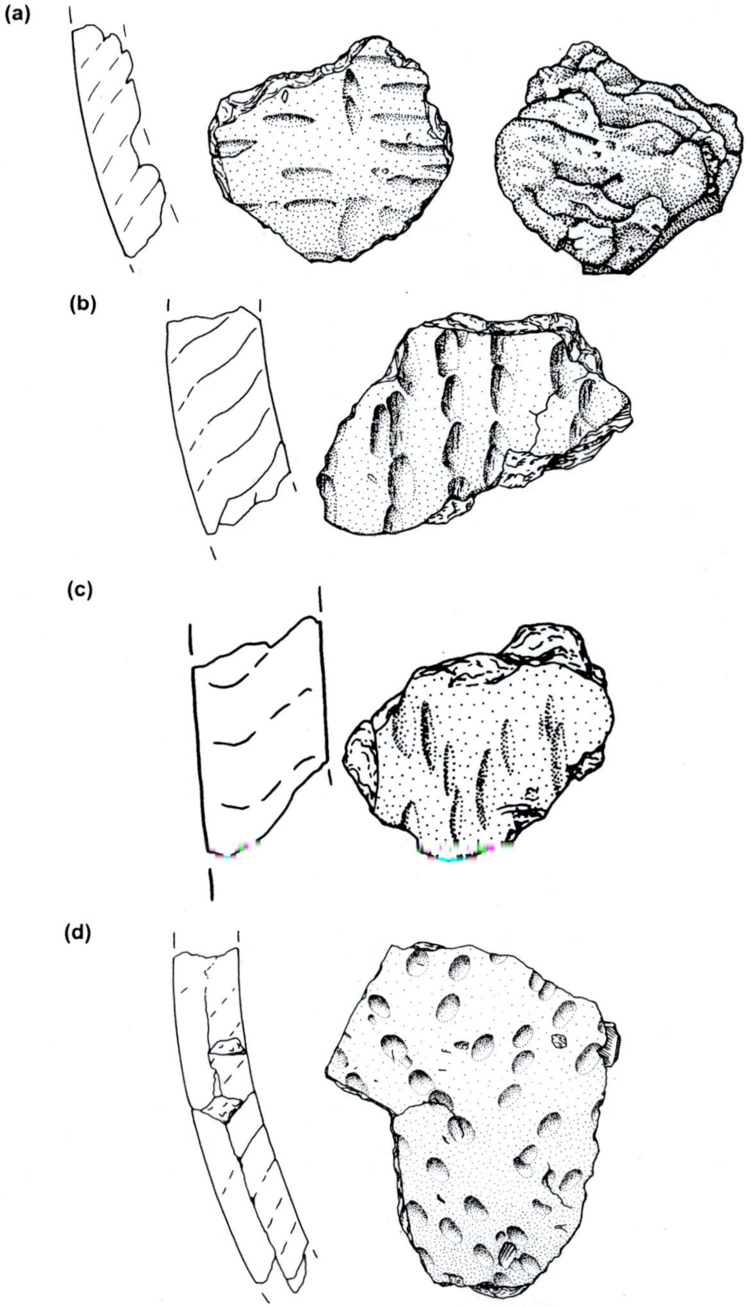

Figure 11.8 a-d Examples of the small, shallow pits ornamenting a large part of the Ertebølle pots on sites in western Scania *(1:1 in 100% reproduction)*.

in variable rows—often vertical—with the narrow ends towards each other and an average distance of four to six mm. In some cases, however, the impressions overlap (Kjellmark 1903: 85; unpublished notes). Judging from larger ornamented fragments, the rows of impressions cover the whole body of the beaker. On Soldattorpet, some sherds have small, almost rectangular impressions. This is most likely due to distortion of originally elliptical impressions by an ensuing smoothing of the outer surface. At Skateholm, exactly the same type—from eliptical to oval—and the same sizes of impressions appear. As in the Soldattorpet material, they are often arranged in rows over the whole body of the vessel, but the rows may be slanting and even chevron-like as well as vertical. Contrary to Soldattorpet, the impressions might also be placed with the broad sides to each other. It is possible that the consistency of shapes and dimensions of these impressions is due to the use of the same tools, which could be different kinds of snail shells, although no suitable objects have yet been found.

At the site of Löddesborg (Jennbert 1984: 58), the same types of ornaments dominate while another smaller group of sherds are ornamented with lines of deep round impressions made with a thin stick (diameter 1.5–3 mm). There are even examples of sherds where the shallow oval/eliptical impressions are overlain by lines of these impressions. Thus, the same people most likely made both types of pottery. One sherd, ornamented with deep round impressions, was found at Skateholm I. This type of ornamentation has its closest parallels in Ertebølle pottery from Jutland (Jennbert 1984: 58).

While it is never possible to reach any definite conclusions about the origin of and reasons behind prehistoric ornaments, it is interesting to speculate. Kjellmark, in a footnote written in 1903, speculated on whether the rows of shallow oval impressions might symbolize or be inspired by footprints in the sand of the raised beach. Even if this may only be a poetic intuition, it signals a belief in the connection between the vessel design and the local environment of the sites where they have been found. This connection may also be evident in the pointed bases, which makes it easier to stand the vessels upright in the sand.

An equally plausible explanation could be that they relate to the shape of the wheat and barley grains being introduced to southern Scandinavia from the LBK groups at around the same time. No imprints of grains were found on ERT pottery either at Soldattorpet (Kjellmark 1903: 100–102) or at Skateholm I, but a few imprints of barley and wheat were identified on Ertebølle sherds from Löddesborg (Jennbert 1984: 93). Ethnographic records tell us that ornaments are often thought to provide protection against impurity (e.g., in the Swahili culture, Donley-Reid 1990: 56) and what would be more important to prehistoric people, than to protect food or indeed a brew made of the precious, imported grains?

Comparable ornamentation consisting of lines of impressions, but made with the fingertips and/or nails or spatula are common on ERT sites in northern Germany (Schwabedissen 1972: 6) and in the Swifterband culture in Holland (De Roever 1979: 16–17; Raemaekers 1999: 53). It is quite possible that the ornaments of the southwest Scanian ERT were just an emulation of this ornamentation, for which the original inspiration may have come from the LBK pottery as did the very idea of making and using ceramic objects.

There are examples of dot-line ornaments on LBK pots for example from Eastern Germany (Behrens 1973: 28, 245). An argument against a direct influence on Scanian ERT is that no attempts, with the exception of the necked globular vessel from Ivetofta (Hulthén 1977: 38–39) and possibly one other similar vessel from Soldattorpet, seem to have been made by ERT people towards copying the shapes of the LBK pottery.

At Soldattorpet, as at most of the sites in southern Scandinavia, Ertebølle pottery was found in the culture layers of a coastal settlement site together with a normal household refuse of flint debris, discarded tools, bones and shell. At Skateholm, some of the pottery was found in similar contexts while the main part of the material occurred in pits of different sizes and within the confines of a small building. The results of the analyses of this material are currently undergoing publication, but it may be mentioned here that there are some differences in the composition of the ceramic finds in three different types of contexts. The differences, which are difficult to interpret, may mean that some of the pottery belongs to a ritual rather than a settlement context. No Ertebølle pottery has been found in any of the contemporary graves excavated in southern Scandinavia.

Thus most contexts place the use of the pots within the domestic sphere which is corroborated by some use-ware traces. The occasional sooting of some sherds and the higher/longer firing of some pointed bases (Hulthén 1977: 42) indicate a use as some kind of cooking or at least food-processing vessel.

Macrofossil and chemical analyses (Arrhenius 1984; Arrhenius and Lidén 1989) have been performed on charred remains from the inside of one of the larger pots from Denmark. The results indicate fermenting as the most probable use for this vessel. Fermenting is a very practical way of preserving many types of organic foodstuffs and pottery vessels are very suitable containers for this process. The method, however, has appeared very seldom in the archaeological discussions concerning vessel function. The analyses of protein, amino acids and trace elements suggested that blood and nuts had been essential components in this particular dish. One important product which comes to mind when talking about fermentation is of course beer. Several scholars (e.g., Jennbert 1984: 147; Katz and Maytag 1991) have pointed to the fact that the most logical use of the small amounts of seed acquired from farming communities would have been for making beer. In fact, this introduction of a fermented drink as a high-status phenomenon at important social gatherings might even be the very reason behind the introduction of farming in northern Europe.

The boat-shaped blubber-lamps, which are found in the same settlement contexts as the pots, could very well have been used as a special tool in connection with fishing for eels (Hulthén 1980). However difficult it is to substantiate this theory to any large degree, it is very convincing. Accepting this functional interpretation reflects of course on the perception of the beakers. If they were introduced as just general ceramic containers, it is remarkable that a special tool such as this lamp was invented within the Mesolithic societies keenly utilizing the specific qualities of ceramic ware and in some cases using an advanced ware composition. Perhaps, we may create a more logical scenario if we hypothesize that both the lamp and the beaker were adopted as special tools for specific purposes: the catching of eels and the brewing of beer.

Conclusions

The complexity and ambiguity of function, ideas of manufacture and meaning in symbolism in early ceramics in prehistoric societies in Scandinavia is well-documented (see, for example, Holm 2006; Hulthén 1977, 1991; Lindahl 2000; Skandfer 2003b). The early ceramics in northern Sweden and Scandinavia cover a long period of time from the earliest indications of Early Comb Ware in the far north in Norway and Sweden during the Late Mesolithic/Early Neolithic to the Middle Neolithic traditions in coastal northern Sweden.

The archaeological contexts of Early Comb Ware obviously indicate hunter-gatherer societies with strong links to the vast area of Comb Ware traditions to the south and southeast and in all probability, as the archaeological context indicates, with less contact with the early agricultural areas of southwestern Scandinavia. The earliest Comb Ware contexts in Scandinavia are coastal and indicate both small amounts of pottery of various types with individual decorations of separate manufacture strategies, and larger amounts which are more formalized. A certain amount of conservatism—with a general idea of consensus about form and decoration techniques like in the vast area of the Comb Ware tradition, and the long-time perspective of the technique—is obvious.

Recent excavations extend the northern limits of the Funnelbeaker tradition to central Scandinavia (Hallgren 1996). There are good reasons to believe that this border of neolithization ideas is extended further north to the Bothnian Coast. Although no Funnelbeaker is recorded here, the archaeological context, including among other things different ceramic traditions of technology and decoration and carbonized grains, indicate developed communications in social networks. These wide networks and communications within and between societies, towards the southwest and south, but also to the east, made this ideological change possible. But even at this time, the economy basically relied on marine resources and hunting-gathering.

The new technology in northern coastal Scandinavia was introduced into hunter-gatherer societies which were based both on terrestrial and marine resources, and continued to be so, often for a long time, after the adoption of ceramics. The northern topographical and ecological conditions were restricting factors for agricultural activity. People involved in the new technologies and ideas had far-reaching contacts and communications responding to, among other things, their wish and need for novelties such as pots. In coastal northern areas, hunter-gatherer societies chose different techniques and distinctive forms of pots in accordance with overall existing and contemporary common concepts and values brought from outside. The introduction of ceramic technology and the underlying ideas, together with social communications that brought alluring objects, including grains, were incorporated into existing hunter-gatherer social and cultural contexts.

Even in the southern part of Sweden, pottery was introduced in an advanced, economically diversified, hunter-gatherer society. Agriculture was not to appear for a long time although smaller amounts of presumably imported grain seem to have been consumed, perhaps as beer. The idea of pottery making was most likely picked up from societies to the southwest and south of the Ertebølle culture area. The general impression presented by the products of the early Ertebølle ceramics craft is one of a clearly defined and delineated entity. As a novelty the pot as a concept was inextricably linked to its shape and ornamentation. It was accepted and used within society, but was not yet a social object, which could be manipulated and given local or individual shapes and ornaments (Magnusson-Staaf 2000). This seems to be in contrast to the individuality of the Early Comb Ware in coastal northern Sweden. The pots from several sites in southwest Scania had a specific ornamentation which differed from the pottery designs both to the east and west, but the small shallow—almost invisible—impressions can only be seen at close range and can hardly be perceived as a regional emblem.

More likely they represent specific regional traits in relation to and in the perception of the pottery, even if this is not apparent in other ways; or it reflects general attitudes in society—for example, a tradition of using ornaments as protection.

The observed changes in the Funnelbeaker pottery of the ensuing Early Neolithic meant that the craft got more standardized while shape and ornaments became more variable and thereby open to social manipulations. The pottery but not necessarily potting was now embedded in the social structure (Stilborg and Bergenstråhle 2001: 38).

In these overviews of the earliest ceramics introduced into hunter-gatherer societies in the northern and southern parts of Sweden, we see very different traditions of shapes, sizes and decorative elements—although the idea of all-over decoration is found in both regions. As a whole, the outlines of different roles for the ceramics appear. In the far north, the Early Comb Ware represents both individualistic schemes of decoration and a general consensus on shapes and decorative techniques over vast areas. The ceramics obviously carried an important symbolic meaning *vis-à-vis* the agricultural areas further to the south. Even the earliest ceramics of the Bothnian Coast carried important social connotations as it symbolized the contacts with the Pitted Ware, Corded Ware and Funnelbeaker groups in middle Sweden. In contrast, the Ertebølle ceramics of southern Sweden, while inspired by ideas of the craft from the south and perhaps even by the idea of fermenting alcoholic beverages, seem to have been rather exclusive objects on the outskirts of society. They were exclusive in the sense that they were used for special purposes and constituted a narrowly defined group in terms of shapes and decoration. These differences in the social roles of the earliest ceramics in different parts of Sweden should be a central topic in the continued research on the introduction of ceramics into Scandinavia.

References

Andersen, S H (1975) 'Ringkloster, en jysk indlandsboplads med Ertebøllekultur', *1973–74 Kuml,* 11–94

——— (1998) 'Ringkloster. Ertebølle trappers and wild boar hunters in eastern Jutland, a survey', 12 (1994–95) *Journal of Danish Archaeology,* 13–59

Andersson, M (2004) 'Domestication and the first Neolithic concept 4,800–3,000 BC', in Andersson, M, Karsten, P, Knarrström, B and Svensson, M, *Stone Age Scania*, Skrifter no. 52, Lund: Riksantikvarieämbetets förlag

Arbman, H (1945) 'Hedningahällan', *Strena Archaeologica Professori A M Tallgren*, XLV *Finska Fornminnesföreningens Tidskrift*, 45–56

Arrhenius, B (1984) 'Analysrapport gällande organiskt material på keramikskärvor funna i Löddesborg', Appendix I, in Jennbert, K, *Den Produktiva Gåvan. Tradition och Innovation i Sydskandinavien for Omkring 5300 år Sedan.* Acta Archaeologica Lundensia Ser. in 4°, No. 16, Stockholm: Almqvist & Wiksell International

Arrhenius, B and Lidén, K (1989) 'Fisksoppa eller vegetabilisk gröt? Diskussion kring matresterna från Tybrind Vig', 3 *Laborativ Arkeologi*, 5–10

Bargatzky, T (1986) 'Innovations and the integration of sociocultural systems', in van der Leeuw, S and Torrence, R (eds), *The Social and Economic Contexts of Technological Change*, The World Archaeological Congress, September 1986, London: Allen & Unwin

Baudou, E (1995) *Norrlands Forntid—Ett Historiskt Perspektiv.* Örnsköldsvik: CEWE

Behrens, H (1973) *Die Jungsteinzeit im Mittelelbe-Saale-Gebiet.* Veröff.des Landesmuseum für Vorgeschichte in Halle Band 27, 1973, Halle: Landesamt für Archäologie Sachsen-Anhalt

Bergenstråhle, I (1995) 'Skateholm, a Late Mesolithic settlement in southern Scania, in a regional perspective', in Bintz, P (ed), *L'Europe des Derniere Chausseurs, 5° Colloque Internationale UISPP Grenoble*, Paris

Bronk Ramsey, C (2003) *OxCal Program v3.9*, Radiocarbon Accelerator Unit, University of Oxford

Brorsson, T (2000) 'Gropkeramiken från Hedningahällan—en glimt av ett hantverk', KFL rapport 00/214, unpublished report

Brown, J A (1986) 'The beginnings of pottery as an economic process', in van der Leeuw, S and Torrence, R (eds), *The Social and Economic Contexts of Technological Change*, The World Archaeological Congress, September, London: Allen & Unwin

De Roever, J P (1979) 'The pottery from Swifterband—Dutch Ertebølle, Swifterband contribution 11', XIX *Helinium*, 13–36

Donley-Reid, L W (1990) 'The power of Swahili porcelain, beads and pottery', in Nelson, S M and Kelhoe, A B (eds), *Powers of Observation: Alternative Views in Archaeology*, Archaeological Papers of the American Anthropological Association no. 2, Arlington: American Anthropological Association

Färjare, A (1996) 'Lillberget–en 6 000 årig boplats väcks till liv', 1995 *Norrbotten*, 84–99

Färjare, Å and Olsson, E (eds) (2000) *Lill-Mosjön—Boplatslämningar och Fångstgropar Från Neolitikum, Äldre Järnålder och Historisk Tid*, Riksantikvarieämbetet, avdelningen för arkeologiska undersökningar, UV Mitt. Dokumentation av fältarbetsfasen 2000: 5, Stockholm: Riksantikvarieämbetets Förlag

Halén, O (1994) *Sedentariness during the Stone Age of Northern Sweden—In the Light of the Alträsket Site, c. 5,000 BC and the Comb Ware Site Lillberget, c. 3,900 BC*, Acta Archaeologica Lundensia, series in 4°, No. 20, Stockholm: Almqvist & Wiksell International

Hallgren, F (1996) 'Sociala territorier och exogamirelationer i senmesolitisk tid. En diskussion utifrån Pärlängsberget, Södermanland', 28 *Tor*, 5–28

——— (2000) 'Lineage identity and pottery design', in Olausson, D and Vandkilde, H (eds), *Form Function and Context: Material Culture Studies in Scandinavian Archaeology*, Acta Archaeologica Lundensia, Series in 8°, Stockholm: Almqvist & Wiksell International

Holback, T, Lindholm, P and Runesson, H (2004) *Bjästamon. Ett Kustbundet Bostadskomplex från Slutet av Neolitikum, Botniabanan, Västernorrland*, Riksantikvarieämbetet, avdelningen för arkeologiska undersökningar, UV Mitt. Dokumentation av fältarbetsfasen 2004: 1, Stockholm: Riksantikvarieämbetets Förlag

Holm, L (2000) 'Pottery and meaning: a communicative aspect', in Olausson, D and Vandkilde, H (eds), *Form Function and Context: Material Culture Studies in Scandinavian Archaeology*, Acta Archaeologica Lundensia, Series in 8°, No. 31, Stockholm: Almqvist & Wiksell International

——— (2003) 'Contacts in the Bothnian region: the Neolithic site Hedningahällan in regional and interregional contexts', in Samuelsson, C and Ytterberg, N (eds), *Uniting Sea. Stone Age Societies in the Baltic Region*, OPIA 33, Uppsala: Department of Archaeology and Ancient History, Uppsala University

Holm, L (2006) *Stenålderskust i Norr. Bosättning, Försörjning och Kontakter i Södra Norrland*, Studia Archaeologica Universitatis Umensis 19, Umeå: University of Umeå

Hulthén, B (1977) *On Ceramic Technology during the Scandinavian Neolithic and Bronze Age*, Theses and Papers in North-European Archaeology 6, Stockholm: Akademilitteratur

——— (1980) 'Ertebøllekulturens lampor', 4 *Ale*, 5–7

——— (1991) *On Ceramic Ware in Northern Scandinavia during the Neolithic, Bronze Age and Early Iron Age*, Archaeology and Environment 8, Umeå: University of Umeå

——— (2000) 'Den neolitiska keramiken från Lill-Mosjön—en ceramologisk analys', in Färjare, A and Olsson, E (eds), *Lill-Mosjön—Boplatslämningar och Fångstgropar Från Neolitikum, Äldre Järnålder och Historisk Tid,* Riksantikvarieämbetet, avdelningen för arkeologiska undersökningar, UV Mitt. Dokumentation av fältarbetsfasen 2000: 5, Stockholm: Riksantikvarieämbetets Förlag

Jennbert, K (1984) *Den Produktiva Gåvan. Tradition och Innovation i Sydskandinavien for Omkring 5300 år Sedan*, Acta Archaeologica Lundensia Ser. In 4°, No. 16, Stockholm: Almqvist & Wiksell International

Karsten, P (2004) 'Peak and transformation of a Mesolithic society', in Andersson, M, Karsten, P, Knarrström, B and Svensson, M (eds), *Stone Age Scania*, Skrifter No. 52, Lund: Riksantikvarieämbetets Förlag

Karsten, P and Svensson, M (2004) 'Introduction', in Andersson, M, Karsten, P, Knarrström, B and Svensson, M (eds), *Stone Age Scania*, Skrifter No. 52, Lund: Riksantikvarieämbetets Förlag

Katz, S and Maytag, F (1991) 'Brewing an ancient beer', July–August 22 (V 44 # 4) *Archaeology*, 26

Kelm, R (1994) 'Mölleholmen. Eine Inselsiedlung des 11. Jahrhunderts im südlichen Schonen, Schweden. 2. Katalog und Abbildungen', Schriftliche Hausarbeit (Dipl.-Prähist), Kiel, unpublished thesis

Kjellmark, K (1903) *En stenåldersboplats i Järavallen vid Limhamn*, Antikvarisk tidsskrift för Sverige del 17 nr. 3, Stockholm: Wahlström & Widstrand

Knutsson, K, Falkenström, P and Lindberg, K-F (2000) 'Appropriation of the past. Neolithisation in the northern Scandinavian perspective', in Larsson, L, Kindmark, H, Knutsson, K, Loeffler, D and Åkerlund, A (eds), *Mesolithic on the Move. Papers Presented at the Sixth International Conference on the Mesolithic in Europe*, Oxford: Oxbow Books

Lindahl, A (2000) 'The idea of a pot ... perception of pottery', in Olausson, D and Vandkilde, H (eds), *Form, Function and Context. Material Culture Studies in Scandinavian Archaeology*, Acta Archaeologica Lundensia, Series in 8°, No. 31, Stockholm: Almqvist & Wiksell International

Magnusson-Staaf, B (2000) Ting och tanke', in Högberg, A (ed), *Artefakter, Arkeologiska Ting. En Bok om Föremål ur ett Arkeologiskt Perspektiv*, Report series no 71, Lund: University of Lund

Raemaekers, D (1999) *The Articulation of a 'New Neolithic': The Meaning of the Swifterband Culture for the Process of Neolithisation in the Western Part of the North European Plain (4,900–3,400 BC)*, Archaeological Studies Leiden University 3, Leiden: Leiden University

Schwabedissen, H (1972) 'Rosenhof (Ostholstein), ein Ellerbek-Wohnplatz am eistigen Ostseeufer', 2 *Archäologisches Korrespondenzblatt*, 10–44

Schierbeck, A (1994) *Hedningahällan—En Undersökning för att Skydda och Vårda,* Riksantikvarieämbetet, UV, Stockholm, Rapport 1994: 31, Stockholm: Riksantikvarieämbetets Förlag

Skandfer, M (2003a) 'Tidlig, nordlig kamkeramikk. Typologi–Kronologi–Kultur', Unpublished dr. art. dissertation in archaeology, University of Tromsø, Norway

——— (2003b) 'The Early Comb-Ceramics in northern Fennoscandia: material expression of cultural identity?' in Bergstøl, J (ed), *Scandinavian Archaeological Practice—in Theory: Proceedings from the 6th Nordic TAG, Oslo 2001*, Oslo archaeological series 1, Oslo: University of Oslo

Stilborg, O and Bergenstråhle, I (2001) 'Traditions in transition: a comparative study of the patterns of Ertebølle lithic and pottery changes in the Late Mesolithic ceramic phase at Skateholm I, III and Soldattorpet in Scania, Sweden', 6, 2000 *Lund Archaeological Review*, 23–42

Taffinder, J (1998) *The Allure of the Exotic: The Social Use of Non-Local Raw Materials during the Stone Age in Sweden*, Aun 25, Uppsala: Department of Archaeology and Ancient History, Uppsala University

Torvinen, M (2000) 'Säräisniemi 1-ware', XVI(1) *Fennoscandia Archaeologica,* 3–35

CHAPTER 12

'ALL CHANGE'? EXPLORING THE ROLE OF TECHNOLOGICAL CHOICE IN THE EARLY NORTHERN COMB WARE OF FINNMARK, ARCTIC NORWAY

Marianne Skandfer

INTRODUCTION

Early Northern Comb Ware (ENCW) is found in a limited area of the easternmost part of Finnmark, arctic Norway, in the present border areas between Russia to the east and Finland to the south. It is the oldest pottery in Norway and among the oldest pottery in Fennoscandia. The pottery has been known since 1910, but has elicited little interest in Norwegian archaeology. The reluctance within archaeological research to engage with this ceramic material contrasts with both the interest shown for Neolithic ceramics in southern Norway and the extensive ceramic research conducted in more southerly areas of Finland.

Despite the lack of more thorough engagement with the details and variability of this ceramic material, the ENCW in Finnmark has been nonetheless taken as a chronological marker for the beginning of the Younger Stone Age in north Norwegian archaeology (Helskog 1980; Olsen 1994; Simonsen 1991a, 1991b). As such, it has been viewed as a significant marker of cultural or ethnic diversity in Stone Age Finnmark (Olsen 1994: 66–67). The discrepancy between the cursory analysis of the ENCW material, and the bold chronological and cultural interpretations attached to it, has been the subject of a series of recent publications which are reviewed and developed here (Skandfer 2003a, 2003b, 2005)[1]. In this chapter, I review the on-going re-interpretation of these hunter-fisher ceramic assemblages, focusing on morphological characteristics, chronology and the prehistoric socio-cultural contexts within which this ceramic technology was introduced, produced, used, understood and finally given up.

ENCW: Säräisniemi 1 Reconsidered

Two main groups of prehistoric pottery have been found in Finnmark: an Early Comb Ware and different variants of Asbestos Tempered Ware. The two groups are not technologically or chronologically related in this area, and they are therefore easily distinguishable from each other. All pottery in Finnmark decorated with comb stamps or similar stamps in combination with rows of pits, and, not being asbestos tempered, is assigned to the earliest Comb Ware of northern Fennoscandia. The pottery is better known as Säräisniemi 1 Ware (Sär 1). It has been presumed to form a typologically distinct group, found in a limited area covering northeastern Russia south of Karelia, north Finland and easternmost Finnmark, Norway (German 1997; Halén 1994; Siiriäinen 1971, 1973; Torvinen 1998, 1999, 2000). In addition to the pottery vessels, the ENCW assemblages also include pieces of what might be clay figurines, although these will not be discussed here.

The Norwegian archaeologist Povl Simonsen first defined the characteristics of Finnmark Comb Ware pottery in a classic paper published in 1957 (Simonsen 1957). This description became known as Sär 1 and was subsequently adopted as the universal definition of the earliest Comb Ware in northern Fennoscandia. Prior to this, studies of Comb Ware in both Finnmark and the rest of northern Fennoscandia had been based on the descriptions and typologies developed for Comb Ware pottery from southern Finland (Ailio 1909; Europäeus-Äyräpää 1930). The aim of these classifications was to establish distinct types of ceramics that had exclusive geographical and chronological distributions. In line with the prevailing culture-historical paradigm of the time, these pottery distributions were believed to reflect distinct ethnic populations or 'cultures'.

More recently, the definition of archaeological types has been regarded as a first step in analysis, rather than a final conclusion. Moreover, the desire to define fixed categories based on geographic distributions of 'typical' artefact types can suppress important regional and localized variations both within and between these categories. Recent works have pointed out that decorative style in finished products often have less information about social identities than does the entire production and use repertoire of the technological tradition (Dietler and Herbich 1989; Pfaffenberger 1992). In this chapter, I try to view the archaeological material not just as a series of diagnostic objects for constructing generalized culture-historical sequences and as 'types' passively reflecting social (cultural) entities, but as the results of technological choices linked together into '*chaînes opératoires*' or production sequences. From a multitude of possible solutions to solve a technological challenge, any given community tends to select only a few (Lemmonier 1986). This shows that technological choices are not so much based on material efficiency or technological rationality as on social structure. Most technological choices are therefore not individual and *ad hoc* selections, but are instead the results of socio-cultural traditions, transmitted through more or less ritualized practices as collective memory within a community (Connerton 1989). The theory of technological choice implies a shift of focus from the finished product ('what') to the production sequence ('how') and thereby the producer ('who'). It constitutes an analytical division between culture as patterns *for* behaviour and social structure as patterns *of* behaviour.

In developing this alternative perspective on the material, I will start by noting that the Sär 1 category of ceramics was originally developed on the basis of a limited quantity of material from a restricted number of sites (Simonsen 1957; Torvinen 2000). In order to discuss technological variations at finer social and spatial scales, and to be able to view the earliest northern Comb Ware in light of broadly similar pottery technologies (other 'types') further to the south and east, I have chosen to avoid the Sär 1 concept. Instead, I introduce the term 'Early Northern Comb Ware' (ENCW) in order to examine the role of local choice in the long-term social reproduction of broader technological tradition.

ENCW ceramics are found at 13 sites in two adjacent regions of eastern Finnmark. Three of the sites are located on the southern coast of the inner part of Varanger fjord and the remaining 10 lie along the Norwegian side of the upper Pasvik River (Figure 12.1). The pottery was introduced to the hunter-fisher communities in these regions through social networks stretching to the southeast and east. Datings show that Early Comb Ware spread rapidly between the hunter-fisher-gatherer groups in Finland and probably north-western Russia up to eastern Finnmark (see below). The technological skill needed for the time-consuming pottery production (which had to be a summer activity in the northern hemisphere) indicates that trained, specialized potters were suddenly present in the groups. This indicates a social network including pottery-producing communities in the Baltic and southwest Russian region.

The current research aims to examine patterns of similarity and difference in materials from all the sites in Finnmark, and to isolate any specific relationships between a range of variables, including vessel size, vessel form, tempers and decoration. These fresh insights can then be compared to the original, and more rigid, Sär 1 definitions of the region's early Comb Ware pottery, in order to draw out more detailed insights into the relationships between the pottery and the past socio-cultural contexts in which it was produced, used and understood.

Early Northern Comb Ware

The collections of Finnmark's ENCW were visually inspected in order to document the morphological variations in the material (Skandfer 2003a,

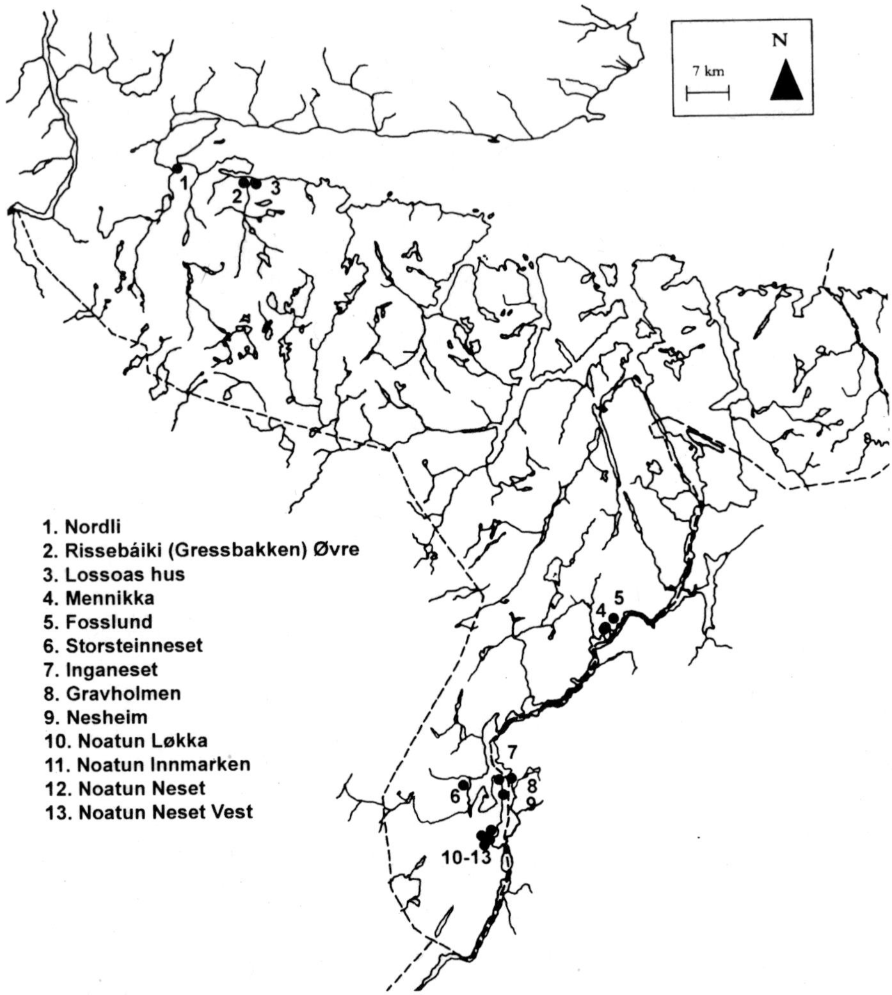

Figure 12.1 Map of ENCW sites in Finnmark, Arctic Norway.

2005). Variations in stamp motifs and other decorative techniques proved to be some of the most useful diagnostic characteristics, a fact noted in a range of other studies of both ENCW (Sär 1), and Comb Ware more generally (Europäeus-Äyräpää 1930; Nummedal 1937, 1938; Simonsen 1957; Solberg 1918; Torvinen 1999, 2000). Statistical analysis of ENCW generated a series of detailed insights into similarities and variations in the ENCW material from Finnmark. The resulting descriptions were far more detailed than earlier descriptions and classifications of ENCW (Sär 1) (e.g., Simonsen 1957; Torvinen 1999, 2000). Interestingly, the results also demonstrate that the ENCW pottery from Finnmark does not fit easily within the general typology of Sär 1 (Skandfer 2003a).

The pottery material consists of a total of 3,847 sherds. The vessels were decorated with particular stamps in combination with pits. Each stamp was produced to decorate only one single vessel. Based on the identification of individual stamps and their combination in patterns, a minimum number of 270 vessels were found. Ninety-five different rims were identified, whilst only three vessel bases have been found in the Finnmark material.

The number of pots from each site varies considerably, with 154 vessels recovered from Noatun Innmarken in upper Pasvik to only one from the small island Gravholmen some kilometres further down the Pasvik River. In general, each site contains between three and 20 vessels. Each identified vessel was subjected to morphological analysis. It was noted that ENCW vessels in Finnmark are similar to other Comb Ware pots in having straight walls slanting towards a round or tapering base. Moreover, there are differences in ENCW vessel-size between larger vessels and smaller cups (Skandfer 2003a: 125, 127–128), a feature also characteristic of Comb Ware pottery assemblages recovered from sites further to the south (Ailio 1909; Europäeus-Äyräpää 1930).

Based on a comparison between ENCW pottery mouth diameters and wall-thickness just below the rim, 27 small cups are recognized as having mouth diameter less than 8 cm, wall thickness of less than 9.5 cm and height from 5 to 7 cm. One flat cup bottom is present. The 243 larger vessels have mouth diameters varying between 13 and 36 cm, wall thickness of 6.8–13 mm and height probably around 40 cm (see Torvinen 2000: 7). In addition to these general correlations between vessel size and wall thickness, the statistical analysis suggested a correlation between wall-thickness and the site where the vessel was found. The indications of a potential relationship between geographic location and wall-thickness are confirmed when the Finnmark material is compared to ceramics from different sites in neighbouring northern Finland (Skandfer 2003a: 162–163). This variation might be interpreted as evidence for on-site ceramic production and use, with each site/community (in practice probably one or a few potters in each community) having its own technological choices regarding the forming of the vessel walls.

All rim profiles are simple, and all rims are straight-walled, with large vessels having a more-or-less rounded edge, and all cups having a rounded edge. Rims thickened inwards, and straight edges are also common among the larger vessels, whilst other rim profiles are rare. In contrast to the patterns of site-based variation in the wall thicknesses noted above, there was no clear link between individual sites and preferred rim types.

The larger vessels were made from coils placed successively on top of each other and then squeezed firmly together to make a joint. They were then smeared to cover the joints and to produce a smooth interior and exterior. The large size of the vessels indicate that they could have been at least partly modelled over or inside a prepared mould, but no impressions of cloth or baskets have been identified on the in/outside. The small cups were made using a different pinching and drawing technique, with a lump of clay placed in the hand and then squeezed and shaped in order to acquire the desired form. The majority of the vessel exteriors appear to have been painted with red ochre. Today, the ochre paint is preserved mostly on the upper parts of the vessel and on the rim edges. It is difficult to say whether they were originally painted all over the exterior surface or only on the upper half. When they were in use, at least the area around the vessel mouths might have been vivid red, with stamp decoration forming patterns of unpainted yellowish/brownish impressions. On the lower half, the paint may have been burnt off during cooking or eroded if the vessels were storage pots placed inside the ground. Pieces and powder of burnt red ochre are commonly observed at the ENCW-sites in Finnmark.

The analyses also revealed between-site variations in the type and quality of the tempers added to the clay pastes. All clay pastes were intentionally tempered, most often with rather coarse pieces of crushed quartz. There are clear differences between the sites with respect to the quality (colour, pureness, coarseness) of the quartz temper, and this appears to relate to the availability of appropriate sources close to the sites. While sand tempering was common at one site in the upper Pasvik Valley and at one site on the Varanger coast, it was practically unknown at other sites in both areas. This probably reflects the fact that few of the sites had beaches in their direct vicinity. Eight vessels at Noatun Innmarken are tempered with chamotte (crushed pieces of burnt clay). This re-use of broken vessels could also suggest that pottery was produced locally here. The interpretation is strengthened by the observation that several sherds from a particular part of Noatun Innmarken have 'clean' horizontal ruptures, indicating that the joints between the coils were broken during the firing. The clay used in the small cups is not tempered.

The colour of the clay in the vessel also varied between sites, from light yellow to dark brick red. These colour differences could stem partly from different mineral characteristics (quantity of iron) but is also affected by firing conditions. The between-site colour variations therefore suggest that either firing techniques differed between the sites or more probably that local clays with a different quantity of iron were selected and used. The analysis is preliminary in this respect. Further examination of local geology vis-a-vis the mineral characteristics in the vessel material would give more information on the questions of provenance.

ENCW pottery is characterized by strictly symmetrical patterns of decoration, with conical pits always combined with comb or comb-like stamp impressions. More than 60% of the vessels are decorated with simple comb-stamps cut with vertical lines (Figure 12.2). However, other decorative motifs do occur, although they are much less common. The decorations were made with differently cut comb-stamps, or include indentations made with pins, nail impressions, shells, twisted cord or even the humerus of small birds. It is interesting to note that only some of these decorative techniques have been argued to define Sär 1 as a distinct 'type' of pottery, while other techniques were omitted from the Sär 1 definition. Sär 1 thereby seemingly contrasts with Early Comb Ware in the south where these techniques were present (Torvinen 2000: 7, 18–19). Closer analysis of the Finnmark material reveals interesting correlations between decoration, vessel size and site. For instance, bird-humerus impressions are a very rare form of decoration when the material is considered as a whole, yet it forms the second most frequent motif on the small cups. Other generally 'rare' motifs, including the pin- or nail-impressions, are quite common on cups. There are also indications that certain sites were characterized by preferences for particular stamp motifs, and motifs that occur in only small quantities in the overall assemblage can be frequent, and even dominate, the site-specific material (Skandfer 2003a: 147–150, 2005).

The quantities of cup-like vessels are small and only some general conclusions about decorative preferences can be drawn. It appears that both pin and bird-humerus impressions are more commonly chosen to decorate

Figure 12.2 ENCW sherd from the site Noatun neset, upper Pasvik River valley (Ts.6128t), decorated with a comb stamp cut with vertical lines in combination with conical pits (photo by A Icagic, Tromsø Museum).

cups rather than vessels, and that cups can be decorated with conical pits only, and finally that cups are more frequently decorated with horizontally organized patterns (Skandfer 2003a).

In general, the stamp motifs are arranged in a repetitive fashion and cover the entire outer surface of the vessel. Sixteen rim edges with decorations were found. Horizontally organized patterns are by far the most common ones, accounting for around 90% of the total pattern variation. However, vertical patterns are not absent in the ENCW (Sär 1) material, as claimed by Torvinen (2000: 8). The decoration patterns mostly form horizontal ribbons, and there is a clear contrast between horizontally and vertically organized patterns in the material. Horizontal and vertical pattern organization is never mixed on the same vessel or cup (Figure 12.3). On the other hand, the stamp impressions can be placed diagonally both in the overall horizontally and vertically organized patterns, thereby 'mediating' the visual impression of either horizontality or verticality in the pattern organization. Again, several between-site differences in the pattern organization can be observed, in what initially appears to be pottery subjected to seemingly homogeneous forms of decoration (Skandfer 2005).

This impression of overall decorative homogeneity has led many researchers to believe that there are distinct decorative motifs and patterns, which together define Sär 1 (Halén 1994; Siiriäinen 1971, 1973; Simonsen 1957; Torvinen 2000). For example, Simonsen (1957) describes

Figure 12.3 Reconstruction of vessel from the site Nordli at the Varanger coast (drawing by E Skandfer).

the comb-stamp motif ending in a conical pit as being a defining feature of Sär 1, and as frequently occuring on vessels from the site of Noatun on the upper Pasvik River, Norway. Simonsen's Sär 1 definition, however, is based on only parts of the Noatun material and no material from the other ENCW sites in Finnmark or northern Finland. This motif has now been documented on only 10 vessels from the other sites in Finnmark. In fact, this particular motif is absent from vessels at over half the number of sites in Finnmark, meaning that the motif is actually rather rare in this region. Similarly, the decoration pattern which Torvinen (1998, 2000) suggests is an additional distinguishing feature of Sär 1 pottery is documented on half the vessels from the three largest and most important north Finland sites in his analyses, but is rare in other parts of the ENCW area. In Finnmark, only six of the 270 vessels have this pattern of diagonal comb-stamps put close together, making the edge of the vertical band look serrated. None of the 13 Finnmark sites have yielded more than one vessel with this decorative pattern (Skandfer 2003a: 137).

My examination of decorative variation in the Finnmark material, and in material from selected sites in northern Finland, indicates clearly that many supposedly 'distinguishing' features of the Sär 1 pottery 'type' are in fact site- and possibly region-dependent decorative variations, rather than being universally present. In summary, analysis of Comb Ware ceramics from Finnmark clearly demonstrates that there is so much between-site and regional variation in pottery characteristics that the crisp classic definition of a distinct Sär 1 pottery is no longer tenable making the whole concept of a 'typical' Sär 1 early Comb Ware ceramic in northern Fennoscandia redundant. Instead, I will argue that richer insights into the social significance of the material can be developed by using the more inclusive term of Early Northern Comb Ware (ENCW) which conceptualizes early Comb Ware pottery as a larger socially reproduced technological phenomenon, subject to long-term reproduction and change through localized traditions of technological choice (Skandfer 2003a, 2005).

Datings

The dating of the material can also be reevaluated. For example, it was long assumed that Comb Ware ceramics in northern Fennoscandia had to be younger than similar pottery found further to the south, and it was not until the first radiocarbon dates from Sär 1 sites became available in the early 1970s that northern Comb Ware was accepted as belonging to the oldest pottery traditions in Fennoscandia (Siiriäinen 1971, 1973). More recently, ENCW pottery in Finnmark has been dated to between 5,630 BC and 4,100 BC (6,570±60 bp–5,515±65 bp), based on 16 datings from eight different sites (Skandfer 2003a, 2005) (Table 1).[2] In addition, a sample of marine shell from beneath the cultural layers on the site Lossoas Hus gives a *terminus ante quem* dating to 6,370–5,570 BC (7,250±100 bp). The datings from Finnmark are among the oldest for Comb Ware from Fennoscandia at large, indicating a rapid spread of the new pottery technology.

All 10 datings derived from carbonized food residues scraped from the pottery lie between 5,630 BC and 4,550 BC. The measured $\delta^{13}C$-values for the food residues show no systematic differences between coastal and inland sites, and it seems that the marine reservoir effect has little or no impact on the datings of food residue from the two coastal sites. All the datings of food residues should therefore be considered reliable. Of the remaining six conventional dates, three are on seal bones and three on charcoal. Two datings, one on seal bone from the site Lossoas hus and one on charcoal from Noatun Innmarken are slightly younger than 4,500 BC. This could mean that they date to an aceramic occupation phase directly succeeding the one

Table 12.1 Radiocarbon datings of ENCW in Finnmark, Arctic Norway (calibrated by OxCal 3.10)

Site	Lab.no.	Dating BP	Dating calBC (2 sigma)	Dated material
Gravholmen	T-238[1]	5,950±300	5,600–4,100	Charcoal
	TUa-5472	6,095±45	5,210–4,900	Charcoal
Inganeset (Kjerringneset IV)	TUa-3025	5,990±55	5,010–4,720	Food residue
Lossoas hus	TUa-3024	6,065±55	5,080–4,830	Food residue
	TUa-3660	5,745±45	4,710–4,490	Seal bone
	T-2468[2]	6,315±90	5,480–5,050	Seal bone
	T-2472[3]	5,865±50	4,850–4,580	Seal bone
Mennikka (Skogfoss)	TUa-3022	5,795±55	4,750–4,510	Food residue
	TUa-3027	5,975±60	5,010–4,710	Food residue
Noatun Innmarken	TUa-3023	6,185±65	5,310–4,980	Food residue
	TUa-3029	5,850±55	4,840–4,550	Food residue
	TUa-2887	5,515±65	4,490–4,240	Charcoal
Noatun Neset	Beta-131296[4]	5,950±90	5,100–4,550	Food residue
Noatun Neset Vest	TUa-3026	6,030±70	5,080–4,720	Food residue
Nordli	TUa-3028	6,570±60	5,630–5,460	Food residue
	TUa-3021	6,330±50	5,390–5,210	Food residue

1. The dating was performed for Povl Simonsen †
2. Published by Knut Helskog (1980). The date was corrected for MRE and fractionation for the author in 2002.
3. Published by Knut Helskog (1980). The date was corrected for MRE and fractionation for the author in 2002.
4. The dating was performed for Ericka Engelstad 2003 (pers. com.).

represented by the pottery, but it could also indicate that ENCW was actually used until around 4,200 BC in Finnmark. One dating of charcoal from the site Gravholmen has a considerable range, between 5,600 BC and 4,100 BC. The dating was done in 1960 and is too vague to determine a specific age for the Gravholmen site, but generally it can be stated that the Comb Ware was used here around 5,000–4,500 BC. To sum up, 13 of the total 16 dates point to the period c. 5,400–4,500 BC as at least the main period in which ENCW pottery was produced in Finnmark, a tradition which may have extended until c. 4,100 BC.

The datings from Finnmark fit well with the majority of datings of ENCW from other parts of northern Finland, where ENCW food residue samples and charcoal from ENCW sites are also dated between 5,900 BC and 3,690 BC, with the majority of datings clustering between 5,500 BC and 4,040 BC (24 out of 27 datings from 12 different sites) (Torvinen 2000: 29). The dates also show that ENCW was being produced and used in the eastern parts of northern Fennoscandia from around 5,500 BC, possibly even a little earlier (Skandfer 2005). However, due to complicated stratigraphy, the datings from Comb Ware sites in northwestern Russia are difficult to relate with certainty to the earliest stage of pottery production and pottery use. Russian archaeologists date the earliest pottery there (different kinds of Pitted and Comb Ware) to around 4,000 BC (German 1997; Gurina 1987: 41, 1997; Gusentsova and Andreeva 1997; Kosmenko 1992; Shumkin 1984: 77), but based on pottery morphology and associated tools and stone technology it is likely that the earliest pottery was also introduced to the Kola-White Sea region around c. 4,700 BC at the latest (Skandfer 2003a: 248).

After around a thousand years of ceramic production being transmitted between generations, the social reproduction of Sär 1 ENCW as a technological phenomenon came to an end around 4,500–4,200 BC in eastern Finnmark, possibly also in northwestern Finland north of the Saarisälkä watershed (Skandfer 2003a, see also Carpelan 1978, figure 2). Further to the south and east, however, we see longer-term continuity in ceramic production, with cumulative technological choices made by successive generations reproducing different manifestations of the Comb Ware tradition, and leading eventually to the emergence of various types of Asbestos Tempered Ware around 2,300 BC (Carpelan 1978; Edgren and Thörnblom 1998: 48, 107; Lavento 2001: 23). Further to the west, Comb Ware was introduced in only a very limited area of the upper Kalix River valley, northern Sweden, around 4,500 BC and was retained until 3,500 BC (Färjare and Wickström 1997; Halén 1994). In the remaining areas of northern Fennoscandia, pottery was not present until Asbestos Ware dispersed into the area from the southeastern or eastern regions around 2,000–1,500 BC. Pottery technology once again fell out of use around AD 300 in the northern regions (Carpelan 1978; Edgren and Thörnblom 1998: 150; Jørgensen and Olsen 1988; Lavento 2001: 90–94) (Figure 12.4).

The Social Reproduction of ENCW as a Long-Term Technological Tradition

Pottery is a rare bird in the archaeological record of Stone Age Norway, in contrast to the situation in neighbouring Finland and Russia. The ENCW pottery in Finnmark has, until recently, been interpreted as evidence for the arrival of a new and foreign 'inland group' of hunter-fisher-gatherers, who practised seasonal movements between large winter house structures on the Pasvik River and small summer camps along the southern shores of the Varanger Fjord (Olsen 1994). This new pottery-making community was argued to have been culturally distinct and geographically separate from an indigenous and long-established coastal group who had no tradition of making ceramics. It has further been suggested that inter-group contacts developed, and that these led to the rather rapid assimilation of the newer arrivals into the traditions of the older coastal group. While the new arrivals adopted fresh practices from the coastal groups, they appear also to have given up their pottery-making traditions as the two cultures fused (Olsen 1994: 66–67).

This interpretation falls in line with the 'culture historical' school in viewing archaeological 'types'—especially pottery—as passive markers of distinct 'cultural' groups. In recent years, new insights suggest that it is a mistake, however, to conclude that pottery presence or absence implies cultural identity differentiation (Skandfer 2003b, 2005). It is now clear that this interpretation has little support in the archaeological material.

First of all, the alleged winter houses on the upper Pasvik River have been demonstrated to be much later features which intrude into the ENCW contexts (Skandfer 2003a; Skandfer and Bruun 2006); we therefore need to develop new perspectives on the settlement geography, as neither inland nor coastal ENCW sites have evidence for substantial dwellings. Given the need for clay, temper material and, not least, stable drying conditions with temperatures above 5°C for several days, the ENCW must have been produced during the summer season. Probably most of the ENCW sites in Finnmark were production sites used in the summer. Additionally, we also see increasing evidence for a series of contemporary sites in Varanger and west Finnmark without pottery, some of them containing small pit-houses (Hesjedal et al. 1996; Skandfer 2003a). At least in Varanger, these could be winter dwellings.

Second, the new analyses of the ENCW material in Finnmark in fact reveal a whole range of systematic local differences within what was

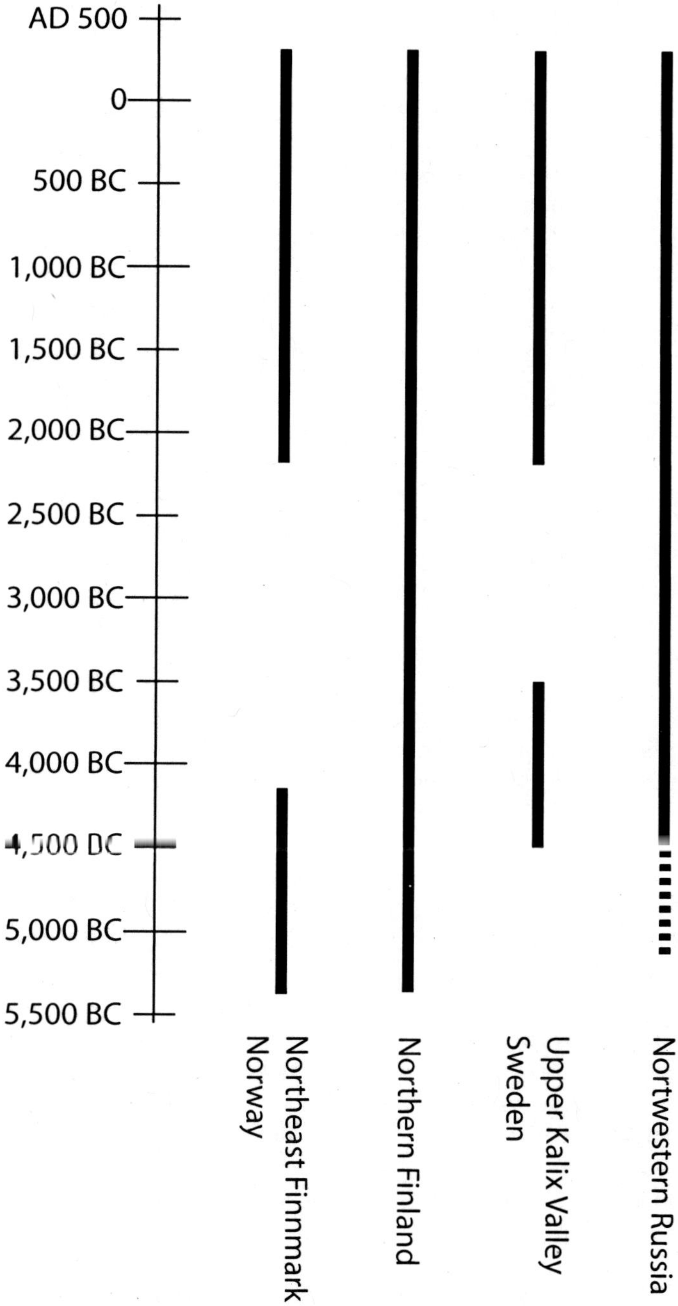

Figure 12.4 Pottery production phases in different parts of northern Fennoscandia during the Stone Age and early Metal Age.

hitherto argued to be 'homogeneous' ENCW pottery. These new insights from Finnmark point towards the practice of localized, on-site pottery production, which generates divergent variations of a wider ceramic technological phenomenon. The observed similarities in the pottery decorative motifs chosen in the upper Pasvik valley and the differences between the Pasvik material and the pottery from the sites on the Varanger coast suggest that the two regions were used by different hunter-fisher groups during summer. An analysis of the pottery from the closest sites on the Finnish side of the present national border shows several similarities with the Pasvik material, but little with the pottery from the Varanger coast. In light of that comparison, however, the Varanger-Pasvik pottery has more in common than does the Pasvik-northern Finland pottery. The full operational sequence of pottery production, from the collection of raw materials, to forming and firing, must have constituted the skilled practices of potters within local site-based communities. Assuming that pottery production skill was transmitted between generations in direct family lines, the systematic use of similar decorative motifs probably indicates socio-cultural and kinship relations on different levels between the ENCW groups in eastern Finnmark and northern Finland. The Pasvik group could then be related to both the Varanger coastal group and to the northern Finland groups by family ties, whereas the Varanger group had more distant, or no direct, family relations with the groups in Finland.

There is further evidence that pottery may have been introduced into Fennoscandia as part of a wider 'package' of technological traditions. These new elements include raw materials and tool types, for example the earliest instances of polished slate and bifacially retouched points. Placing the focus on the bifacially retouched points, a different picture of regional variation than the seeming contrast in Finnmark between east (ceramic) and west (aceramic) emerges.

Unlike pottery, bifacially retouched points with a pointed base and various tools of polished slate are found all over northern Fennoscandia from the time ENCW was introduced to the eastern parts of the region. However, two different reduction sequences for the bifacially retouched points have been documented in Finnmark: one in the eastern Varanger-Pasvik area and one in western Finnmark (Skandfer 2005: 16–17). The Varanger-Pasvik reduction sequence is quite easily recognized from its standardized preforms, which were formerly believed to represent different types of bifacially retouched tools (Gjessing 1942; Helskog 1980: 48–49; Nummedal 1937, 1938; Olsen 1994: 52; Simonsen 1991a: 222, figures 113, 114), whereas the western Finnmark sequence lacks such distinctive preforms. The Varanger-Pasvik sequence is also observed at ENCW sites in northern Finland, something which underlines the close relationship between ENCW sites in the region. Furthermore, this 'eastern' reduction sequence is found in pit-houses at later dwelling sites such as Gropbakkeengen on the Varanger coast (Figure 11.5). This technological link indicates a direct relation between the ENCW groups and later aceramic occupation on the coast of eastern Finnmark, and at the same time shows that the different reduction sequences in eastern and western Finnmark were maintained even after the pottery went out of use.

It seems that new and distinctive lithic toolkit traditions, which had no earlier history in the region, may have been brought into eastern Finnmark at the same time as early Comb Ware pottery. While the new tools are found all over Fennoscandia, pottery is found only at select sites in the easternmost areas, but the new lithic technologies are shared by both pottery producing and aceramic groups. The new and common repertoire of material culture is quite different from the older Mesolithic material culture. Moreover, the material culture from sites without pottery shows no closer resemblance to previous Mesolithic technological traditions than do sites where the ENCW is found. Instead, it seems that some late Mesolithic technological traditions, such as the use of the oblique point, are continued in both ceramic and aceramic societies equally around 5,400–4,500 BC. Here, I must add that these points are present at all ENCW sites along the Varanger coast and are known also from ENCW sites in northern Finland, but they are neither found on aceramic sites along the Varanger coast nor in the Pasvik River valley. It follows that a functional differentiation between specialized sites can again be suggested.

In the literature (Helskog 1980; Olsen 1994; Simonsen 1991b), the arrival of pottery in northern Fennoscandia has been given greater emphasis as an indicator of cultural and social change, whereas changes in stone tool technology have tended to be downplayed. As a result, there has been a tendency to interpret pottery as a more important archaeological class of evidence. There may be several reasons for this, but the fact that stone tools are reduced to practical instruments (in line with a western view of tools), whereas pots are seen as more creative, cultural and 'artistic', is probably one factor, combined (often unconsciously) with the concept of progress where pottery is interpreted as sign of a higher—or as in Finnmark of an at least quite different—cultural setting. The fact that pottery is easily recognizable and at the same time relatively rare compared to stone artifacts and debris, is probably an additional reason as to why pottery has been given greater emphasis. Material culture acts as a reservoir of information about the past cultural contexts in which the artifacts were produced, used and deposited. Pottery may be a more 'visible' class of artefact, but this visibility should not colour our interpretations of the past, which should be based on analysis of wider datasets, including stone tool assemblages.

Pottery was introduced into Finnmark around 5,400 BC, but then disappeared again some time after 4,500 BC. As a result, the ENCW has been

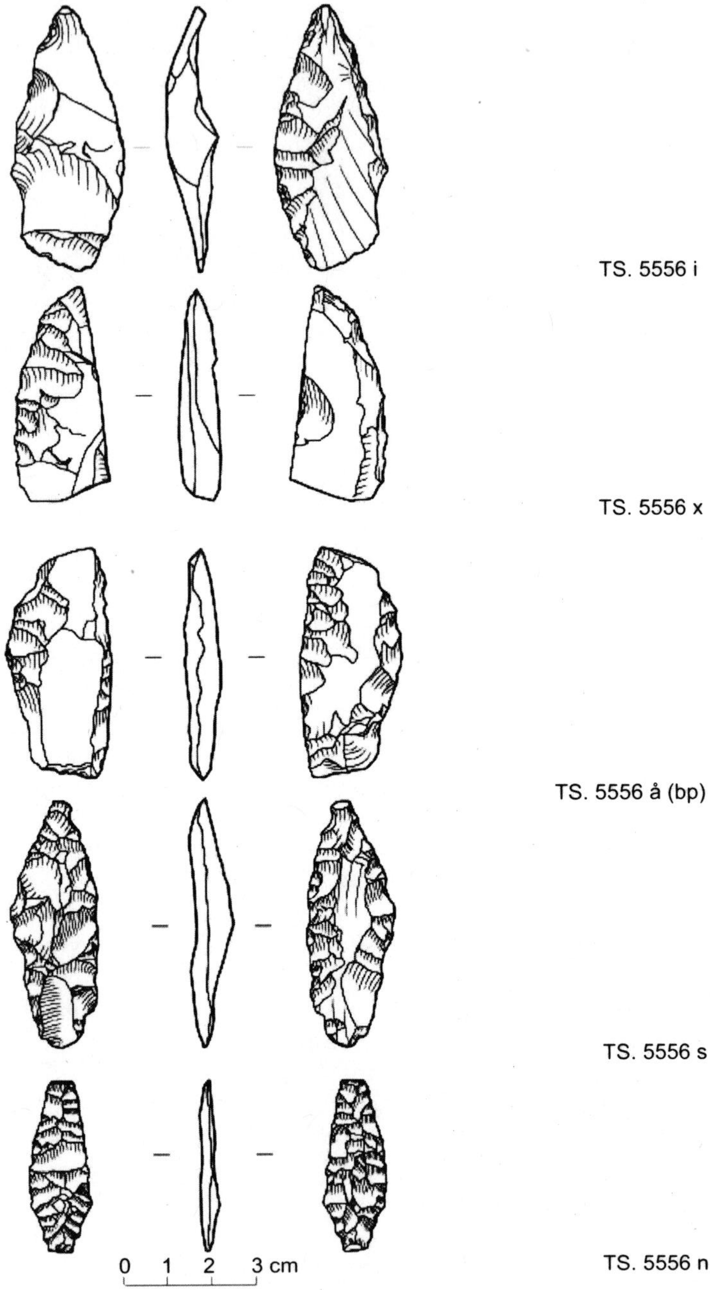

Figure 12.5 Bifacially retouched points exhibiting the 'eastern' (Varanger-Pasvik) reduction sequence.

commonly understood as a cultural and technological 'dead end' in the wider social history of the region. However, as noted above, the introduction and disappearance of pottery has been regarded as perhaps the most important technological change defining this period, and perhaps the degree of attention given to this chronologically and geographically delimited category of material culture has been undue. In contrast, several other long-lasting technological traditions, including the use of polished slate tools and the settling in pit-houses, also appear in the ENCW-phase, but then persist for several millennia after the demise of pottery making in the whole of northern Fennoscandia. In the larger parts of northern Finland and northwestern Russia, the Comb Ware tradition continued alongside the new stone technology and pit-houses. These insights suggest that the ENCW should perhaps be viewed not as a dead end in the prehistory of Finnmark but as a catalyst, in the sense of an entity that initiates something without itself being part of the end product. It seems that many other dimensions of the new technological traditions which were established around 5,400–4,500 BC were reproduced in wide, long-lasting and stable socio-cultural networks in the next four millennia.

Locating Prehistoric Technologies in their Socio-Cultural Context

The ENCW groups were part of a wider hunter-fisher-gatherer economy which prevailed in Fennoscandia at large. On a regional scale, the evidence from the Varanger-Pasvik region of northern Fennoscandia points towards small, mobile hunter-fisher-gatherer groups which share the same distinctive production technologies across the eastern region, e.g., raw materials, reduction sequences and pots. These distinct ceramic and stone tool technological traditions, which must have been reproduced from generation to generation amongst the Varanger-Pasvik hunter-fishers, seem to single them out as being culturally distinct and geographically separate from other hunter-fisher groups in western Finnmark. In this way, the Varanger-Pasvik groups were part of a widespread technological (and cultural) phenomenon: the early Comb and Pitted Wares in north-eastern Fennoscandia, around the inner Gulf of Bothnia, and further east into Russia. I suggest that this division of technological traditions can be traced further, with a distinct 'eastern' stone technology extending eastwards into Finland and Russia, and a 'western' tradition which is found further southwards along the Atlantic coast, as well as into Sweden to the south-west. Of course there would have been contacts—as clearly indicated by the shared bifacially retouched tool form—but my suggestion is that the traditions of learning and production were being transmitted locally, and that they were not shared across the whole wide region. Instead, there seems to have been a technology frontier in several classes of artefact between 'eastern' and 'western' Fennoscandia. It has not been investigated whether this frontier can be traced in the preceding Mesolithic period or if it is a new geographical division, established with the introduction of pottery.

However, similar regional traditions can probably be traced in the later distribution of rock carvings from around 4,500 BC. Rock carvings have been found only in the western region of Finnmark, further south along the Atlantic coast and in northern Sweden to the southeast, but not in eastern Finnmark or in Finland. The distribution of the rock carvings could be a continuation of the older 'east'/'west' technological division. But further to the east, the picture changes since rock carvings dating back to around 4,500 BC or a little later are also found in the northeasternmost ENCW region in the White Sea–Lake Onega area (Lobanova 1995; Savvateev 1990), possibly in combination with later Comb Ware production. This coincides with the disappearance of pottery in Finnmark and possibly northernmost Finland and the occurrence of Comb Ware in upper Kalix, northernmost Sweden. We might here see a retructuring of networks, expressed through the dynamics of material culture.

I have suggested that the hunter-fisher societies shared many common features and that they interacted. In this socio-cultural setting, early Comb Ware should be viewed as a technological phenomenon participated in by several of these hunter-fisher groups occupying the easternmost region. The presence or absence of pottery at a site in this region should not, however, be used alone as evidence for a socio-cultural or ethnic dualism cutting across eastern Finnmark. Instead, the stone technology shared among all the sites indicates that the pottery was used only at specialized or seasonal camps, probably summer dwellings. In addition, other common elements found all over Finnmark and Fennoscandia as a whole indicate a shared sense of common culture, an impression which is modified by the observed different technological choices in stone tool reduction, which coincide with the 'east'/'west' geographical ceramic/aceramic division (Skandfer 2003b; 2005: 19–20).

How can we develop an understanding of the role of pottery in these past societies? Pots can be seen as a kind of material culture which signals meaning and identity. There are different kinds of social identity such as age, status and gender. On a larger scale, social identity can also mean ethnicity or group identity. Ethnic or cultural identity is made up of contrasts towards the Other, as well as of collective templates transferred by birth and through childhood within a specific society (Barth 1969, 2000; Eriksen 1993; Hodder 1982; Jenkins 1997; Jones 1998). As many anthropologists and archaeologists have shown, material culture is and has been used to signal identity at the frontier, but also within the societies. There has been an emphasis on the social processes taking place along culture boundaries, but in recent years enculturation within specific societies and in specific historical settings has gained more attention (Jenkins 1997).

Under given historical conditions, groups can consider it important to actively express ethnic differences. Ethnicity or cultural difference can be expressed through the use of culture-specific objects or—perhaps more commonly—as material correlates of culture-specific behaviour (Hodder 1982; McGuire 1982). We may conclude that the categories of material culture which two or more ethnic groups have in common, but which they produce or use differently, are often the most potent resources for signalling identity. Such objects are particularly powerful symbols of identity because although they form elements in the same broad material 'language', this commonality brings into sharp and powerful focus the details of production or use that speak loudly about inter-community differences, much more so than would the use of very different artefacts that cannot be recognized, compared and contrasted.

It is important to consider components of the wider material culture assemblage that were potentially involved in these inter-ethnic identity negotiations and the characteristics of such inter-ethnic relations, that is, (1) how groups were organized seasonally and spatially over the resource landscape; (2) how, when, with whom and where they interacted; (3) what kinds of relationships grew out of these interactions; (4) what kinds of identity needed to be signalled and at what social scales; and (5) what kinds of material culture were used in communicating identity. Variations in the ceramic material should, in my opinion, be seen as:

1. Local or regional differences in technological choices, reflecting local learning traditions. The observed site-dependent and local or regional variation in pot decoration, form and size indicate that the pots were made locally, possibly by only a few selected potters in each community. The pots could have been seen as symbols for different communities/groups within the wider group of ENCW, Comb Ware and Pitted Ware producers.
2. Showing kinship and signalling family ties. Here I take as an underlying assumption that potting was passed on from generation to generation within a family setting. Similarities in preferred decoration would then express family ties. Regional similarities indicate kinship on different levels, whereas clear differences indicate more distant kinship relations.
3. Related to social regulation of land use. The pottery material as well as the associated stone tools show local and regional differences which might indicate a kind of 'territoriality'. 'Territoriality' and 'boundary' are, however, complex concepts, embedded with

massive cultural entailments. Territoriality most commonly refers to exclusive use of a defended area with clear borders, an understanding which focuses on only one aspect of a range of possibilities to regulate resource use under different conditions among hunter-gatherers (Kelly 1995: 163; see also Barth 2000). Instead, several anthropologists suggest that the term *land tenure* is used for the regulation of relations between people and land (Ingold 1987; Kelly 1995). Additionally, the recognition of variation in material culture often leads to an assertion of a boundary as a particular conceptual construct that people in given situations impress on the world. However, as Barth (2000: 17) claims, 'making a [categorical] distinction does not necessarily entail drawing a border'. Land tenure, gift exchange and sharing are all parts of the permission-granting behaviour whereby hunter-gatherers regulated access to resources (Ingold 1987; Kelly 1995: 164). In the vast and sparsely populated area of northern Fennoscandia, rich with varied resources, defending areas was probably not necessary, whereas on the other hand marriage/mating partners must have been one of the most critical resources. The social regulation of access to natural resources could have been performed at arranged meetings between entire groups of different hunter-fisher-gatherer communities or selected members thereof. In such settings, variations in material culture could have served as meaningful socio-cultural symbols. The observed variations in the material culture suggest that large areas were shared by several groups/communities related through kinship, and that these groups used different dwelling sites in the region. More research is needed, but so far the material could indicate that different kindred communities used different regions, although other groups probably had access as well.

The present material indicates that eastern Finnmark became the outlier of the earliest Stone Age pottery production in Fennoscandia, probably because the region already was part of the resource area of hunter-fisher groups that also used areas further to the southeast, and that were exposed to pottery production through long-distance social networks across the Baltic Sea (see Hallgren 2004; see also Nũnez 1990). The mobile groups with open kinship systems and low levels of territoriality quickly spread the new technology. In contrast, the lack of pottery in western Finnmark, as also in western parts of northern Fennoscandia at large, suggests that the groups living there had a different settlement pattern, possible already with more sedentary settlements as indicated by the earliest pit-houses along the Atlantic coast, and that they had a different (type of) contact or kinship network.

In summary, the differences in the archaeological material between eastern and western Finnmark are interpreted as arising from mainly cultural factors. Rather than representing the sharp and large-scale forms of ethnic identity usually associated with the 'culture historical' school, I would like to argue that these general divergences signal only that groups in the different areas belonged to different social and cultural networks which brought them into different forms of contact. Pottery was probably used as an active symbol in the negotiation of social identities, but likely only within the eastern network and not between the different social networks. At small social and spatial scales, the finer grained variability in pot or tool shape, decoration or design spoke loudly about the maker or users' origin, skills, status and identity. In contrast to the simplistic and coarse-grained ethnicity model, which maps broad material culture distributions onto broad groups, the present study suggests that patterns of similarity and difference within a shared material culture point towards more local negotiations within a wider network of contact, mutual knowledge and shared technological traditions that linked ENCW groups in eastern Fennoscandia both to ceramic-producing societies further south and east as well as to the aceramic groups in the west. This broader social focus means that presence or absence of pottery should not be regarded as the sole means with which identities were made, but just one resource amongst a much wider shared group of technological traditions.

Contrary to traditional culture-evolutionistic theory in archaeology, the technology of pottery did not spread gradually among the hunter-fisher groups in northern Fennoscandia after the 6th millennium BC. Instead, pottery went out of use from around 4,500–4,200 BC in the northernmost region. At the same time, we find indications of more sedentary societies, relying to a greater extent on marine resources, and providing exactly the kind of socio-economic conditions ethnographically documented to be favourable for ceramic production and use. How can this be? Let me first give a short presentation of the archaeological material.

Shortly after pottery technology ceased, semi-subterranean house structures were built in large settlements along the coast of eastern Finnmark, also close to the former ENCW sites. Such houses are dated from around 5,000 BC in western Finnmark (Hesjedal et al. 1996). Around 4,500 BC, as Comb Ware was abandoned in northeasternmost Fennoscandia, pottery was taken up in the technological repertoire in neighbouring upper Kalix River valley to the west. Similar pit-houses are found at these later Comb Ware sites in upper Kalix, northern Sweden and northern Finland, as well as on non-ceramic sites in interior northern Sweden at least from around 4,200–4,000 BC (Halén 1994; Lundberg 1985, 1986, 1997; Pesonen 2002; Torvinen 2000). The widespread practice of building semi-subterranean houses could mark a more sedentary lifestyle for the hunter-fisher societies of northern Fennoscandia around 4,500–4,200 cal BC. The new house structures suggest that the hunter-fisher groups answered to a change in living conditions by introducing, for instance, more spatially restricted settlement patterns and thereby reduced mobility. Coastal settlements with large numbers of pit-houses along the Barents coast, like the Gropbakkeengen and the Slettnes sites, suggest a stronger reliance on marine resources. Several families engaging in specialized fishing or sea-mammal hunting probably dwelt here for at least parts of the year. At the same time, interior northernmost Finland and the upper Pasvik Valley seem practically uninhabited.

Better understanding of the historical conditions under which ENCW technology and practices were introduced and—from an archaeological viewpoint—discarded shortly thereafter in northernmost Fennoscandia, demands much more research. So far, very little has been done on the transition phase from small, mobile communities with mostly open dwelling sites before c. 4,500 BC towards seemingly larger and more sedentary communities with 'villages' of semi-subterranean houses and more reliance on marine resources in the succeeding phases (see however Damm 2006 for a suggestive model). The pottery is introduced and used by mobile hunter-fisher groups, and then discarded as the lifestyle became more sedentary, quite in contradiction to current theories for ceramic adoption.

The disappearance of ceramic technology from Finnmark and possibly also northernmost Finland around c. 4,500–4,000 BC suggests that pottery tradition should not be universally linked to degrees of sedentism and the practicality of pottery in less mobile groups, but instead to the active socio-cultural networks in which these groups were taking part. Further investigation of how these changing networks were expressed materially is needed, but the fast-spread tradition of semi-subterranean houses in both western and eastern parts of Fennoscandia suggest that the building of these houses, possibly as a result of increased sedentism and more localized interaction, was a shared technology in the whole area. At the same time, more spatially restricted settlement patterns would lead to challenges for the groups to survive both biologically and culturally. In Finnmark, this could have lead to a reorientation away from the coast- inland axis and to the formation of new coastal networks, breaking the ties to the Comb Ware groups further southeast.

Concluding Remarks

The Comb Ware pottery is a long-known, much discussed, yet poorly researched element of the archaeological record of Finnmark. The reluctance within archaeology to engage with this ceramic material can probably be partly understood in the light of the culture-historical 'otherness' in the region, particularly the inland areas, which form part of the native Saami settlement area. As a result, the 'Saami past' was long regarded as topic for ethnographic rather than archaeological study, with the excavation of the first ENCW material in the upper Pasvik River Valley, Finnmark, carried out by an ethnologist (Solberg 1918). A further, and possibly related reason, is that ENCW materials—that is hunter-fisher-gatherer pottery—fits awkwardly within broader schemes of social evolution which assert a direct link between ceramics and early farming. In contrast, ceramic technology has a very deep, though discontinuous, history in Finnmark while farming—in practice cattle and sheep husbandry—only arrived in large areas in the 20th century.[3]

In the present analysis I have sought to break open these categories and engage in a finer grained study of the variability within the ENCW material, and then return to its social, cultural and symbolic context, the contexts in which it was produced, used and understood as an active element in the material and technological repertoire.

In order to achieve these insights, I have focused on the material not merely as representing a selection of finished tools or artefacts which can then be subjected to large formal classifications, but sought to investigate them as individual outcomes of a socially embedded technological process. The artefacts form the material result of these sequences of decisions made during production, and so variability depends on both personal choice and local raw materials, but more on inherited and acquired technological skill and the socio-culturally sanctioned setting. As a result, the interpretations presented here are based not so much on the presence or absence of archaeological 'types' as on identified technological variations in both pottery and lithic material, which result from choice.

Although more research is needed, the analyzed material from Finnmark suggests a division between 'eastern' and 'western' hunter-fisher cultural networks in northern Fennoscandia in the centuries before 5,000 BC. The 'eastern' groups chose to take up and participate in the new but already widespread socio-cultural technological phenomenon of Comb and Pitted Ware production, while the 'western' groups did not and remained aceramic.

Additionally, the analyses of stone tools show that in eastern Finnmark/Varanger, the presence/absence of pottery at sites should not be interpreted as a simple indicator of crude ethnic divisions based on the use or absence of pottery, but as the product of a single mobile hunter-fisher group that only produced and used pottery at some of their habitations.

After one thousand years—which is a short time in prehistory, particularly in a hunter-gatherer context—the new tradition of pottery making went out of use in Finnmark and possibly in northernmost Finland. The ENCW is therefore commonly presented as a dead end in the cultural history of the region. However, based on the present more resolutely social and symbolic analysis of the ENCW and its corresponding material (Skandfer 2003a), I have suggested that the short use of pottery should instead be investigated further as a catalyst for profound and long-lasting changes in northern Fennoscandia during the subsequent millennia.

Notes

1. This chapter is an abridged version of my 2005 paper in *Fennoscandia Archaeologica* XXII.
2. The datings have been recalibrated by OxCal 3.10 for the purpose of this chapter, thereby diverging slightly from the dates (calibrated by OxCal 3.5) given in my 2005 paper in *Fennoscandia Archaeologica* XXII.
3. Sheep husbandry in combination with fisheries and hunting can probably be dated back to early Medieval times in the coastal areas, and some reindeer husbandry (pastoralism) could have been taken up even earlier, also in the interior. In most parts of Finnmark, cattle husbandry and cultivation was not introduced until the 20th century, and it has never been more than a minor industry in the region.

References

Ailio, J (1909) *Die Steinzeitlichen Wohnplatzfunde in Finland* 2(1), Helsingfors: Finnische Altertumsgesellschaft: in Kommmission bei der Akademischen Buchhandlung

Barth, F (1969) 'Introduction', in Barth, F (ed), *Ethnic Groups and Boundaries,* Oslo: Universitetsforlaget

——— (2000) 'Boundaries and connections', in Cohen, A P (ed), *Signifying Identities. Anthropological Perspectives on Boundaries and Contested Values,* London and New York: Routledge

Carpelan, C (1978) 'Om asbestkeramikens historia i Fennoskandien', 85 *Finskt Museum*, 5–35

Connerton, P (1989) *How Societies Remember*, Cambridge: Cambridge University Press

Damm, C B (2006) 'Interregional contacts across northern Fennoscandia 6,000–4,000 BC', in Arneborg, J and Grønnow, B (eds), *Proceedings from Dynamics of Northern Societies—a SILA and NABO conference, May 2004,* Copenhagen: SILA

Dietler, M and Herbich, I (1989) 'Tich Matek: the technology of Lou pottery production and the definition of ceramic style', 21 *World Archaeology,* 148–164

Edgren, T and Thörnblom, H (1998) *Finlands Historia*, Esbo: Schildts Förlags Ab

Eriksen, T H (1993) *Små Steder—Store Spørsmål*, Oslo: Universitetsforlaget

Europäeus-Äyräpää, A (1930) 'Die relative chronologie der steinzeitlichen keramik in Finland', *Acta Archaeologica*, 166–190

Färjare, A and Wickström, C (1997) 'En sälidol från Överkalix. Diskussion om en djurfigurins funktion och betydelse på den kamkeramiska boplatsen Lillberget', in Åkerlund,

A, Bergh, S, Nordbladh, J and Taffinder, J (eds), *Till Gunborg. Arkeologiska Samtal*, Stockholm: Stockholm Archaeological Reports, University of Stockholm

German, K E (1997) 'Neoliticheskaia keramika v Severnoi Karelii (Säräisniemi I)', 1 *Arkheologiia Severa. Sbornik Nauchnykh Statei*, 63–73

Gjessing, G (1942) *Yngre Steinalder i Nord-Norge*, Oslo: Instituttet for sammenlignende kulturforskning

Gurina, N N (1987) 'Main stages in the cultural development of the ancient population of the Kola Peninsula', IV *Fennoscandia Archaeologica*, 35–48

——— (1997) 'Istoriia kul'tury drevnego naseleniia Kol'skogo poluostrova', 32 *Arkheologicheskie Izyskaniia*, Saint Petersburg: Institut Istorii Material'noi Kul'tury RAN

Gusentsova, T M and Andreeva, N A (1997) 'Pamiatniki s keramikoi Sperrings v Severo-Vostochnykh raionakh Leningradskoi Oblasti', 1 *Arkheologiia Severa. Sbornik Nauchnykh Statei*, 57–62

Halén, O (1994) *Sedentariness during the Stone Age of Northern Sweden in the Light of the Alträsket Site, c. 5,000 BC, and the Comb Ware Site Lillberget, c. 3,900 BC: Source Critical Problems of Representativity in Archaeology*, Acta Archaeologica Lundensia 4(20), Lund: Lund University

Hallgren, F (2004) 'The introduction of ceramic technology around the Baltic Sea in the 6th millennium', in Knutsson, H (ed), *Coast to Coast Arrival: Results and Reflections. Proceedings of the Final Coast to Coast Conference 1–5 October 2002 in Fallköping, Sweden*, Uppsala: Uppsala University

Helskog, K (1980) 'The chronology of the younger Stone Age in Varanger, North Norway, revisited plus comments', 13 *Norwegian Archaeological Review*, 47–54

Hesjedal, A, Damm, C, Olsen, B and Storli, I (1996) *Arkeologi på Slettnes. Dokumentasjon av 11.000 års Bosetning*, Tromsø: Tromsø Museums Skrifter XXVI

Hodder, I (1982) *Symbols in Action*, New Studies in Archaeology, Cambridge: Cambridge University Press

Ingold, T (1987) *The Appropriation of Nature. Essays on Human Ecology and Social Relations*, Iowa City: University of Iowa Press

Jenkins, R (1997) *Rethinking Ethnicity: Arguments and Explorations*, London: Sage Publications

Jones, S (1998) *The Archaeology of Ethnicity*, London: Routledge

Jørgensen, R and Olsen, B (1988) *Asbestkeramiske Grupper i Nord-Norge*, Tromura Kulturhistorie 13, Tromsø: University of Tromsø

Kelly, R L (1995) *The Foraging Spectrum. Diversity in Hunter-Gatherer Lifeways*, Washington: Smithsonian Institution Press

Kosmenko, M G (1992) *Mnogosloinye Poseleniia Iuzhnoi Karelii*, Petrosavodsk: Karelia

Lavento, M (2001) 'Textile ceramics in Finland on the Karelian Isthmus', 109 *Finska Forminnesföreningens Tidsskrift*, Helsinki: Suomen Muinaismuistoyhdistys

Lemmonier, P (1986) 'The study of material culture today: toward an anthropology of technical systems', 5 *Journal of Anthropological Archaeology*, 147–186

Lobanova, N (1995) 'Economy and mode of life in the Neolithic population on the eastern shore of Lake Onega', XXI *Fennoscandia Archaeologica*, 103–111

Lundberg, Å (1985) 'Villages in the inland of northern Sweden 5,000 years ago', in Backe, M et al. (eds), In *Honorem Evert Baudou,* Archaeology and environment 4, Umeå: University of Umeå

——— (1986) 'Skärvstensvallar i Norrland', 8 *Acta Bothensia Occidentalis. Skrifter i Västerbottnisk Kulturhistoria*, 81–99

——— (1997) *Vinterbyar—Ett Bandsamhälles Territorier i Norrlands Inland 4,500–2,500 f.Kr*, Studia Archaeologica Universitatis Umensis 8, Umeå: University of Umeå

McGuire, R (1982) 'The study of ethnicity in historical archaeology', 1 *Journal of Anthropological Archaeology*, 159–178

Nummedal, A (1937) 'Yngre stenaldersfund fra Nyelven og Karlebotn i Øst-Finnmark', 1935–36 *Universitetets Oldsakssamlings Årbok*, 69–128

——— (1938) 'Yngre stenaldersfund fra Nyleven og Karlebotn (II)', 1937 *Universitetets Oldsakssamlings Årbok*, 1–26

Nũnez, M (1990) 'On Subneolithic pottery and its adoptation in late Mesolithic Finland', VII *Fennoscandia Archaeologica*, 27–52

Olsen, B (1994) *Bosetning og Samfunn i Finnmarks Forhistorie*, Oslo: Universitetsforlaget

Pesonen, P (2002) 'Semisubterranean houses in Finland—a review', in Ranta, H (ed), *Huts and Houses. Stone Age and Early Metal Age Buildings in Finland*, Helsinki: National Board of Antiquities

Pfaffenberger, B (1992) 'Social anthropology of technology', 21 *Annual Review of Anthropology*, 491–516

Savvateev, J A (1990) *Kamennaia Letopis' Karelii: Petroglify Onezhskogo Ozera i Belogo Moria*, Petrozavodsk: Karelia

Schanche, K (1988) 'Mortensnes, en boplass i Varanger: en studie av samfunn og materiell kultur gjennom 10.000 år', unpublished masters thesis, University of Tromsø

Shumkin, V (1984) 'The Stone Age of Eastern Lappland', unpublished PhD thesis, University of Saint Petersburg [translated into English by B Grudinko]

Siiriäinen, A (1971) 'Shoreline dating of the Säräisniemi 1-ceramics in Finland', 78 *Suomen Muinaismuistoyhdistys*, 9–19

——— (1973) 'Studies relating to shore displacement and Stone Age chronology in Finland', *Finskt Museum*, 5–22

Simonsen, P (1957) 'Bopladserne ved Noatun i Pasvikdalen', 58 *Suomen Muinaismuistoyhdistyksen Aikakauskirja*, 234–267

Simonsen, P (1991a) *Veidemenn på Nordkalotten, Vol. 1,* Stensilserie B 17, Tromsø: Institute of Social Science, Tromsø: University of Tromsø

——— (1991b) *Veidemenn på Nordkalotten, Vol. 2.* Stensilserie B 17, Tromsø: Institute of Social Science, Tromsø: University of Tromsø

Skandfer, M (2003a) 'Tidlig, nordlig kamkeramikk. Typologi–kronologi–kultur', unpublished PhD thesis, University of Tromsø

——— (2003b) 'The early Comb-ceramics in northern Fennoscandia: material expression of cultural identity?' in Bergstøl, J (ed), *Scandinavian Archaeological Practice—In Theory,* Oslo: University of Oslo

——— (2005) 'Early Northern Comb Ware in Finnmark: the concept of Säräisniemi 1 reconsidered', XXII *Fennoscandia Archaeologica*, 3–23

Skandfer, M and Bruun, I M (2006) 'De eldste husene i Pasvik', 26 *Varanger Årbok*, 106–117

Solberg, O (1918) 'Mennikka-fundet', 1916 *Oldtiden,* 1–11

Torvinen, M (1998) 'Sär 1-keramiikaa käyttänyt väesto–etnisiä kysymyksia', 4 *Muinaistukija*, 38–45

——— (1999) 'Sär 1—tutkielma luoteisen varbaiskeramiikan alalta', unpublised PhD thesis, University of Helsinki

——— (2000) 'Säräsniemi 1-ware', VI *Fennoscandia Archaeologica*, 3–35

CHAPTER 13

'TINY ISLANDS IN A FAR SEA'—ON THE SEAL HUNTERS OF ÅLAND, AND THE NORTHWESTERN LIMIT IN THE SPREAD OF EARLY POTTERY

Fredrik Hallgren

> We do not now, he said, nor did we ever live imprisoned on 'tiny islands in a far sea,' ... The sea is our home, as it was to our ancestors. The ancestors' world 'was anything but tiny. They thought and recounted their deeds in epic proportions.' They lived in great associations of islands ... linked by the sea, not separated by it. (Sahlins 2000: 180)

INTRODUCTION

Around 5,000 cal BC, hunter-gatherers living on the eastern shores of the Baltic Sea, from Finland in the north to Poland in the south, adopted ceramic technology and started producing and using pottery. While most of these examples of hunter-gatherer ceramics share some similarities in vessel shape—mostly unprofiled pots with a pointed or round bottom—they can be divided into several more or less distinct craft traditions based on aspects of design and technology, for example, type and composition of decoration, use (or not) of colour pigment, choice of temper etc. The earliest pottery of southern Finland, and Russian Karelia and Ingermanland, is of a distinct type named Sperrings 1, so named after a settlement site in Esbo, close to Helsinki (Europaeus 1927: 49–55). Sperrings 1 pottery is also known by the more cumbersome name Early Older Comb Ware, abbreviated Ka I:1. This pottery occurs as far west as the offshore Åland islands of Finland, but was not adopted among the hunter-gatherers on the western shores of the Baltic.

In most of Sweden and Norway, ceramic technology was not introduced until millennia later.

The theme of this chapter is the Sperrings 1 pottery of the seal hunters in Åland. While the small archipelago of Åland at 5,000 cal BC could be characterized as 'tiny islands in a far sea', the hunter-gatherers living there should not be considered as living in isolation. Rather they lived on islands 'linked by the sea, not separated by it'. These quotes from Marshall Sahlins refer to islands of the Pacific, but have relevance to the conditions of the Baltic Sea of the 6th millennium cal BC.

The Sperrings 1 Pottery of the Seal Hunters of Åland

Åland is a group of islands in the middle of the Baltic Sea, situated in the strait separating central Sweden and southern Finland (Figure 13.1). Around 5,000 cal BC, most of present-day Åland was under the sea, leaving a group of small islands and skerries (Figure 13.2), separated by some 100 km of open water from the nearest islands on both the Finnish and Swedish side (Stenbäck 2003: 85). The islands were dominated by low bedrock outcrops, hewn smooth by glaciers of the ice age, and shingle, washed by the sea, that covered the islands following the melting of the ice. Today the Baltic Sea is brackish, but around 5,000 cal BC, there were rapid fluctuations in the salinity of the Baltic, ranging from fresh to brackish water, the latter still less saline than the present day conditions (Emeis et al. 2003). Åland is situated at 60° latitude, and while the summers are warm, the sea around Åland is partly covered by fast and drifting ice during winter. There do not seem to be estimates of ice conditions during the Stone Age, but the abundance of ringed seal at several of the Åland sites indicate ice-covered seas during winter, as ringed seal litter on the ice.

The ten or so known settlements with Sperrings 1 pottery on Åland represent the oldest known occupation of the islands. The finds on these sites are dominated by pottery, but also contain knapped ryolite, stone axes, polishing stones and bones from seal (Alhonen and Väkeväinen 1981; Andersson 1990; Dreijer 1941; Nuñez 1986; Nuñez and Storå 1992; Stenbäck 2003; Väkeväinen 1979). It is likely that seal hunting was one of the main reasons for people to visit these remote islands. The fauna assemblages from various Sperrings 1 sites are dominated by different seal species. At Östra Jansmyra, only ringed seal was found, while at Vargstensslätten, bones from grey seal dominated the assemblage (Väkeväinen 1979). In reference to Storå's discussion of seasonal seal-hunting patterns in the Baltic during the Middle Neolithic, this could suggest occupation during late winter at Östra Jansmyra and during summer at Vargstensslätten (cf Storå 2001: 31, 46). Osteological material from Sperrings 1 sites on mainland Finland also tend to be dominated

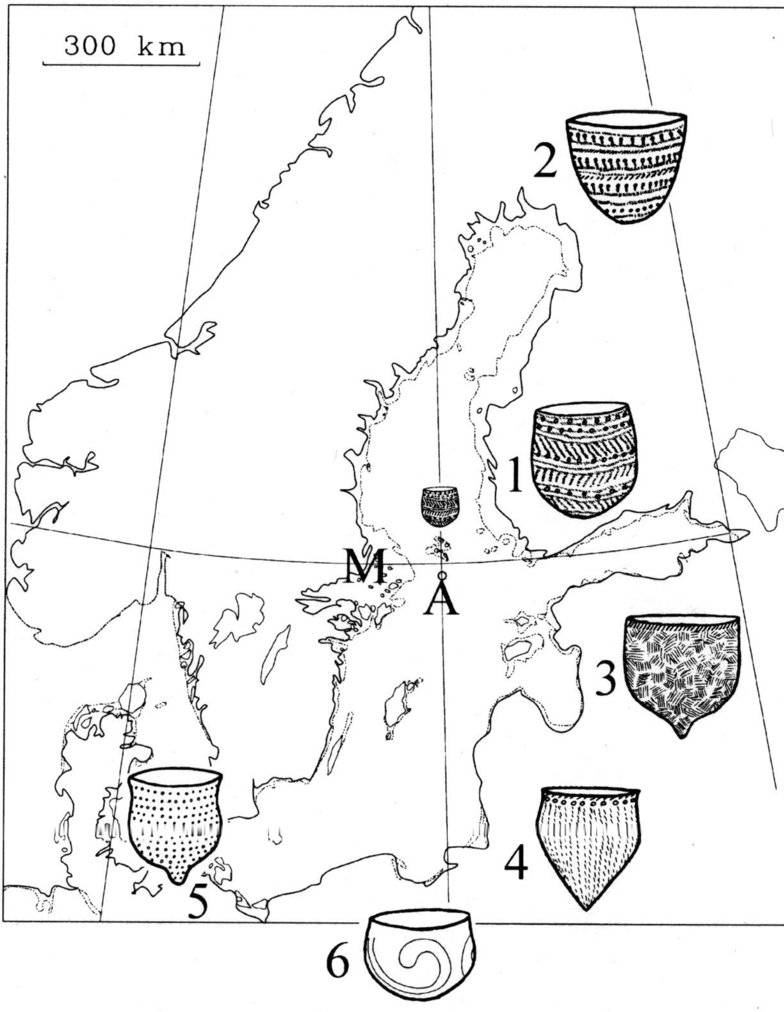

Figure 13.1 Early ceramic traditions around the Baltic Sea around 5000 cal BC. Legend: (M) Mälardalen; (Å) the Åland archipelago; (1) Sperrings 1 pottery; (2) Säräisniemi 1 pottery; (3) Narva pottery; (4) Neman pottery; (5) Ertebølle pottery; (6) Linear Band pottery (Säräisniemi 1 pottery has a further distribution beyond the map, cf figure 13.8). Shoreline map adopted from Munthe (1940).

by seal, but in addition, contain fish and terrestrial game. There are no traces of domesticated plants or animals at Sperrings 1 sites (Nuñez 1986: 25; Nuñez and Storå 1992: 147; Zvelebil 1993: 153). It is reasonable to suggest that the early sites on Åland were hunting stations, used by task-groups from the Finnish mainland (Alhonen and Väkeväinen 1981: 67; Nuñez 1994: 114;

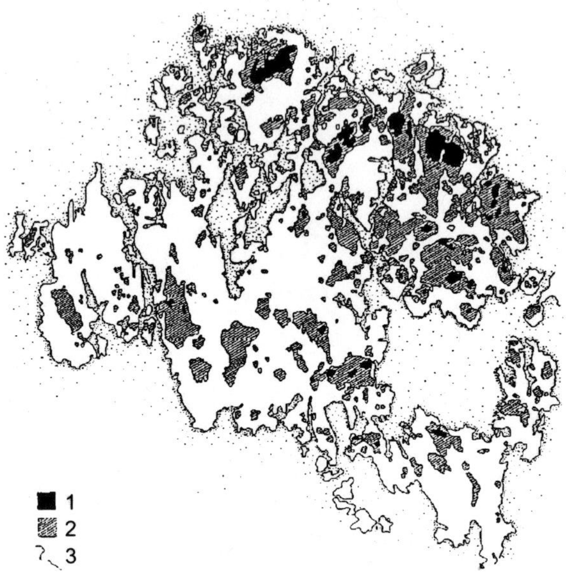

Figure 13.2 Åland during the Stone Age. Most known Sperrrings 1 sites are located on the ancient shores of the large island (3 km across) in the eastern part of the archipelago, close to the modern village Långbergsöda, Saltvik parish. Legend: (1) land at 6,000 BP; (2) land at 3,500 BP; (S) Saltvik parish. From Nuñez (1994), with modifications.

Welinder 1977). This interpretation finds support in the fact that the pottery is made of a clay that was not available on Åland. Thus, either the ceramic vessels, or the clay for pottery manufacture, were transported from the mainland to Åland (Alhonen and Väkeväinen 1981: 72).

A case study of two Sperrings 1 sites on Åland, Östra Jansmyra I and Vargstensslätten II included pottery (Hallgren 2004). The sites in question were located on one of the larger islands in the eastern part of the Stone Age archipelago (close to the modern village of Långbergsöda), Östra Jansmyra, sheltered by a small lagoon, and Vargstensslätten II, an exposed beach facing the open sea (Figure 13.3). Both sites contained what can be classified as typical Sperrings 1 pottery, sherds of unprofiled pots with a round or pointed bottom, that have surface covering decoration from rim to bottom (Alhonen and Väkeväinen 1981; Äyräpää 1956; Dreijer 1941; Edgren 1992; Väkeväinen 1979). The decoration is alternately applied in horizontal, vertical, or diagonal position, forming horizontal or vertical zones around the vessel (Figure 13.4). Common decorative elements are cord-stamp, drag and stab, incised lines, and various stamp impressions. As a rule, all vessels also have pit-impressions applied on top of the other decoration (Figure 13.5a–f). Some of the pottery from Vargstensslätten has

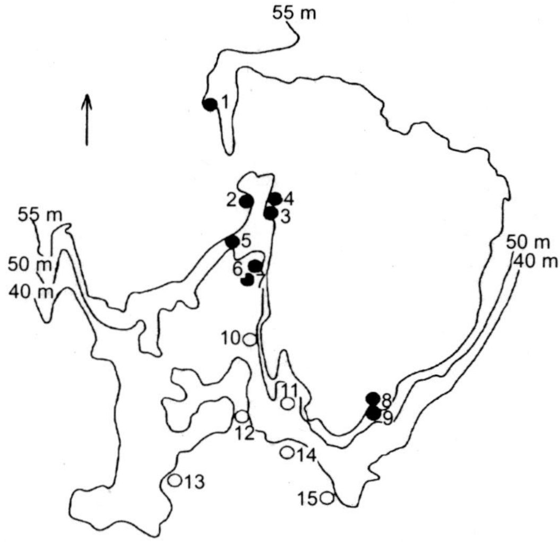

Figure 13.3 The Stone Age sites at Långbergsöda, Saltvik parish, Åland, related to ancient shore lines. The settlements discussed in this chapter are: (3) Östra Jansmyra I and (8) Vargstensslätten II. The distance between Östra Jansmyra I and Vargstensslätten II is approximately 1.5 km. Legend: black dots—Sperrings 1 and 2 sites (Ka I:1 and Ka I:2); circles—younger Comb Ware sites and Pitted Ware sites. From Alhonen and Väkeväinen (1981: 69), with modifications.

Figure 13.4 A Sperrings 1 vessel from Vargstensslätten II, Åland. From Alhonen and Väkeväinen (1981: 73).

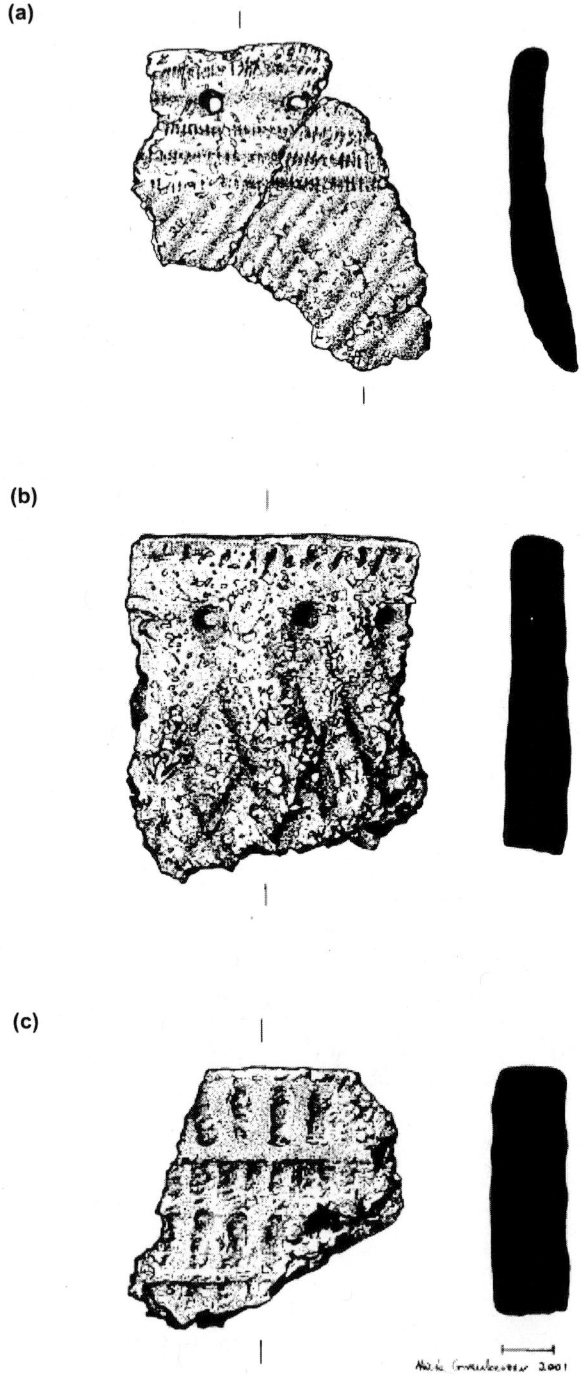
(a)
(b)
(c)

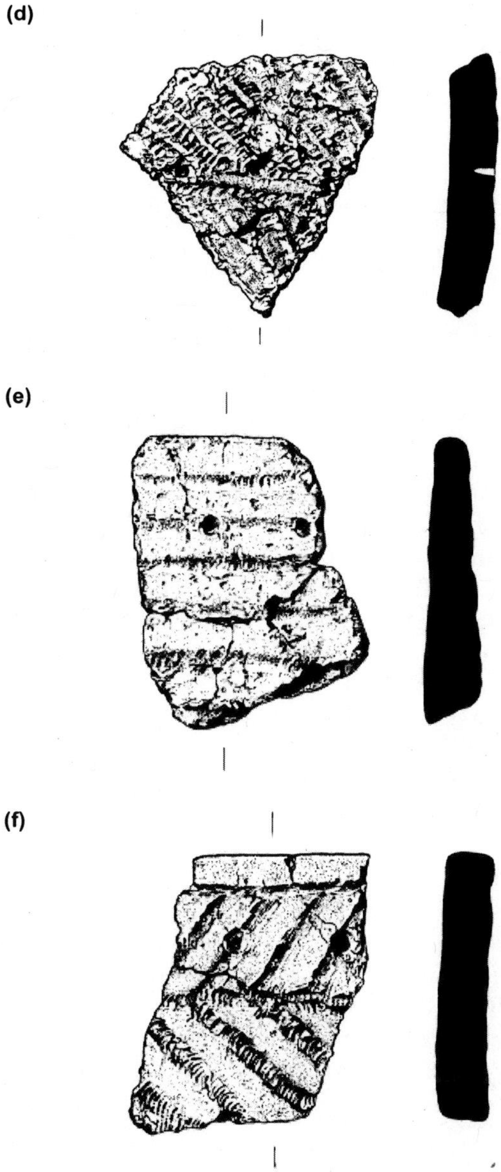

Figure 13.5 The 14C-dated Sperrings 1 sherds from Östra Jansmyra I (a–c) and Vargstensslätten II (d–f), Långbergsöda, Saltvik parish, Åland. (a) Östra Jansmyra I, Ua-17854 6,100±75 BP; (b) Östra Jansmyra I, Ua-17855 6,065±80 BP; (c) Östra Jansmyra I, Ua-17856 6,185±120 BP; (d) Vargstensslätten II, Ua-17857 5,990±90 BP; (e) Vargstensslätten II, Ua-17858 5,990±90 BP; (f) Vargstensslätten II, Ua-17859 6,165±75 BP. Drawings by Alicja Grenberger. (Scale 1:1).

been painted red, a practice also known from the eponymous site Sperrings (Europaeus 1927: 50). According to analysis performed by Alhonen and Väkeväinen, the vessels are made of Ancylus- or freshwater-clay from mainland Finland, in some cases perhaps with an addition of clay mud (Alhonen and Väkeväinen 1981). The clay has been tempered with crushed rock with a grain size of up to 10 mm. Sometimes, rather large quantities of temper have been used, as can be seen in the drawing of the weathered sherd in Figure 13.5b.

A palynological study has shown that the basin adjacent to the site Östra Jansmyra I became isolated from the sea around 6,000 bp due to isostatic uplift (Andrén et al. 1996). However, as the dated geological samples consisted of bulk sediment, the results are problematic when trying to correlate them to an archaeological chronology (cf Hedenström and Possnert 2001; Olsson 1986: 290). To confirm the age of the Sperrings 1 tradition in Åland, organic remains ('food-crusts') on six decorated potsherds were selected for ^{14}C dating, three each from the sites Östra Jansmyra I and Vargstensslätten II (Figure 13.5a–f). As the sites in question are seal-hunting camps, the organic remains are likely to contain marine products and the ^{14}C dates may therefore be affected by the marine reservoir effect. During the brackish late Litorina stage of the Baltic Sea (Middle Neolithic), the reservoir effect has been estimated to be about 100 years for the coast of nearby eastern central Sweden (Segerberg et al. 1991; cf Eriksson 2004: 151–152), while in the freshwater Ancylus stage (Early Mesolithic), the reservoir effect is considered to have been much lower (Lindqvist and Possnert 1997: 73, 1999: figure 17). As mentioned above, there were rapid fluctuations in the salinity of the Baltic Sea around 5,000 cal BC, ranging from fresh to brackish water, with the latter being less saline than present day conditions (Emeis et al. 2003). This suggests that the reservoir effect may have varied from close to zero to perhaps 100 years during this time. The ^{13}C values of the dated crusts range between –24.8 and –26.8 ‰ vs PDB, values that do not indicate a pronounced saline environment (cf Eriksson 2004: tables 3, 4; Katzenberg et al. 2000: 5–8; Lindqvist and Possnert 1999: 79; van Klinken et al. 2000: 47). Taken together, this suggests that the ^{14}C dates are fairly accurate. The datings gave the expected age of around 5,000 cal BC, in accordance with ^{14}C datings of Sperrings 1 sites on mainland Finland and in Russian Karelia (Figure 13.6).

Early Pottery Among Hunters and Farmers on the Eastern Shores of the Baltic Sea

Sperrings 1 is only one of the several ceramic traditions that appeared on the eastern shores of the Baltic Sea around or slightly before 5,000 cal BC

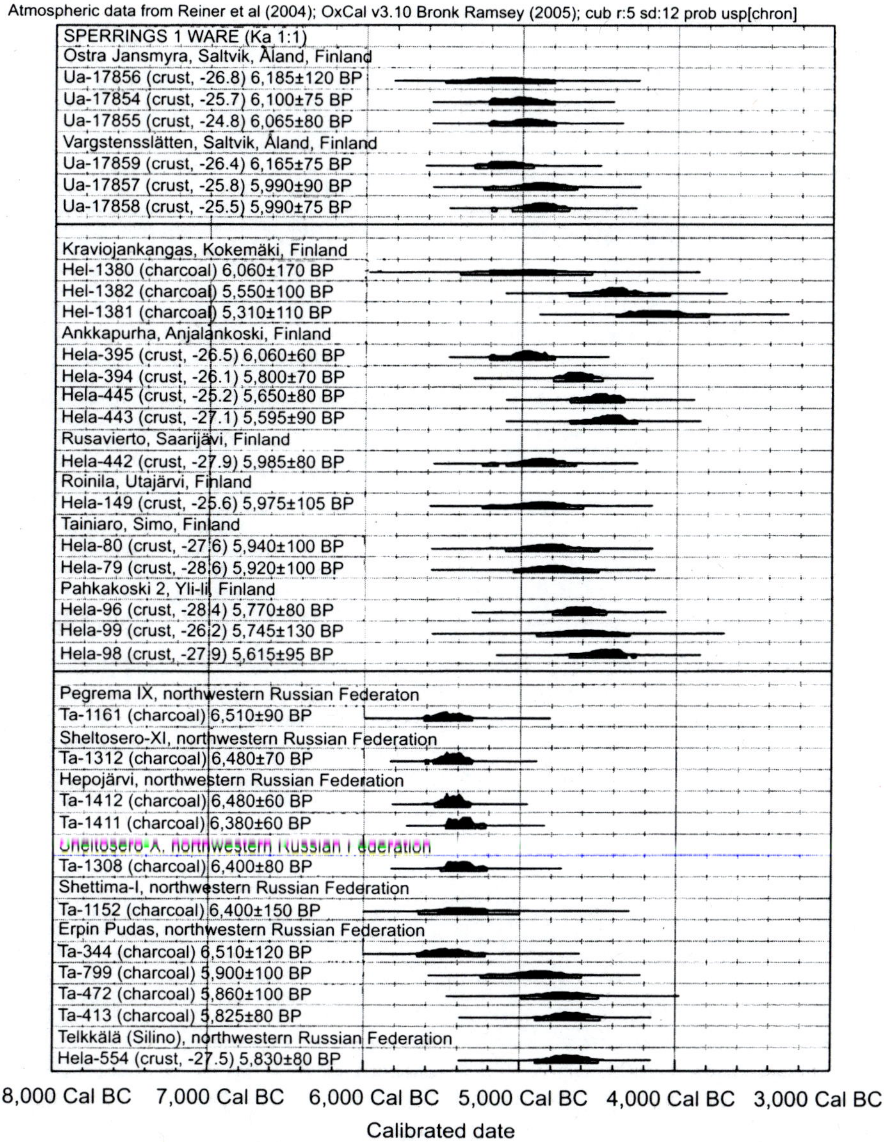

Figure 13.6 The 14C-dates from Östra Jansmyra I and Vargstensslätten II, Åland, compared to 14C-datings from Sperrings 1 contexts from mainland Finland, and from Karelia and Ingermanland in western Russia. (Dolukhanov et al. 2005; Leskinen 2002; Nuñez 1990; Schulz 2004; Takala and Sirviö 2003; Timofeev and Zaitseva 1997; Carpelan, personal communication 2005).

(Figure 13.1). Säräisniemi 1 pottery prevailed in the north, material very similar to Sperrings 1, differing in decoration (Carpelan 2004: 26–29; Skandfer 2003; Torvinen 2000). The traditions of Narva, Neman and Ertebølle prevailed in the south, all with more or less pointed bottom vessels (Brazaitis 2002; Černjawskij 1987; Gebauer 1995; Hulthén 1977, 1995; Ilkiewicz 1997; Jennbert 1984; Koch Nielsen 1987; Loze 1988; Persson 1999; Piličiauskas 2002; Timofeev 1998a, 1998b). Within both Narva and Ertebølle traditions, flat lamps or plates were found, displaying further similarity between these traditions. In contrast, the early Linear Band Pottery of the LBK farmers in northern Poland differed from all these variants of hunter-gatherer pottery in both vessel shape and style of decoration (Bogucki and Grygiel 1993: 413; Kirkowski 1990; for a more detailed description of each of these traditions, and fuller complete references, see Hallgren 2004).

Figure 13.7 shows a selection of ^{14}C dates from these early ceramic traditions around the Baltic Sea. Without going into specifics on each of the dated samples, some of the charcoal and wood dates in the graph may be slightly too old due to the age of the wood, and some of the organic crust dates may be affected by a reservoir effect (Hallgren 2004). Significantly, the graph suggests broad contemporaneity between the oldest phase of each of these different ceramic traditions. Within a rather short time span, ceramic technology was adopted in a region stretching from the Barents Sea in the north, to the southern Baltic in the south.

While it can be said that the large geographic scale of the phenomenon is remarkable, the fact that the spread of ceramic technology came to a grinding halt on the borders of the Scandinavian Peninsula is equally remarkable. The Ertebølle pottery characteristic of northern Germany, Denmark, and northwestern Poland, was also adopted by the hunter-gatherers of Scania, the southern tip of Sweden (Hulthén 1977; Jennbert 1984; Stilborg and Bergenstråhle 2001). In contrast, the hunter-gatherers of southern central Sweden did not adopt ceramics (Jennbert 1984: Figure 75; Price 2000: 265–269). In northern Fenno-Scandinavia, the same phenomenon can be observed (Figure 13.8). Säräisniemi 1 pottery is largely found in northern Finland and on the Kola Peninsula, with a northwestern limit on the southern shores of the Varangerfjord in northeastern Norway (Skandfer 2003; Torvinen 2000). Pottery was absent on the northern side of the same fjord, as well as further west and south in Norway (Glørstad 2005: 40–43; Skandfer 2005: 98). Another spatial limit in the distribution of Säräisniemi 1 ceramics is found in the Kalixälven river valley in northeasternmost Sweden (Halén 1994: 147–150; Torvinen 2000; Carpelan, personal communication). Pottery is absent in the neighbouring Luleälven river valley slightly further to the south. Still another sharp limit in the spread of early hunter-gatherer pottery is the strait between Åland and Mälardalen of eastern central Sweden.

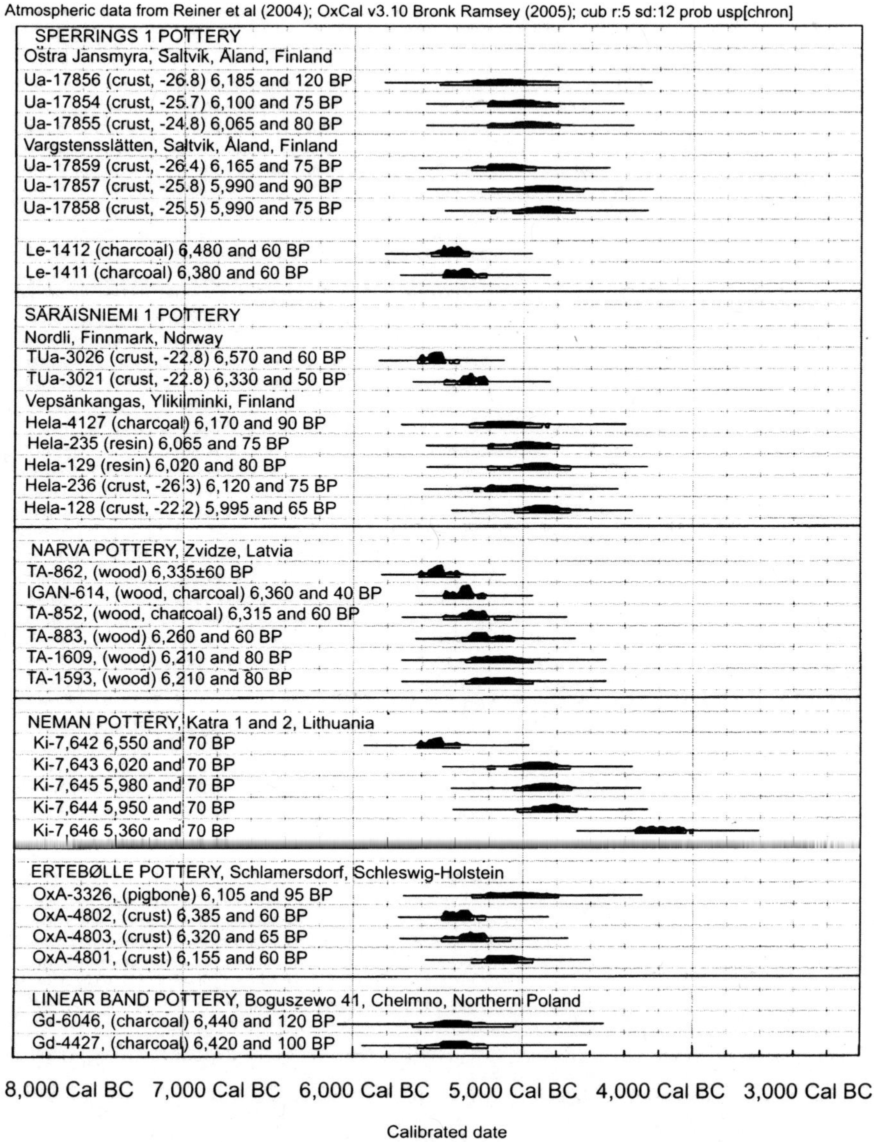

Figure 13.7 The 14C-dates from Östra Jansmyra I and Vargstensslätten II, Åland, compared to 14C-dates from different early ceramic traditions around the Baltic Sea (Dolukhanov et al. 2005; Hartz et al. 2000; Kirkowski 1990; Loze 1988; Persson 1999; Piličiauskas 2002; Skandfer 2003;Timofeev and Zaitseva 1997; Torvinen 2000; Zhilin 2000). [The cultural attribution of the pottery from Schlamersdorf and the validity of the associated 14C-dates are debated—Tomas Terberger, personal communication, August 2005.]

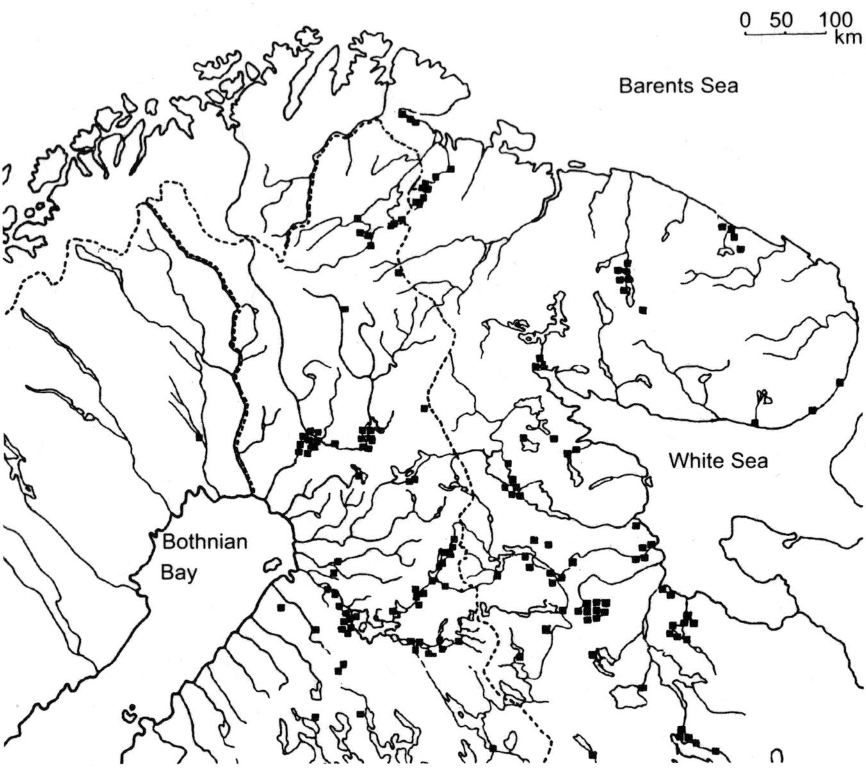

Figure 13.8 Find spots with Säräisniemi 1 pottery in northern Fenno-Scandinavia, and adjacent parts of northwestern Russia. (Note that 'the Bothnian Bay' is the northern part of the Baltic Sea.) The distribution of Säräisniemi 1 ceramics has a sharp western limit in the region between the Varangerfjord (Norway) in the north and the Kalixälven river valley (Sweden) to the south. Pottery was absent further west in northern Sweden and Norway. Map from Torvinen (2000: 32), with modifications.

The 'Aceramic' Hunter-Gatherers of Mälardalen

To the west of Åland, lies about 100 km of open water before the vast Stone Age archipelago of what is known today as the region of Mälardalen, in eastern central Sweden (Figure 13.1). The distance is equal to the strait that separated Åland from the closest islands on the Finnish side to the east (Stenbäck 2003: 85), and could be traversed by canoe in one day in fair weather (Colbing 2004: 104; Danielsson 1999; Hallgren 2003: 596). Just like their neighbours in Åland, the hunter-gatherers of Mälardalen were seal hunters and fishermen, using hunting stations in similar locations in the outer archipelago (Åkerlund 1996; Knutsson et al. 1999; Larsson et al. 1997). The material culture, on the other hand, differed substantially.

Pottery was absent in Mälardalen, and whereas the lithic assemblages on Åland were totally dominated by knapped ryolite, the contemporary sites in Mälardalen were dominated by quartz (Lindgren 2004).

In both Åland and Mälardalen, the ringed seal was an important prey during the Stone Age. In historical and recent times, this seal species was hunted in prolonged expeditions on the ice of late winter. Even though ice may cover the whole of the northern Baltic Sea, the hunters were dependent on boats tugged along on the ice, as cracks and open leads of water were frequent, and the ice could break up or start drifting at any moment (Sundfeldt and Johansson 1964). Drifting ice can travel at a speed of 20–30 km a day. Once the ice started drifting, the hunters would frequently end up on a foreign shore. In 1551 AD, seal hunters from the northern coast of Finland (then part of Sweden) petitioned the Swedish King, complaining that when they drifted with the ice to Sweden or Åland, the locals would not let them continue to hunt seal. The King graciously granted them right to hunt seal wherever the ice took them (Sundfeldt and Johansson 1964: 33).

During the 6th millennium BC, the seal hunters would have to settle such negotiations. Long distance exogamous marriage connections, and associated visiting and feasting, is one common way to cement alliances between different groups in tribal societies (e.g., Chernela 1993; Fuglestvedt 1999: 29–31; Lawrence 1994; Walthall and Koldehoff 1998: 268–269; Wiessner 2001). The importance of 'intertribal' marriages has been stressed by Moore, in a study based on Boas' unpublished population statistics of American Indians (Moore 2001). Moore's study presents marriage statistics of 9,200 individuals from 52 different tribes, pertaining to the period 1888–1903. The average frequency of intertribal marriage in this sample was 5% (marriages with Europeans not included). While Moore emphasized that such a high figure has consequences for the understanding of the spread of culture, language and genes, this is by no means a high rate of intertribal marriage—the corresponding figure for parts of the Amazon where 100% tribe (language-group) exogamy is practised. (Chernela 1993).

Of the 471 intertribal marriages in Moore's sample, most (64%) connected spouses from neighbouring tribes of the same language family though 'most often...languages that were not mutually intelligible' (Moore 2001: 43), but marriages between neighbouring tribes of different language families were also frequent (26%). The more distant marriages connected spouses of the same—as well as of different—language families (6% and 4% respectively). Moore concludes that 'these data show a continual, significant movement of people, and hence their genes, language, and culture, from society to society across hundreds and thousands of miles of territory' (Moore 2001: 43). However, while these kinds of contacts may have facilitated the *possibility* of the spread of cultural practices like craft traditions, it is not self-evident that these traditions were accepted in the new social and cultural setting.

Identity in Cultural Practice

How we live, what we do, defines who we are (Wenger 1998: 145). The activities we take part in as members of a community, shape and create our identity:

> We become who we are by being able to play a part in the relations of engagement that constitute our community. Our competence gains its value through its very partiality. As an identity, this translates into a form of individuality defined with respect to community. It is a certain way of being part of a whole through mutual engagement (Wenger 1998: 152).

Material culture, artefacts, and the practice of the crafts that produce them, plays a key role in the formation of this identity (Wenger 1998: 184–187). While material culture is created within a social context, the design of artefacts, and the practice of their production, is also a means to connect to the history of that context (Lave and Wenger 1991: 101). By participation in the practice of a craft, the individual becomes a part of that practice, and part of the community that reproduces it: 'We recognize the history of a practice in the artefacts, actions, and language of a community. We can make use of that history because we have been part of it and it is now part of us, we do this through a personal history of participation' (Wenger 1998: 153). However, our identity is not only formed by what we do, but also what we do not do: 'We not only produce our identities through the practices we engage in, but we also define ourselves through practices we do not engage in. Our identities are constituted not only by what we are but also by what we are not' (Wenger 1998: 164). Therefore, while there may have been, and probably were, frequent contacts across the Baltic Sea between members of different communities, with a different cultural repertoire (cf Zvelebil 2006), that need not necessarily have led to the further spread of pottery craft.

This point can be illustrated by a quote from Kroeber, describing the far-reaching travels by the Mohave Indians of the lower Colorado River, travels that took them as far as to the Yokuts of central California, and the Chumash of the Californian coast.

> Visits carried parties of Mohave as far as the Chumash and Yokuts. Sheer curiosity was their main motive, for the Mohave were little

> interested in trade. They liked to see lands, timidity did not discourage them, and they were as eager to know manners of other peoples as they were *careful to hold aloof from adopting them....* These journeyings brought with them friendships and alliances as well as enmities (Kroeber 1925: 727, emphasis added).

So why did the hunter-gatherers of Mälardalen not adopt pottery craft? Because they were not people that practised the craft of pottery!

Paraphrasing the quote from Marshal Sahlins at the start of this chapter, we have to imagine the hunter-gatherers with Sperrings 1 pottery on Åland—or the hunter-gatherers of the archipelago of Mälardalen without pottery, for that matter—not as imprisoned on tiny islands in a far sea, but as people living in a world that was 'anything but tiny', as a people who 'thought and recounted their deeds in epic proportions', and that 'lived in great associations of islands linked by the sea...not separated by it.' (Sahlins 2000: 180). Incidentally, the communities thus linked, at times practised quite different traditions, cultural differences that were not necessarily diminished by the exposure to alternatives (Bergsvik 2003; Jenkins 1997: 77; Jones 1997: 95).

Acknowledgements

Thanks to Christian Carpelan, Helsinki University, who has provided extensive comments on my previous paper on the Sperrings 1 of Åland (Hallgren 2004).

References

Åkerlund, A (1996) *Human Responses to Shore Displacement. Living by the Sea in Eastern Middle Sweden during the Stone Age*, Stockholm: Riksantikvarieämbetet

Alhonen, P and Väkeväinen, L (1981) 'Diatom analytical studies of Early Comb Ware vessels from Åland', 1980 *Suomen Museo*, 67–77

Andersson, J (1990) 'Undersökning av ett skadat område på stenåldersboplatsen Östra Jansmyra i Saltvik Långbergsöda Fornlämning Saltvik 202 Långbergsöda', unpublished report, Mariehamn: Ålands museum

Andrén, E, Veski, S and Storå, J (1996) 'Biostratigraphical investigations in connection with archaeological dwelling sites at Jansmyra, northeastern Åland', 50 *Pact*, 355–365

Äyräpää, A (1956) 'Den yngre stenålderns kronologi i Finland och Sverige', LXII *Finskt Museum*, 5–52

Bergsvik, K A (2003) 'Mesolithic ethnicity—too hard to handle?' in Larsson, L, Kindgren, H, Knutsson, K, Loeffler, D and Åkerlund, A (eds), *Mesolithic on the Move,* Oxford: Oxbow Books, pp 592–599

Bogucki, P and Grygiel, R (1993) 'The first farmers of Central Europe: a survey article', 20 *Journal of Field Archaeology*, 399–426

Brazaitis, D (2002) 'Narviðkos keramikos stiliai rytø Lietuvoje', 23 *Lietuvos Archeologija*, 29–40

Carpelan, C (2004) 'Environment, archaeology and radiocarbon dates. Notes from the Inari region, northern Finnish Lappland', 13 *Iskos*, 17–45

Černjawskij, M M (1987) 'The Neolithic in north-western Byelorussi', in Wiślańskiego, T (ed), *The Neolithic and Early Bronze Age in the Chelmno Land*, Toruń: Uniwersytetu Mikołaja Kopernika w Toruniu, pp 429–439

Chernela, J (1993) *The Wanano Indians of the Brazilian Amazon, A Sense of Space*, Austin: University of Texas Press

Colbing, S (2004) *Skärgårdsliv*, Stockholm: Wahlström & Widstrand

Danielsson, J (1999) *148 Dagar i Havskajak. En Paddlares Ensamfärd i 6.379 Kilometer*, Stockholm: Bokförlaget T Fischer & Co

Dolukhanov, P, Shukurov, P, Gronenborn, D, Sokoloff, D, Timofeev, V, Zaitseva, G (2005) 'The chronology of Neolithic dispersal in Central and Eastern Europe', 32 *Journal of Archaeological Science*, 1441–1458

Dreijer, M (1941) 'Ålands äldsta bebyggelse', 1940 *Finskt Museum*, 1–66

Edgren, T (1992) 'Den förhistoriska tiden', in Edgren, T and Törnblom, L (eds), *Finlands Historia 1*, Esbo: Schildts

Emeis, K-C, Struck, U, Blanz, T, Kohly, A and Voss, M (2003) 'Salinity changes in the central Baltic Sea (NW Europe) over the last 10,000 years', 13(3) *The Holocene*, 411–421

Eriksson, G (2004) 'Part-time farmers or hard-core sealers? Västerbjers studied by means of stable isotope analysis', 23(2) *Journal of Anthropological Archaeology*, 135–162

Europaeus, A (1927) 'Stenålderskeramik från kustboplatser i Finland', XXXVI *Finska Fornminnesföreningens Tidskrift*, 45–77

Fuglestvedt, I (1999) 'Interregional contact in the Late Mesolithic: the productive gift extended', in Boaz, J (ed), *The Mesolithic of Central Scandinavia*, Oslo: Universitetets Oldsaksamling, pp 27–38

Gebauer, A B (1995) 'Pottery production and the introduction of agriculture in southern Scandinavia', in Barnett, W K and Hoopes, J W (eds), *The Emergence of Pottery Technology and Innovation in Ancient Societies,* Washington DC: Smithsonian Institution Press, pp 99–112

Glørstad, H (2005) 'Tangen—En neolittisk boplass fra Kragerø kommune i Telemark. Noen betraktninger omkring boplassens kulturmiljø og traktbegerkulturens vestgrense', 2005 *Viking*, 25–54

Halén, O (1994) *Sedentariness during the Stone Age of Northern Sweden,* Stockholm: Almqvist & Wiksell International

Hallgren, F (2003) 'My place or yours?' in Larsson, L, Kindgren, H, Knutsson, K, Loeffler, D and Åkerlund, A (eds), *Mesolithic on the Move,* Oxford: Oxbow Books, pp 592–599

——— (2004) 'The introduction of ceramic technology around the Baltic Sea in the 6th millennium', in Knutsson, H (ed), *Coast to Coast—Arrival, Results and Reflections,* Uppsala: Uppsala University, pp 123–142

Hartz, S, Heinrich, D and Lübke, H (2000) 'Frühe bauern an der küste. Neue ^{14}C-daten und aktuelle aspekte zum neolithisierungsprozess im Norddeutschen Ostseeküstengebiet', 75 *Praehistorische Zeitschrift*, 129–152

Hedenström, A and Possnert, G (2001) 'Reservoir ages in Baltic Sea sediments—a case study of an isolation sequence from the Litorina Sea stage', 20 *Quaternary Science Reviews*, 1779–1785

Hulthén, B (1977) *The Ceramic Technology during the Scandinavian Neolithic and Bronze Age,* Stockholm: Stockholm University

——— (1995) 'Ceramics artefacts—a key to ancient society', in Vincenzini, P (ed), *The Ceramics Cultural Heritage*, Faenza: Techna, pp 1–7

Ilkiewicz, J (1997) 'From studies on Ertebølle type cultures in the Kozalinian coastal area (Dabki 9, Koszalin-Dzierzecino 7)', in Król, D (ed), *The Built Environment of Coast Areas during the Stone Age,* Gdańsk: Regional Centre for Studies and Preservation of Built Environment, pp 50–65

Jenkins, R (1997) *Rethinking Ethnicity: Arguments and Explorations*, London: Sage Publications

Jennbert, K (1984) *Den Produktiva Gåvan*, Malmö: Liber Förlag

Jones, S (1997) *The Archaeology of Ethnicity: Constructing Identities in the Past and Present*, London: Routledge

Katzenberg, M A, Saunders, S R, Abonyi, S (2000) 'Bone chemistry, food and history: a case study from 19th century Upper Canada', in Ambrose, S H and Katzenberg, M A (eds), *Biogeochemical Approaches to Paleodietary Analysis*, New York: Kluwer Academic/Plenum, pp 1–22

Kirkowski, R (1990) 'Boguszewo, gmina Gruta, województwo Toruńskie, stanowisko 41, obiekty 3 i 5', in Jankowskiej, D (ed), *Z Badań nad Chronologią Absolutną Stanowisk Neolitycznych z Ziemi Chełmińskiej,* Toruń: Uniwersytetu Mikołaja Kopernika w Toruniu, pp 9–14

Knutsson, K, Lindgren, C, Hallgren, F and Björck, N (1999) 'The Mesolithic in eastern Central Sweden', in Boaz, J (ed), *The Mesolithic of Central Scandinavia*, Oslo: Universitetets Oldsaksamling, pp 87–123

Koch Nielsen, E (1987) 'Ertebølle and Funnel beaker pots as tools', 57 *Acta Archaeologica*, 107–120

Kroeber, A L (1925) *Handbook of the Indians of California,* Washington DC: Smithsonian Institution Press

Larsson, M, Lindgren, C and Nordqvist, B (1997) 'Regionalitet under Mesolitikum—från senglacial tid till senatlantisk tid i Syd- och Mellansverige', in Larsson, M and Olsson, E (eds), *Regionalt och Interregionalt Stenåldersundersökningar i Syd- och Mellansverige*, Stockholm: Riksantikvarieämbetet, pp 13–55

Lave, J and Wenger, E (1991) *Situated Learning Legitimate Peripheral Participation*, Cambridge: Cambridge University Press

Lawrence, D (1994) 'Customary exchange across the Torres Strait', 34(2) *Memoirs of the Queensland Museum*, 241–446

Leskinen, S (2002) 'The late Neolithic house at Rusavierto', in Ranta, H (ed), *Huts and Houses: Stone Age and Early Metal Age Buildings in Finland,* Helsinki: National Board of Antiquities, pp 147–169

Lindgren, C (2004) *Människor och Kvarts*, Stockholm: Stockholm University

Lindqvist, C and Possnert, G (1997) 'Om reservoareffekt-problemet', in Burenhult, G (ed), *Ajvide och Den Moderna Arkeologin*, Falköping: Natur och Kultur, pp 73–74

Lindqvist, C and Possnert, G (1999) 'The first seal hunter families on Gotland. On the Mesolithic occupation in the Stora Förvar cave', 7 *Current Swedish Archaeology*, 65–87

Loze, I (1988) *Poseleniia Kamennogo Veka Lubanskoi Niziny: Mezolit, Rannii i Srednii Neolit*, Riga: Zinatne

Moore, J H (2001) 'Ethnogenetic patterns in native North America', in Terrell, J E (ed), *Archaeology, Language, and History*, Westport: Bergin & Garvey, pp 31–56

Munthe, H (1940) *Om Nordens, Främst Balticums Senkvartära Utveckling och Stenåldersbebyggelse*, Stockholm: Kungliga Svenska Akademien

Nuñez, M (1986) 'Om bosättningen på Ålandsöarna under stenåldern', *Åländsk Odling*, 13–28

——— (1990) 'On Subneolithic pottery and its adoption in Late Mesolithic Finland', 7 *Fennoscandia Archaeologica*, 27–52

——— (1994) 'Discovery and occupation of the Finnish Stone Age archipelago in the 4th millennium bc', XXXVII *Kontaktstencil*, 109–124

Nuñez, M and Storå, J (1992) 'Shoreline chronology and economy in the Åland archipelago 6,500–4,000 BP', 36 *Pact*, 143–161

Olsson, I U (1986) 'Radiometric dating', in Berglund, B E (ed), *Handbook of Holocene Palaeoecology and Palaeohydrology*, Chichester: Wiley, pp 273–312

Persson, P (1999) *Neolitikums Början. Undersökningar Kring Jordbrukets Introduktion i Nordeuropa*, Göteborg: University of Göteborg

Piličiauskas, G (2002) 'Dubičių tipo gyvenvietės ir Neolitinė Nemuno kultūra pietų Lietuvoje', 23 *Lietuvos Archeologija*, 107–136

Price, T D (2000) 'The introduction of farming in northern Europe', in Price, T D (ed), *Europe's First Farmers*, Cambridge: Cambridge University Press, pp 260–300

Sahlins, M (2000) 'Sentimental pessimism and ethnographic experience; or, why culture is not a disappearing object', in Daston, L (ed), *Biographies of scientific objects*, Chicago: University of Chicago Press, pp 158–202

Schulz, E-L (2004) 'Ankkapurhan arkeologisen aineiston radiohiiliajoitukset', in Uino, P (ed), *Ammoin Ankkapurhassa. Kymenlaaksossa Kivikaudella*, Hesinki: Museovirasto, p 46

Segerberg, A, Possnert, G, Arrhenius, B and Lidén, K (1991) 'Ceramic chronology in view of ^{14}C datings', 5 *Laborativ Arkeologi*, 83–91

Skandfer, M (2003) 'Tidlig, nordlig kamkeramikk. Typologi, kronologi, kultur', unpublished PhD thesis, University of Tromsø, Tromsø

——— (2005) 'Fra eldre til yngre steinalder? Kronologiske og begrepsmessige utfordringer sett fra Nord-Norge', 8 *Primitive Tider*, 97–105

Stenbäck, N (2003) *Människorna vid Havet. Platser och Keramik på Ålandsöarna Perioden 3,500–2,000 f.Kr.*, Stockholm: Stockholm University

Stilborg, O and Bergenstråhle, I (2001) 'Traditions in transition. A comparative study of the patterns of Ertebølle lithics and pottery. Changes in the Late Mesolithic ceramic phase at Skateholm I, III, and Soldattorpet in Scania, Sweden', 6 *Lund Archaeological Review*, 23–41

Storå, J (2001) *Reading Bones. Stone Age Hunters and Seals in the Baltic*, Stockholm: Stockholm University

Sundfeldt, J and Johansson, T (1964) *Färdmän från isarna*, Stockholm: Forum

Takala, H and Sirviö, T (2003) 'Telkkälä, Muolaa—a multi-period dwelling site on the Karelian Isthmus', XX *Fennoscandia Archaeologica*, 55–78

Timofeev, V I (1998a) 'The east-west relations in the late Mesolithic and Neolithic in the Baltic region', in Kosko, A (ed), *Beyond Balkanization*, Poznan: Adam Mickiewicz University, pp 44–58

——— (1998b) 'The beginning of the Neolithic in the eastern Baltic', in Zvelebil, M, Domańska, L and Dennell, R (eds), *Harvesting the Sea, Farming the Forest. The Emergence of Neolithic Societies in the Baltic Region*, Sheffield: Sheffield Academic Press, pp 225–236

Timofeev, V I and Zaitseva, G I (1997) 'Some aspects on the radiocarbon chronology of the Neolithic cultures in the forest zone of the European part of Russia', 11 *Iskos*, 15–22

Torvinen, M (2000) 'Säräisniemi 1 Ware', XVI *Fennoscandia Archaeologica*, 3–35

Väkeväinen, L (1979) 'Nya rön angående den äldre stenåldern på Åland', 1978 *Åländsk Odling*, 49–52

van Klinken, G J, Richards, M P and Hedges, R E M (2000) 'An overview of causes for stable isotopic variations in past European human populations: environmental, ecophysiological, and cultural effects', in Ambrose, S H and Katzenberg, M A (eds), *Biogeochemical Approaches to Paleodietary Analysis*, New York: Kluwer Academic/ Plenum, pp 39–63

Walthall, J A and Koldehoff, B (1998) 'Hunter-gatherer interaction and alliance formation: Dalton and the cult of the long blade', 43 *Plains Anthropologists*, 257–273

Welinder, S (1977) 'Åländsk fångststenålder', 1976 *Åländsk Odling*, 46–58

Wenger, E (1998) *Communities of Practice: Learning, Meaning, and Identity*, Cambridge: Cambridge University Press

Wiessner, P (2001) 'Of feasting and value. Enga feasts in a historical perspective (Papua New Guinea)', in Dietler, M and Hayden, M (eds), *Feasts: Archaeological and Ethnographic Perspectives on Food, Politics and Power*, Washington DC: Smithsonian Institution Press, pp 115–143

Zhilin, M (2000) 'Chronology of the transition from the Mesolithic to the Neolithic in the forest zone of eastern Europe', 19 *Lietuvos Archeologija*, 287–297

Zvelebil, M (1993) 'Hunters or farmers? The Neolithic and Bronze Age societies of north-east Europe', in Chapman, J and Dolukhanov, P M (eds), *Cultural Transformations and Interactions in Eastern Europe*, Aldershot: Avebury, pp 147–162

——— (2006) 'Mobility, contact, and exchange in the Baltic Sea basin 6,000–2,000 BC', 25(2) *Journal of Anthropological Archaeology*, 178–192

Chapter 14

The Pitted Ware Culture in Eastern Middle Sweden: Material Culture and Human Agency

Mats Larsson

Introduction

The focus of this chapter is to develop a more social, symbolic and agency-oriented history of the rich ceramic Pitted Ware Culture (PWC) traditions of the Middle Neolithic period in eastern Middle Sweden. Although this is a much-discussed archaeological phenomenon, the PWC has tended to be regarded as an unchanging and homogeneous category of material culture. These ceramics were produced over large areas of southern Scandinavia. This established, but static perspective, on the long-term PWC technological tradition is increasingly open to criticism as it has tended to ignore the dynamic roles of those who created, used, exchanged and understood these vessels at more local and community-based scales.

In order to move away from these more established explanations of the PWC, it seems appropriate to develop more resolutely social and symbolic insights into the role of PWC as a system of material communication that was reproduced and transformed by the negotiated intervention of human agency. The argument here is that through socially embedded reproduction of the PWC ceramic craft, wider public meanings and interpretations were constantly negotiated and contested. Through reproduction and transformation in this long-term technological practice, individuals and groups were able to maintain the means for communication through material culture despite profound and cumulative changes in other domains of practice and perception.

From this it is possible to develop a social, symbolic and communicative perspective on ceramic 'practice' by focusing on the site of Åby in eastern Sweden where repeated excavations have yielded large volumes of PWC ceramics (Figure 14.1). Detailed analyses have revealed that rather than being homogeneous and uniform, these ceramic assemblages exhibit distinct characteristics, which include repeated decorative compositions employing 'hanging triangles'.

An Overview of Research into Pitted Ware Culture

The point of departure in this chapter is to present a general overview of the much-discussed PWC craft as a cultural phenomenon manifest in time and space (i.e., a regional overview) and then present a summary review of the history of research into PWC. Briefly, attention is drawn to the diverse

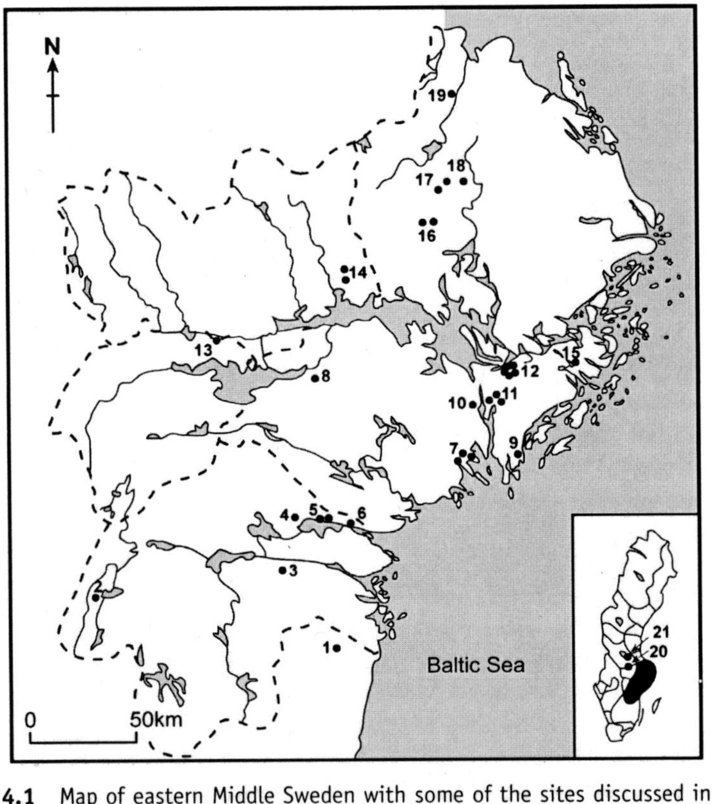

Figure 14.1 Map of eastern Middle Sweden with some of the sites discussed in the text marked: (1) Tryserum; (2) Alvastra; (3) Össby; (4) Åby; (5) Svintuna, Fagervik; (6) Säter; (7) Djupvik, Överråda; (8) Moäng; (9) Nynäshamn; (10) Kvedesta; (11) Kyrktorp, Korsnäs; (12) Häggsta; (13) Körartorp; (14) Fågelbacken, Äs; (15) Återvall; (16) Åloppe; (17) Sotmyra, Vadbro; (18) Tibble; (19) Torslunda; (20) Korsnäset; (21) Orsand (after Edenmo et al. 1997).

models that have been used to explain the emergence of the PWC. These accounts can be broadly divided into (a) those that understand the origins of PWC as an outcome of the convergence of two ethnic groups, and (b) that variations in the material culture reflect social variations within a single large-scale ethnic entity.

One of the most challenging themes in Scandinavian Stone Age archaeology over the last few decades has been the origins, inter-relations and cultural affinities of three archaeological 'cultures': Funnel Beaker Culture (FBC), Battle Axe Culture (BAC) and Pitted Ware Culture (PWC).

Despite these extended debates, the precise forms, social, symbolic and economic contents, and the relationships between these cultures during the Middle Neolithic in southern Sweden remains a question of baffling complexity. For example, the three cultures have been argued to represent the cultural manifestations of quite different economic strategies: the farming economy of the FBC has been contrasted with the hunting/gathering economy of the PWC. However, the PWC has alternatively been seen as a hunting/gathering aspect of the FBC in the west or as a homogeneous hunting society in the east (Edenmo et al. 1997). It has also been seen as regression towards a hunter-gatherer economy, with pottery-producing farmers reverting to an economic life of foragers (Browall 1991).

Here, I propose to review these debates with the aim of uncoupling the history of PWC ceramics from the economic history of the region and discussing the social history of the period.

The Historical Perspective on Pitted Ware Culture Research

Studies of PWC on the Swedish mainland started early in the last century with the discovery by Oscar Almgren (1906) in 1901 of the site Åloppe in Uppland (eastern Middle Sweden). This was followed by the discovery of the classic site Säter (Nerman 1911). The culture thus discovered initially had the name 'the East Swedish settlement culture' (Östsvenska boplats kulturen). The decoration of the pottery has always been an important part of the chronology of the PWC and this was the case from the beginning (Bagge 1938, 1951; Bagge and Kjellmark 1939). Early on, the PWC was also divided into porous and coarse ware (Edenmo et al. 1997: 169). When the hypothesis of the regressive change in the shore displacement was first formulated, this was a major breakthrough in the discussion concerning chronology. Based on these new studies, Nerman (1927) developed the Säter chronology. The site is situated on the shores of Slätbaken in the county of Östergötland. The division into stages Säter II–IV was in use until Axel Bagge's investigations in 1935–1936 at Fagervik, close to Säter.

The older definitions of the PWC include variables like material culture, economy and grave form. Stone technology and axes (Meinander 1957; Welinder 1978), the amount of flint (Welinder 1973a) and funerary rites with the use of red ochre (Wyszormirska 1984) have all been used. Several different traits have been utilized from single artefact types like arrowheads or ceramics. In southern Sweden and Denmark, its definition has been based on arrowheads and cylindrical blade cores. It is necessary, however, to avoid stressing these cultural markers too strongly, since the same types of flint and stone artefacts are found in both the FBC and the PWC.

In eastern Middle Sweden, where we find most of the large settlement sites, the definition has been based on ceramics, divided into the five Fagervik stages (Bagge 1951; Edenmo et al. 1997) (Figures 14.2 and 14.3). Axel Bagge based his chronology on the shore displacement curve and added one early and one late stage, Fagervik I and V respectively (Bagge 1951). The last stage belongs to the BAC. It is important to remember that the correlation between Säter and Fagervik is not complete. Säter II is actually a mixture of Fagervik I and II and Säter IV partly corresponds to Fagervik IV–V (Bagge 1951: 62, 72). Bagge believed the above to be different settlement stages and that each was characterized by a specific pottery assemblage and shore displacement (Bagge 1938: 151, 1951: 60, 64). According to Bagge, they did not actually correspond to different styles of decoration, instead they should be viewed as heterogeneous. For example, he stated that the earliest Fagervik stages (I–II) belonged to the Funnel Beaker Culture and its middle Swedish equivalent the Vrå Culture (1951: 84–95). Fagervik I displays a whole range of features that are shared with the FBC like cord decoration. Later, Stig Welinder (1976) also viewed Fagervik I as a transitional stage between the FBC and PWC. Bagge (Bagge 1951: 97–98; Bagge and Kjellmark 1939: 179) had already discussed the difference between Fagervik II and III. These differences have led to the conclusion that Fagervik II is to be seen as precluding the 'real PWC', that is Fagervik III (Becker 1954: 84–86; Ebbesen 1975: 200–201; Table 14.1).

Central to the discussion is the Alvastra pile dwelling in the district of Östergötland in eastern Middle Sweden. A megalithic grave is also situated not far from the pile dwelling (Janzon 1984), and is associated with Middle Neolithic pottery. The same kind of pottery was also found inside the pile dwelling (Browall 1991: 114; Malmer 2002: 102). Objects of flint are much more frequent here than on other PWC sites in central Sweden. There is no great change in the economy. The people practised farming and kept cattle and pigs. Interestingly enough, the bone assemblage is dominated by cattle followed by pig (Malmer 2002: 109). They also hunted, especially red deer. The transformation from FBC to PWC seems to have been gradual and peaceful, not connected to any economic changes in other parts of eastern Sweden, for example on the island of Gotland where recently published data

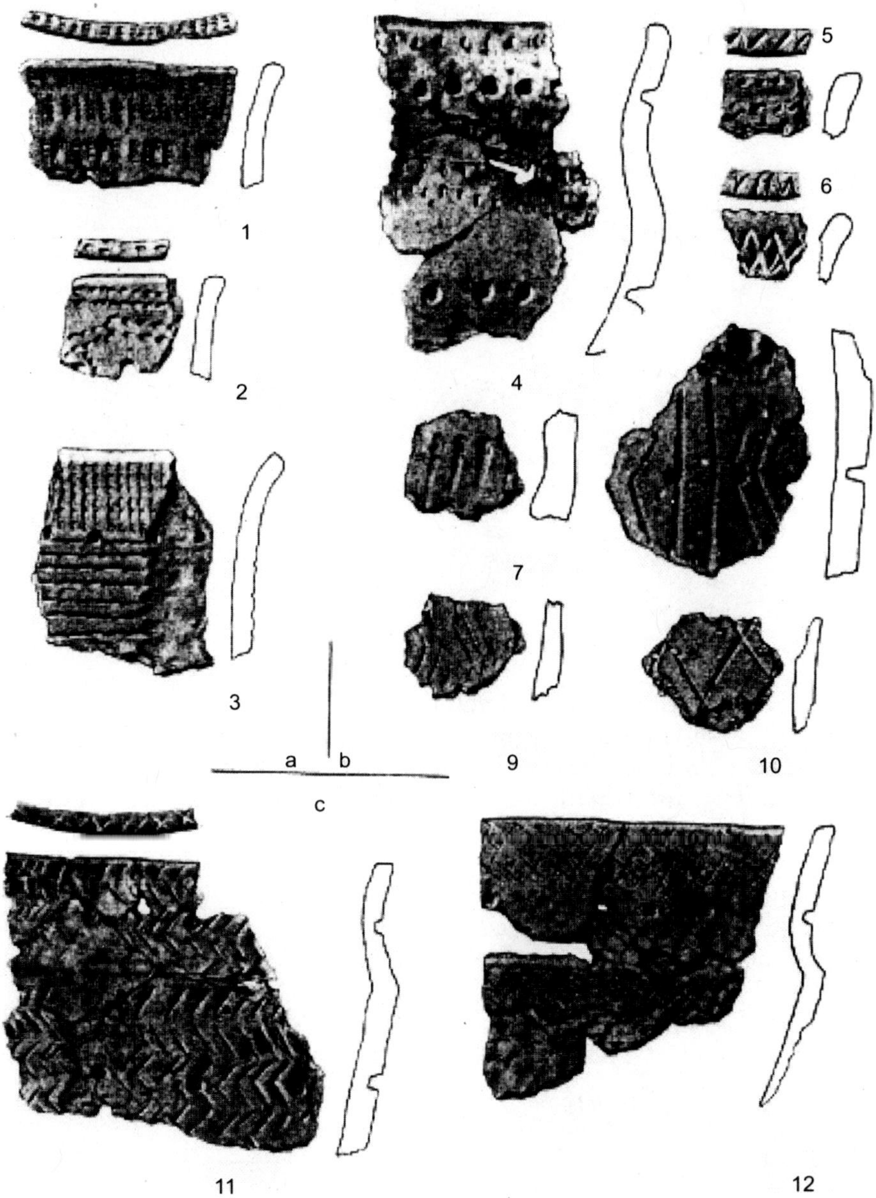

Figure 14.2 Characteristic decoration from the Fagervik stages: (a) Fagervik I; (b) Fagervik II; (c) Fagervik III (after Edenmo et al. 1997) (1:2).

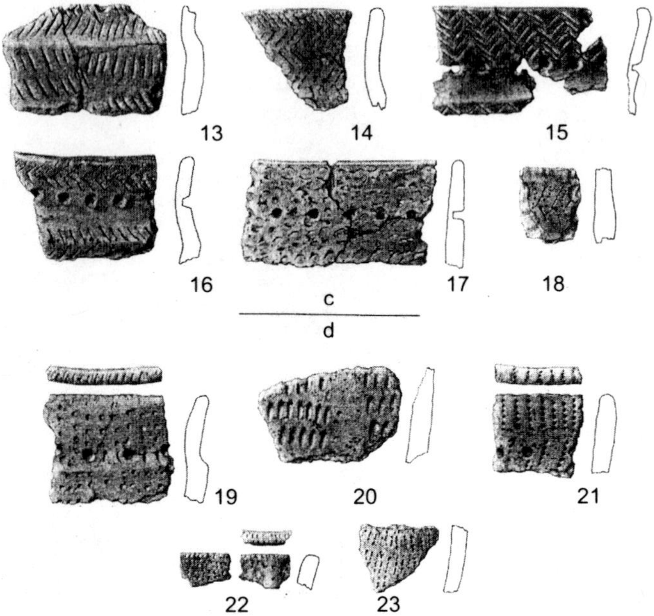

Figure 14.3 Sherds from (a) Fagervik III; (b) Fagervik IV (after Edenmo et al. 1997) (1:2).

from Västerbjers clearly shows that the people here were almost completely dependent on seals for food (Eriksson 2004: 31).

Niklas Stenbäck does not see the development of the PWC as a reaction against the FBC and what it stood for. He sees the PWC as a return to old hunter-gatherer values, even if they do not entirely abandon farming and husbandry (Stenbäck 2003: 194). The pits on the pottery could have been a way of dealing with these changes; a sort of 'de-neolithization'. What we see is actually different economic and ceramic histories.

Alexander Gill (2003) has a completely different view of how the PWC developed. In about the same way as Anders Carlsson (1998), he stresses the importance of the beaches. He sees these as crucial for our understanding of

Table 14.1 Simplified table showing the different cultures and their chronology, economy and geographical distribution.

Culture	Chronology	Timespan BC	Economy	Geographic region
FBC (MN A)	MN I-V	3300-2700	Farming	S. Scandinavia
BAC (MN B)	1-6	2700-2350	Farming	Scandinavia
PWC	Fagervik I-V	3300?-2700	Hunting/gathering (evidence for farming)	S. Scandinavia/ E. Sweden

the PWC. Gill discusses the change from the FBC 'gardens' to the 'beaches'. During the Middle Neolithic the 'beaches' were used for funerary rites, and they were not actual settlements (Gill 2003: 129–134). Lars Sundström (2003) discusses the development of the PWC from an Early Neolithic perspective and views the PWC as an outcome of social stress in the FBC. He interprets the PWC as an answer to a discrepancy between the social ideal and the real situation (Sundström 2003: 217).

From the above, it is obvious that it is very hard to actually define the PWC. Mats Malmer (1973: 60) once defined the PWC as the Scandinavian Neolithic culture that does not fit the definitions for Ertebølle, FBC, BAC and late Neolithic culture. To avoid a discussion of what constitutes 'culture', terms like 'societies' or 'tradition' have been proposed (Andersson 2003: 288, 2004; Edenmo et al. 1997; Gustafsson et al. 2003: 6).

Recently Strinnholm (2000: 123) thoughtfully propounded: 'There might not be a story to tell about the Pitted Ware Culture, but there are many important tales to tell about the Middle Neolithic' (translation by author).

Towards a Social History of the PWC: A Case Study of the Åby Settlement

During the last couple of decades, some of the most influential works regarding the Neolithic have been written in the UK (Barrett 1994; Bradley 1998, 2000, 2002; Hodder 1982, 1990; Thomas 1991, 1996; Tilley 1996). These works have had a huge impact in Sweden. However, one of the problems relating to these works is that they are mostly interpretative, sometimes lacking any deeper connection to the archaeological material. The writers do not base their interpretations on the use of large amounts of pottery or chronological/stylistic studies. This does not in any way diminish their importance, but it goes to show how difficult dealing with pottery is. This shows the necessity for sharpening arguments, in this case based on a study of the Åby site in eastern Middle Sweden, to develop a social history of the PWC (Figure 14.4).

The site is highly significant because of the many detailed excavations undertaken over the years; the evidence from this site is ideal for relocating pottery into a more social and symbolic practice-based community context. The Åby site has, during the last couple of years, been discussed in detail (Larsson 1999, 2003, 2004, 2006) so only a short summary of the site and the main results will be discussed.

This large Neolithic site is situated to the north of the town of Norrköping, eastern Middle Sweden. The site was discovered in 1928 and an area of 200 x 40 m in which pottery as well as flint and stone implements was delimited. The height above sea level varies between 26 and 29 m. Axel

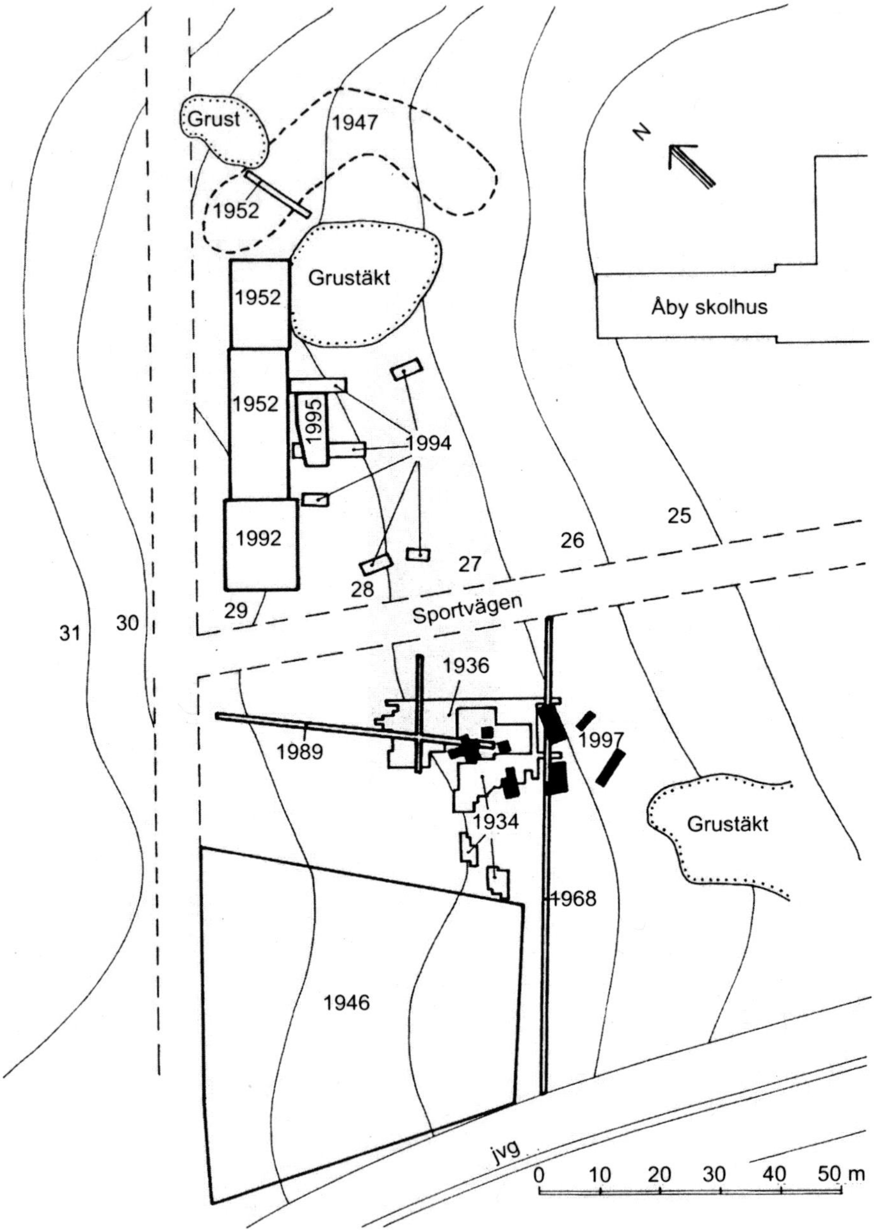

Figure 14.4 The site Åby with all the different excavations marked.

Bagge undertook the first comprehensive excavation between 1934 and 1936. Accounts of these have never been fully published, and accordingly, it is necessary to rely on his site reports. According to Bagge, the finds included mostly pottery of Säter III, together with smaller numbers of Säter II and IV.

Except for possibly two hut structures, Axel Bagge also found two probable graves. They both had an oval shape but no skeletal parts were found, just some flint flakes. During the following years, a large number of mainly small excavations were carried out (Einarsson 2002; Gruber and Larsson 1999; Larsson 1999, 2006).

The Investigations in 1995 and 1997

During the spring of 1995, a 63 sq m area was excavated (Gruber and Larsson 1999; Larsson 2004). Beneath the culture layer, features such as pits, post-holes and what has been interpreted as a sunken house structure (A3) were found. A small oblong pit (A21) has been interpreted possibly as a child's grave. The largest part of the find material is pottery; the number of sherds found in the small area amassed to almost 12,000 and weighed approximately 46 kg. Most of the ceramic material is made up of PWC typical chalk tempered pottery. The amount of decorated pottery is 41% (Figure 14.5).

A closer look at the decoration on the pottery reveals that the PWC characteristic motifs prevail. The most common motif is pits of various sizes (43%) as well as herringbone motifs (7.5%) using tooth stamp technique. Short vertical tooth stamp imprints (6%) are also found directly on the rim. Due to the fragmentary state of the pottery, it is difficult to see how the different motifs combine, but it is obvious that pits are important. Pits are found together with other motifs on 48% of the pottery. The rim decoration is something that proved to be important in the interpretation of the site (see below). This element was significant when compared to the results of Bagge's and later excavations (Einarsson 2002; Gruber 1995: 30; Larsson 2003: 126, 2004, 2006).

Compared to the amount of pottery, stone, quartz and flint make up a minor part of the material. The weight of the stone material is almost 9.5 kg. Most of the flint is made up of debris but a couple of tanged arrowheads, a transverse arrowhead, scrapers and parts of ground flint axes were found. In the stone material, evidence for both thick-butted and thin-butted axes was found.

In 1997, an additional 69 sq m large area was excavated. Eight trenches were dug. The trenches were mostly situated in the area where Axel Bagge's excavations were undertaken in the 1930s. The ultimate aim was to delimit the settlement area to the south and also try and connect to Bagge's old trenches.

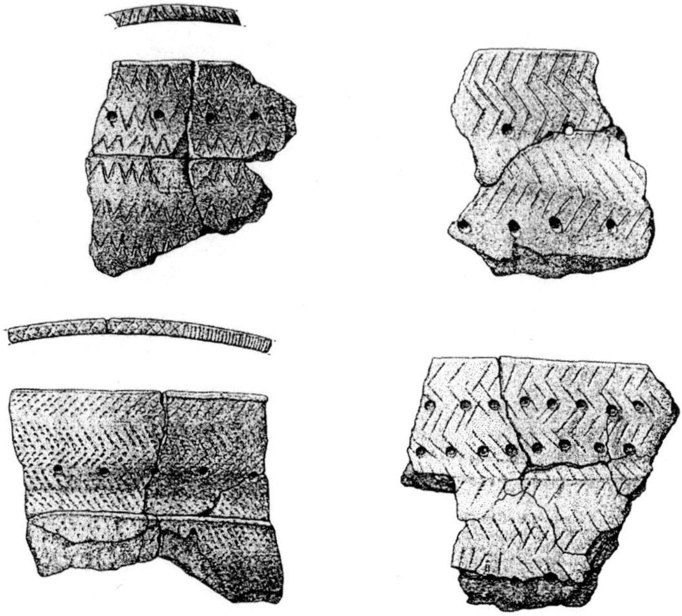

Figure 14.5 Sherds from Åby (R Holmgren) (1:2).

A large amount of finds were located, for example almost 100 kg of pottery. As was the case with the excavation in 1995, the amount of stone and flint was much smaller, with a few typical flint arrowheads of different sizes and types, all blades. Scrapers as well as some fragments of thick-butted and thin-bladed axes were found. In a rather delimited area, both burnt and unburnt bone were found and human bone as well as animal bone was identified. Together with the bone material, some fire-damaged fragments of flint axes were identified, some fragments of double-edged battleaxes and a couple of hollow-edged small chisels. These were made from either greenstone or volcanic rock.

The dating of the Åby site is somewhat problematic. Based on the distribution of the different types of decoration, it is obvious that the site ought to be dated to the stage Fagervik III according to Axel Bagge's scheme (1951). As discussed above, the large number of radiocarbon dates obtained during the last decade clearly shows that the five stages are more or less contemporary. The radiocarbon values from the Åby site itself clearly show this. Until now, six dates have been obtained, all based on charred residue inside pottery. These are placed in the interval between 3,600 cal BC and 2,040 cal BC (Larsson 2006). All values, except one, come from decorated sherds (Table 14.2). The decoration on the pottery may be placed in the Fagervik III with some traces of Fagervik IV elements.

Table 14.2 Radiocarbon dates from the site Åby (calibration after Stuiver et al. 1993).

Lab nr	Date BP	cal BC 1 sigma
Ua-7459	4550+/-100	3495-3052
Ua-7460	4455+/-95	3343-2920
Ua-7461	4665+/-60	3620-3357
Ua-7462	4430+/-60	3305-2921
Ua-9451	4105+/-60	2880-2490
Ua-9166	3800+/-85	2400-2040

One of the sherds has a 'hanging triangle motif (Figure 14.6). This also points to the latest date.

The site could be interpreted in many ways: chronologically different, one large settlement, several smaller ones over a wider area or small sites that were contemporaneous. In the interpretation here, the latter seems the most likely explanation. The evidence may not be that clear, considering the radiocarbon dates, but during the excavations and the following

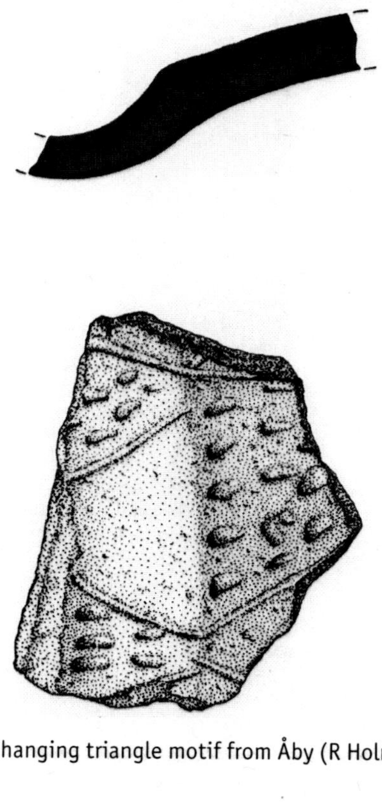

Figure 14.6 Sherd with hanging triangle motif from Åby (R Holmgren) (1:1).

analysis it was possible to delimit the site. Through analysis of the ceramics, at least two, maybe three, settlements were clearly detected in the area. People lived for several centuries at the site, they built different types of huts and they buried their dead. Although it is impossible to be precise, it seems each settlement continued for a long time. There was no incentive to move because they lived in an ecologically favourable area. They had the sea, small inland lakes, a lot of sandy soil for small-scale agriculture—if they actually practised this, which is subject to debate—and they kept pigs. The people who lived on the site had long-distance contacts judging from the flint and quartzite. The flint probably came in the shape of complete axes and arrowheads (Figure 14.7). They made highly decorated pottery that was used for storage and cooking. The size of the pots differed, from miniature pots to very big ones. Evidence from the excavation in 1997 clearly indicated that they were placed in pits in the ground. From the same excavation, there is also evidence for what has been interpreted as a ritual

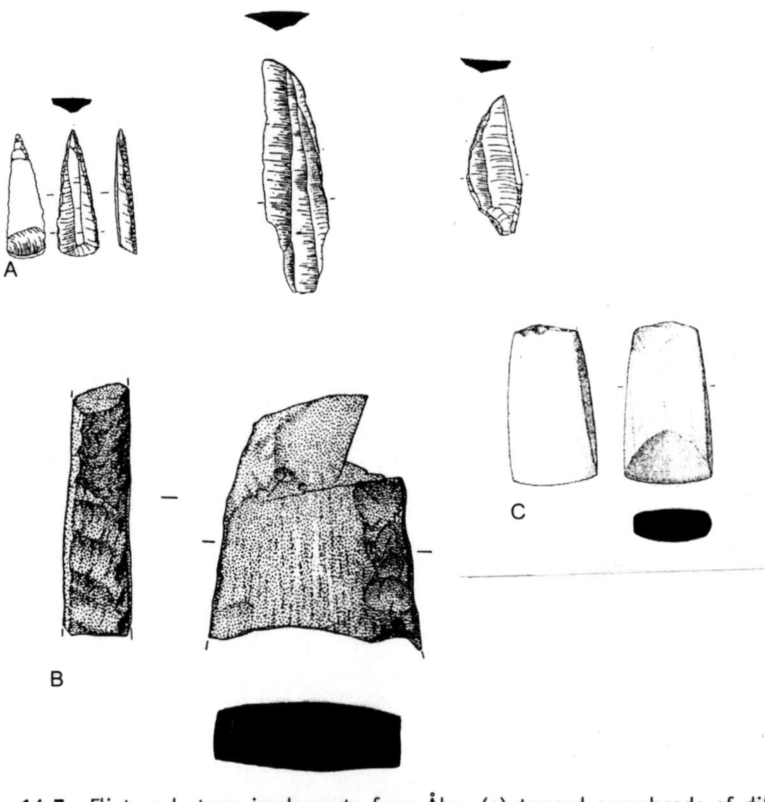

Figure 14.7 Flint and stone implements from Åby: (a) tanged arrowheads of different types; (b) thin-bladed flint axe, fire damaged; (c) chisel from green stone (R Holmgren) (1:2).

area in which human bone, both cremated and uncremated, was found. In this area several complete arrowheads and axes, as well as burned flint, were recovered.

The evidence shows that we have a complex settlement area with examples of different aspects of life in the Middle Neolithic. People lived their lives with different aspects of the material culture. They shaped their own collective and individual identity and their history evolved over the centuries.

Hanging Triangles, Seals and Human Agency

It is now possible to develop an interpretation of these assemblages. The aim is to move away from dry definitions of large-scale 'cultures' and to explore the routines and contents of daily lives—in other words to explore how time, place and material conditions gave form, content and meaning to ancient lifeways. Understanding transformation in these lives and associated conditions has become the defining theme of a new generation of archaeological research that concerns itself with the relationships between day-to-day practice and the reproduction and transformation of technological traditions, social relations and symbolic meanings.

In using the concept of agency (cf Barrett 2000), it is possible to focus on the concept of culture as meaningfully constituted interactions. The concept of agency entered the repertoire of archaeological theory largely as a reaction to the perceived sterility of the functional and adaptive perspectives on long-term culture change advocated by New Archaeology. Agency/practice theory concerns itself with the ways in which cultural forms are reproduced through routinized actions. Agency is the means by which objectives are achieved and theories of agency recognize that men and women are situated in webs of symbolic meanings and make choices, hold intentions and take action to realize their goals. Agents are people whose practices and forms of engagement serve as the driving force, keeping things in check, moving and/or changing (Dobres 2000: 133). As discussed by John Barrett, several agency studies take their basic analytical framework from Giddens' structuration theory—a society should be seen as a group of individuals that are part of a social network and through this have recurrent, and in time, lasting relations. Bourdieu's *habitus* is more about how social practices seem to be ordered across time and space (Barrett 2001: 148–149).

Interest in agency has, in the last couple of years, intensified in at least four distinct areas of archaeological inquiry: gender, material culture, phenomenology and inequality (Dobres and Robb 2000: 7–13). Both Dobres and Barrett are careful not to see just the *individuals* as agents but stress that agents can be a collective (Barrett 2000: 61; Dobres 2000: 133).

It is also important to stress, as John Barrett (2000: 61–62) has stated, that action, time, space and agency work together and carry the past into the future. This is *habitus* at work. In both this and a previous article (Barrett 1999: 29), he introduces the concept of inhabitation or inhabiting a landscape that involves an understanding of the landscape through earlier experiences. In the following discussion, the focus will be on the collective but including the individual, as Alasdair Whittle (2003) has done in a recently published work. In a chapter entitled 'Difficult Individuals', he discusses the individual in the context of the community and household (Whittle 2003: 50–77). In a case study based on Hungarian settlements and cemeteries from the later Neolithic, he suggests that variation in this period could be examined as the effect of particular individuals or at least small-scale social groupings (Whittle 2003: 63). This notion, together with John Barrett's (2000: 67) writings some years ago regarding how the structuring principles which secured a community's identity were worked out and eventually changed, are important issues in this interpretation. In other words, how group identities were constructed was crucial. In the following section, the concern for the identity of larger groups and its mediation and expression via material culture will be elaborated.

Recently (Larsson 2004), ethnicity was discussed using a scheme by Jones (1997) that proved useful in an interpretation of the changes in the Middle Neolithic. Ethnic identity is, according to her, that aspect of a person's self-conceptualization which results from identification with a broader group in opposition to others, based on perceived cultural differentiation and/or common descent. The concept of ethnicity has, over the years, of course been strongly debated. Lynn Meskell (2001: 189–213), in a review of the discussion, writes that there has never been a consensus in terminology, and ethnicity has been used to denote the individual versus the group. Ethnic identity, she writes, is only one social determinate, that can cut across status, occupation, gender etc and that allows contact between groups (Meskell 2001: 189). In this context it is important to remember, as Knut Bergsvik (2003: 300) firmly states, that ethnicity is a socially constructed phenomenon primarily concerned with how people think about themselves as groups and how they set themselves apart from other groups. It is a social construction of 'us' and 'them' which is marked in cultural terms.

Julian Thomas (1999: 96) has argued, in a highly influential work, that to shift or hold on to a specific style of pottery must be seen as very important in social reproduction during the Neolithic. Highly important for social reproduction, the *habitus*, were ritual expressions. The way in which objects are placed in the ground and/or water, have lately been called structured deposits by some archaeologists (Bradley 2000; Thomas 1991). Richard Bradley (2000: 122) has recently suggested that the placing of material in the ground invoked a whole series of references to the origin of the object. The flint brought to the site came from far away and was immersed in almost mythical history in which myths, stories about ancestors and events long ago were brought together. At sites like Åby, Häggsta, Bollbacken and Siretorp, there is evidence for what have been interpreted as ritual activities (Larsson 2006). At Häggsta and Bollbacken, evidence for buildings were found (Artursson 1996). At Åby, there is no clear picture of how the items were deposited but the combination of human bone, axes, arrowheads and decorated pottery makes it possible to interpret the area as a site of formal deposition.

Niklas Stenbäck (2003) has proposed an interesting aspect of a change in structuring principles. In a study of the PWC at Åland in the Baltic, he stresses the importance of the sea and a 'seal ideology' in the same way as Jan Storå (2001) does. Marek Zvelebil (2003: 1–26), in a very stimulating article, connects animals like elk with shamanism and discusses the very long time span during which these symbols were in use. Human agency played an important role in this as ritual specialists were used for guidance and connection with non-humans (Zvelebil 2003: 20). Liliana Janik (2003: 116) discusses interestingly the possibility of different animals as markers of cultural identity. That the importance of different animals changed in the PWC is of great significance in the context of this study.

Through ritual, people could achieve power and influence. It is important to remember that people, through deliberate changes and manipulation of symbols as knowledgeable social actors, through contact and innovation, provided the *agency* for incorporating innovation within the existing traditions (Zvelebil 2003: 4). We can see these ritual-structured depositions as a way for a group of people to handle changes in society. In turning to the past and linking up with the ancestors, ritual was a means for the people to adjust to a new situation. In this way, a community's identity was secured through agency.

An important site in this context is the Alvastra pile dwelling in eastern Sweden. This was initially an FBC ritual site of the earliest Middle Neolithic (Browall 2003; Malmer 2002). These people also built at least one megalithic grave in close vicinity to the pile dwelling. The site metamorphosed into a PWC ritual centre. As discussed above, the economy did not change at all and neither did the actual rituals performed there. Farming and cattle breeding was still very important. Why did this change occur? It would seem that Alvastra is a good example of agency. Humans made their own history, individual and communal contacts and influences eventually changed at least the pottery designs but nothing much else. To elaborate further, material culture items were used by agents, both individuals and small-scale societies, to keep society going and at the same time make people feel secure.

Briefly summing up this interpretative framework, the main concern here is how people used material culture to make individual choices as a reaction against the communities' standard norms. These were human actions that each generation learnt. At the same time, they learnt how to behave in culturally specific ways from the example set by previous generations and thus linked to the ancestors as well. The people of the PWC used routinized tasks like ritual depositions both as a way of handling changes and invoking a new identity. In doing this, they connected back to their ancestors but they also looked forward. They expressed certain cultural values that belonged to history in the making. Pottery was produced, used and sometimes broken and deposited. People re-enacted the production of pottery and the eventual destruction of it. By inscribing certain types of decoration—both individual and communal—on the walls of the pots, these became part of their society with a meaning that was clear to everyone. Thus, a biography of the pot was created but also of society as a whole. Production, use and deposition were implicated in the generational reproduction of different forms of identity.

Returning to archaeology, the Åby site and the ceramics, what was the role of ceramics in the Middle Neolithic? From the arguments, it is clear that ceramics had many different roles depending on how and where it was used, and probably also, by whom. The pottery of the PWC was highly decorated and ought to have been very important in different social circumstances as well as being used in various historical settings. Indeed, the history of the PWC might be a good example of how group identity changed. Why, how and via what mechanisms is, of course, the issue here.

As discussed above, it is obvious that during the last hundred years or so, the view of the PWC has changed massively. Today, the focus is the perspective on material culture which has been developed above—not dry categories but artefacts as 'symbols in action'. In routines, ritual and symbolic contexts, the use of material culture is an expression, outcome and medium via which *habitus* was re-enacted over time.

Returning to the site of Åby in eastern Sweden, the site can be seen as two more or less contemporary settlements. Chronologically, there is probably no great difference between the two areas. Thorough analyses of the pottery have shown that there is a marked difference in the way that rim decoration was used between the two areas. At the site excavated in 1952 and 1995, rim decoration of type 85 (vertical lines on the rim \\\\\) clearly dominates. This type of rim decoration is present only on very few sherds in the area excavated in the 1930s and in 1997 (Einarsson 2002: figures 4–8; Gruber 1995: 30; Larsson 2004). Within this interpretation, people might show their social belonging, or ethnic individuality, using different patterns of ceramic decoration.

Hanging Triangles, Seals and Human Agency

The title of this section might seem enigmatic but the following explanation should show what, in particular, hanging triangles have to do with the wider narrative of the social history of the PWC.

In the pottery from the Åby site, the decoration has another element which, though rare, is very interesting in this context—the 'hanging triangles' (Figure 14.6). The 'hanging triangle' motif is rather late in the chronological sequence in south Scandinavia and is common in the late Funnel Beaker groups like Stävie and Karlsfält as well as the Globular Amphorae culture (Edenmo et al. 1997: 141–152) (Figures 14.8 and 14.9). One sherd with this motif from Åby has been radiocarbon dated (2,880–2,490 cal BC (Ua-9451) and is in good accordance with other dates from the period (Larsson 2006).

The use of this kind of motif could be said to be the visible manifestation of memory that is linked to the use of it on the surface of the pot (Jones 2002: 131). Andrew Jones (2002: 130) also writes, based on studies of Grooved Ware pottery at Barnhouse on Orkney, that social relations are inscribed on the surface of the vessel. The production of Grooved Ware pottery at the Barnhouse settlement was also closely allied to the expression of settlement-specific identities and to the consumption of cattle products (Jones and Richards 2003: 48). John Chapman (2000: 187), referring to late Neolithic Hungary, writes that different households are each dependant upon varied inter-community and some extra-community relations and made strategic use of material/symbolic resources to construct their own social order. In this way, public meanings and interpretations were negotiated and contested through human agency. In times of rapid change, people used, re-invented and re-used both new and old principles while still being able to communicate through the medium of material culture.

In this context, the 'hanging triangles' have their place. Individuals or groups of people re-enacted both old and new structuring principles, respecting some of the old ones. John Barrett (2000: 67) wrote some years ago that agents, as they move through time and space and given structural conditions, modify existing material culture. What Barrett is actually doing is to emphasize the distinction between daily routines and what people actually thought and valued. Structure, in this context, is a set of social rules and resources located outside space and time, existing only as memory traces (Whittle 2003: 11).

The 'hanging triangles' enabled people to connect with each other in disparate regions. In other words, they were the visible signification of a social change. In this context Bergsvik's (2003: 298) comment on ethnicity is noteworthy: 'In a context of ethnicity, one might argue that the more people have in common in terms of values, technology or subsistence practice, the

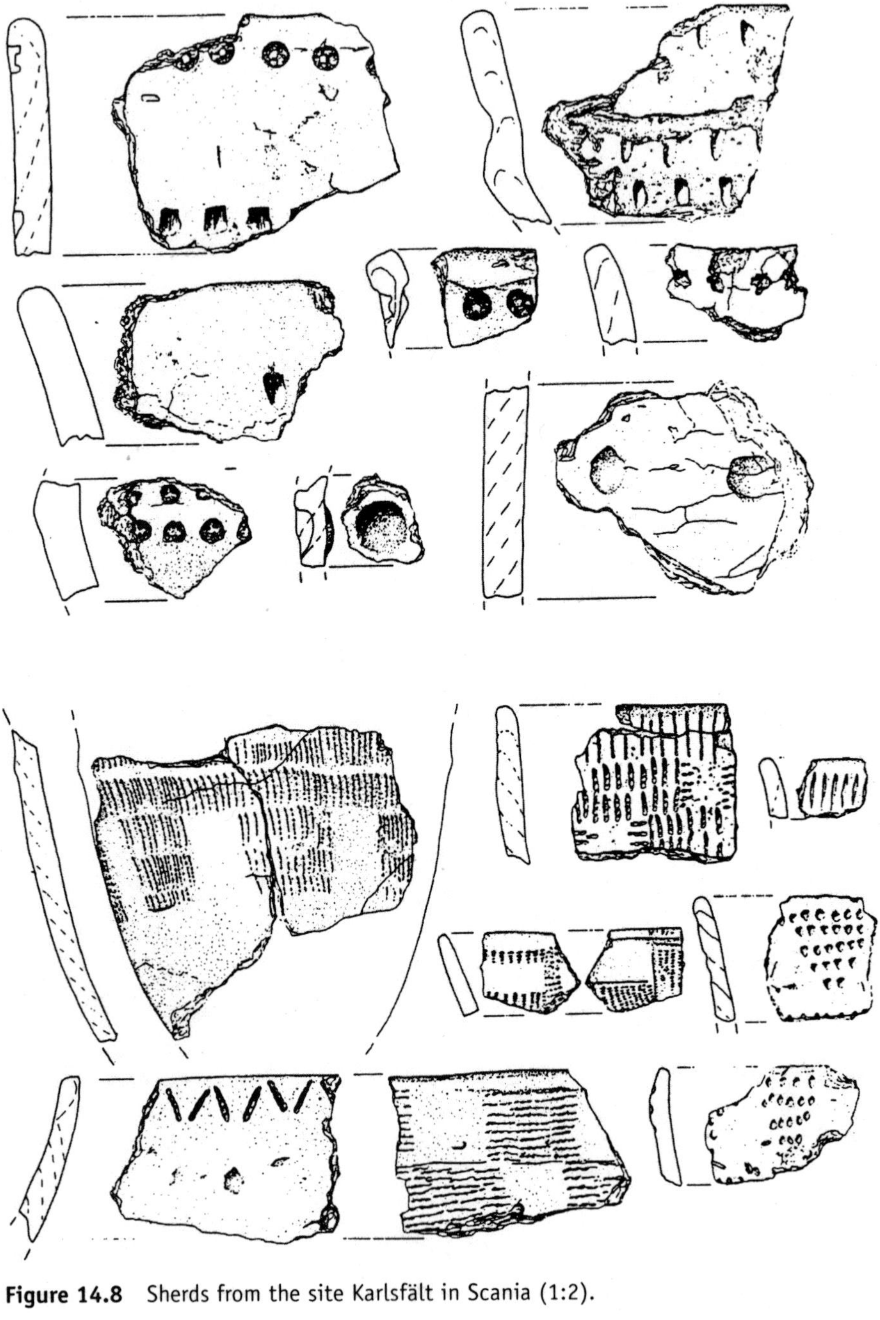

Figure 14.8 Sherds from the site Karlsfält in Scania (1:2).

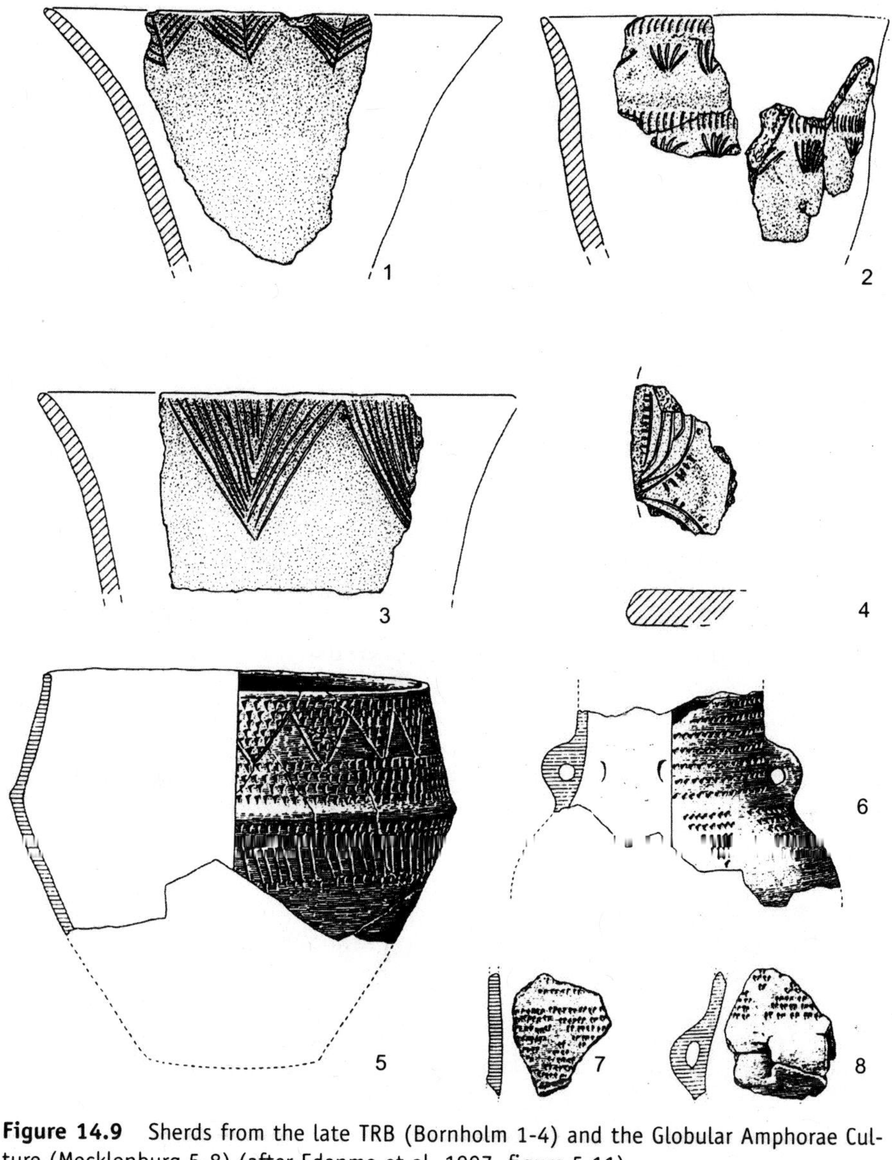

Figure 14.9 Sherds from the late TRB (Bornholm 1-4) and the Globular Amphorae Culture (Mecklenburg 5-8) (after Edenmo et al. 1997, figure 5.11).

more they are likely to recognise and acknowledge other people as similar to themselves.'

So, the FBC in eastern Middle Sweden evolved into the PWC, through peoples' actions and involvement with each other—human agency. By modifying ritual and looking at animals in a new way, for example seals, the ritual specialists made a connection back in time to hunting-gathering days while still keeping some of the FBC rituals like depositions on settlement sites and at special sites like the Alvastra pile dwelling. Human agents were deeply knowledgeable about the surrounding world; they interacted with other individuals and communities as part of an ongoing process of change in which rituals formed an important element.

They retained their material culture as well, like different types of axes. Ritual specialists carried the past into the future. As we have seen, in the material culture of the PWC there is a strong connection over time to the earlier traditions of the FBC both in the decoration of the pottery and in the flint and stone tools. This is especially true of the pottery in the early stages Fagervik I–II. The later stages show a similarity with the FBC, visualized through the 'hanging triangles'.

Their history and new connections helped the people of the PWC to create a new structure in society, but this was a structure that only existed in the lives it sustained. A structure enabled practitioners to find the effectiveness of their place in the world. These structuring principles, which secured a community's identity, were eventually changed in the PWC. Therefore, the PWC people thus created their own history. In using a motif like the 'hanging triangles', and through human agency, old relations and connections could be upheld, and deepened. In using material culture, humans—through agency—could completely change the means by which the world could be seen as meaningful and understandable.

In their social role as material infringements in the social field, artefacts are designed to influence how people interact and come to know about each other and themselves (Wobst 2000: 47). This is how people in the central FBC areas in southern Sweden and Denmark came to know of this new way of being in the world. These influences are eventually seen in the late FBC groups like Stävie and Karlsfält. All humans, both as individuals and as part of a social group, have the power to create and alter structure or tradition through action.

Conclusions and Outlook

Presented here is a new way of looking at the PWC. In using human agency, the PWC can be interpreted as the outcome of changing structuring principles in both ritual and material culture. In this, we can also see how agency was used in helping people to achieve things. Accordingly, there is no reason to label the PWC as a 'new people' or 'surviving hunter-gatherers—the PWC were humans with a history that developed out of the early Neolithic FBC. They kept features like thick-butted axes and double-edged battleaxes, and incorporated items of material culture like blade arrowheads of different types and pottery decoration, such as the 'hanging triangles' that had developed elsewhere. These all belong to the late FBC groups in south Scandinavia and it is especially in the transition between the earlier and later periods of the Middle Neolithic that these affiliations are particularly strong.

In using human agency theory and structuration theory, the problem of the cultural diversity at the transition between the earlier to the later phases of the Middle Neolithic is not 'solved', but the interpretation has been enriched. In discussing material culture, ritual, human agency and social identity, some aspects are easier to understand than before. Knowingly, human agents used the past to connect with the future and used rituals to change the structure of society.

People demonstrated their ethnic belonging, particularly through the use of different patterns of ceramic decoration. Material culture had in this way an active role in linking together different ethnic groups that were socially constituted. Mixing old and new elements made it possible for people to create 'a new world' (Thomas 1996: 37). Public meanings, especially of ethnic identity and interpretations, were negotiated and contested. In times of rapid change, people used, re-invented and re-used both new and old principles while still being able to communicate, as we have seen, through the medium of material culture.

This approach is not restricted to the Middle Neolithic—it could also be used, for example, for neolithization in south Scandinavia. To study the late Mesolithic, the last hunter-gatherers and the transition to farming is a challenging task but in using human agency theory and structuration theory, this historical period can be viewed in a new way.

My conclusion is that, over many generations, the communities inhabiting the site employed hanging triangles as core cultural symbols within the long-term reproduction of a technological tradition. These material practices served as a powerful mechanism for maintaining local forms of routinized social relations across time and space, but also served as a means of signalling how local community identity and worldview were distinct from more distant groups.

Acknowledgements

The study is part of the research programmes 'Us and them: cultural identity during the middle Neolithic' and 'Millennium to millennium: 3,000–2,000 BC in southern Britain and southern Sweden' presently conducted by the Archaeological Research Laboratory at Stockholm University, the University of Kalmar, both Sweden, and the Department of Archaeology, University of Sheffield, UK. The project is sponsored by The Swedish Research Council and STINT (Swedish Foundation for International Cooperation in Research and Higher Education). Sections of this article were written while I spent some weeks at the Department of Archaeology at the University of Sheffield, UK, between 2003 and 2005.

I would like to thank Dr Peter Jordan, Department of Archaeology, University of Aberdeen, and Professor Marek Zvelebil, Department of Archaeology, University of Sheffield, for many valuable comments on the original manuscript and for the invitation to be a part of the present volume. Any inadequacies are, of course, mine.

References

Almgren, O (1906) 'Uppländska stenåldersboplatser', 1 *Fornvännen*, 1–19

Andersson, M (2003) *Skapa Plats i Landskapet. Tidig- och Mellanneolitiska Samhällen Utmed två Västskånska Dalgångar*. Acta Archaeologica Lundensia 8, Malmö

——— (2004) *Making Place in the Landscape*, Stockholm: Riksantikvarieämbetet

Artursson, M (1996) *Bollbacken: En sen Gropkeramisk Boplats och ett Gravfält från Äldre Järnålder*. RAÄ 258; Tortuna socken; Västmanland. Rapporter från Arkeologikonsult. Slutundersökningsrapporter, MBM402. ATA Dnr: 421–4805–1996

Bagge, A (1938) 'Stenåldersboplatsen vid Fagervik i Krokeks socken, Östergötland', Meddelanden från Östergötlands fornminnes- och museiförening 1937–1939, Linköping, pp 151–159

——— (1951) 'Fagervik', XXII *Acta Archaeologica*, Köpenhamn, 57–118

Bagge, A and Kjellmark, K (1939) *Stenåldersboplatserna vid Siretorp*, Stockholm: KVHAA

Barrett, J (1994) *Fragments from Antiquity*, Oxford: Blackwell

——— (1999) 'Chronologies of landscape', in Ucko, P J and Layton, R (eds), *The Archaeology and Anthropology of Landscape: Shaping Your Landscape*, London: Routledge, pp 21–30

——— (2000) 'A thesis on agency', in Dobres, M-A and Robb, J (eds), *Agency in Archaeology*, London: Routledge, pp 61–69

——— (2001) 'Agency, the duality of structure, and the problem of the archaeological record', in Hodder, I (ed), *Archaeological Theory Today*, Oxford: Blackwell, pp 141–164

Becker, C J (1954) 'Die mittelneolitischen Kulturen in Sudskandinavien', XXV *Acta Archaeologica*, 49–150

Bergsvik, K A (2003) 'Mesolithic ethnicity—too hard to handle?' in Larsson, L, Kindgren, H, Knutson, K, Loeffler, D and Åkerlund, A (eds), *Mesolithic on the Move: Papers Presented at the Sixth International Conference on the Mesolithic in Europe, Stockholm 2000*, Oxford: Oxbow Books, pp 290–302

Bradley, R (1998) *The Significance of Monuments: On the Shaping of Human Experience in Neolithic and Bronze Age Europe*, London: Routledge

——— (2000) *An Archaeology of Natural Places*, London: Routledge

——— (2002) *The Past in Prehistoric Societies*, London: Routledge

Browall, H (1991) 'Om förhållandet mellan Trattbägarkultur och Gropkeramisk kultur', in Browall, H, Persson, P and Sjögren, K-G (eds), *Västsvenska Stenåldersstudier*. GOTARC Serie C, Arkeologiska Skrifter No. 8, Göteborg, pp 111–142

——— (2003) *Det Dorntida Alvastra*, Stockholm: Statens Historiska Museum

Carlsson, A (1998) *Tolkande Arkeologi och Svensk Forntidshistoria. Stenåldern*, Stockholm Studies in Archaeology 17, Stockholm

Chapman, J (2000) 'Tension at funerals: social practice and the subversion of community structure in later Hungarian prehistory', in Dobres, M-A and Robb, J (eds), *Agency in Archaeology*, London: Routledge, pp 169–196

Dobres, M-A (2000) *Technology and Social Agency*, Oxford: Blackwell

Dobres, M-A and Robb, J (2000) 'Agency in archaeology: paradigm or platitude', in Dobres, M-A and Robb, J (eds), *Agency in Archaeology*, London: Routledge, pp 3–19

Ebbesen, K (1975) *Die Jungere Trichterbecherkultur auf den Dänischen Inseln*, Copenhagen: Akedemisk Forlag

Edenmo, R, Larsson, M, Nordqvist, B and Olsson, E (1997) 'Gropkeramikerna. Fanns de?' in Larsson, M and Olsson, E (eds), *Regionalt och Interregionalt. Stenåldersundersökningar i Syd- och Mellansverige*, Arkeologiska Undersökningar Skrifter Nr. 23, Stockholm: Riksantikvarieämbetet, pp 135–213

Einarsson, N (2002) *Materiell Kultur som Kommunikationsmedel. En Gropkeramisk Boplats i Åby, Östergötland*, CD-uppsats Högskolan i Kalmar

Eriksson, G (2004) 'Part time farmers or hard-core sealers? Västerbjers studied by means of stable isotope analysis', 23 *Journal of Anthropological Archaeology*, 135–162

Gill, A (2003) *Stenålder I Mälardalen*, Stockholm Studies in Archaeology 26, Stockholm: Jannes Snabbtryck Kuvertproffset HB

Gruber, G (1995) 'Åby', *Elva Undersökningar Senare... En Gropkeramisk Boplats för Flera Mindre Hushåll?* Uppsats i påbyggnadskurs i arkeologi vid Stockholms Universitet Ht 1995

Gruber, G and Larsson, M (1999) *Åby- En Gropkeramisk Boplats*. Rapport UV Öst 1999: 19, Linköping: Riksantikvarieämbetet

Gustafsson, P, Lindholm, P and Runesson, H (2003) 'Gropar och keramik. Ett försök att med statistik och rumslig analys belysa den inre strukturen på den gropkeramiska boplatsen vid Lindskrog', *Landningsplats-Forntiden*. Arkeologiska fördjupningsstudier kring yngre stenålder, järnålder och historisk tid inom det område som tas i anspråk för den tredje landningsbanan vid Arlanda Flygplats, Arkeologiska undersökningar Skrifter Nr. 49, Stockholm: Riksantikvarieämbetet, pp 15–60

Hodder, I (1982) *Symbols in Action*, Cambridge: Cambridge University Press

——— (1990) *The Domestication of Europe*, Oxford: Blackwell

Janik, L (2003) 'Changing paradigms: food as a metaphor for cultural identity', in Parker-Pearson, M (ed), *Food, Culture and Identity in the Neolithic and Early Bronze Age*, British Archaeological Reports International Series 1117, Oxford: Archaeopress, pp 113–125

Janzon, G (1984) 'A Megalithic grave at Alvastra in Östergötland, Sweden', in Burenhult, G (ed), *The Archaeology of Carrowmore*, Stockholm: Theses and Papers in North-European Archaeology 14, Institute of Archaeology, University of Stockholm

Johnson, M (2000) 'Self-made men and the staging of agency', in Dobres, M-A and Robb, J (eds), *Agency in Archaeology*, London: Routledge, pp 213–232

Jones, A (2002) *Archaeological Theory and Scientific Practice*, Cambridge: Cambridge University Press

Jones, A and Richards, C (2003) 'Animals into ancestors: domestication, food and identity in Late Neolithic Orkney', in Parker-Pearson, M (ed), *Food, Culture and Identity in the Neolithic and Early Bronze Age*, British Archaeological Reports International Series 1117, Oxford: Archaeopress, pp 45–51

Jones, S (1997) *An Archaeology of Ethnicity: Constructing Identities in the Past and Present*, London: Routledge

Kihlstedt, B, Larsson, M and Nordqvist, B (1997) 'Neolitiserngen i Syd-Väst- och Mellansverige-social och ideologisk förändring', in Larsson, M and Olsson, E (eds), *Regionalt och Interregionalt. Stenåldersundersökningar i Syd- och Mellansverige*, Arkeologiska Undersökningar Skrifter Nr. 23, Stockholm: Riksantikvarieämbetet, pp 135–213

Larsson, M (1999) 'Den gropkeramiska kulturens "mikrorum". Kring boplatsen Åby i Östergötland', *Forskaren i Fält. En Vänbok till Kristina Lamm*, Arkeologiska Undersökningar Skrifter Nr. 27, Stockholm: Riksantikvarieämbetet, pp 43–53

——— (2003) 'People and sherds—the Pitted Ware site Åby in Östergötland, eastern Sweden', in Samuelsson, C and Ytterberg, N (eds), *Uniting Sea. Stone Age societies in the Baltic Sea Region. OPIA 33*. Uppsala: Department of Archaeology and Ancient History, pp 117–131

——— (2004) 'Living in cultural diversity. The Pitted Ware Culture and its relatives', 14 *Journal of Nordic Archaeological Science*, 61–69

——— (2006) *A Tale of a Strange People: The Pitted Ware Culture in Southern Sweden*, Lund: University of Lund

Malmer, M P (1973) 'Om den gropkeramiska kulturens väsen', in Simonsen, P and Stamsö Munch, G (eds), *Bonde-Veideman. Bofast—Icke Bofast I Nordisk Forhistorie*, Tromsö Museums Skrifter XIV, Tromsö: Tromsö Museum, Universitetsforlaget, pp 59–61

——— (2002) *The Neolithic of South Sweden. FBC, GRK and STR*, Stockholm: KVHAA

Meinander, C F (1957) 'Kolsvidja', 58 *Finska Fornminnesföreningens Tidskrift*, 186–213

Meskell, L (2001) 'Archaeologies of identity', in Hodder, I (ed), *Archaeological Theory Today*, London: Polity Press, pp 187–214

Nerman, B (1911) 'Östergötlands stenålder', *Meddelanden från Östergötlands Fornminnesförening*, 1–39

——— (1927) 'Ett bidrag till fågn om gånggriftstidens havsnivå i Östergötland', 22 *Fornvännen*, 247–250

Stenbäck, N (2003) *Människorna vid Havet. Platser och Keramik på Ålandsöarna Perioden 3,500–2,000 f Kr*, Stockholm Studies in Archsaeology 28, Stockholm: Jannes Snabbtryck Kuvertproffset HB

Storå, J (2001) *Reading Bones: Stone Age Hunters and Seals in the Baltic*, Stockholm Studies in Archaeology 21, Stockholm: Jannes Snabbtryck Kuvertproffset HB

Strinnholm, A (2000) *Bland Säljägare och Fårfarmare. Struktur och Förändring i Västsveriges Mellanneolitikum. Coast to Coast Book 4*. Uppsala: Department of Archaeology and Ancient History

Stuiver, M, Long, A and Kra, R S (eds) (1993) Calibration 1993, 35(1) *Radiocarbon*, 1–244.

Sundström, L (2003) *Det Hotade Kollektivet. Neolitiseringsprocessen ur ett Östmellansvenskt Perspektiv*, Uppsala: Department of Archaeology and Ancient History

Thomas, J (1991) *Rethinking the Neolithic*, Cambridge: Cambridge University Press

——— (1996) *Time, Culture and Identity*, Cambridge: Cambridge University Press

Thomas, J (1999) *Understanding the Neolithic*, London: Routledge

Tilley, C (1982) *An Assessment of the Scanian Battle-Axe Tradition: Towards a Social Perspective*, Lund: Gleerup

——— (1996) *Ethnography of the Neolithic. Early Prehistoric Societies in Southern Scandinavia*, Cambridge: Cambridge University Press

Welinder, S (1971) 'Överåda—a Pitted Ware Culture Site in eastern Sweden', 1969–70 *Meddelanden från Lunds Universitets Historiska Museum*, 5–98

Welinder, S (1973a) 'Om den gropkeramiska kulturen', in Simonsen, P and Stamsö Munch, G (eds), *Bonde-Veideman. Bofast—Icke Bofast I Nordisk Forhistorie*, Tromsö Museums Skrifter XIV, Tromsö: Tromsö Museum, Universitetsforlaget, pp 54–59

——— (1973b) 'The radiocarbon age of the Pitted Ware Culture', 1971–72 *Meddelanden från Lunds Universitets Historiska Museum*, 107–119

——— (1976) 'The economy of the Pitted Ware Culture in eastern Sweden', 1975–76 *Meddelanden från Lunds Universitets Historiska Museum*, 20–30

——— (1978) 'The acculturation of the Pitted Ware Culture in eastern Sweden', 1977–78 *Meddelanden från Lunds Universitets Historiska Museum*, 98–110

Whittle, A (2003) *The Archaeology of People: Dimensions of Neolithic Life*, London: Routledge

Wobst, M (2000) 'Agency in (spite) of material culture', in Dobres, M-A and Robb, J (eds), *Agency in Archaeology*, London: Routledge, pp 40–51

Wyszormirska, B (1984) 'Figurplastik och gravskick hos nord- och östeuropas fångstkulturer', 4 *Acta Archaeologica Lundensia*, 18

Zvelebil, M (2003) 'People behind the lithics: social life and social conditions of Mesolithic communities in temperate Europe', in Bevan, L and Moore, J (eds), *Peopling the Mesolithic in a Northern Environment*, British Archaeological Reports International Series 1157, Oxford: Archaeopress, pp 1–26

Chapter 15

Pitted Ware Culture Ceramics: Aspects of Pottery Production and Use at Ottenby Royal Manor, Öland, Sweden

Ludvig Papmehl-Dufay

Introduction

Cultural relations in the Middle Neolithic (around 3,300–2,300 BC) of southern and eastern Sweden have been a major issue of discussion among Scandinavian archaeologists for decades. Diversity in the material remains has resulted in a variety of interpretations concerning ethnic or cultural identity, typically reflected in the three-group model, the Funnel Beaker Culture (FBC), the Pitted Ware Culture and the Boat Axe (or Battle Axe) Culture (Malmer 2002). The Pitted Ware Culture, appearing in the Middle Neolithic with a decidedly 'non-Neolithic' character, has been intensely studied and debated (Edenmo et al. 1997). For a long time, it was regarded as a coastal fisher-hunter-gatherer culture with little or no farming, with sites yielding large amounts of characteristic pottery (Björck 1998; Edenmo et al. 1997; Olsson et al. 1994). During the last decade several authors have suggested alternative interpretations, arguing material differences to reflect ideological patterns and symbolic statements rather than cultural identities (Carlsson 1998; Gill 2003). This chapter will focus on the role of pottery in the Middle Neolithic Pitted Ware Culture in eastern Sweden, exemplified by the recently excavated site at Ottenby Royal Manor on the southernmost part of the island of Öland in the Baltic Sea. The overall characteristics of the Pitted Ware Culture are not dealt with in detail (see Edenmo et al. 1997; Malmer 2002; Stenbäck 2003; Larsson, this volume).

Pottery as a cultural object is intimately connected to a wider social network in which a number of other factors are involved (Gosselain 1999). When studying archaeological ceramic assemblages, it is easy to become fixated with the potsherds themselves and not address the craft tradition they represent. Ceramics have held a key position in the definition and study of the Pitted Ware Culture for about 100 years (Almgren 1906; Bagge 1951; Löfstrand 1974; Malmer 1969), but only rarely have researchers aimed at understanding the social significance of the craft of ceramics (for a similar argument on Beakers, see Boast 2002). The often very large ceramic assemblages have resulted in limited analytical treatment of the category and in Scandinavian research, typology and ornamental styles have been the focus of thought. Fredrik Hallgren argued that at the FBC site Skogsmossen in Västmanland pottery, design has been heavily governed by social factors (Hallgren 2000: 184–187). In a study of Iron Age pottery from Uppåkra and Hötofta in Scania, southern Sweden, Ole Stilborg argued that a number of factors are responsible for variations in a ceramic assemblage, including the social function of the site, the area and durability of the site, the number of potters engaged in ceramic production at the site, and so on (Stilborg 2001: 124). It seems this applies to all time periods, and thus chronology and ethnicity are not the only relevant factors in explaining variation in Pitted Ware ceramic assemblages.

Pitted Ware Ceramics

The term 'pitted' refers to the cylindrical pits that appear as decoration on Neolithic ceramics over a very large geographical area covering most of eastern Scandinavia (Figure 15.1). During the 1960s and 1970s, the pits were often argued to have served a practical rather than a decorative

Figure 15.1 Pitted Ware pits. Potsherd from Ottenby Royal Manor, Öland, Sweden (photo by the author).

function (Almgren 1967; Löfstrand 1968), but today they are most often regarded as symbolically loaded decoration (Stenbäck 2003: 192). The term 'Pitted Ware' has been criticized by several authors, who argue that it is normative in the sense that it implies that cylindrical pits should occur on all Pitted Ware vessels in opposition to all other vessels lacking such pits (e.g., Meinander 1957: 199; Wyszomirska 1984: 35–36), which, of course, is not the case.

The use of the term 'ware' in this context may also be criticized. In ceramic technology studies, the term 'ware' is commonly used to characterize technological aspects of ceramic products. It has been defined as 'a specific combination of technological features, characterised by one specific set of surface properties, and one fabric, or a set of closely related fabrics... entirely independent of any considerations of vessel shape, decoration, or vessel size' (Nordström 1972: 34–35). Thus vessels of a particular ware have in common a range of technological features, regardless of formal and decorative design. Pitted Ware ceramics, however, are a category more or less completely based on formal and decorative features. The use or not of calcareous tempering has been ascribed a chronological significance (e.g., Bagge 1951), but otherwise 'ware' in the common usage of the term is not a defining feature of Pitted Ware ceramics. Thus ceramics that, from their formal and decorative appearance, without doubt would be described as Pitted Ware may be of almost any physical composition as regards clay and temper.

Pitted Ware sites often display enormous amounts of pottery. Assemblages of 200–400 kg are common (e.g., Olsson et al. 1994: 25), and at Ajvide on Gotland, the collected assemblage exceeds 3,000 kg (Burenhult 2002: 32; Norderäng 2002, 2003, 2004). The practical function of the vessels has sometimes been discussed, a common suggestion being that the vessels were used for cooking marine foods, production of seal train-oil and storage (e.g., Segerberg 1999: 62). Large vessels (rim diameter approximately 25–40 cm) with profiled shoulder/carination and a rounded base tend to be the dominating form, and decoration is usually confined to the upper part of the vessel (Figure 15.2). Decorated sherds commonly make up 30 to 50% of an assemblage (Bagge and Kjellmark 1939: 108–132; Olsson et al. 1994: 28; Papmehl-Dufay 1999: 27, 2004: 39, 2005: 17). By far the most common decorative element is the eponymous cylindrical pit, typically arranged in one or several horizontal rows above the shoulder and further down on the body. The pits are often combined on the neck and shoulder with horizontal rows of vertically oriented geometric patterns of comb or line stamp. A chronological tendency can be seen in that earlier Pitted Ware ceramics have less surface-covering decoration than have the later (Bagge 1951), and in contrast, pits are more common on early pottery than later (Gustafsson et al. 2003). Several recent studies have suggested that variations in decoration within a site should be given

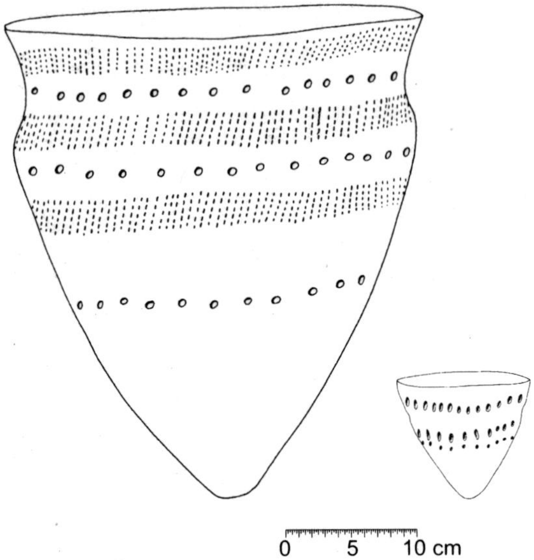

Figure 15.2 Pitted Ware ceramic vessels from Siretorp, Blekinge, southeast Sweden (after Stilborg 2002: 74).

a social rather than a chronological explanation, for example at Åby in Östergötland where the occurrence of various decorative patterns on the edge of the rim was shown to differ between two neighbouring areas and interpreted as reflecting different social units (Einarsson 2002; Gruber 1995; Larsson 2004).

The most influential work on Pitted Ware pottery in Sweden is undoubtedly the essay on the site of Fagervik by Axel Bagge (1951). In this work the enormous ceramic material of 170,000 potsherds from Fagervik in Östergötland, eastern Sweden, was considered. Bagge distinguished five phases of occupation, Fagervik I–V, situated at different levels above the present sea level and seemed to reflect a chronological sequence from early Middle Neolithic FBC through the Pitted Ware, up to the late Middle Neolithic Boat Axe Culture (Bagge 1951: 67–84). The different phases were interpreted as corresponding to variations in pottery style, including decoration, vessel shape and tempering. The Fagervik phases were described in some detail but not properly defined, which has been subject to heavy criticism (e.g., Edenmo et al. 1997; Löfstrand 1974; Malmer 1969). Bagge was careful to emphasize that Fagervik I–V should be understood as phases of occupation on a particular site and not universal styles of pottery decoration (Bagge 1951: 62, 64), a note that later researchers seem to have forgotten. Despite criticism the attribution of newly uncovered pottery assemblages to one or several of the Fagervik phases, based mainly on occurring decorative patterns, has become more or less a routine practice in Swedish archaeology (Olsson et al. 1994: 28; Werthwein 2001: 15).

In his work on Fagervik, Bagge distinguished three types of vessel profile to which he attributed a chronological significance (Bagge 1951: 67) (Figure 15.3). The A-profile was described as a funnel beaker and thus not regarded as a type of the Pitted Ware Culture; the B-profile was described as a smoothly S-shaped 'degeneration' of the A-profile, and was regarded as the earliest vessel shape of proper Pitted Ware ceramics connected to Fagervik II but reappearing in Fagervik IV. Finally the C-profile, having a concave neck and a sharply marked shoulder, represented the typical Pitted Ware vessel and was associated with Fagervik III and IV. Recently Mats Malmer has proposed the addition of a fourth type, the D-profile, where the vessel wall is slightly convex from rim to base (Malmer 2002: 77).

Ottenby Royal Manor

In August and September 2004 a limited excavation just south of Ottenby Royal Manor, situated in Ås parish on the southernmost part of the island of Öland in the Baltic Sea, was undertaken (Papmehl-Dufay 2005, 2006). This large island is extremely rich in prehistoric remains, but so far very few Pitted Ware sites have been archaeologically investigated (Figure 15.4). At Köpingsvik a number of excavations have taken place and

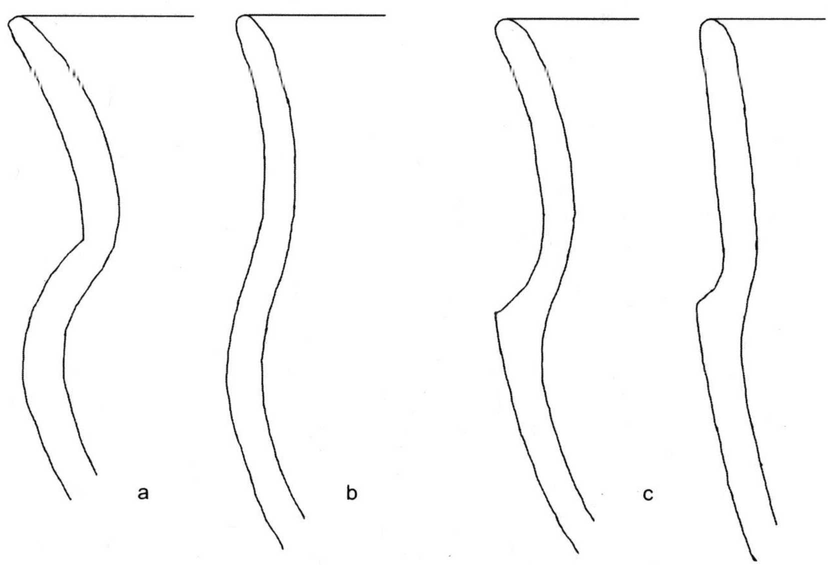

Figure 15.3 Profile types as distinguished by Bagge at Fagervik (after Bagge 1951: 82).

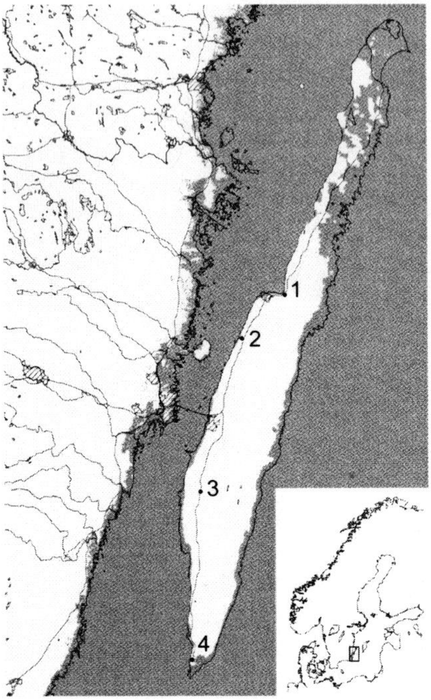

Figure 15.4 The island of Öland with estimated shoreline around 3,000 BC, sites mentioned in the text indicated; (1) Köpingsvik; (2) Halltorp; (3) Resmo; (4) Ottenby Royal Manor (map adapted from Claesson and Mikaelsson 2001).

extensive Pitted Ware material recovered, including more than 20 burials, but the documentation is generally poor and the treatment of the site problematic (Papmehl-Dufay 2006: 84–118; Schulze 2004). At Halltorp a single sunken feature has been excavated producing small amounts of finds comprising both Pitted Ware and Boat Axe Culture pottery (Alexandersson 2002). The site at Ottenby Royal Manor was located in the late 1980s and very briefly investigated in 1991 (Papmehl-Dufay 2005; Petersson 2001; Rosberg 1993). In Resmo parish on the western part of the island, four megalithic tombs are situated, and field-walking in recent years has revealed an abundance of Stone Age settlement sites on the island (Alexandersson et al. 1996).

The Pitted Ware site at Ottenby Royal Manor is situated at approximately 8–10 m above the present sea level, in a flat sandy terrain some 4 km from the present-day southernmost point of the island. During the Middle Neolithic, the site was located near the shore on the very tip of the island. The excavation in 2004 revealed a finds-producing layer of varying thickness, generally ranging from 10 to 50 cm. The finds recovered are strongly dominated by pottery, but also include approximately 3 kg of lithic material and 3.5 kg of burnt and unburnt animal bones (Table 15.1). On a general stylistic level the pottery is in accordance with what has been previously recovered in the neighbouring province of Blekinge to the west (e.g., Bagge and Kjellmark 1939; Hagelqvist and Widenborg 1977), but significant similarities can also be noted with sites like Köpingsvik on Öland and Humlekärrshult in Småland (Bagge 1936; Papmehl-Dufay 2004, 2006). Sunken features were documented in a limited part of the investigated area and included small postholes and a section of a large circular channel of unknown function. The faunal assemblage from the site is dominated by marine species such as fish and seal, but pig and sheep/goat are also present (Reuterdahl 2005). Six bones from domesticated animals as well as one hazelnut shell and one sample of charred food remains on a potsherd were submitted to ^{14}C-dating. The results indicated that, while the hazelnut shell and the food crust represent the Pitted Ware occupation at around 3,100–2,800 cal BC, all the dated bones which were from domesticated animals were later. The food crust was analyzed by FT-IR and GC-MS as well (Isaksson 2006), and the results indicate a mixture of terrestrial animals and vegetables rich in carbohydrates. No marine input could be detected, and

Table 15.1 Summarizing presentation of the finds from the excavation at Ottenby Royal Manor 2004.

material	note	weight g	number
pottery	Stone Age pottery	23,294.8	6,809
unburnt bone		3,157.1	8,426
burnt bone		393.7	2,992
ordovician flint		805.6	147
kristianstad flint		380.8	94
southscandinavian flint		115.8	81
flint undetermined		42.3	50
quartz		60.0	16
quartzite		420.0	14
porphyr		146.1	6
mica-slate	axe	491.7	1
greenstone	axe	218.9	1
basalt	axe fragment	76.7	1
sandstone	grinding stone	393.5	1
burnt clay	bead fragment	0.4	1
hazelnut shell		0.1	1
iron		26.3	8
brick		17.9	1
slag		8.7	2
sintered clay		6.0	3
glass	bead	1.3	2
bronze	pin	0.4	1
pottery	Iron Age pottery	4.1	1
Total		30,062.2	18,659

thus the reservoir effect does not have to be accounted for in this case. Pottery use at the site will be further discussed below.

The ceramic assemblage from the 2004 excavation amounts to approximately 23 kg, including 1,838 sherds and 4,971 fragments. A detailed treatment of this assemblage, as well as details on definitions and the recording process, has been dealt with elsewhere (Papmehl-Dufay 2006, see also Papmehl-Dufay 2005). Summarized data on the assemblage is presented in Table 15.2. The sherds have been categorized as vessel part categories, and since several categories may be present on one sherd, the sum of the vessel part categories exceeds the total number of sherds. A number of reconstructed vessel details are shown in Figure 15.5.

Wall thickness in Pitted Ware assemblages is usually presented as the mean value of measured maximum thickness, thus more correctly 'mean maximum thickness' (e.g., Hulthén 1998). The distribution of maximum thickness measured on each sherd in the Ottenby assemblage is shown in Figure 15.6. The mean value of the maximum thickness is 8.3 mm. This stands in some contrast with results from other Pitted Ware assemblages, e.g., Hagestad in Skåne where mean maximum thickness was 12 mm (Hulthén 1977: 127), Alvastra in Östergötland where mean maximum thickness ranged between 12 and 14 mm (Hulthén 1998: 20, 35), and Vendel in Uppland where mean maximum thickness was just below 10 mm (Papmehl-Dufay 1999: 22–30). In the assemblage of approximately 18 kg from the excavation in 1950 at Köpingsvik on Öland (Figure 15.4), the corresponding figure was 9.3 mm. Thus, without any calculations on statistical significance, it seems that Pitted Ware ceramics from Ottenby Royal Manor are relatively thin-walled.

All vessels are handmade, and the means of joining the coils generally correspond to that commonly referred to as N-technology (Andersen 1975:

Table 15.2 Summarizing data on the ceramic assemblage from Ottenby 2004 (fragments not included).

vessel part	n	mean weight g	mean max thickness mm	decorated % of n	pits % of n	linestamp % of n	combstamp % of n	other decoration % of n
rim	261	12.5	7.9	83	42.5	43	19.5	17
neck	535	12.1	8.3	81.5	52	38.5	17.5	4
shoulder	190	15.5	8.9	90	47	52	20	2
body	1,525	8.6	8.3	47	32.5	12.5	5	3.5
base	32	23.6	12.2	3	0	0	0	3
total	2,543							
average		9.4	8.3	50.5	34	16.5	6.5	5.5

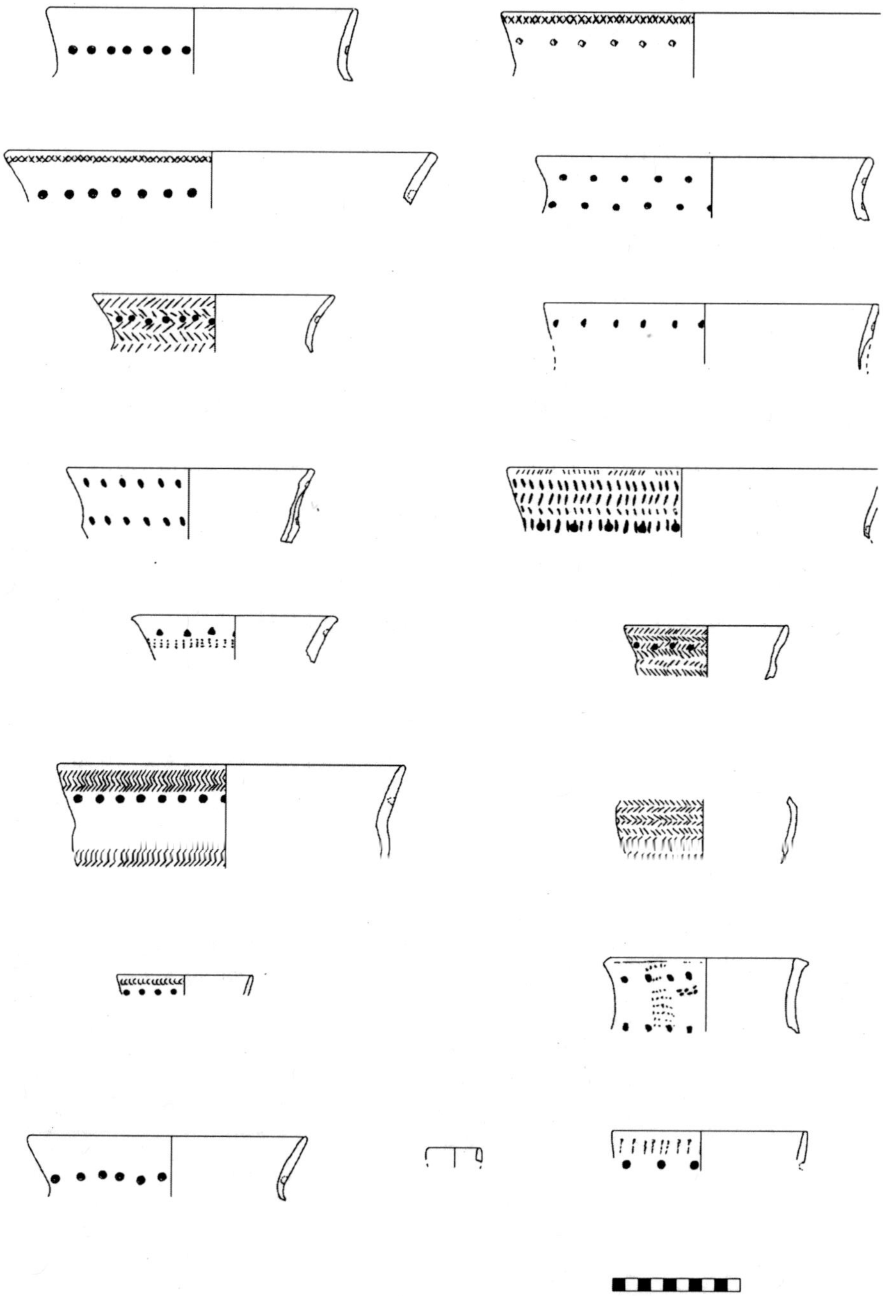

Figure 15.5 Reconstructed vessel details from Ottenby Royal Manor (drawing by the author).

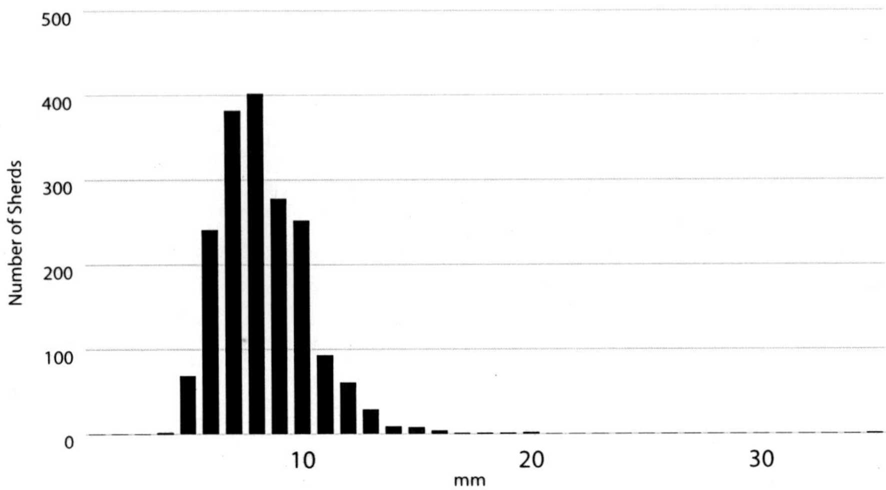

Figure 15.6 Maximum and minimum sherd thickness in the Ottenby Royal Manor assemblage.

57; Hulthén 1977), that is the coils are joined diagonally resulting in a relatively thin wall thickness. This is the common vessel building technique for Pitted Ware ceramics in general. Some vessels are apparently modelled or 'pinched', which enables an even thinner wall to be produced. On a few sherds it has been possible to detect the mixing of different techniques, usually in the form of modelling the base part and then coiling the rest of the vessel from the base and up.

Just over half of the sherds in the Ottenby Royal Manor assemblage are decorated. The single most common decorative element is the cylindrical pit, occurring on 834 sherds and fragments. On 432 sherds and fragments line stamping has been noted, the most common arrangement of which is one or several horizontal rows of vertically oblique impressions. Comb stamping has been noted on 167 sherds and fragments, for which the most common arrangement is one or several horizontal rows of vertical impressions. From Table 15.2 it is evident that the frequency of decoration varies over sherd categories, and no less than 81.5% of all rim sherds and 90% of the shoulder sherds display decoration. For body sherds the figure is 47%. These figures tell us firstly that if all vessels are assumed to have generated rim sherds, the majority of the vessels were decorated. Secondly, the even higher portion of decorated shoulder sherds indicate that an even larger portion, if not all, of vessels with a pronounced shoulder were decorated. However, the mean weight of the decorated shoulder sherds is higher than that of the decorated rim sherds and thus the difference may be a result of variation in fragmentation. Irrespective of these arguments it can be concluded that the majority of the vessels represented in the Ottenby assemblage have been decorated on their upper part, and quite a large portion of these have also been decorated further down.

Only 56 rim sherds allowed for an estimation of rim diameter, the results of which are shown in Figure 15.7. The figures presented should be regarded as an approximation. It is clear, however, that vessels of several different sizes are present in the assemblage. Hypothetically at least four different size classes can be identified: 10–16 cm (11 sherds), 18–23 cm (13 sherds), 25–30 cm (17 sherds) and 32–36 cm (12 sherds). Apart from these groups one vessel has been identified with a rim diameter of only 4 cm, and two measurements indicate vessels with a rim diameter of approximately 40 cm. Thus the majority of the vessels seem to fall in the interval 18–30 cm and a greater number if the interval is extended to 36 cm. The occurrence of different size classes of vessels in Pitted Ware assemblages has been observed previously on several sites, for example, Bollbacken where the distribution was similar to that observed here (Artursson et al. 1996: 224).

The vessel with a rim diameter of 4 cm mentioned above, and partly reconstructed in Figure 15.5, deserves particular attention. Occurring in sizes of about a fingernail, more or less tiny and often undecorated vessels are commonly found in small numbers on Pitted Ware sites all over the distribution area, both in burials and in cultural layers (Bagge and Kjellmark 1939; Janzon 1974: 104; Lindholm 2003; Stenberger et al. 1943: 100) (Figure 15.8). In some cases the vessels clearly imitate larger versions both regarding shape and decoration, and again in other cases they seem to be

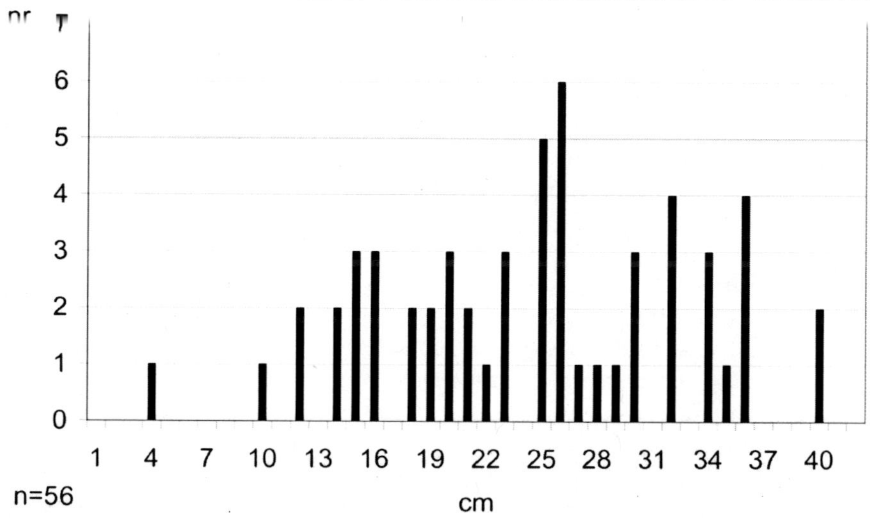

Figure 15.7 Estimated rim diameter on 56 sherds from Ottenby Royal Manor.

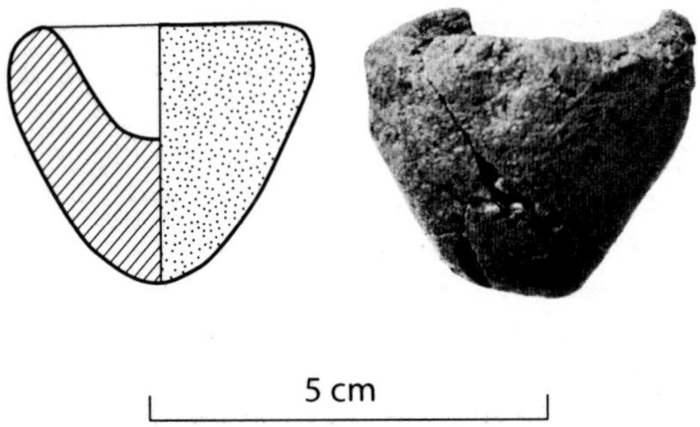

Figure 15.8 Minature vessel from the cultural layer at Västerbjers, Gotland (after Stenberger et al. 1943, figure 61).

pinched from a small lump of clay without much effort put into the actual shaping of the object. In Pitted Ware studies these small vessels are usually referred to as mini-vessels or miniature vessels. Gunborg Janzon (1974) preferred the former term, arguing that these vessels have served a practical rather than a symbolic function. She used the height of the vessels to define two types of mini-vessels: type A:1 with a maximum height of 5 cm, and type A:2 with a height more than 5 but less than 10 cm (Janzon 1974: 104). Margareta Leijonhufvud suggested another definition, where the rim diameter was used to distinguish between miniature vessels (diameter of less than 8 cm) and cups (rim diameter 8–10 cm) (Leijonhufvud 1989: 79). This distinction reflects an idea that the very small vessels must be of mainly symbolic significance, whereas the 'larger small vessels' may have served a practical function. Judging from the widespread occurrence and varied execution of vessels of less than 10 cm in rim diameter, some with carefully shaped profile and decoration and some without any distinctive formal attributes, it seems likely that the categories of Janzon and Leijonhufvud encompass vessels produced for different purposes, some most probably representing 'true' miniatures and some being small vessels designed primarily for a practical purpose. If this is accepted, it has important implications for our understanding of the social significance of ceramic production within the Pitted Ware Culture. It seems reasonable to assume that the production of miniatures in general has been confined to certain categories of socially, economically or in some other way important objects (Arrhenius 1961: 150–158). Here, it is argued that the miniature pots should be regarded as objects treasured not for their precious material or high quality of execution, but for the associations they evoked.

During the excavation at Ottenby Royal Manor a total of 15 sherds were selected for lipid residue analysis by means of Gas Chromatography/Mass Spectrometry (GC/MS). The sampling strategy was to target as many different vessel types as possible; however, for practical reasons sherd size and identifiable diagnostic features also became a factor. Lipid residues were extracted from the ceramic ware by means of a chloroform/methanol solvent and sonification (cf. Evershed et al. 2001). These lipids most probably derive from the last few uses of the pot (Craig et al. 2004). A detailed record of the analytical procedure and the results has been presented elsewhere (Papmehl-Dufay 2006: 163–165, 220–227), and in the following only a brief review of the results is offered. The lipid residue analysis was performed by the author in close collaboration with Dr Sven Isaksson, at the Archaeological Research Laboratory, Stockholm University, Sweden.

The results of the lipid residue analysis reveal a striking variability in the amount of residues present in the ceramic ware (Figure 15.9). Three sherds were more or less clear of lipid residues, and five sherds contained amounts below 10 µg/g. Three sherds fell in the range 25–60 µg/g, which is regarded as low. The two sherds OT7 and OT8 contained large amounts of lipids (216 and 314 µg/g respectively), and finally sherd OT9 stands out with an exceptional amount of 5,842.3 µg/g, almost 6 mg lipid residues per gm potsherd (see Figure 15.10).

Each sample was checked for the presence of a number of biomarkers, and the resulting protocol enabled interpretation of the lipid residues in terms of origin, for example animals vs vegetables (cholesterol vs phytosterols) as well as terrestrial animals vs marine animals/fish and vegetables (ratio C18:0/C16:0 (S Isaksson 2000: paper VII: 12–19; Olsson 2004: 23–24), isoprenoidic fatty acids, ω-(o-alkylphenyl) fatty acids (Hansel et al. 2004)). The interpretations are indicated as abbreviations in Figure 15.9.

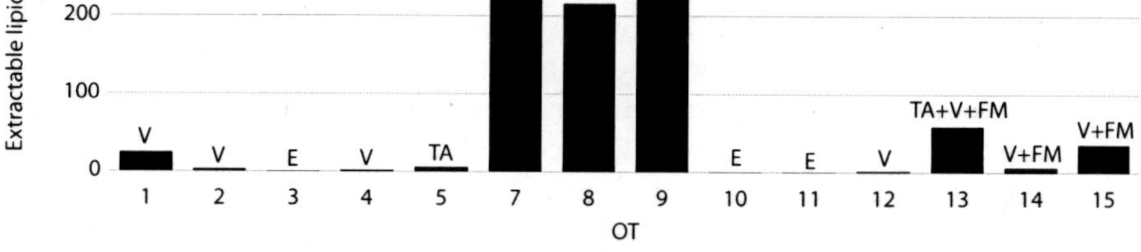

Figure 15.9 Results of the lipid residue analysis, expressed as amount of extracted lipid residues in µg per g potsherd. Interpretations are indicated above each block. Abbreviations: (E) Empty; (V) Vegetables; (TA) Terrestrial animals; (FM) Fish or marine animals; (BW) Beeswax.

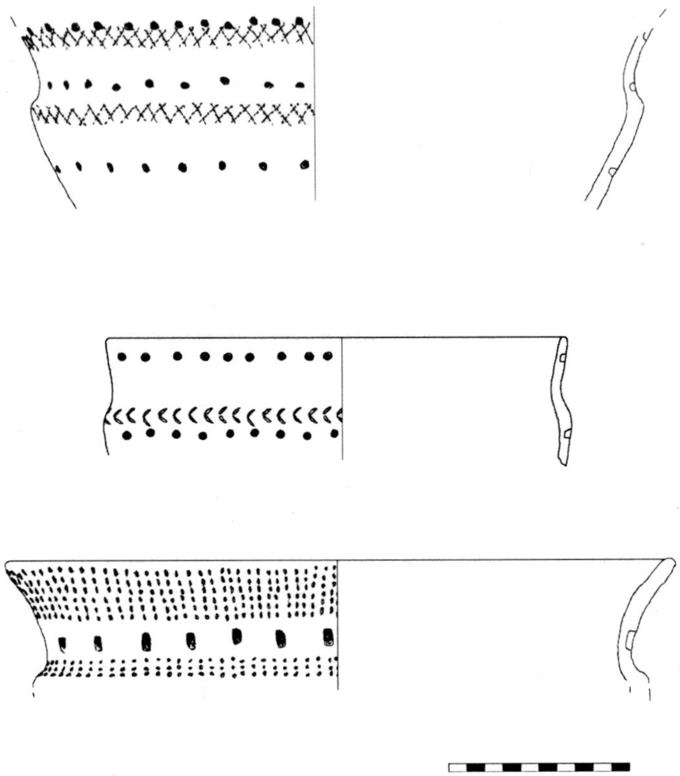

Figure 15.10 Reconstruction drawings of the three analyzed vessels: OT7, OT8 and OT9 (drawings by the author).

The initially assumed dominance of marine animal and fish markers was not evidenced, and instead vegetables and terrestrial animals are indicated in several of the samples. Ten of the samples have been interpreted as containing traces of vegetables, four as containing traces of marine animals or fish and four samples as containing traces of terrestrial animals. A mixture of animal and vegetable fats is evidenced in five samples, two of which contain fat from marine animals or fish and two of which contain fat from terrestrial animals. The sample OT13 most likely contains a mixture of both terrestrial and marine animals/fish and vegetables. Direct traces of vegetables in the form of phytosterols were detected in four of the samples, three of which also contained wax residues.

The exceptionally rich sample OT9 deserves special comment. The residues in this sample include extremely well preserved as well as partly degraded acyl lipids, and the biomarkers noted indicate a mixture of terrestrial animals, vegetables and beeswax. The latter is indicated through the presence of *n*-alkanes, *n*-alcohols and palmitic wax esters characteristic of this substance (Copley et al. 2005: 529; Evershed et al. 2001: 333–334). The presence of cholesterol indicates the presence of animal fats. The ratio C18:0/C16:0 is just within the range of values indicating the presence of vegetables or marine animals/fish. However, other biomarkers indicating the presence of marine animals/fish are absent, and so the results point towards a mixture of substances derived from both terrestrial animals and vegetables. Phytosterols are not present, nor the TMTD isoprenoid, which are associated with marine animals/fish. However, there are small amounts of the isoprenoid fatty acid TMHD, which indicates the presence of green plants. Wax residues noted in the sample most probably represent degradation products from beeswax. The presence of beeswax together with terrestrial animal commodities may indicate the use of honey, although there is also the possibility that the beeswax had been used for sealing the pot (Copley et al. 2005: 529).

To conclude the lipid residue analysis, the sample is too small to indicate any correlations between vessel design and patterns of use. However, the results clearly suggest a complex multi-purpose pattern of vessel use, rather than single purpose hunting-related cooking pots. The presupposed dominance of fats from marine catch is not at all evidenced, and instead vegetables and terrestrial animals predominate. The presence of beeswax in one sample, possibly representing the utilization of honey, constitutes the earliest evidence so far in Sweden for this product.

Based on the principal of recording the complete assemblage, a sample of sherds was analyzed in detail for their mineralogical properties. A brief characterization of ware and inclusions by means of the polishing of sections and use of binocular microscope on 33 sherds was followed by thin section analysis on 24 of these sherds. In addition four clays collected from three locations in the vicinity of the site were analyzed (Fig 15.11). The clay samples were fired as briquettes at 600°C for about 15 minutes, the thin sections of which were analyzed along with the ceramic sections from the site. The microscopic analyses were performed by Dr Ole Stilborg at the Laboratory for Ceramic Research, Lund University, Sweden (Stilborg 2006).

The analysis of the raw clay samples showed that all four collected clays are highly calcareous, that is, not particularly suited for ceramic production (Shepard 1957: 22, 30). On the geological survey map of Öland, clay is documented in only a few small locations, none of which is situated in proximity to Ottenby Royal Manor. Thus all four samples derive from what has been mapped as clayey till, and the high calcium content in the clays reflects the limestone bedrock in the area. Several other locations were tested for presence of suitable clay, without success.

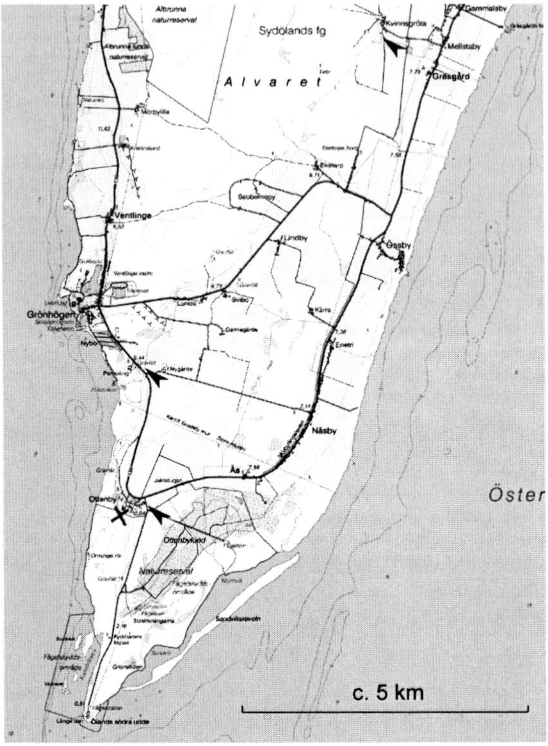

Figure 15.11 Map of southernmost Öland with the location of the raw clay samples (black arrows) and the Pitted Ware site at Ottenby Royal Manor (cross) indicated (© Lantmäteriverket Gävle 2005; permission I 2005/2034).

The microscopic analysis of the ceramics shows some variability concerning type and amount of temper. All clays in the ceramic wares analyzed are non-calcareous, and thus no match could be made with any of the sampled local clays. However, judging from the rock types noted among the natural inclusions, most clays represented by the pottery are to be regarded as local. A large portion of the ceramics analyzed displayed 'temper', consisting of what has been interpreted as naturally occurring mineral inclusions, that is, the clays contained sufficient amounts of non-plastics and no further material has been added. Further, in a number of samples, the temper material has been interpreted as consisting of added unsorted sand, which is indeed an odd choice. Some of these may well represent naturally occurring inclusions though, and thus the dominating fabric at Ottenby Royal Manor consists of naturally tempered wares. One sherd displays temper of slate, and crushed limestone was noted in another sample. Four sherds stand out as distinctly granite-tempered, and in three of these, the clay is distinctly different from that noted in the other samples from the site. This clay is extremely fine and contains only small amounts of silt and fine sand, and based on the results from raw clay surveys on the island as well as extensive analyses of ceramic samples (Papmehl-Dufay 2006; Stilborg 2006), these sherds are interpreted as consisting of non-Öland clays. Interestingly, two of these sherds display other deviant features as well; one of the non-local wares is identical to the exceptional sample OT9 in the lipid residue analysis above with very large amounts of lipid residues and evidence among other things for beeswax, and another displays a kind of decoration not noted on any other vessel on Öland but with close parallels in contemporary material from western Scania (Figure 15.12) (see Lidén 1940: 160; Papmehl-Dufay 2005: 32).

Concluding Reflections

> It is not possible to analyse and explain specific aspects of ceramic technology without considering the practical and societal function of the ceramic products. (Stilborg 1997: 25)

This study has attempted to approach an assemblage of ceramic sherds and fragments at a Pitted Ware site from a number of different angles, targeting aesthetic as well as technological and functional aspects of the pottery. To reach some kind of understanding about the people involved in ceramic craft tradition represented in an archaeological potsherd assemblage is indeed a difficult task. In the introduction to this chapter, it was argued that most previous studies of Pitted Ware ceramics have stayed

Figure 15.12 One of the non-local wares in the Ottenby Royal Manor assemblage. The decoration on this vessel has parallels in southwestern and western Sweden, e.g., Jonstorp (photo by the author).

mainly at a level of potsherd presentation, discussing in great detail the development of various decorative patterns but only rarely considering social aspects of the craft tradition that once produced them. However, given the very large assemblages recovered along the eastern coast of Sweden and the apparent dominance of decorated vessels, an understanding of the social implications for ceramic craft seems important in any discussion concerning this culture and time period. Some ideas on this topic have emerged during the work on Pitted Ware pottery from the island of Öland, which are summarized here.

Despite many criticisms in recent years as to the prehistoric reality reflected in the culture groups concept, from the evidence of stable isotope studies (Eriksson 2004; Fornander 2006) and differences in burial practice (Janzon 1974; Sjögren 2004), the Pitted Ware Culture in eastern Sweden should be seen as representing a specific cultural entity with a clearly identifiable way of living, myths of origin, perceptions on life and death and, not least, a socially embedded ceramic tradition. It is probable that within this entity numerous local identity groups were included. As to the pottery aesthetically, as well as technologically, there are regional differences within the overall tradition. From the raw material analysis presented briefly above, together with the results of several previous studies (e.g., Hulthén 1977, 1998), it can be concluded that the bulk of the pottery was locally produced, most likely at the same site where it was later discarded. Evidence for ceramic production at the sites is generally scanty or non-existent, but considering the relatively simple methods used not much could be expected after 5,000 years (see Lindahl and Matenga 1995: 35). However, as was seen at Ottenby Royal Manor above, there is also evidence for a certain mobility of ceramic artefacts at these sites, possibly representing pots as well as single sherds being circulated within and between areas. This may well reflect social networks such as marriage patterns or identity groups (e.g., Hallgren 2000: 186–187; Hulthén 1997: 143–144) but could also reflect the exchange of commodities. The deviant usage pattern recorded in one of the non-local pots at Ottenby Royal Manor could indicate the use of the pot as receptacle for the transport of foodstuff, but the presence on many Pitted Ware sites of one or a few potsherds of deviant stylistic character could also indicate the circulation of pots and sherds as commodities in their own right (e.g., Stenbäck 2003: 199).

The frequent appearance in Pitted Ware assemblages of small amounts of potsherds representing other ceramic traditions (e.g., Boat Axe Culture ceramics) is interesting in this respect (see Hulthén, in Artursson et al. 1996: 236). At Ottenby Royal Manor, apart from some 7,000 sherds and fragments of Pitted Ware pottery, five sherds have been identified as Boat Axe Culture ceramics. One of these was analyzed in thin section, and proved to be tempered with grog of second generation. The clay in this sherd was not of deviant character, although the distinct tempering tradition was not present in any other sample of Pitted Ware analyzed from the island. Together with the completely different design (bowl with decoration of twisted cord and whipped cord impressions), this clearly suggests that this vessel was produced by a potter of a different ceramic tradition.

Several authors have argued that the Pitted Ware Culture is conceptually closely connected to the coastal environment; seal hunting and fishing obviously were important and probably all year round activities (Storå 2001), and studies of stable isotopes in human bone and teeth clearly shows that people depended completely on marine food resources (Eriksson 2004; Fornander 2006). Plastic clay figurines, supposedly of symbolic significance, are most often interpreted to depict seals or even 'seal-humans' (Janzon 1974; Storå 2001: 50–51), and on many sites (including Köpingsvik on Öland) cemeteries have been placed close to the shore. It may not be far-fetched to suggest that the settlement and burial pattern, the subsistence economy and diet are all parts of an extensive and well-defined 'way-of-life' concept, a cultural understanding of what life should be like, and it is not difficult to draw parallels with late Mesolithic cultures in southern Scandinavia (e.g., Knutsson 2000).

These observations, one might argue, do not agree with the existence of a lively and socially important ceramic craft tradition. However, there seems no reason why production and use of ceramics could not have been an important activity at these sites and within this culture. As evident in many chapters throughout this volume, the commonly perceived connection of pottery production with sedentary agricultural societies cannot be maintained. The Pitted Ware Culture in this sense is yet another example of a hunter-fisher-gatherer culture in which ceramic tradition played an important role. A number of traits make the ceramic tradition of this particular hunter-fisher-gatherer culture unique; a great uniformity in pottery design can be seen over a large area and for a long time period, and very large amounts of decorated pots have been produced and deposited in quite specific landscape settings, almost always in connection with the sea or, more rarely, a lake. Certain elements of the pottery design, such as the horizontal rows of cylindrical pits and the sharply marked shoulder, are remarkable in their frequency and geographic distribution. In his study of Neolithic pottery from the Åland islands, Niklas Stenbäck showed that the amount of pottery on Pitted Ware sites by far exceeds that on both earlier and later sites in the region (Stenbäck 2003: 98–100). This pattern seems to be true for Pitted Ware sites all over eastern Sweden (Segerberg 1999: 61). The social position of the potter is another interesting issue in this context. Given the large amounts of pottery produced it is reasonable to assume that, at least during the part of the year when the average temperature was above 0°C, production of pottery has been a visible and quite dominant activity at these sites. The raw material studies on Öland have shown that suitable clay is extremely difficult to find on the island, and thus the clay quarry sites may also have served as places of special significance. Whoever made the pots recovered from the Pitted Ware sites, it is clear that these potters have been skilful and deeply immersed in their own craft tradition, and it is possible that they have enjoyed a certain social status from their skills as well.

The analysis of pottery use presented briefly above showed, contrary to what could be expected judging from the faunal evidence, that at least at this site the pottery has not been used first and foremost for processing of marine products. Instead vegetables dominate in the observed biomarkers, and terrestrial animals seem to have been of as much importance as the marine resources. Unfortunately so far no burials have been found on the site and thus no direct data on dietary habits at Ottenby Royal Manor are yet available. It does seem, however, that seal and fish were indeed of great importance but that the pottery has mainly been used for processing of terrestrial/vegetable products. The sample analyzed is small and numerous source-critical factors should be considered, however the possibility emerges that the use of the pottery has been culturally restricted to certain types of commodities, or rather that certain kinds of substances were prohibited from being prepared or stored in the vessels.

Pitted Ware pottery is no doubt one of the most discussed and debated aspects of the Middle Neolithic in eastern Sweden. The large amount of ceramics on Pitted Ware sites however have resulted in a homogenization of the category, where variation and the unusual have been seen as something undesirable (M. Isaksson 2000). The resulting image of Pitted Ware pottery as rough and homogeneous is a gross generalization; the variations observed at Ottenby Royal Manor (and other sites for that matter, e.g., Papmehl-Dufay 2004) cover the whole range from coarse thick-walled to smooth and thin wares. The potsherds represent a long-lasting and socially important craft tradition, and thus variations are not explained by one single factor. Context of production, context of use, the individual potter, as well as chronology and cultural identity are all relevant in the discussion. The present study has approached this often-treated material category from a different angle, targeting the ceramic craft and the use of its product rather than the archaeological potsherds and fragments of decorative elements. Through a detailed archaeological analysis focussing on variation in a ceramic assemblage, and by utilizing chemical as well as mineralogical analyses on smaller samples in order to achieve further knowledge of the production and use of the vessels at the site, a more dynamic image of the ceramic craft within the Pitted Ware Culture in eastern Sweden is starting to appear.

Acknowledgements

I am grateful to a number of people who, by their expert knowledge, patience and invaluable comments, have helped me in producing this paper: Professor Kerstin Lidén and Dr Sven Isaksson at the Archaeological Research Laboratory in Stockholm, Dr Ole Stilborg at the Laboratory for Ceramic Research in Lund. Financial support was provided by the Swedish Research council, Berit Wallenbergs Stiftelse and Helge Ax:son Johnsons Stiftelse. Last but not least, I am happy and honoured to have been invited to the conferences in St Petersburg 2003 and Sheffield 2005 and to participate in this volume.

References

Alexandersson, K (2002) *En Mellanneolitisk Boplats vid Halltorp. Fornlämning 146, Högsrum Socken, Borgholms Kommun, Öland. Arkeologisk Förundersökning 2001*, Arkeologisk undersökning rapport, April 2002: 6, Kalmar: Kalmar Läns Museum

Alexandersson, K, Gurstad-Nilsson, H and Källström, M (1996) 'Stenålder i järnåldersland. Om den pågående specialinventeringen på Öland', 2–3 *Bulletin för Arkeologisk Forskning i Sydsverige*, 4–17

Almgren, B (1967) 'Zur technik der neolithischen keramik', XI *Tor*, 230–251

Almgren, O (1906) 'Uppländska stenåldersboplatser', 1906: 1 *Fornvännen*, 1–19, 101–118

Andersen, S H (1975) 'Ringkloster. En jysk inlandsboplads med Ertebøllekultur', *Kuml* 1973–74, 11–108

Arrhenius, B (1961) 'Vikingatida miniatyrer', VII *TOR*, 139–164

Artursson, M, Hadevik, C and Hulthén, B (1996) 'Keramiken på den gropkeramiska boplatsen', in Artursson, M (ed), *Bollbacken: En sen Gropkeramisk Boplats och ett Gravfält från Ääldre Järnålder Raä 258, Tortuna sn Västmanland,* Tryckta rapporter från Arkeologikonsult, Upplands Väsby: Arkeologikonsult AB, pp 194–240

Bagge, A (1936) 'Stenåldersboplatsen vid Humlekärrshult, Oskarshamn', 24 *Kalmar Läns Fornminnesförening Meddelanden*, 55–104

——— (1951) 'Fagervik. Ein rückgrat für die periodeneinteilung der ostschwedischen wohnplatz- und bootaxtkulturen aus dem mittelneolithikum. Ein vorläufige mitteilung', XXII *Acta Archaeologica*, 57–118

Bagge, A and Kjellmark, K (1939) *Stenåldersboplatserna vid Siretorp i Blekinge,* Stockholm: Kungl Vitterhets Historie och Antikvitets Akademien

Björck, N (1998) *Fräkenrönningen –En "by" för 5000 år Sedan. En Gropkeramisk Boplats, RAÄ 399, Valbo Socken, Gästrikland*, Gävle: Rapport Länsmuseet i Gävleborgs län 1998: 14, Länsmuseet Gävleborg

Boast, R (2002) 'Pots as categories: British beakers', in Woodward, A and Hill, J D (eds), *Prehistoric Britain: The Ceramic Basis*, Prehistoric Ceramics Research Group, occasional publication 3, Oxford: Oxbow books, pp 96–105

Burenhult, G (2002) 'The grave-field at Ajvide', in Burenhult, G (ed), *Remote Sensing,* vol II, Theses and papers in North-European Archaeology 13: b, Stockholm: Stockholm Department of Archaeology, pp 31–167

Carlsson, A (1998) *Tolkande Arkeologi och Svensk Forntidshistoria. Stenåldern*, Stockholm Studies in Archaeology 17, Stockholm: Stockholms Universitet

Claesson, T and Mikaelsson, J (2001) 'Berg och jord', in Forslund, M (ed), *Natur och Kultur på Öland,* Kalmar: Länsstyrelsen i Kalmar Län, pp 15–41

Copley, M S, Berstan, R, Mukherjee, A J, Dudd, S N, Straker, V, Payne, S and Evershed, R P (2005) 'Dairying in antiquity. III. Evidence from absorbed lipid residues dating to the British Neolithic', 32 *Journal of Archaeological Science*, 523–546

Craig, O E, Love, G D, Isaksson, S, Taylor, G and Snape, C E (2004) 'Stable carbon isotopic characterisation of free and bound lipid constituents of archaeological ceramic vessels released by solvent extraction, alkaline hydrolysis and catalytic hydropyrolysis', 71(2) *Journal of Analytical and Applied Pyrolysis*, 613–634

Edenmo, R, Larsson, M, Nordqvist, B and Olsson, E (1997) 'Gropkeramikerna –fanns de? Materiell kultur och ideologisk förändring', in *Regionalt och Interregionalt. Stenåldersundersökningar i Syd- och Mellansverige*, Riksantikvarieämbetet Arkeologiska undersökningar, Skrifter nr 23, Stockholm: Riksantikvarieämbetet, pp 135–213

Einarsson, N (2002) 'Materiell kultur som kommunikationsmedel. En gropkeramisk boplats i Åby Östergötland', unpublished seminar paper, University College, Kalmar

Eriksson, G (2004) 'Part time farmers or hard-core sealers? Västerbjers studied by means of stable isotope analysis', 23 *Journal of Anthropological Archaeology,* 135–162

Evershed, R P, Dudd, S N, Lockhart, M J and Jim, S (2001) 'Lipids in archaeology', *Handbook of Archaeological Science*, Chichester: Wiley

Fornander, E (2006) 'The wild side of the Neolithic. A study of Pitted Ware diet and ideology through analysis of stable carbon and nitrogen isotopes in skeletal material from Korsnäs, Grödinge parish, Södermanland', unpublished seminar paper, Stockholm: Archaeological Research Laboratory, Stockholm University

Gill, A (2003) *Stenålder i Mälardalen*, Stockholm Studies in Archaeology 26, Stockholm: Stockholms Universitet

Gosselain, O P (1999) 'In pots we trust. The processing of clay and symbols in Sub-Saharan Africa', 4(2) *Journal of Material Culture*, 205–230

Gruber, G (1995) 'Åby elva undersökningar senare... En gropkeramisk boplats för flera mindre hushåll?' unpublished seminar paper, Stockholm: Stockholm University

Gustafsson, P, Lindholm, P and Runeson, H (2003) 'Gropar och keramik. Ett försök att med statistik och rumslig analys belysa den inre strukturen på den gropkeramiska boplatsen vid Lindskrog', in Anund, J (ed), *Landningsplats Forntiden*, Riksantikvarieämbetet arkeologiska undersökningar Skrifter nr 49, Stockholm: Riksantikvarieämbetet, pp 15–60

Hagelqvist, T and Widenborg, M (1977) 'Björkärr. En boplats från yngre stenåldern i östra Blekinge. Ett undersökningsobjekt i kronologi och ekologi', unpublished seminar paper, Lund: Lund University

Hallgren, F (2000) 'Lineage identity and pottery design', in Olausson, D and Vandkilde, H (eds), *Form, Function and Context: Material Culture Studies in Scandinavian Archaeology,* Acta Archaeologica Lundensia, series in 8° No. 31, Lund: Almqvist & Wiksell International, pp 73–191

Hansel, F A, Copley, M S, Madureira, L A S and Evershed, R P (2004) 'Thermally produced ω-(o-alkylphenyl) alkanoic acids provide evidence for the processing of marine products in archaeological pottery vessels', 45 *Tetrahedron Letters*, 2999–3002

Hulthén, B (1977) *On Ceramic Technology during the Scanian Neolithic and Bronze* Age, Theses and Papers in North-European Archaeology 6, Stockholm: Akademilitteratur

Hulthén, B (1997) 'Stenålderskeramiken från Ire på Gotland', in Åkerlund, A (ed), *Till Gunborg: Arkeologiska Samtal*, SAR 33, Stockholm: Institutionen för Arkeologi, Stockholms Universitet, pp 131–145

——— (1998) *The Alvastra Piledwelling Pottery: An Attempt to Trace the Society behind the Sherds*, Stockholm. Monographs 5, Stockholm/Lund: The Museum of National Antiquities

Isaksson, M (2000) 'Forskning i Platons skuggvärld? Om kulturteori och variation i neolitiskt källmaterial utifrån analys av tidigare forskning och organiska beläggningar på gropkeramik från Jonstorp, Jonstorp sn, Skåne', in *CD-Uppsatser i Laborativ Arkeologi 99/00 Del 1*, Stockholm: Archaeological Research Laboratory, Stockholm University, pp 3–40

Isaksson, S (2000) *Food and Rank in Early Medieval Times*, Theses and Papers in Scientific Archaeology 3, Stockholm: Archaeological Research Laboratory, Stockholm University

——— (2006) 'Analys av organisk beläggning på skärva Ottenby 219. Arkeologiska forskningslaboratoriet, uppdragsrapport nr 31', unpublished analysis report, Stockholm: Archaeological Research Laboratory, Stockholm University

Isaksson, S, Olsson, M and Hjulström, B (2005) 'De smorde sina krås. Spår av vegetabilisk olja i keramik från yngre järnålder', 100 *Fornvännen*, 179–191

Janzon, G O (1974) *Gotlands Mellanneolitiska Gravar*, Acta Universitatis Stockholmiensis, Studies in North-European Archaeology 6, Stockholm: Stockholms Universitet

Knutsson, H (2000) 'Two technologies—two mentalities', in Knutsson, H (ed), *Halvvägs Kust till Kust. Stenålderssamhällen i Förändring*, Coast to Coast Book 2, Uppsala: Department of Archaeology and Ancient History, Uppsala University, pp 11–41

Larsson, M (2004) 'Living in cultural diversity. The Pitted Ware Culture and its relatives', 14 *Journal of Nordic Archaeological Science*, 61–69

Leijonhufvud, M (1989) 'Keramiken på Västerbjers, Gothem sn. En studie i boplats- och gravkeramik från stenåldern', unpublished seminar paper, Stockholm: Stockholm University

Lidén, O (1940) *Sydsvensk Stenålder Belyst av Fynden på Boplatserna i Jonstorp. 2. Gropkeramikskulturen*, Lund: Gleerupska Universitets-Bokhandeln

Lindahl, A and Matenga, E (1995) *Present and Past: Ceramics and Homesteads, an Ethnoarchaeological Project in the Buhera District, Zimbabwe*, Studies in African Archaeology 11, Uppsala: Department of Archaeology and Ancient History, Uppsala University

Lindholm, P (2003) 'Gravspråk på olika nivåer—den gropkeramiska graven vid Lindskrog', in Anund, J (ed), *Landningsplats Forntiden*, Riksantikvarieämbetet Arkeologiska undersökningar, Skrifter nr 49, Stockholm: Riksantikvarieämbetet, pp 61–82

Löfstrand, L (1968) 'Die grübchen der grübchenkeramik', XII *Tor* 1967–68, 128–132

——— (1974) *Yngre Stenålderns Kustboplatser. Undersökningarna vid Äs och Studier i den Gropkeramiska Kulturens Kronologi och Ekologi*, AUN 1, Uppsala: Institutionen för Arkeologi, Uppsala Universitet

Malmer, M P (1969) *Gropkeramiksboplatsen Jonstorp RÄ*, Antikvariskt Arkiv 36, Stockholm: Kungl Vitterhets Historie och Antikvitets Akademien

——— (2002) *The Neolithic of South Sweden. TRB, GRK and STR*, Stockholm: Kungl Vitterhets Historie och Antikvitets Akademien

Meinander, C F (1957) 'Kolsvidja', 58 *Finska Fornminnesföreningens Tidskrift*, 185–213

Norderäng, J (2002) 'Ajvideboplatsen. Rapport från arkeologisk undersökning 2001 av fornlämning nr. 171 på fastigeten Ajvide 2:1 i Eksta socken, Gotland', Högskolan på Gotland, unpublished report

——— (2003) 'Ajvideboplatsen. Rapport från arkeologisk undersökning 2002 av fornlämning nr. 171 på fastigeten Ajvide 2:1 i Eksta socken, Gotland', Högskolan på Gotland, unpublished report

——— (2004) 'Ajvideboplatsen. Rapport från arkeologisk undersökning 2003 av fornlämning nr. 171 på fastigeten Ajvide 2:1 i Eksta socken, Gotland', Högskolan på Gotland, unpublished report

Nordström, H-Å (1972) *Cultural Ecology and Ceramic Technology. Early Nubian Cultures from the Fifth and the Fourth Millennia B.C.*, Acta Universitatis Stockholmiensis, Studies in North-European Archaeology 4, Stockholm: Almqvist & Wiksell International

Olsson, E, Granath Zillén, G and Mohr, A (1994) *Korsnäs –En Gropkeramisk Grav- och Boplats på Södertörn*, Riksantikvarieämbetet UV Stockholm, Rapport 1994: 63, Stockholm: Riksantikvarieämbetet

Olsson, M (2004) 'Skärvor från det förflutna. En FTIR- och GCMS-analys av keramik från Tuna i Alsike, Uppland', unpublished seminar paper, Stockholm: Archaeological Research Laboratory, Stockholm University

Papmehl-Dufay, L (1999) 'Myten om gropkeramikerna. Klassificering och analys av neolitisk keramik från Vendel 1:1, Vendel sn, Uppland', in *CD-Uppsatser i Laborativ Arkeologi 98/99. Del 1*, Stockholm: Archaeological Research Laboratory, Stockholm University, pp 1–40

——— (2004) 'Kulturell identitet på Öland under mellanneolitikum. Om keramikens betydelse inom den gropkeramiska kulturen', in Werner, T and von Hackwitz, K (eds), *Aktuell Arkeologi* VIII, Stockholm: Arkeologiska Institutionen, Stockholms Universitet, pp 31–46

——— (2005) *Mellanneolitikum vid Ottenby Kungsgård. Arkeologisk Undersökning av Raä 40, Ås Socken, Ölands Sydspets Aaugusti–September 2004*, Rapporter från Arkeologiska forskningslaboratoriet 2, Stockholm: Archaeological Research Laboratory, Stockholm University

——— (2006) *Shaping an Identity. Pitted Ware Pottery and Potters in Southeast Sweden.* Theses and papers in Scientific Archaeology 7, Stockolm: Archaeological Research Laboratory, Stockholm University

Petersson, M (2001) 'Keramiska in- och uttryck', in Magnusson, G (ed), *Möre. Historien om ett Småland,* Kalmar: Kalmar Läns Museum, pp 165–184

Reuterdahl, M (2005) 'Osteologisk analys, Ottenby Kungsgård, Öland. Boplatsundersökning på en gropkeramisk lokal, 2004', in Papmehl-Dufay, L (2005) *Mellanneolitikum vid Ottenby Kungsgård. Arkeologisk Undersökning av Raä 40, Ås Socken, Ölands Sydspets Augusti–September 2004,* Rapporter från Arkeologiska forskningslaboratoriet 2, Stockholm: Archaeological Research Laboratory, Stockholm University, pp 60–111

Rosberg, A (1993) 'Ottenby Kungsgård—rötter i förhistorien', 78 *Kalmar Län 1993. Årsbok för Kulturhistoria och Hembygdsvård,* 106–109

Schulze, H (2004) *Köpingsvik på Öland—30 Undersökningar 1970–1994. Arkeologiska Undersökningar i Köpingsvik, Utförda av Riksantikvarieämbetet och Kalmar Läns*

Museum. Raä 215, 216 m fl, Köpings sn, Borgholms kn, Öland, Arkeologiska enheten Rapport 2004: 2, Kalmar: Kalmar Läns Museum

Segerberg, A (1999) *Bälinge Mossar. Kustbor i Uppland Under Yngre Stenåldern,* AUN 26, Uppsala: Department of Archaeology and Ancient History, Uppsala University

Shepard, A O (1957) *Ceramics for the Archaeologist,* Publication 609, Washington DC: Carnegie Institution of Washington

Sjögren, K-G (2004) 'Megalithic tombs, ideology and society in Sweden', in Knutsson, H (ed), *Coast to Coast—Arrival: Results and Reflections,* Coast to Coast-book 10, Uppsala: Department of Archaeology and Ancient History, Uppsala University, pp 157–182

Stenbäck, N (2003) *Människorna vid Havet. Platser och Keramik på Ålandsöarna Perioden 3500–2000 f.Kr.,* Stockholm studies in Archaeology 28, Stockholm: Stockholms Universitet

Stenberger, M, Dahr, E and Munthe, H (1943) *Das Grabfeld von Västerbjers auf Gotland,* Kungl, Vitterhets Historie och Antikvitets Akademien, Stockholm: Kungl Vitterhets Historie och Antikvitets Akademien

Stilborg, O (1997) *Shards of Iron Age Communications: A Ceramological Study of Internal Structure and External Contacts in the Gudme-Lundeborg Area, Funen during the Late Roman Iron Age,* Monographs on Ceramics, Lund: Laboratory for Ceramic Research, Lund University

——— (2001) 'Keramiken i centrum. Keramik som indikator på centralplatsstatus', in Larsson, L (ed), *Uppåkra: Centrum i Analys och Rapport*, Uppåkrastudier 4, Stockholm: Almqvist & Wiksell International, pp 123–141

——— (2002) 'Mellanneolitikum. Gropkeramisk kultur ca 3100–ca 2600 f.Kr.', in Lindahl, A, Olausson, D and Carlie, A (eds), *Keramik i Sydsverige. En Handbok för Arkeologer*, Riksantikvarieämbetet UV Syd Rapport 2002: 6, Lund: Keramiska Forskningslaboratoriet, Lunds Universitet, pp 71–74

——— (2006) Clays and tempering in wares from Köpingsvik and Ottenby. Appendix 4, in: Papmehl Dufay, L *Shaping an Identity: Pitted Ware Pottery and Potters in Southeast Sweden,* Theses and papers in Scientific Archaeology 7, Stockholm: Archaeological Research Laboratory, Stockholm University, pp 296–323

Storå, J (2001) *Reading Bones. Stone Age Hunters and Seals in the Baltic*, Stockholm Studies in Archaeology 21, Stockholm: Stockholms Universitet

Werthwein, G (2001) *Boplatslämningar, Hålvägar och en Skärvstensvall i Igelsta. Arkeologisk Särskild Undersökning av Fornlämning Raä 279 och 280, Östertälje Socken, Södertälje Kommun, Södermanland*, Rapport 2001: 7, Stockholm: Stockholms Läns Museum

Wyszomirska, B (1984) *Figurplastik och Gravskick hos Nord- och Nordösteuropas Neolitiska Fångstkulturer,* Acta Archaeologica Lundensia 18, Lund: LiberFörlag/Gleerup

Part 3

Early Pottery in Forager–Farmer Interaction Zones

Chapter 16

Hunter-Gatherers and Early Ceramics in Poland

Marek Nowak

Introduction

The aim of this chapter is to examine evidence for the early use of ceramic vessel technology by prehistoric communities in Poland. The task is challenging due to the fact that the post-glacial prehistory of central Europe has traditionally been analyzed in terms of discrete chronological epochs, and focused on the analysis of separate archaeological 'cultures', each marked by a different material and economic base. Constrained by established disciplinary convention, attempts to understand how early ceramics were dispersed into the region are made more complex due to two lingering assumptions: that Mesolithic hunter-gatherer societies were radically different from the farming societies of the Neolithic, and that pottery was unknown to the former, while it was a universal feature of the latter.

A closer contextual reading of the evidence—one which emphasizes social process over broad-scale classification—highlights the existence of two separate 'histories' concerning the arrival of farming and pottery. The first 'history' involved a very gradual and piecemeal spread of the agro-pastoral economy into new areas of Poland, which left extended periods during which farming communities were surrounded by, and interacting with, pre-existing hunting and gathering communities. The second 'history' involved a gradual take-up of pottery technology by these different communities, at different rates, at different times and in different places.

While the two dispersal processes may have had some broad general association, they are nonetheless distinct. The most striking insights generated by the present study are that many groups of hunter-gatherers appear to have had long histories of producing and using pottery long before they chose to take up farming. Moreover, while the arrival of farming populations brought new pottery traditions that the hunter-gatherers readily copied and made their own, they were also actively assimilating quite different ceramic styles from other hunter-gatherer communities who lived further to the north and east.

Mesolithic Hunter-Gatherer Cultures and the Arrival of Farming

Communities leaving material traces which archaeologists nowadays consider to be typical of the Mesolithic appeared on the territories of present-day Poland in the 9th millennium BC (Galiński 2002: 156, 175, 345; Kobusiewicz 1999: 102, 116; Sulgostowska 2003). Starting from the first half of the 7th millennium BC, the tool inventory of the European Mesolithic underwent typological and technological transformations, which consisted of the ongoing standardization of flint industries. With time, the process was reinforced and either obliterated or diminished the hitherto typological (and functional) diversity of Mesolithic inventories. Typical of this convergence process are mainly trapezes and truncations made on regular blade blanks as well as end-scrapers and side-scrapers (Galiński 2002; Kozłowski 1989). On the other hand, that does not mean the late Mesolithic inventories were absolutely standardized. One can still observe fairly distinct archaeological units.

Between 6,000 BC and 4,000 BC, these units were the post-Maglemosian Complex, late Komornica Culture, and Janisławice Culture (Figure 16.1). Due to their unique character, the Janisławice inventories continue to arouse much interest as well as much controversy. Certainly, they are not reminiscent of any previous unit of the central European Mesolithic. A number of authors point to their potential links with the Pontic regions (Domańska 1990; Kozłowski 1989: 163, 200–201) or, recently, with the regions of southern Scandinavia (Galiński 2002: 356–360).

By the middle of the 6th millennium BC, we see the first arrival of farming communities on the territories of central Europe. These early Neolithic settlements form part of the Linear Band Pottery Culture (LBK), and are characterized by their clustered distribution, which is generally limited to areas with fertile soils appropriate for an agricultural economy. It should be noted that increasing numbers of LBK sites have recently been found outside the 'typical' enclaves of early Neolithic settlements. There are single sites in the mountains, in the 'non-loess' parts of the uplands and in the lowlands, sometimes considerably remote from the clearly delineated areas of LBK settlement (Figure 16.1). While these more recently discovered sites

Figure 16.1 General map of Polish territories in the period 6,000–4,800 BC. (A) enclaves of LBK settlement; (B) single sites of LBK outside aforementioned enclaves; (C) Janisławice Culture; (D) Late Komornica Culture; (E) post-Maglemose complex; (F) Mesolithic sites within LBK enclaves: Glanów (G) and Śniejowice (S) (Chochorowski 2001; Pazdur et al. 2004); possibly transitional, Mesolithic/Neolithic sites: Podgaj (P) and Dęby (D) (Domańska 1990).

are exceptions to the general rules of distribution of LBK settlements, they nonetheless point towards 'linear' penetrations of the surrounding areas occupied by Mesolithic hunter-gatherer peoples.

In the context of the LBK settlement in Kujawy, we encounter a very interesting phenomenon represented by sites such as Podgaj 32 (Czerniak 1994: 54–58; Domańska 2003). These are four sites situated, significantly, not on fertile marsh black soils, but on sands. The inventories from these sites include a flint industry with some Mesolithic features, and LBK-like ceramics. The character of these ceramics is currently under discussion, which is hindered by the inadequate number of relevant publications. Early opinions were that these ceramics were 'barbarized' LBK ceramics. At present a more widespread view is that of Czerniak (1994), who states these are indeed LBK ceramics, but only the kinds of coarse wares used in domestic cooking contexts. The interpretation of this phenomenon is not easy since there could be two possibilities. First, that the ceramics represent traces of LBK people who either inhabited or penetrated the sand zone and who for some reasons (perhaps practical ones) used the Mesolithic type of flint tool inventories in these environments. Second, the ceramics may represent evidence for the early use of LBK ceramic cooking vessels amongst Mesolithic hunter-gatherer peoples.

Around 4,800–4,700 BC, the LBK culture disappeared, and was followed by the Stroke Band Pottery Culture (SBK) in western Poland with units that are classified as belonging to the Lengyel-Polgar complex (Figure 16.2). Around 4,500 BC, the SBK culture declined. The areas settled by SBK people became integrated with the expanding Lengyel-Polgar complex,

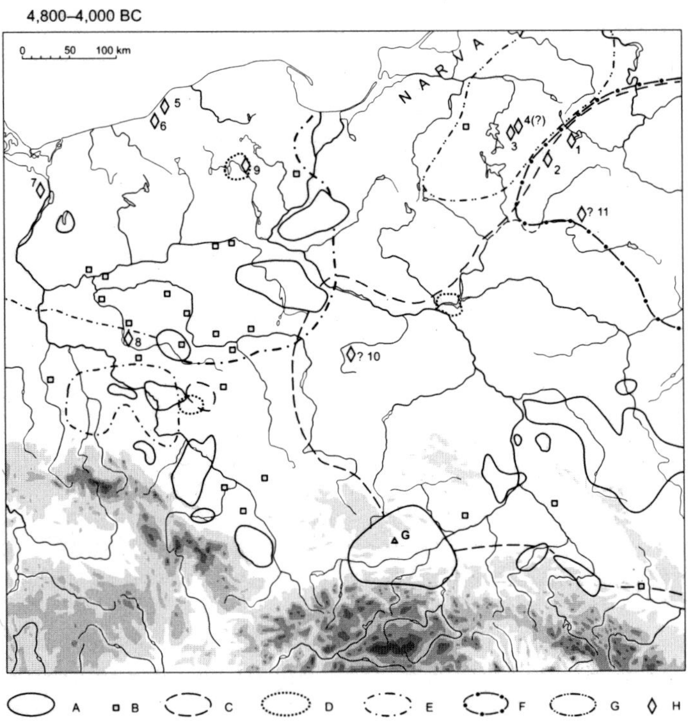

Figure 16.2 General map of Polish territories in the period 4,800–4,000 BC. (A) enclaves of SBK and Lengyel-Polgar settlement; (B) single sites of SBK and Lengyel-Polgar complex outside aforementioned enclaves; (C) Janisławice Culture; (D) Late Komornica Culture; (E) post-Maglemose complex; (F) Prypet-Neman Culture; (G) early stage of Zedmar Culture; (H) selected 'foraging' sites with pottery: (1) Augustów-Wójtowskie Włóki, (2) Woźna Wieś, (3) Szczepanki, (4) Dudka, (5) Dąbki, (6) Koszalin-Dzierżęcino, (7) Tanowo 3, (8) Chobienice, (9) Męcikał 6, (10) Witów, (11) Nowodworce.

which survived on the Polish territories for as long as the middle of the 4th millennium BC. The settlement of these groupings continued to be concentrated in limited areas, the majority of which were the same fertile areas that previously had been occupied by the LBK. These post-LBK sites were slightly more numerous with small enclaves located in Greater Poland and Lower Silesia. Moreover, as with the LBK, recent years have seen the discovery of increasing numbers of post-LBK sites. Most of these belong to the Brześć Kujawski group of the Lengyel-Polgar complex and often lie at a considerable distance from the more typical early kinds of Neolithic enclaves that are mentioned above. These new sites include settlements which were discovered in eastern Pomerania and the Mazurian Lake District (Paner et al. 2003; Rybicka and Wysocki 2004). In summary, it should be noted that most areas of present-day Poland were not permanently settled by Neolithic peoples but remained occupied by Mesolithic groups of hunter-gatherers.

In the literature, there are many debates surrounding the late Mesolithic in Poland, which starts from the 5th millennium BC (Bagniewski 1979, 1982, 1998; Czerniak 1994: 9–10; Galiński 1991; Kobusiewicz 1999; Kozłowski 1989; Kukawka 1997; Schild 1998). One major problem is the apparent homogeneity of the materials recovered from Mesolithic sites, caused by geological and geomorphological factors that at most sites hinder considerably the sequence of deposition of natural and anthropogenic sediments as well as archaeological artefacts (Schild 1989). Despite these difficulties in demarcating compact archaeological complexes, it is argued here that lands outside the densely settled early agricultural enclaves were occupied by populations of hunter-gatherers until at least the end of the 3rd millennium BC, and possibly even longer. In other words, farmers co-existed with hunter-gatherers for at least 3,500 years. This hypothesis is supported by radiocarbon dates, the typology of flint materials and the presence of ceramic vessels in hunter-gatherer contexts (Nowak 1999, 2001). Cartographic analysis shows that the main concentrations of these late hunter-gatherer settlements were in lowland areas, including the Pomeranian and Mazurian Lake Districts, some areas of northeastern Mazovia, Greater Poland, Lower Silesia and central Poland (Nowak 2001: 586).

The northern edges of the Polish territories were influenced by the cultures of southern Scandinavia. In the 5th millennium BC, and possibly in the early 4th millennium BC, southern Scandinavia was settled by communities of hunters and gatherers who formed the Ertebølle Culture, a highly-structured cultural complex (Larsson 1990; Rowley-Conwy 1998, 1999; Tilley 1996: 9–69). This culture deserves special attention because of the pottery that is frequently recovered from these sites. Interestingly, these ceramics appear to have been neither an import nor an imitation of the 'southern' Neolithic pottery produced by the LBK groups. These distinctive 'hunter-gatherer' ceramic wares are also present on some 'Polish' sites that were connected with the Ertebølle culture, for example, Dąbki (Ilkiewicz 1989), Koszalin-Dzierżęcino (Ilkiewicz 1998), Tanowo (Galiński 1992) and Chobienice (Kobusiewicz and Kabaciński 1998) (Figure 16.2).

The Dąbki settlement is situated partly on sands, partly on peat and appears to date to c. 5,200–4,000 BC (Pazdur 1991: 34). The excavations recovered very rich flint, bone, horn and ceramic assemblages (the latter amounting to 1,500 fragments). The typology of the flint industry was post-Maglemosian, with a few features of the Ertebølle culture (Ilkiewicz 1989: 28; Kabaciński 2001). The ceramics are clearly linked to the pottery of the Ertebølle culture, although many vessels have their own unique features too. According to Ilkiewicz (1989: 34), the first pottery appeared around c. 4,600–4,500 BC and the technology remained in use as long as the settlement was occupied. The vessels were made of fat clay 'containing considerable amounts of organic substances, mixed with iron compounds, unevenly distributed, badly mixed sandy and rocky materials, sharp-edged quartz and bits of granite. Sometimes, ceramic breakages were used as admixture' (Ilkiewicz 1989: 28). The vessel bases were formed from a single lump of clay. The walls were of uneven thickness and were made from flattened bands, with straight or rounded joints set obliquely in relation to the walls. The vessel surface was smeared with fatty irony silt and then polished. The pots were fired in temperatures up to 650°C, in a weakly reductive atmosphere. According to Ilkiewicz and Wirska-Parachoniak,

> the production of vessels was of a very low standard, which, presumably, was due to the lack of potters' tradition, and also a short-lived development of this branch. The skill of making vessels was adopted without a sound knowledge, particularly of the proper choice of clay, its preparation and selection of the appropriate admixture. The skill was presumably worked out unaided, thus inventing an original tradition of ceramics. (Ilkiewicz 1989: 31; Wirska-Parachoniak 1991: 47–50)

The basic vessel form was a beaker with a flared neck, giving the pots a more or less S-shaped profile, which tapered down to circular, concave or sharply pointed base (Figures 16.3: 11, 16.4: 3, 5). A range of other forms were also produced, including oval bowls (thick-bottomed ones, perhaps lamps) (Figure 16.3: 6, 7), conical bowls (Figure 16.3: 13), sack-like vessels and compound vase-shaped vessels (Figure 16.3: 1, 2, 10). Only 4% of these ceramic fragments bear traces of ornamentation (Ilkiewicz 1989: 31). Decorations were produced using various techniques, including imprinting, perforating or incising. Some plastic ornaments were also added, usually below the rim (Figure 16.3: 4, 5). In summary, most of the features of

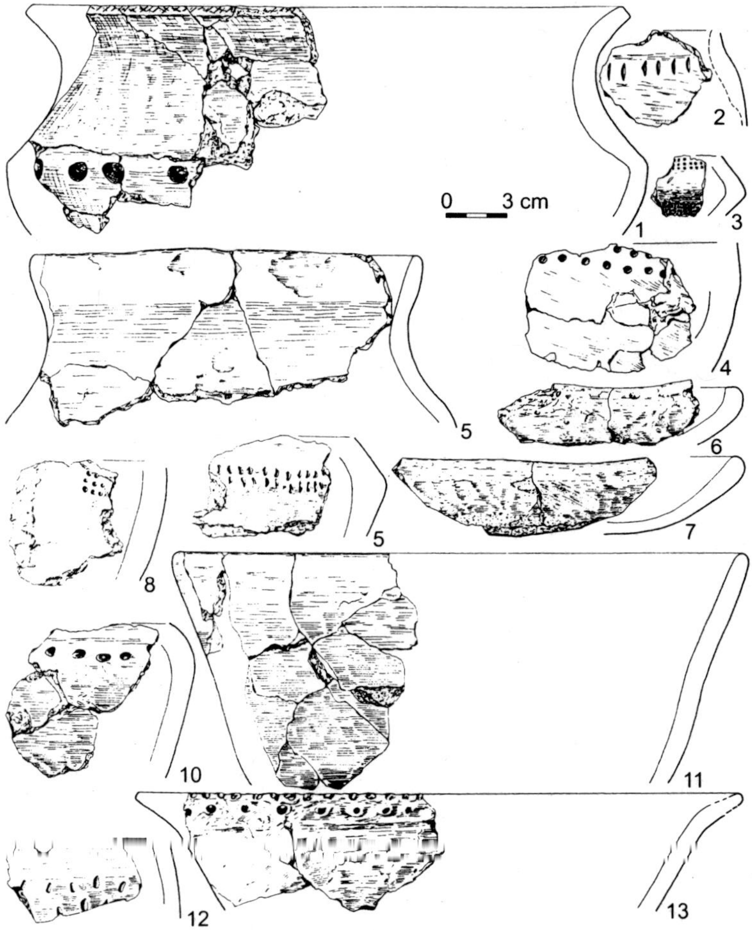

Figure 16.3 Ertebølle pottery from site Dąbki 9 (source: Ilkiewicz 1989, figure 7).

the Dąbki ceramic complex resemble those of the Ertebølle culture. Nevertheless, it is interesting to note the presence of other features, which were not derived from Ertebølle pottery styles. For example, some vessels have a more prominent profile with distinct necks and embossed or stroked ornamentation on and below the rim, and along the maximum protrusion of the belly (Figure 16.3: 2–4, 8, 9, 10, 12, 16.4: 5) features more reminiscent of the pottery of the Brześć Kujawski group of the Lengyel-Polgar complex (Ilkiewicz 1989: 30–33).

There are other possible indications that the hunter-gatherers at the Dąbki settlement had 'Neolithic' contacts. Examination of the animal bones recovered from the site (analyses conducted by Sobociński, after

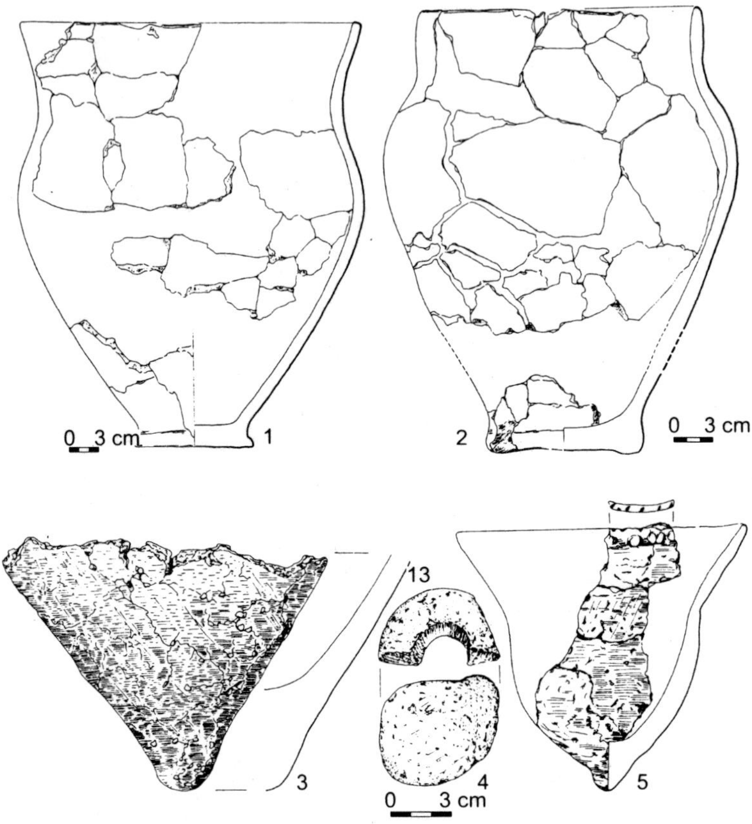

Figure 16.4 Linin pottery from Chwalim (1, 2) and Ertebølle pottery from Dąbki 9 (3–5) (source: Ilkiewicz 1989, figure 6: 21–23; Szmyt 1993, figures 24, 25).

Ilkiewicz 1989: 21–25, 1997: 51) suggests that the community may also have been keeping a small but gradually increasing number of domesticated cattle and pigs, although the poor stratigraphy of the site makes this conclusion tentative.

The pottery from the sites of Koszalin-Dzierżęcino and Tanowo (Figure 16.5) is similar to the pottery from Dąbki, but it is possible to detect larger technological diversity in the Tanowo pottery assemblages (Galiński 1992: 63–65, 70–75, 78–79, 85–87, 2005: 71–72, 78–85). Interestingly, Ertebølle type ceramics have also been found in the Lubuskie Lake District, which is a significant distance from the sea coast. The site in question is Chobienice (Czerniak and Kabaciński 1997: 76; Kobusiewicz 1999: 160–161; Kobusiewicz and Kabaciński 1998: 95–96). This Ertebølle-type pottery also has certain technological similarities to Lengyel-Polgar pottery, according to Kobusiewicz and Kabaciński (1998: 95).

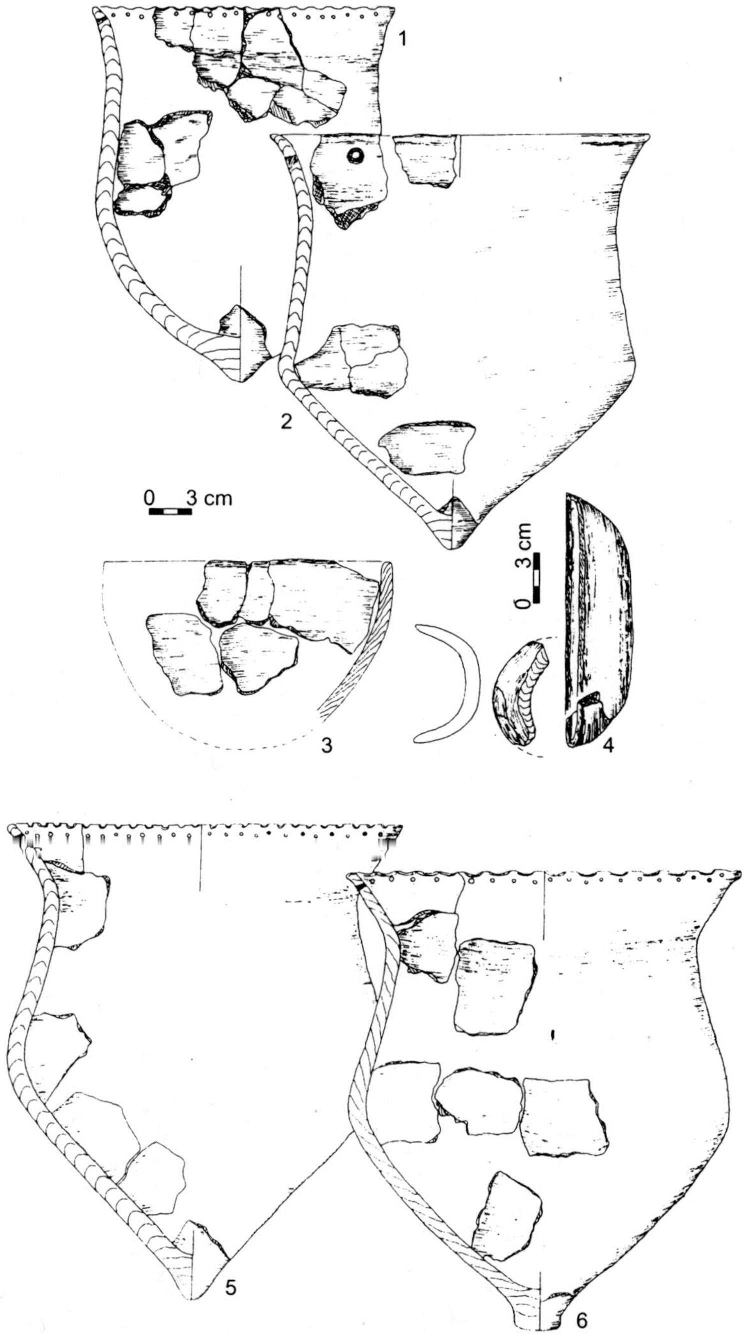

Figure 16.5 Ertebølle pottery from Tanowo 3 (source: Galiński 1992, figures 2.25, 2.26).

The pottery-using hunter-gatherers of the Ertebølle culture inhabited the northern fringes of central Europe, but their distinctive ceramics were only one of several kinds of 'non-Neolithic' pottery that were produced by hunter-gatherer communities in eastern parts of central Europe (Figure 16.2). As noted in the introduction, the use of pottery by prehistoric hunter-gatherers undermines the classic distinction between the Mesolithic and Neolithic, and has spawned a series of adapted terms, which include the para-Neolithic, proto-Neolithic, sub-Neolithic, Forest Neolithic, Comb-Pitted Pottery Culture and the Ceramic Mesolithic (Gumiński 1999: 59; Józwiak 2003: 11–12; Werbart 1998), not to mention the less popular ones. For the purposes of this chapter, the term 'pottery-using hunter-gatherers' is used.

It is now widely acknowledged that pottery-using hunter-gatherer communities inhabited vast regions of eastern Europe which is here defined as the European territories east of present-day Poland. These communities came into view along the southern borders of eastern Europe at the turn of the 7th and 8th millennia BC, and then subsequently spread over the entire territory of eastern Europe (Dolukhanov et al. 2005; Józwiak 2003: 43–46). These forager communities remained in the region through several millennia and were only eclipsed by transition to the agricultural economy in the Bronze Age. The economy of these communities was based mainly on hunting and gathering, but in some areas there was some limited familiarity with agriculture (Gumiński 1998; Gumiński and Michniewicz 2003; Rimantiené 1998: 217). The production and use of pottery was very widespread, and the vessels had distinctive technology, morphology and decoration that were quite different to the strictly Neolithic ceramics made by farming communities. With some exceptions, this hunter-gatherer pottery is very different to other Neolithic pottery in central Europe, but does have some broad resemblance to the ceramics of the Ertebølle culture, as noted above. In addition, the hunter-gatherer flint industries also have their own unique features, but in general tend to be similar to those of the preceding Mesolithic.

In the Baltic Countries, the first use of ceramics by hunter-gatherers may have occurred as early as the middle of the 6th millennium BC (Liiva and Loze 1993, 1994; Timofeev 1998: 226) and in Belarus begins around 5,000 BC (Józwiak 2003: 58). As a result, it is possible that pottery may have been present in the northeastern part of today's Poland as early as the first half of the 5th millennium BC. In archaeological terminology, the first use of pottery has been designated as the Dubiciai stage of Neman Culture (Kempisty and Sulgostowska 1991), although more recently it has been suggested that a distinction should be made between this stage and the Neman Culture proper. Consequently, the term 'Prypet-Neman Culture' was introduced instead of 'Dubiciai stage of the Neman culture' (Józwiak

2003: 57–63). Materials of this kind have been identified at only a few sites in the northeastern part of Poland. These sites include Woźna Wieś and Augustów-Wójtowskie Włóki (Kempisty and Sulgostowska 1991) and ceramic assemblages include thick-walled, scantily decorated vessels with uneven bases. The clay pastes contain considerable quantities of fibrous plant admixture (Figure 16.6: 1, 3–6). While the pottery is a new innovation, many of the flint artefacts show strong continuity from the earlier Mesolithic periods, a scenario repeated in the Belarus, Lithuanian and Volhynian sites of this culture (Józwiak 2003: 60). This suggests that it was the local foraging population who either invented or adopted the pottery.

Around 4,500–4,400 BC, we see the emergence of the Zedmar Culture in the Mazurian Lake District. In the 1970s and 1980s, it was suggested that this culture formed a southern extension of the Narva Culture, which comprised pottery-using hunter-gatherers and was generally located in the Baltic countries (Kempisty 1983), although this hypothesis has been firmly rejected following more recent research by Gumiński (1999). He

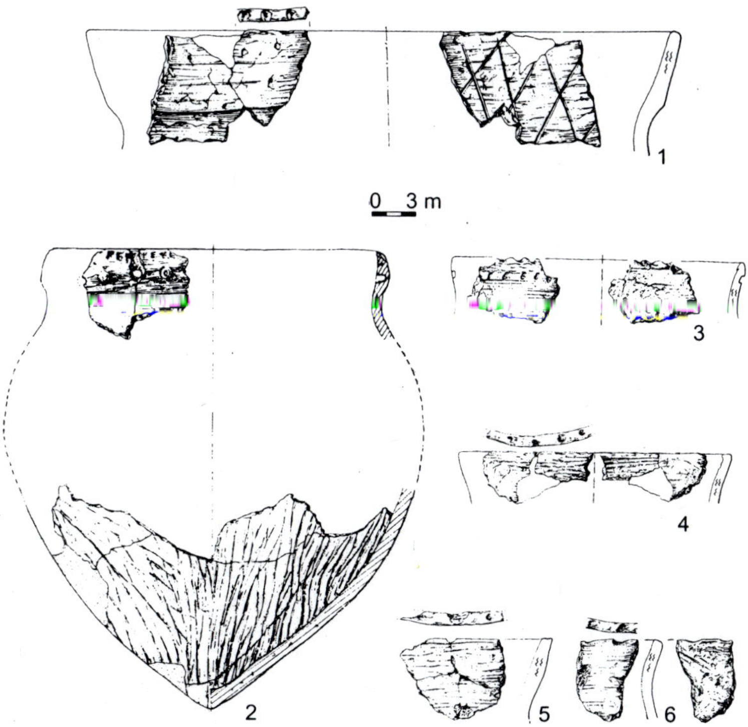

Figure 16.6 Prypet-Neman Culture pottery from Woźna Wieś (1, 3–6) and Neman Culture pottery from Sośnia (2) (source: Kempisty and Sulgostowska 1991, plate XXXIV; Kempisty and Więckowska 1983, plate XVIII: 1).

demonstrated that Zedmar Culture should be linked to the Ertebølle or Swifterbant cultures rather than to the pottery-using hunter-gatherer cultures of eastern Europe. One reason for this is the use of very different organic and mineral tempers, sometimes in tandem, in the clay pastes of the distinctive Zedmar Culture pottery. The pottery is poorly fired, making it fragile and crumbly. The pots or cups have S-shaped profiles with everted necks, and beakers have clearly delineated and protruding rims. However, in contrast to all other hunter-gatherer pottery styles which have pointed bases, the Zedmar vessels are flat-bottomed. Decoration was also used sparingly, and is found on only 10% of the sherds. The ornamentation was applied by the impression of a finger, nail, stick or stamp. The pots are also incised and furrowed, usually below the rim, and notched rims were also common.

Gumiński's work (1999, 2003) demonstrated conclusively that the Zedmar ceramics were not just markedly different from Narva Culture pottery, but were also quite different from all other hunter-gatherer pottery styles. The analysis also revealed that Zedmar ceramics shared certain traits that belonged to the pottery of Neolithic cultures. These included notched rims, rims with eaves, straight and slanted lines of imprints, all of which were typical of Lengyel-Polgar pottery of the Brześć Kujawski group during the second half of the 5th millennium BC (Figure 16.11: 1, 4).

Changing Forager–Farmer Interactions: 4,000–2,000 BC

The 5th and 4th millennia BC marked the start of pivotal changes in the patterns of interaction between farmers and hunter-gatherers in the basins of the Vistula and the Oder Rivers. After this point, it is possible to see the increasing spread of the Neolithic over much larger areas, as the agro-pastoral economy is taken up beyond the narrow enclaves occupied by the early farmers. This wider transition to agriculture accelerates so that by the third quarter of the 4th millennium BC, the Neolithic was clearly apparent in the majority of the Polish territories (Figure 16.7). These transformations can be understood as the second stage of neolithization, with the first being the initial arrival and settlement of Neolithic pioneers of the LBK culture. This wider expansion of farming appears to have been connected with the emergence and later expansion of the Funnel Beaker Culture (FBC). The origins of this culture are extremely complex, and aside from the characteristic FBC pottery, regional variants of the culture include both local Mesolithic (in the lowlands) and early Neolithic elements (principally in the uplands). As to the topic of FBC genesis, it has to be emphasized that recently published pottery data from Tanowo (Galiński 2005) demonstrate that pottery-using hunter-gatherers in western Pomerania had significant impact on the

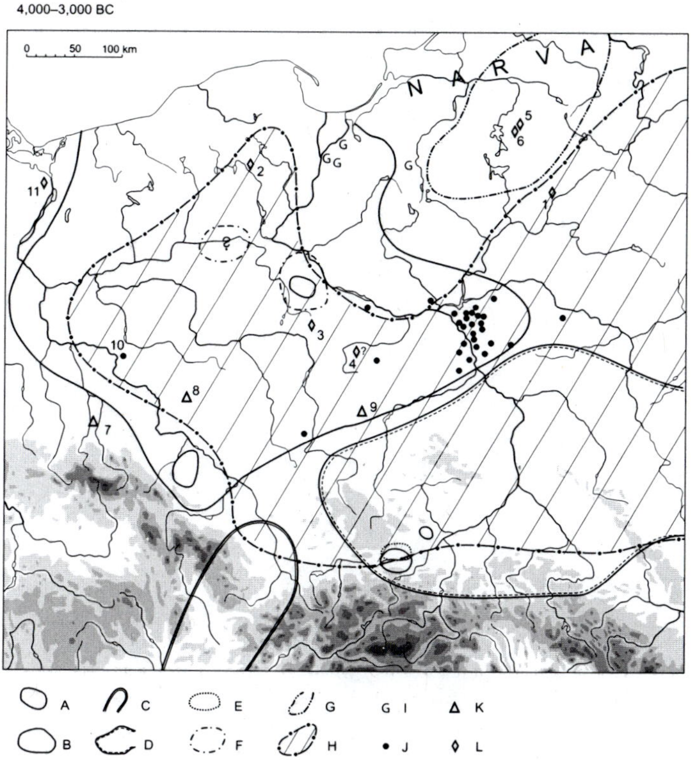

Figure 16.7 General map of Polish territories in the period 4,000–3,000 BC. (A) enclaves of the latest Lengyel-Polgar settlement; (B) Eastern Group of Funnel Beaker Culture (FBC); (C) Silesian-Moravian Group of Funnel Beaker Culture (FBC); (D) South-Eastern Group of Funnel Beaker Culture (FBC); (E) Baden Culture; (F) early Globular Amphorae Culture; (G) Zedmar Culture; (H) Neman Culture; (I) Comb Pottery Culture; (J) sites with Linin type pottery (10—Chwalim); (K) late Mesolithic sites without pottery; (L) selected 'foraging' sites with pottery: (1) Sośnia, (2) Męcikał 11, (3) Korzecznik, (4) Witów, (5) Dudka, (6) Szczepanki, (7) Dąbrowa Krępnica, (8) Pobiel, (9) Mokracz, (10) Chwalim, (11) Tanowo.

development of FBC pottery style, perhaps similar to the situation in the Lower Elbe area.

Despite the considerable expansion of FBC culture, it never spread over the full territory of Poland, leaving a number of areas in which communities of hunter-gatherers were existing alongside neighbouring farming groups (Figure 16.8) (Bagniewski 1996, 1998; Kanwiszer and Trzeciak 1991: 120–121; Kobusiewicz 1999: 153–175). Most of these 'late' hunter-gatherer groups appear to have both made and used ceramics (Cyrek et al. 1983; Galiński 1991). This phenomenon is particularly interesting because their ceramic technology appears to have been inherited from earlier hunter-

3,000–2,000 BC

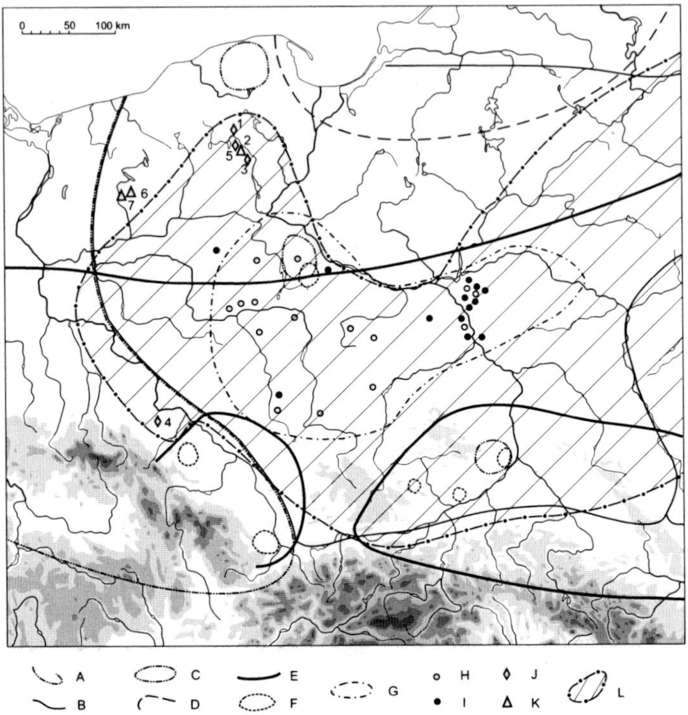

Figure 16.8 General map of Polish territories in the period 3,000–2,000 BC. (A) Western Group of Globular Amphorae Culture; (B) Central and Eastern Groups of Globular Amphorae Culture; (C) the latest Funnel Beaker Culture (FBC); (D) Rzucewo/Pamariu Culture; (E) regions with concentrations of Corded Ware sites (Little Poland, Silesia, Northern Poland); (F) Bell Beaker Culture; (G) concentration of sites with Linin type 4 pottery; (H) sites with Linin type 3 pottery; (I) pottery with Linin type 2 pottery; (J) selected 'foraging' sites with pottery: (1) Zbrzyca, (2) Męcikał 6, (3) Turowiec 3, (4) (K) Mesolithic sites without pottery, (5) Turowiec 1, (6) Kalisz Pomorski, (7) Prostynia; (L) Neman Culture.

gatherer pottery traditions rather than being adopted from the expanding Neolithic groups of the FBC (Józwiak 2003: 71). These distinctive hunter-gatherer ceramics are mainly concentrated in the northeastern areas of Poland, but have also been found in most other areas, apart form the mountainous and sub-mountainous regions (Józwiak 2003; Kempisty 1983) (Figure 16.7). These distributions are commonly interpreted as a reflection of the westward expansion of pottery-using east European hunter-gatherer communities into ecologically similar enclaves, and the ceramics have been argued to represent part of the Neman Culture (Józwiak 2003) (Figures 16.7 and 16.8). In northeastern Poland, there are also limited finds of ceramics which appear to have been derived from other pottery-using hunter-gatherer cultures, including the Narva Culture and the Comb Pottery Culture (Józwiak 2003: 69, 82). Besides, the Zedmar Culture still existed in this region. In summary, until the beginning of the Bronze Age, there were a number of hunter-gatherer groups using pottery derived from other foragers to the East, in particular, in northeastern Poland. Alongside these groups were also hunter-gatherer populations who did not use pottery and can be classed as Final Mesolithic in character, that is, with highly standardized flint industry that makes it difficult to use cultural classifications typical for Mesolithic industries dated to the Atlantic period.

From the archaeological evidence, this is nevertheless quite difficult to understand. Often it is not easy to define the point at which ceramics are adopted, because many sites are located on sands which make the stratigraphy difficult to reconstruct. Moreover, there is strong continuity in the flint industries before and after the adoption of pottery (trapezes, blade truncations, side-scrapers), which only sometimes were supplemented with 'Neolithic' elements (points, retouched inserts, stone axes, bifacial, flat retouches, lamellar retouches). Therefore, contrary to previously quoted allochthonous views, regional variations of the Neman Culture do appear to have been a continuation of older indigenous Mesolithic groups, with only the adoption of pottery a cultural tradition adopted from the East.

Let us examine in more detail the early 'non-farmer' ceramics that date to the 4th and 3rd millennium BC, and partly to the 2nd millennium BC. The majority of this hunter-gatherer pottery is characterized by the use of grog and granite temper, occasionally with organic admixtures. Most vessels are thin-walled and the surfaces have been smoothed with bunches of grass. The pottery is found at a number of sites but its extremely fragmented nature makes it difficult to reconstruct the original forms. Mostly, the vessels appear to have been S-shaped, and include bowls, vases, cups and pots, with both even and uneven bases (Figure 16.9). The predominant ornamentation is made up of furrowed imprints, various stamps and knobs placed under the rims, which are also grooved and incised (Józwiak 2003: 74, 93–121; Kempisty and Więckowska 1983: 76–78).

This very general description of 'non-Neolithic' ceramics refers mainly to variations of the classic Neman Culture. Another very interesting type of ceramic used by prehistoric hunter-gatherers bears significant resemblance (including morphology, some technological features and decoration) to the pottery made by Neolithic farming communities who inhabited the eastern regions of central Europe. Kempisty defined this pottery as 'Linin Type' (1983), although more recently Józwiak (2003) included it into Neman Culture as an 'unusual' element which forms the 'Linin horizon'. These ceramic traditions persisted for a considerable period, from the 2nd half of the 4th millennium BC to the early 2nd millennium BC, and included the adoption of several forms similar to those of subsequent Neolithic archaeological

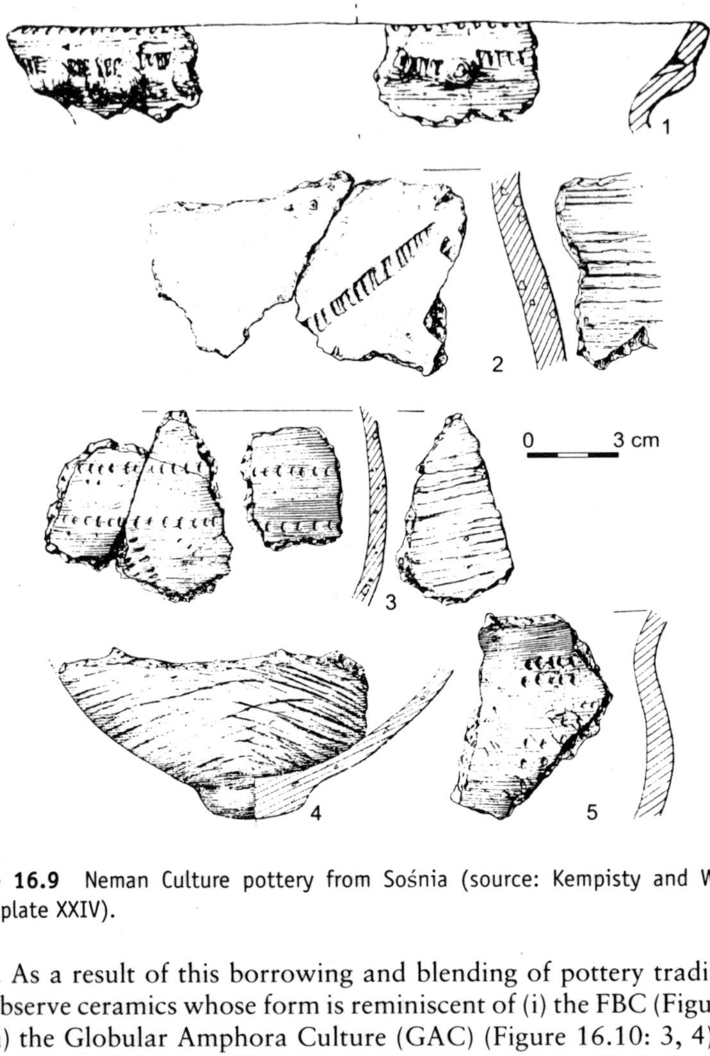

Figure 16.9 Neman Culture pottery from Sośnia (source: Kempisty and Więckowska 1983, plate XXIV).

units. As a result of this borrowing and blending of pottery traditions, we can observe ceramics whose form is reminiscent of (i) the FBC (Figure 16.10: 2); (ii) the Globular Amphora Culture (GAC) (Figure 16.10: 3, 4); (iii) the Corded Ware Culture (CWC) (Figure 16.10: 1); (iv) the Early Bronze Iwno Culture with local elements of the Bell Beaker Culture (Figure 16.10: 5, 6) (respectively, the Linin horizon 1, 2, 3, and 4, according to Józwiak). The latter type of Linin pottery is the most frequent (Józwiak 2003: 204–209). We must also note that various manifestations of the Linin pottery tradition have been found in association with 'proper' examples of Neman Culture ceramics. However, these finds were made at sandy sites where the stratigraphy is often difficult to interpret.

One of the best researched sites containing Linin ceramics is Chwalim (Figure 16. 7: No. 10) (Kobusiewicz and Kabaciński 1993), where a series

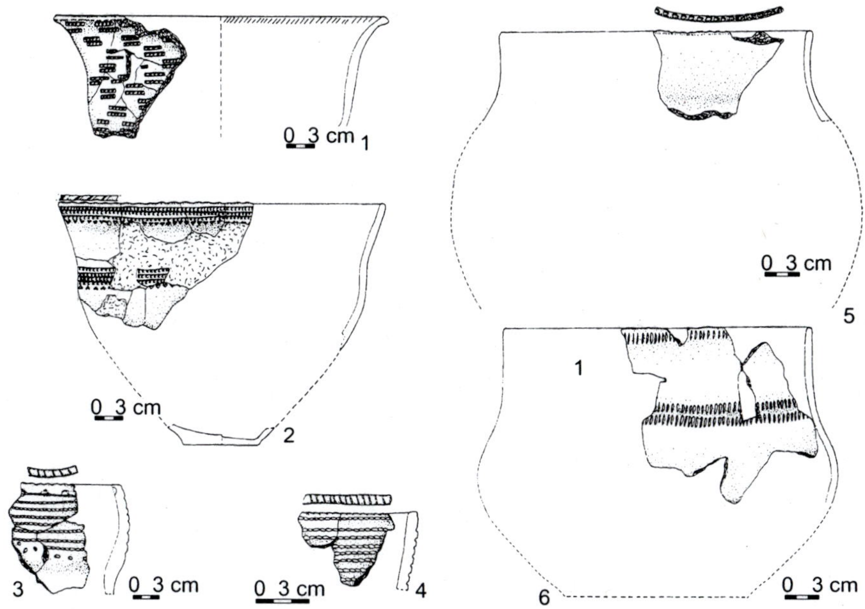

Figure 16.10 Linin pottery; type 1—2; type 2—3, 4; type 3—1; type 4—5, 6 (types after Józwiak 2003). 1, 3, 4—Linin, 2—Działy Czarnowskie, 5, 6—Korzecznik (source: Kempisty 1983: 185–187; Olszewski 1987, figures 18, 19).

of radiocarbon dates enables a more precise understanding of the material. These dates enable us to establish the occupation of the site to the second half of the 4th millennium BC (Kobusiewicz and Kabaciński 1993: 77–78). The assemblage included 73 ceramic fragments and two fairly complete vessels (Figure 16.4: 1, 2). According to Szmyt (1993), the ceramic technology employed the same formulae as pottery of the GAC, including the use of a coarse temper of crushed stone, plus the addition of smaller quantities of organic material and sand to the clay paste. The majority of ceramics have walls of medium thickness. In some cases, the vessel surface had been smoothed with a bunch of grass. None of the vessels has a pointed base or an S-shaped profile. Instead, we see flat-bottomed shapes more reminiscent of the GAC, and perhaps even the FBC. Only four of the sherds had been decorated. Ornamentation methods included the use of double cord, twisted cord, rectangular stamps and plastic applications.

In the first half of the 4th millennium BC, we can note the existence of the next stage of Zedmar Culture in the Mazurian Lake District, which was still characterized by hunter-gatherer communities with pottery (Figure 16.7). As in the preceding period, these ceramics display similarities with Neolithic pottery, though not the with the Brześć Kujawski group, but with the pots of the lowland FBC groups (Figure 16.11: 2, 3, 5). In particular, we

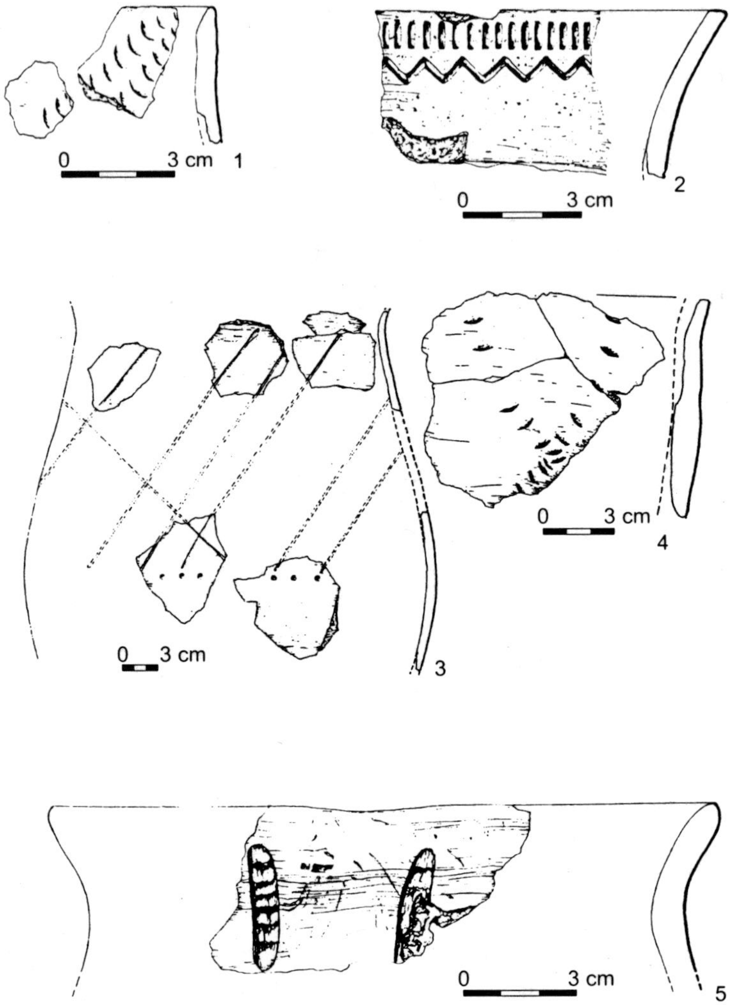

Figure 16.11 Zedmar Culture pottery from Dudka (source: Gumiński and Fiedorczuk 1988, figures 6, 7, 8).

can observe the following features (Gumiński and Fiedorczuk 1988): (i) the vessels have flat bottoms and no handles; (ii) the necks are short and bent distinctly outwards; (iii) the decoration is scant; (iv) the necks bear single rows of stamp prints.

Finally, it is worth noting that in some parts of Poland, there is evidence that might be interpreted as Neolithic farmers adopting traits characteristic of hunter-gatherer pottery. The phenomenon has been observed in the pottery of the FBC Culture in Kujawy (Kośko 1981, 1988), the Chełmno Land

(Kukawka 1990, 1991) and at Bronocice (personal communication Janusz Kruk) in the western parts of Little Poland.

Conclusions: Foragers, Farmers and Early Ceramics in Prehistoric Poland

The traditional approach to the post-Pleistocene prehistory of central Europe focuses on the detection and documentation of discrete archaeological cultures. As a result, discussions relating to change resort to descriptions of how such homogeneous cultural 'units' appear, develop and disappear. A broader form of categorization involves the allocation of these cultures to larger epochs that make up stages in the general evolutionary progress of mankind. When analyzed in these terms, the archaeology of the Mesolithic and Neolithic is sharply divided by the arrival of the agricultural revolution: the Mesolithic ends as suddenly as the Neolithic begins, with each period characterized by very different social, economic and technological features.

In reality, central European prehistory in particular was far more complex—over 3,500 years farming communities interacted with adjacent groups who had continued to practise hunting and gathering (Zvelebil 1981, 1993). Although some of these forager groups did acquire practical knowledge of agriculture and pastoralism from adjacent farmers, most did not and survived by exploiting only local wild resources. In contrast, many hunter-gatherer groups did appear to take up traits from Neolithic groups, one of which was pottery, incorporating elements of this 'new' technology into their existing ceramic traditions and practices. However, the history of early ceramics in the region is more complex than a simple process of adoption from newly arrived farming populations, to be followed later by the eventual take up of farming (Figure 16.12). In fact, there is evidence for hunter-gatherers using pottery as early as the 5th millennium BC (Figure 16.2), but the characteristics of this pottery are somewhat dualistic. Some vessels resemble the pottery of other foraging groups in eastern Europe, and may have been acquired from them. Other features appear to have been derived from local Neolithic farming cultures who had migrated into the area as part of the general advance of the agricultural revolution.

These 'dualistic' features of local pottery continued in the 4th and 3rd millennium BC (Figures 16.7 and 16.8). The degree to which 'proper' Neolithic traits were incorporated into the pottery varied from region to region across present-day Poland, but they were generally significant enough to distinguish local hunter-gatherer pottery from the pottery made by other forager groups further to east or northeast. However, only after the turn of the 3rd and 2nd millennium BC is it possible to identify a steady and constant increase in Early Bronze pottery traits, which rise along with the adoption

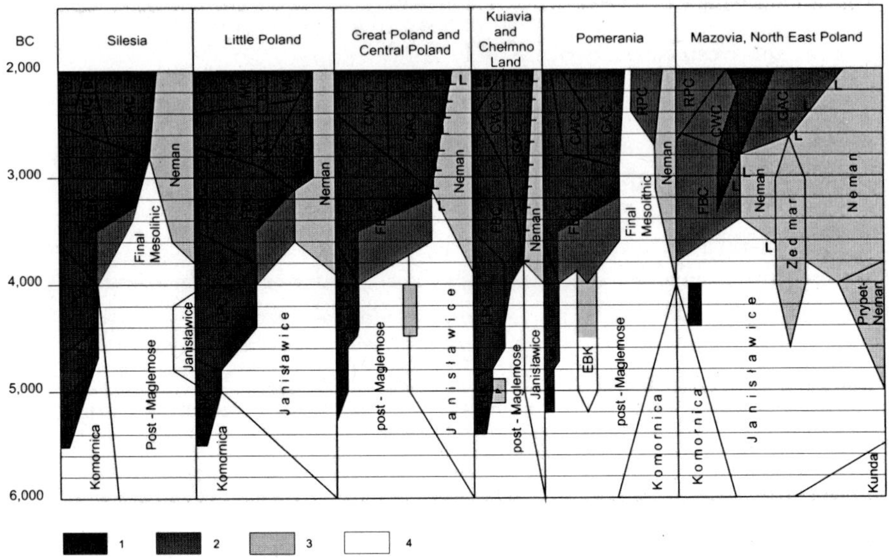

Figure 16.12 Archaeological cultures and main socio-economic formations in Polish territories between 6,000 and 2,000 BC. (1) agro-pastoral and pastoral Neolithic; (2) agro-pastoral Neolithic with significant contribution of hunting and gathering; (3) pottery-using hunter-gatherers (para- and proto-Neolithic); (4) hunter-gatherers (Mesolithic). LBK—Linear Band Pottery Culture; SBK—Stroke Band Pottery Culture; LPC—Lengyel-Polgar Complex; FBC—Funnel Beaker Culture; GAC—Globular Amphorae Culture; CWC—Corded Ware Culture; U—Unetice Culture; BB—Bell Beakers; BC—Baden Culture; ZC—Złota Culture; MC—Mierzanowice Culture; P—sites of Podgaj 32 type; IC—Iwno Culture; DG—Dobre Group; RPC—Rzucewo/Pamariu Culture; L—pottery of Linin type.

of elements of the agro-pastoral economy (Figure 16.12). The overall situation changed radically with the formation of the Trzciniec Culture around 1,900–1,800 BC, which was characterized by the transition to farming, but nonetheless included a significant cultural input from preceding pottery-using hunter-gatherer populations (Józwiak 2003: 180–183, 204–209).

The extended period through which populations of hunter-gatherers and farmers co-existed must have been characterized by different forms and intensities of culture contact and exchange. In these contexts, it is difficult to envisage how the only outcomes of these interactions would have been the selective adoption and modification of farmer ceramics by hunter-gatherers. Cartographic analyses bear witness to the complex nature of these contacts. For example, after 4,000 BC (Figures 16.7 and 16.8), in both the lowlands and certain areas of the uplands, forager and farmer populations were so thoroughly intermingled that it is difficult to demarcate separate zones populated by Neolithic farming cultures (the FBC, the GAC and the CWC) and pottery-using hunter-gatherers (the Neman Culture and the Zedmar

Culture). This evidence suggests that the advance of the agricultural frontier (Dennel 1985; Zvelebil 1998, 2001; Zvelebil and Rowley-Conwy 1984, 1986) halted in the region in the 4th and the 3rd millennium BC leaving a wide forager–farmer interaction zone extending from the northwest to the southeast. This zone must have included the exchange of marriage partners between groups, as well as trade in other goods and services. Only after the turn of the 3rd and the 2nd millennium BC did the co-existence of farmers and hunter-gatherers veer towards a full transition to agriculture, which came about as part of wider radical transformations in hunter-gatherer society and culture. Nonetheless, for the preceding period to have lasted so long, the extended forager–farmer interactions must have been mutually advantageous, probably equating to 'Availability Phase I: Co-operative' in Zvelebil's transition to farming model (1998: 16–17).

Classical, anthropological and archaeological understandings of farmers and foragers tend to emphasize the profound differences between two idealized 'categories' of society and economy. From our modern perspective, one which is heavily tainted by concepts of social 'progress', we often find it hard to understand why some prehistoric forager communities refused to take up farming when, from our biased vantage point, we assume that it must clearly have represented a degree of desired advancement. But we have to remember that in the Neolithic, the differences between communities representing these two types of economy may not have been so dramatic (Zvelebil 2002: 383–384, 2003: 49–50). Moreover, the foraging lifestyle may have had several inherent advantages and, in comparison to farming, ensured long hours of leisure, a highly varied and nutritious diet, and relatively good health, so long as appropriate ecological conditions were present. Most probably, the forager way of life also contained internal self-regulatory social mechanisms that prevented or thwarted any moves towards the intensification of production (Cohen 1989: 18–23, 75–100; Sahlins 1974: 1–100).

If hunting and gathering did remain a viable strategy throughout extended periods of Poland's prehistory, how is it possible to explain and understand the adoption and use of ceramics by these forager societies? As described above, we cannot argue that the uptake of pottery was directly connected with the transition to agriculture, nor with any associated social transformations that made ceramics essential technology for hunter-gatherer societies in a state of transition. Again, it must be noted that pottery does not have to be treated as an indispensable attribute of the Neolithic under all circumstances, and that the use of ceramics by prehistoric foragers had its own distinct technological history.

So how and why did hunter-gatherers use pottery? Pottery does make it easier to store and process food and would have generated a series of clear practical advantages to the forager societies who used it (Rice 1999: 6–10;

Tite 1999: 212). However, given the understanding of the hunter-gatherer communities who inhabited the basins of the Vistula and Oder Rivers in the 4th and 3rd millennium BC, it is suggested here that the symbolic dimension of pottery use, as well as the prestige attached to it, must have been of greater importance. Since these groups had already adopted pottery from 'kindred' hunter-gatherer groups to the east, actively incorporating the new technological practice into their repertoire of traditions, they felt no sense of cultural inferiority when encountering new pottery styles that came with the new 'Neolithic' farmers and breeders in their midst. As a result, we do not see a rapid and 'one-way' take up of a novel technology as foragers sought prestige through acquisition of exotic goods. In fact, the long period of forager–farmer interactions led to both sides incorporating elements of each other's pottery styles, with these exchanges no doubt mediated with a sense of mutual respect. Many of Poland's prehistoric hunter-gatherer communities were in no rush to adopt the new kind of pottery technology, and nor did they feel any immediate requirement to take up agro-pastoral farming. The different communities must have initially met, interacted and co-existed as equals.

The profound social transformations that took place between the Neolithic and the Bronze Age overturned this period of mutual accommodation. The majority of remaining forager societies were drawn rapidly into a new set of social and economic relations out of which emerged a new stratum of leaders, a general intensification of production and the widespread transition to farming and pastoralism (Czebreszuk 2001; Kadrow 2001: 153–169). In these new social and symbolic contexts, it comes as no surprise that around 1,900–1,800 BC, the once widespread production of distinctive hunter-gatherer ceramics goes into rapid decline, and eventually ceases.

Acknowledgements

I would like to express my sincere gratitude to Marek Zvelebil and Peter Jordan for the invitation to submit a paper to this volume as well as for their helpful comments on an earlier draft and stylistic suggestions. All faults and oversights remain my own. Thanks are due to Marta Kapera for translating the text as well as to Elżbieta Pohorska-Kleja for preparing illustrations. T Galiński, W Gumiński, J Ilkiewicz, J Kabaciński, M Kobusiewicz, Z Sulgostowska, and M Szmyt kindly gave their permissions to make use of the pottery illustrations. All are gratefully acknowledged.

References

Bagniewski, Z (1979) *Społeczności Myśliwsko-Rybackie w Okresie od IX do III Tysiąclecia p.n.e. na Terenie Polski Południowo-Zachodniej*, Wrocław: Ossolineum

Bagniewski, Z (1982) 'Społeczności myśliwsko-rybackie w okresie od IX do III tysiąclecia p.n.e. na terenie Polski południowo-zachodniej, część II', 11 *Studia Archeologiczne*, 41–116

——— (1996) 'Mezolit Pojezierza i Równiny Drawskiej', 28 *Studia Archeologiczne*, 5–202

——— (1998) 'Later Mesolithic settlement in central and eastern Pomerania', in Zvelebil, M, Domańska, L and Dennell, R (eds), *Harvesting the Sea, Farming the Forest: The Emergence of Neolithic Societies in the Baltic Region*, Sheffield: Sheffield Academic Press, pp 111–119

Chochorowska, E (2001) 'Obozowisko mezolityczne w Ściejowicach pod Krakowem w świetle datowania radiowęglowego', 32 *Materiały Archeologiczne*, 39–52

Cohen, M N (1989) *Health and the Rise of Civilisation*, New Haven and London: Yale University Press

Cyrek, K, Grygiel, R and Nowak, K (1983) 'Podstawy wydzielania mezolitu ceramicznego na Niżu polski', in Malinowski, T (ed), *Problemy Epoki Kamienia na Pomorzu*, Słupsk: Wyższa Szkoła Pedagogiczna w Słupsku, pp 85–110

Czebreszuk, J (2001) *Schyłek Neolitu i Początki Epoki Brązu w Strefie Południowo-Zachodniobałtyckiej (III i Początki II tys. Przed Chr.). Alternatywny Model Kultury*, Poznań: Adam Mickiewicz University

Czerniak, L (1994) *Wczesny i Środkowy Okres Neolitu na Kujawach 5400–3650 p.n.e.*, Poznań: Institute of Archaeology and Ethnology, Polish Academy of Sciences

Czerniak, L and Kabaciński, J (1997) 'The Ertebølle culture in the southern Baltic coast', in Król, D (ed), *The Built Environment of Coast Areas during the Stone Age*, Gdańsk: Archaeological Museum in Gdańsk, pp 70–79

Dennel, R (1985) 'The hunter-gatherer/agricultural frontier in prehistoric temperate Europe', in Greek, S W and Pearlman, S M (eds), *The Archaeology of Frontiers and Boundaries*, San Diego: Academic Press, pp 113–139

Dolukhanov, P, Shukurov, A, Gronenborn, D, Sokoloff, D, Timofeev, V and Zaitseva, G (2005) 'The chronology of Neolithic dispersal in central and eastern Europe', 32 *Journal of Archaeological Science*, 1441–1458

Domańska, L (1990) *Kaukasko-Nadczarnomorskie Wzorce Kulturowe w Rozwoju Późnomezolitycznych Społeczeństw Niżu Strefy Pogranicza Europy Wschodniej i Środkowej*, Inowrocław: Adam Mickiewicz University

——— (2003) 'Hunter-gatherers and farmers: neighbors in north-eastern Kuiavia, Poland', 30 *Documenta Praehistorica*, 93–98

Galiński, T (1991) 'Uwagi na temat mezolitu ceramicznego i neolitu strefy leśnej na Niżu polskim', 36 *Archeologia Polski*, 5–72

——— (1992) 'Obozowisko mezolityczne i protoneolityczne na stanowisku w Tanowie badane w latach 1989–1991', 38 *Materiały Zachodniopomorskie*, 53–122

——— (2002) *Społeczeństwa Mezolityczne: Osadnictwo, Gospodarka, Kultura Ludów Łowieckich w VIII—IV Tysiącleciu p.n.e. na Terenie Europy*, Szczecin: National Museum in Szczecin

——— (2005) 'Nowe materiały tzw. fazy wczesnopucharowej osadnictwa protoneolitycznego na Pomorzu', 13/14 *Folia Praehistorica Posnaniensia*, 71–90

Gumiński, W (1998) 'The peat-bog site Dudka, Masurian Lakeland: an example of conservative economy', in Zvelebil, M, Domańska, L and Dennell, R (eds), *Harvesting the*

Sea, Farming the Forest: The Emergence of Neolithic Societies in the Baltic Region, Sheffield: Sheffield Academic Press, pp 103–109

Gumiński, W (1999) 'Kultura Zedmar a kultura Narva. Razem czy osobno', 42 *Światowit*, 59–69

——— (2003) 'Szczepanki 8. Nowe stanowisko torfowe kultury Zedmar na Mazurach', 46 *Światowit*, 53–98

Gumiński, W and Fiedorczuk, J (1988) 'Badania w Dudce, woj. suwalskie, a niektóre problemy epoki kamienia w Polsce północno-wschodniej', 33 *Archeologia Polski*, 113–150

Gumiński, W and Michniewicz, M (2003) 'Forest and mobility. A case from the fishing camp site Dudka, Masuria, north-eastern Poland', in Larsson, L, Kindgren, H, Knutsson, K, Leoffler, D and Åkerlund, A (eds), *Mesolithic on the Move: Papers Presented at the Sixth International Conference on the Mesolithic in Europe, Stockholm 2000*, Oxford: Oxbow Books, pp 119–127

Ilkiewicz, J (1989) 'From studies on cultures of the 4th millennium BC in the central part of the Polish coastal area', 36 *Przegląd Archeologiczny*, 17–56

——— (1997) 'Wczesnoneolityczny inwentarz rogowy i kościany ze stanowiska Dąbki 9, gmina Darłowo', 21 *Koszalińskie Zeszyty Muzealne*, 163–198

——— (1998) 'Osiedla proto- i wczesnoneolityczne na stanowisku 7 w Koszalinie-Dzierżęcinie', 1 *Acta Archaeologica Pomoranica*, 41–52

Józwiak, B (2003) *Społeczności Subneolitu Wschodnioeuropejskiego na Niżu Polskim w Międzyrzeczu Odry i Wisły*, Poznań: Adam Mickiewicz University

Kabaciński, J (2001) 'The Mesolithic-Neolithic transition in the southern Baltic Coastlands', 39 *Fontes Archaeologici Posnanienses*, 129–162

Kadrow, S (2001) *U Progu Nowej Epoki. Gospodarka i Społeczeństwo Wczesnego Okresu Epoki Brązu w Europie Środkowej*, Kraków: Institute of Archaeology and Ethnology, Polish Academy of Sciences

Kanwiszer, A and Trzeciak, P (1991) 'Łódź radiocarbon dates III', 33 *Radiocarbon*, 115–130

Kempisty, E (1983) 'Neolityczne kultury strefy leśnej w północnej Polsce', in Malinowski, T (ed), *Problemy Epoki Kamienia na Pomorzu*, Słupsk: Wyższa Szkoła Pedagogiczna w Słupsku, pp 175–199

Kempisty, E and Sulgostowska, Z (1991) *Osadnictwo Paleolityczne, Mezolityczne i Paraneolityczne w Rejonie Woźnej Wsi, woj. Łomżyńskie*, Warszawa: Institute of Material Culture History, Polish Academy of Sciences

Kempisty, E and Więckowska, H (1983) *Osadnictwo z Epoki Kamienia i Wczesnej Epoki Brązu na Stanowisku 1 w Sośni, woj. Łomżyńskie*, Wrocław: Ossolineum

Kobusiewicz, M (1999) *Ludy Łowiecko-Zbierackie Północno-Zachodniej Polski*, Poznań: Poznań Scientific Society

Kobusiewicz, M and Kabaciński, J (eds) (1993) *Chwalim. Subboreal Hunter-Gatherers of the Polish Plain*, Poznań: Institute of Archaeology and Ethnology, Polish Academy of Sciences

——— (1998) 'Some aspects of the Mesolithic-Neolithic transition in the western part of the Polish Lowlands', in Zvelebil, M, Domańska, L and Dennell, R (eds), *Harvesting the Sea, Farming the Forest: The Emergence of Neolithic Societies in the Baltic Region*, Sheffield: Sheffield Academic Press, pp 95–102

Kośko, A (1981) *Udział Południowo-Wschodnio-Europejskich Wzorców Kulturowych w Rozwoju Niżowych Społeczeństw Kultury Pucharów Lejkowatych*, Poznań: Adam Mickiewicz University

——— (1988) *Osady Kultury Pucharów Lejkowatych w Inowrocławiu-Mątwach woj. Bydgoszcz, Stanowisko 1*, Inowrocław: Adam Mickiewicz University

Kozłowski, S K (1989) *Mesolithic in Poland. A New Approach*, Warszawa: Warsaw University

——— (2001) 'Eco-cultural/stylistic zonation of the Mesolithic/Epipalaeolithic in central Europe', in Kertész, R and Makkay, J (eds), *From the Mesolithic to the Neolithic. Proceedings of the International Archaeological Conference held in the Damjanich Museum of Szolnok, September 22–27, 1996*, Budapest: Archaeolingua, pp 261–282

Kukawka, S (1990) 'Udział komponentu północno-wschodniego w wytwórczości ceramicznej kultury pucharów lejkowatych na Ziemi Chełmińskiej (na marginesie pracy A. Kośko 'Osady kultury pucharów lejkowatych w Inowrocławiu-Mątwach, woj. Bydgoszcz, st. 1. Inowrocław 1988')', 35 *Archeologia Polski*, 139–160

——— (1991) *Kultura Pucharów Lejkowatych na Ziemi Chełmińskiej*, Toruń: Toruń Scientific Society

——— (1997) *Na Rubieży Środkowoeuropejskiego Świata Wczesnorolniczego. Społeczności Ziemi Chełmińskiej w IV Tysiącleciu p.n.e.*, Toruń: Mikołaj Kopernik University

Larsson, L (1990) 'The Mesolithic of southern Scandinavia', 4 *Journal of World Prehistory*, 257–299

Liiva, A and Loze, I (1993) 'Mesolithic and Neolithic habitation of the eastern Baltic', 35 *Radiocarbon*, 503–506

——— (1994) 'Tartu radiocarbon dates XIII', 36 *Radiocarbon*, 153–158

Nowak, M (1999) 'Drugi etap neolityzacji ziem polskich w świetle danych archeologicznych i palinologicznych', in Wasylikowa, K (ed), *Rośliny w Dawnej Gospodarce Człowieka. Warsztaty Archeobotaniczne '97*, Polish Botanical Studies, Guidebook Series 23, Kraków: Institute of Botany, Polish Academy of Sciences, pp 39–78

——— (2001) 'The second phase of Neolithization in east-central Europe', 75 *Antiquity*, 582–592

Olszewski, P A (1987) *Osadnictwo Epimezolityczne w Korzeczniku, woj. Konińskie, Stanowisko 6/7*, Inowrocław: Adam Mickiewicz University

Paner, H, Fludzinski, M and Godon, K (2003) 'Ratownicze badania archeologiczne w obrębie autostrady A1 na terenie województwa pomorskiego w latach 2001–2002', 19 *Pomorania Antiqua*, 37–63

Pazdur, M (1991) 'Radiocarbon chronology of the site Dąbki', 38 *Przegląd Archeologiczny*, 33–34

Pazdur, M, Fogtman, M, Michczyński, A, Pawlyta, J and Zając, M (2004) '^{14}C chronology of Mesolithic sites from Poland and the background of environmental changes', 46 *Radiocarbon*, 809–826

Rice, P M (1999) 'On the origins of pottery', 6 *Journal of Archaeological Method and Theory*, 1–54

Rimantienė, R (1998) 'The first Narva culture farmers in Lithuania', in Zvelebil, M, Domańska, L and Dennell, R (eds), *Harvesting the Sea, Farming the Forest: The*

Emergence of Neolithic Societies in the Baltic Region, Sheffield: Sheffield Academic Press, 213–218

Rowley-Conwy, P (1998) 'Cemeteries, seasonality and complexity in the Ertebølle of southern Scandinavia', in Zvelebil, M, Domańska, L and Dennell, R (eds), *Harvesting the Sea, Farming the Forest: The Emergence of Neolithic Societies in the Baltic Region*, Sheffield: Sheffield Academic Press, 193–202

——— (1999) 'Economic prehistory in southern Scandinavia', in Coles, J, Bewley, R and Mellars, P (eds), *World Prehistory. Studies in Memory of Grahame Clark*, Oxford: British Academy, pp 125–160

Rybicka, M and Wysocki, J (2004) 'Materiały kultury późnej ceramiki wstęgowej z Równiny Dolnej, st. III, gm. Korsze, woj. warmińsko-mazurskie (wyniki badań w 2001 roku)', 42 *Prace i Materiały Muzeum Archeologicznego i Etnograficznego w Łodzi*, 79–107

Sahlins, M (1974) *Stone Age Economics*, London: Tavistock

Schild, R (1989) 'The formation of homogeneous occupation units ("Kshemenitsas") in open-air sandy sites and its significance for the interpretation of Mesolithic flint assemblages', in Bonsall, C (ed), *The Mesolithic in Europe. Papers Presented at the Third International Symposium 1985 Edinburgh*, Edinburgh: Edinburgh University Press, pp 89–94

——— (1998) 'The perils of dating open-air sandy sites of the north European Plain', in Zvelebil, M, Domańska, L and Dennell, R (eds), *Harvesting the Sea, Farming the Forest: The Emergence of Neolithic Societies in the Baltic Region*, Sheffield: Sheffield Academic Press, pp 71–76

Sulgostowska, Z (2003) 'Mesolithic colonisation of south-eastern Subbalticum', in Larsson, L, Kindgren, H, Knutsson, K, Leoffler, D and Åkerlund, A (eds), *Mesolithic on the Move: Papers Presented at the Sixth International Conference on the Mesolithic in Europe, Stockholm 2000*, Oxford: Oxbow Books, pp 47–51

Szmyt, M (1993) 'Pottery', in Kobusiewicz, M and Kabaciński, J (eds), *Chwalim. Subboreal Hunter-Gatherers of the Polish Plain*, Poznań: Institute of Archaeology and Ethnology, Polish Academy of Sciences, pp 61–72

Tilley, C (1996) *An Ethnography of the Neolithic. Early Prehistoric Societies in Southern Scandinavia*, Cambridge: Cambridge University Press

Timofeev, V I (1998) 'The beginning of the Neolithic in the eastern Baltic', in Zvelebil, M, Domańska, L and Dennell, R (eds), *Harvesting the Sea, Farming the Forest: The Emergence of Neolithic Societies in the Baltic Region*, Sheffield: Sheffield Academic Press, pp 225–236

Tite, M S (1999) 'Pottery production, distribution, and consumption: the contribution of the physical sciences', 6 *Journal of Archaeological Method and Theory*, 181–233

Werbart, B (1998) 'Subneolithic: what is it? "Subneolithic" societies and the conservative economies of the Circum-Baltic region', in Zvelebil, M, Domańska, L and Dennell, R (eds), *Harvesting the Sea, Farming the Forest: The Emergence of Neolithic Societies in the Baltic Region*, Sheffield: Sheffield Academic Press, pp 37–44

Wirska-Parachoniak, M (1991) 'Wyniki dotychchczasowych badań analitycznych wczesnoneolitycznej ceramiki ze stanowiska Dąbki, woj. Koszalin', 38 *Przegląd Archeologiczny*, 35–51

Zvelebil, M (1981) *From Forager to Farmer in the Boreal Zone. Reconstructing Economic Patterns through Catchment Analysis in Prehistoric Finland*, British Archaeological Reports International Series 115, Oxford: Archaeopress

Zvelebil, M (1986) 'Foragers and farmers in Atlantic Europe', in Zvelebil, M (ed), *Hunters in Transition. Mesolithic Societies of Temperate Eurasia and their Transition to Farming*, Cambridge: Cambridge University Press, pp 67–93

——— (1993) 'Hunters or farmers? The Neolithic and Bronze Age societies of north-east Europe', in Chapman, J and Dolukhanov, P (eds), *Cultural Transformations and Interactions in Eastern Europe*, Basingstoke: Avebury, pp 146–163

——— (1998) 'Agricultural frontiers, Neolithic origins, and the transition to farming in the Baltic Basin', in Zvelebil, M, Domańska, L and Dennell, R (eds), *Harvesting the Sea, Farming the Forest: The Emergence of Neolithic Societies in the Baltic Region*, Sheffield: Sheffield Academic Press, pp 9–28

——— (2001) 'The agricultural transition and the origins of Neolithic society in Europe', 28 *Documenta Praehistorica*, 1–27

——— (2002) 'Demography and dispersal of early farming populations at the Mesolithic-Neolithic transition: linguistic and genetic implications', in Bellwood, P and Renfrew, C (eds), *Examining the Farming/Language Dispersal Hypothesis*, Cambridge: McDonald Institute for Archaeological Research, pp 379–394

——— (2003) 'Who were we 6,000 years ago? In search of prehistoric identities', in Jones, M (ed), *Traces of Ancestry: Studies in Honour of Colin Renfrew*, Cambridge: McDonald Institute for Archaeological Research, pp 41–60

Zvelebil, M and Rowley-Conwy, P (1984) 'Transition to farming in northern Europe: a hunter-gatherer perspective', 17 *Norwegian Archaeological Review*, 104–128

——— (1986) 'Foragers and farmers in Atlantic Europe', in Zvelebil, M (ed), *Hunters in Transition. Mesolithic Societies of Temperate Eurasia and their Transition to Farming*, Cambridge: Cambridge University Press, pp 67–93

Chapter 17

Early Pottery in Hunter-Gatherer Societies of Western Europe

Philippe Crombé

Introduction

This chapter will mainly focus on the characteristics, chronology and geographic origins of early pottery made by hunting, fishing and gathering groups situated in western Europe, focusing on northern France, Belgium and the Netherlands. The area under discussion is characterized by a broad delta-front (North Sea) coastline, followed by a flat coversand landscape bordering the western limit of the central European loess belt. Since the melting of the ice sheets at the end of the Pleistocene, various hunter-gatherer cultures had occupied this zone, and practised a broad spectrum economy.

During the second half of the 6th millennium BC, the loess area of western Europe was rapidly colonized by central European pioneer farming communities belonging to the *Linienbandkeramische Kultur* or Linear Pottery Culture (LBK) (Lüning et al. 1989; van Berg 1990) (Figure 17.1). As well as an agro-pastoral economy, these communities introduced to western Europe a completely new material culture, including highly decorated pottery.

For many years it was commonly accepted that knowledge of pottery manufacturing carried by these early farming communities was spread even further to the west as various forms of interaction between pioneering LBK farmers and the older hunter-gatherer communities began to develop. These processes of culture contact and exchange eventually led to the initial adoption of derivative pottery technologies by the indigenous foragers, and later to the more general transition from a hunting, fishing and gathering economy into a farming one. In this way, the histories and origins of pottery

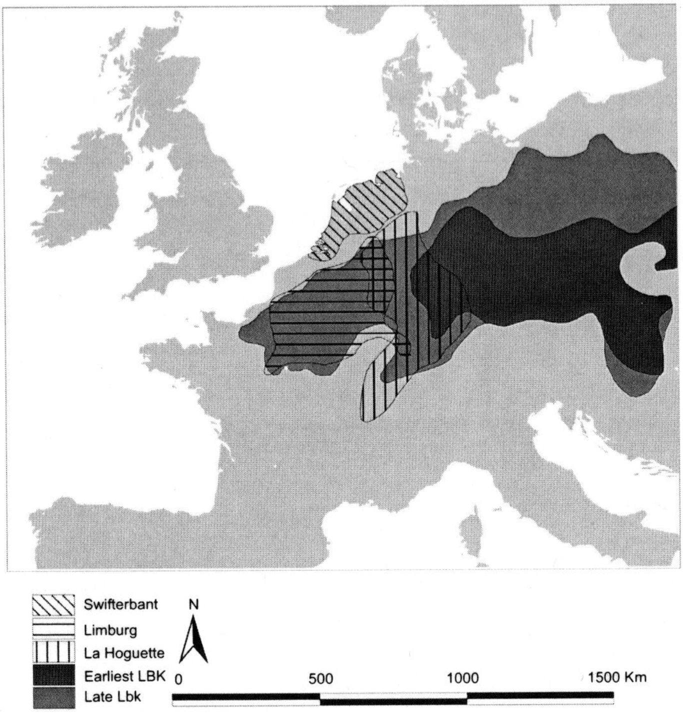

Figure 17.1 Distribution of the main pottery traditions and cultures in western Europe discussed in this chapter.

and the transition to farming in these older hunter-gatherer societies are closely interlinked, with one dispersal followed quickly by another, with both inventions tracing their origins back to central Europe and ultimately to the Near East.

Today, these scenarios are being increasingly questioned by a range of alternative diffusion models. In these scenarios, it is long-distance interactions between various pottery-making hunter-gatherer communities that play a crucial role in the dispersal of ceramic innovations to these areas, rather than the LBK colonizations and subsequent forager–farmer interactions. Pottery knowledge might, for example, have entered western Europe via the Baltic coast from as far as eastern Europe or Siberia.

This chapter presents a critical overview of these competing ceramic origin models and concludes that in western Europe the picture is very complex due to technical problems (e.g., dating and associations problems) and the multiplicity of ceramic styles and traditions.

However, irrespective of these debates it is clear that pottery use amongst northwest European hunter-gatherers was far more widespread than previously thought, and that pots and farming did not necessarily co-exist.

EARLIEST HUNTER-GATHERER POTTERY IN WESTERN EUROPE

In the loess areas colonized by LBK farmers, three kinds of ceramics are often found which are considerably different, both morphologically and technologically, from the traditional LBK pottery. These pottery styles are better known as La Hoguette pottery (Jeunesse 1987) (Figure 17.2), Limburg pottery (Modderman 1970, 1974) (Figure 17.3), and *Begleitkeramik* (Brounen 1999; Jeunesse 1994) (Figure 17.4). All three have in common the frequent (sometimes exclusive) use of an organic temper, consisting of either burnt bone (Limburg and La Hoguette) or shell (La Hoguette), and a predominance of open vessel forms, such as large bowls (Limburg) and egg-shaped vessels (La Hoguette), some of them made with a conical bottom.[1] The principal differences between these potteries are within the decoration. La Hoguette pottery and *Begleitkeramik* are decorated by mainly horizontal undulating bands of impressions associated with applied bands or cordons.

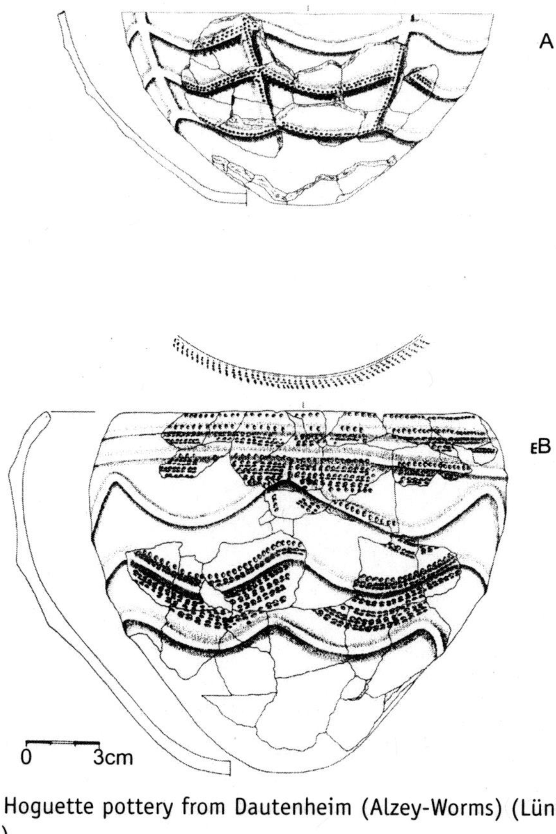

Figure 17.2 La Hoguette pottery from Dautenheim (Alzey-Worms) (Lüning et al. 1989, figures 26 and 28).

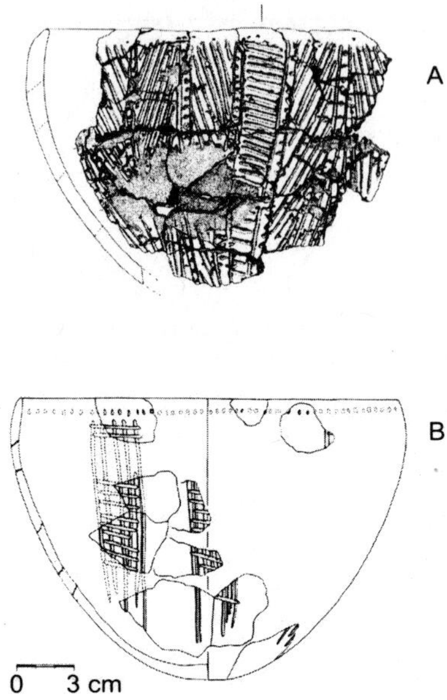

Figure 17.3 Limburg pottery from: (A) Weiler-la-Tour (Jadin et al. 1991); (B) Cuiry-lès-Chaudardes (Constantin 1985, figure 77.1).

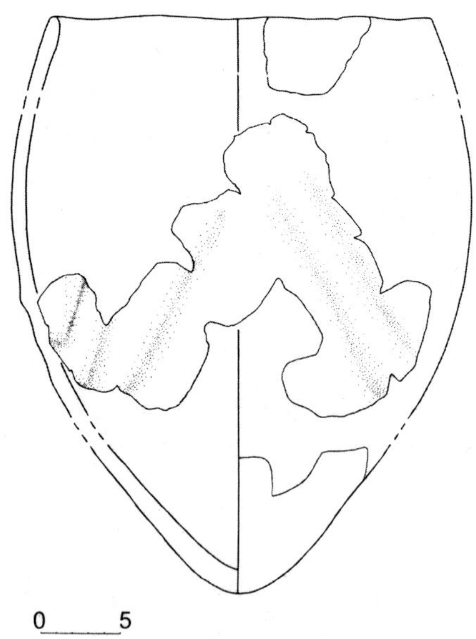

Figure 17.4 Begleitkeramik from Ede-Frankeneng (Schut 1988, figure 4).

On La Hoguette vessels, these impressions were made by means of a spatula with two, three, four, or six points, whereas on *Begleitkeramik* they generally consist of small oblique lines. On some *Begleitkeramik* vessels the decoration is only applied bands without impressions. Limburg pottery, on the other hand, is decorated mainly with vertical panels filled with horizontal and/or oblique incised lines as well as cross-hatchings.

In addition to these stylistic differences, these pottery groups occupy different, partly overlapping, distribution areas (Figure 17.1). The distribution of Limburg pottery is mainly confined to the loess area between the Loire and the Rhine valleys (Paris basin, Belgian and Dutch loess belt), while La Hoguette pottery has a more eastern to southeastern distribution with a main concentration in the southwest German Neckar valley and the upper Rhine valley. Most of the *Begleitkeramik* finds are currently known from the area in which Limburg and La Hoguette pottery occur together, that is, between the Meuse and the Rhine valley.

Until now most of these potteries have been recovered from LBK settlements, where they occur in close spatial connection (postholes, foundation trenches, refuse pits, etc) with LBK settlement waste. Only incidental finds have been reported outside LBK settlements, either inside or outside LBK territories, both at open-air locations (e.g., Haelen, Stuttgart, Liège, Himeling) and in caves/rock shelters (e.g., Bavans, Baulmes, Oberlag).[2] These mainly consist of isolated finds or small assemblages including only a few potsherds or a few vessels. It is often stated that the limited number of finds and findspots outside LBK settlements is mainly due to site taphonomy, in particular the absence of contemporaneous pit structures which served as 'depositional traps' for fragile pottery. However, the open-air site of Haelen 'Broekweg' (see below) situated in acid sandy soil proves that *Begleitkeramik* can perfectly survive even outside features. In addition, none of the mentioned rock shelters and caves have yielded substantial amounts of pottery.

So far the chronology of these three pottery groups is little understood due to lack of reliable radiocarbon dates. The few dates currently available are not associated clearly enough to be useful. Even dates from stratified contexts, such as the cave site of Bavans (Aimé 1989), are problematic and controversial (cf Lanting and Van der Plicht 1999–2000: 46; Perrin 2002). Only the site of Stuttgart-Bad-Cannstatt yielded a closely associated date of 6,353±45 bp obtained from a food crust sample from a La Hoguette potsherd, which coincides with the younger phases of the LBK culture (Kalis et al. 2001). Further dating is based on the association with LBK material. In Germany, La Hoguette pottery is found in LBK contexts belonging to the oldest LBK phase, dated around 5,600–5,500 BC. Some scholars (Gronenborn 1990; Jeunesse 2000: 371) are convinced of an earlier appearance prior to the arrival of the LBK culture, suggesting a dating between 5,800 and 5,600 BC. However, this seems not to be based on firm evidence. The end of La Hoguette may have differed regionally: in the eastern part of its distribution area, it seems to have disappeared almost completely after the earliest/early LBK phase, while in the west it probably continued until the end of the LBK (around 4,900 BC). Limburg pottery, on the other hand, seems to have appeared slightly later than La Hoguette, probably not before the older LBK phase, dated around 5,300 BC. However, an earlier appearance cannot be fully excluded, as settlements of the oldest LBK phase are not known within the distribution area of Limburg ceramics. The end of this pottery tradition was probably synchronic with the western La Hoguette pottery. The relative dating of *Begleitkeramik* is even less clear, but it is most likely that it is similar to that of Limburg pottery.

Since the first discoveries of Limburg and La Hoguette pottery, numerous scholars have debated the origin of these early ceramics, trying to identify their producers. In the 1970s, Modderman (1970, 1974) interpreted Limburg pottery as belonging to Mesolithic hunter-gatherers who were inspired by LBK potters. In contrast, Constantin (1985: 127–128) stated that Limburg pottery might have been made by the LBK farmers themselves, as a kind of special purpose pottery. Today, both interpretations have been rejected by most scholars on the basis of the absence of any similarity in style, technology and decoration between Limburg and LBK pottery. Still, there is an almost general consensus that local hunter-gatherers made La Hoguette/ Limburg/*Begleitkeramik* pottery, but the direction of influence is now sought in a more southerly direction. Based on similarities mainly in the decorative pattern with Cardial and Epicardial pottery, it is believed that Limburg and La Hoguette pottery traditions are rooted in the earliest farming groups from the western Mediterranean (southern France and northeastern Spain) (Jeunesse 1987; Lichardus-Itten 1986; Lüning et al. 1989; van Berg 1990). According to this model, the knowledge of pottery manufacturing would have reached the indigenous occupants of western and central Europe through the Rhône and Saône valleys of southeastern France. Some scholars (Gronenborn 1999: 140–141; Jeunesse 2000: 364–365) believe that not only was pottery brought from the south but the same occurred for the basic principles of small-scale farming and stock herding (see below).

A minority of scholars, however, argue against a southern connection, stating that the entire model is based only on stylistic similarities. Indeed, the morphological (absence of pointed bases and large bowls) and technical (absence of bone and shell temper) differences between La Hoguette/ Limburg pottery and Cardial pottery[3] are as important as with LBK pottery (van Berg 1990: 118). Some (Constantin and Vachard 2004) radically contest the southern connection, claiming that the stylistic parallels with Cardial ware are far too limited and of too general a nature. In 1985, Constantin (1985: 134–139) carefully pointed out the stylistic similarities between Limburg pottery and eastern European 'Forest Neolithic' pottery, an idea which was elaborated further by van Berg (1997: 232–234). In an extensive paper, the latter emphasized the marked similarities in form (conical to pointed bottoms), technology (frequent use of shells and bone as temper) and decoration with pottery from various indigenous traditions along the southern and southeastern Baltic coastline such as the cultures of Narva (Estonia, Latvia, Lithuania), Neman (NE Poland, southern Lithuania) and even Ertebølle (northern Germany, northern Poland, southern Scandinavia). These Baltic cultures would have obtained their pottery technology from even more easterly traditions (China, Japan, Siberia), which possessed ceramics from a much earlier date. However, the theory of a (north)eastern connection is hampered by the, so far, complete lack of pottery finds pre-dating the appearance of La Hoguette/ Limburg/*Begleitkeramik* pottery (>5,500–5,300 BC) along the coast of the North Sea or on the German-Polish plain. The oldest pottery in this area belongs to the Ertebølle and is dated to approximately 5,300–5,100 BC. It was found in northern Germany at the site of Schlamersdorf LA5 (Hartz et al. 2002: 330–331). It should, however, be emphasized that these early dates were obtained on food crust samples of potsherds, which might have been affected by a reservoir age due to the presence of freshwater fish (Fischer and Heinemeier 2003) (see below).

Summing up, we can state that although there is still debate about the precise chronology and genesis (influence from east, south or north), there is an almost overall agreement that La Hoguette/Limburg/*Begleitkeramik* pottery was produced by indigenous groups living along the western periphery of the LBK territory. There is also a general consensus that these pottery-producing hunter-gatherers (perhaps also small-scale farmers?), although partly occupying the same territory as the LBK farmers, were living largely independently from the latter. The presence of native pottery on numerous LBK settlements would only reflect occasional contact and exchange between both population groups.

However, both assumptions still remain largely speculative as they are not sufficiently supported by objective archaeological evidence. Although La Hoguette/Limburg/*Begleitkeramik* pottery is occasionally found on sites with Late or Final Mesolithic lithic artefacts, the association between both categories of finds is usually doubtful or unclear (cf Manen and de Keroualin 2003; Perrin 2002). The context of Bavans, which is generally referred to because of the presence of numerous trapezes and Montbani blades as well as La Hoguette potsherds, was clearly subject to post-depositional disturbances. So far the only findspot with a clear spatial association is the travertine-covered site of Stuttgart-Bad-Cannstatt, which was unfortunately only excavated over a very small surface (4 sq m). Here, six sherds of La Hoguette pottery were found, with a small assemblage of typical Late Mesolithic artefacts, including two trapezes, one Bavans-point, a few regular blades with use retouch, and two fragments of antler harpoons (Strien and Tillmann 2001).

Even if we accept that local hunter-gatherers produced these three kinds of pottery, it still needs to be proved that they formed an autonomous group next to the LBK farmers. In fact there are indications which rather point to the opposite. For example, all pottery finds outside LBK settlements consist of very small, often specialized assemblages (see above), supporting the idea of ephemeral special purpose sites rather than true base camps. Until now, no sites are known which yielded substantial amounts of La Hoguette/Limburg/*Begleitkeramik* pottery. On the other hand, a few LBK sites have produced large amounts of Limburg pottery, for example, the site of Aubechies 'Coron Maton' (Constantin 1985: 105). On this site, Limburg pottery represents almost one quarter of all pottery finds. It is difficult to believe that these were LBK sites which kept more intensive contacts with the natives than other LBK sites. In addition, the almost complete absence of La Hoguette/Limburg/*Begleitkeramik* pottery further west and northwest from the LBK territories and loess frontier does not favour a local indigenous origin independent of the LBK. Except for two outliers of La Hoguette/*Begleitkeramik* pottery—the site of Fontenay-le-Marmion (Calvados) along the French Atlantic coast (Jeunesse 1986) and the site of Ede 'Frankeneng' along the lower Meuse course (Schut 1988)—the distribution of these 'non-LBK' potteries is restricted to the area immediately surrounding the LBK territories. It is most unlikely that this distribution pattern is biased by taphonomic factors, such as differential preservation conditions due to differences, for example, in soil acidity, or inter-regional differences in research frequency (Crombé et al. 2005: 52).

Based on these observations, an alternative explanation can be proposed in which La Hoguette/Limburg/*Begleitkeramik* pottery was manufactured by local indigenous groups, and were assimilated by and partly integrated into the incoming LBK communities. This model fits better with the presence (sometimes in large quantities) of 'non-LBK' pottery on LBK sites, the apparent absence of La Hoguette/Limburg base camps and the clustering of sites in the vicinity of LBK settlements and territories. Today there is increasing evidence that some of the local hunter-gatherers along the northwest limit of the LBK culture were acculturated and/or incorporated into the LBK societies. Among this evidence, there is the typological and technological affiliation between the LBK arrowheads and the Late Mesolithic trapezes (Allard 2005; Crombé et al. 2005; Crombé in press; Gronenborn 1990; Jeunesse 2002; Löhr 1994), the continued exploitation of wild resources (wild game, fish, and nuts) (Arbogast 2001; Crombé et al. 2005; Jeunesse and Arbogast 1997) and of typical 'Mesolithic' raw materials such as Wommersom quartzite (Lodewijckx and Bakels 2000: 38) as well as the existence within some LBK sites (eg the French site of Cuiry-lès-Chaudardes) of hunters as a distinct social 'class' (Hachem 2000).

If we accept the above model, the small assemblages of La Hoguette/Limburg/*Begleitkeramik* pottery found outside LBK settlements can be interpreted as the remains of temporary camps occupied by assimilated hunter-gatherers in the context of transhumance. These incorporated hunter-gatherers may have been 'used' by LBK farmers to travel around with the LBK cattle within as well as outside the LBK territories, leaving behind discrete remains. The small site of Stuttgart-Bad-Cannstatt with its specialized toolkit (see above), situated next to a freshwater pond in the Neckar valley, fits this model particularly well. Besides evidence of hunting (red deer and roe deer), limited evidence of stock-herding (sheep/goat) has been found on this site, which was visited only during spring (April) and late summer (September/October) (Kalis et al. 2001). Based on the similarities in composition, the Dutch site of Haelen (Bats et al. 2002), although less well preserved than Stuttgart, might have had a similar function. On this site, situated in the flat coversand area approximately 15 km from the nearest LBK settlement, a small lithic and ceramic assemblage has been found, including an LBK-like point, eight imported regular blades with intense wear traces (e.g., from hide-working) and 76 potsherds clearly belonging to one (so far unrefitted) vessel of *Begleitkeramik*.

If we accept the above reasoning, the question of why these acculturated and incorporated hunter-gatherers produced and used a different kind of pottery from their LBK fellow-villagers still needs to be answered. Without denying a possible external influence from southern farming and/or (north)eastern hunting groups, other factors may have played a significant role in the creation of La Hoguette/Limburg/*Begleitkeramik* pottery. The egg-shaped conical form might reflect certain functional choices; ethnographical studies (de Roever 2004: 137; van Berg 1997) have shown that point-based vessels, compared to vessels with a rounded bottom, are much more suited for transport as they are easier to carry on the back, the hip or to transport in a canoe. Furthermore, a conical end makes it much easier to force vessels into soil, certainly in unsheltered environments, for example in temporary camps. On the other hand, the creation of a separate local pottery style might have been induced by social considerations. It is not unlikely that pottery in these societies was used as a social marker (symbolic objects), a means to make a distinction between the LBK immigrants and the acculturated locals living in the same community. Many ethnographical studies (Gosselain 1998; Gosselain and Livingstone Smith 2005) have demonstrated that, besides techno-functional constraints, traditions and symbolic conceptions may play a major role in the production of pottery. There are, for instance, many African ethnographical situations where the selection of a clay temper is determined by social and cultural prescriptions.

The fact that the decoration of La Hoguette/Limburg/*Begleitkeramik* pottery presents affinities with far-reaching areas to the north (Baltic cultures) and to the south (Cardial) does not necessarily indicate contact and/or influence between these areas. Indeed, some decorative motifs might simply have been taken from the immediate environment of these hunter-gatherers, for example from basketry, (fishing) nets, etc.

The Swifterbant Pottery

In the Belgian and Dutch lowland coastal area north of the loess frontier, beyond the distribution area of La Hoguette/Limburg/*Begleitkeramik* pottery, the earliest pottery tradition belongs to the Swifterbant culture. Despite its dating problems, it is clear that this pottery post-dates the LBK and La Hoguette/Limburg/*Begleitkeramik* pottery traditions. The exact date of appearance of Swifterbant pottery is still a subject of discussion. A series of radiocarbon dates from sites such as Hardinxveld (Louwe Kooijmans 2001a, 2001b), Almere (Hogestijn and Peeters 2001) and Doel (Crombé 2005; Sergant et al. 2006), suggest an early date at the start of the 5th millennium BC (5,000–4,900 BC), but many of these dates were retrieved from food crusts on potsherds. Recent stable isotope analysis and gas chromatography mass spectrometry analysis on samples from one of these sites—Doel 'Deurganckdok sector B'—have indicated the presence of fish organisms, most likely of freshwater origin, in some dated samples, suggesting that some radiocarbon dates might be affected by a reservoir age (Craig 2005; Craig et al. 2007; Crombé et al. in press). As a matter of fact the food crust dates do not match with other dates from the same site obtained on burnt bones, charcoal, carbonized seeds and hazelnut shells. The latter are on an average 200 to 400 bp years younger. Based on this evidence, at least at Doel, it can be concluded that ceramics did not appear earlier than 4,550–4,500 BC (Van Strydonck and Crombé 2005). Whether the food crust samples of the Hardinxveld and Almere sites were also affected by a fish effect remains to be investigated further (Lanting and Van der Plicht 1999–2000: 53–55). In January 2006, a four-year research project was initiated at the Brussels ^{14}C laboratory, in collaboration with the University of Ghent, in order to tackle this matter.[4]

The earliest Swifterbant pottery (de Roever 2004; Raemaekers 1999: 106–112) (Figure 17.5) is characterized mainly by S-shaped vessels and cylindrical to ovoid vessels with an upstanding shoulder/rim. Bowl-shaped pots also occur but mainly in the southernmost sites (e.g., the Belgian Doel sites) (Crombé et al. 2002; Crombé and Vanmontfort in press). Bases vary from rounded over sagging to pointed. Decoration is limited mostly to impressions or incisions on the top of the rim (*Randkerbung*) and small

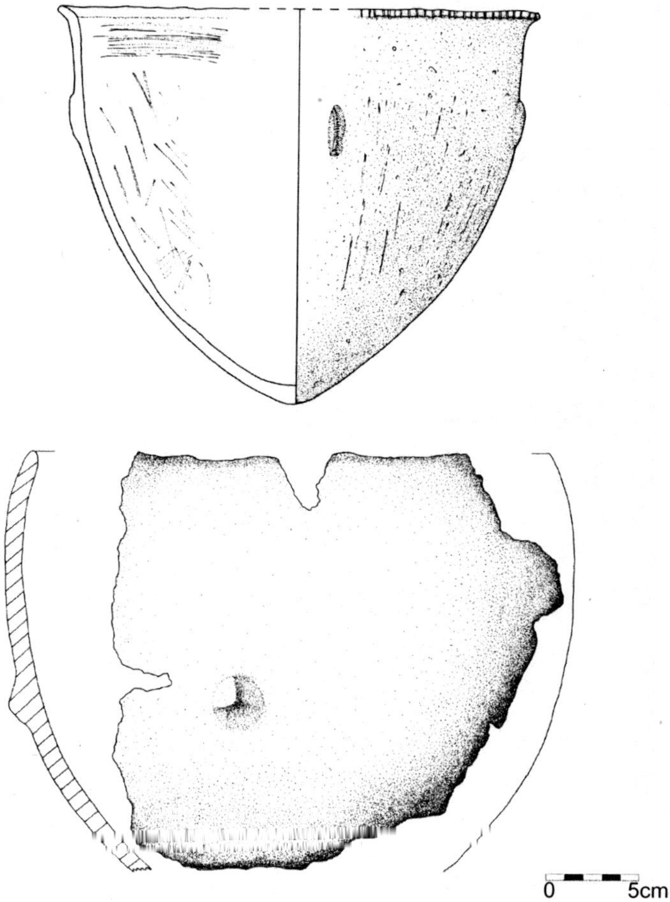

Figure 17.5 Early Swifterbant pottery from Doel, Deurganckdok' (Crombé et al. in press).

circular to oval knobs, the latter only perforated incidentally. Other decoration motifs, such as perforations under the rim or body decoration, are rather rare. This early pottery is built out of coils; the tempering apparently varying regionally from predominantly grog in the south (Belgian sites) to plant and grit in the coastal area of the western Netherlands. During the second half of the 5th millennium BC, Swifterbant pottery changed both in its morphological, technical and decorative senses (Figure 17.6). The morphology became more pronounced, especially at the neck/shoulder transition and at the point of maximum diameter. Organic material became the dominant temper, although grit and grog remained in use. The most profound changes occurred at the level of decoration with an important increase in body decoration. The latter consists of horizontal rows of fingertip

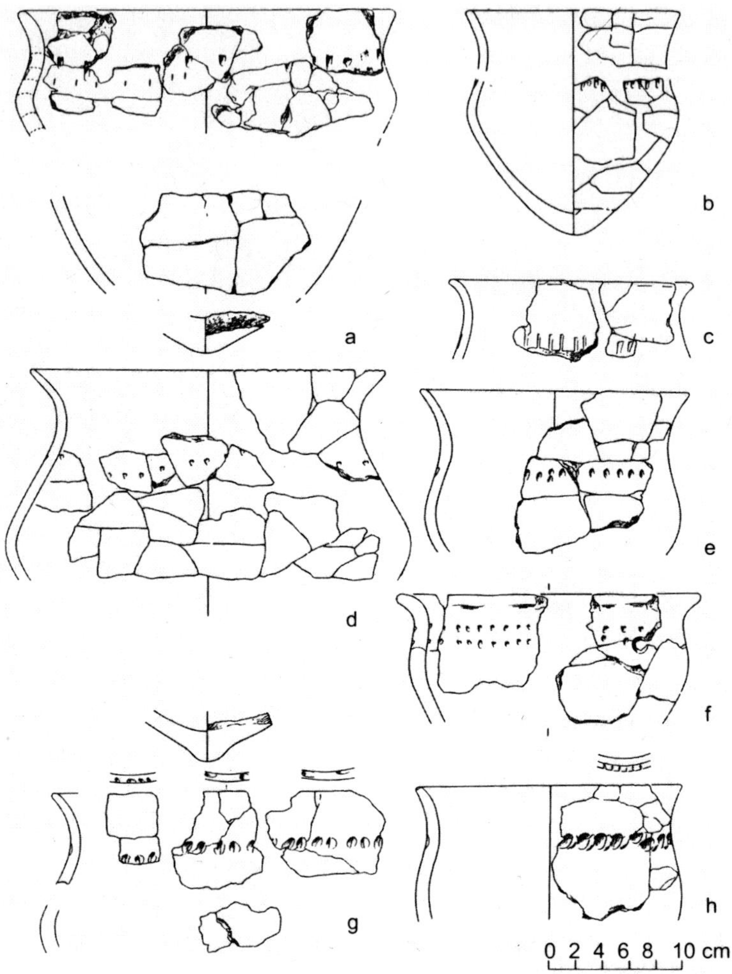

Figure 17.6 Late Swifterbant pottery (second half of 5th millennium cal BC) from Swifterbant site S3 (de Roever 2004, figure 14).

or fingernail impressions, hollow spatulas or grooves which were applied on the neck and shoulder or on the inside of the rim. Decoration covering the entire outer wall also occurred, but mainly in the southern Swifterbant area. Knobs, on the other hand, disappear almost completely.

In contrast to La Hoguette/Limburg/*Begleitkeramik* pottery, there is direct and irrefutable proof that Swifterbant pottery was produced by local 'Mesolithic' groups. Analysis of the finds and features found closely associated with Swifterbant pottery on numerous wetland sites in Belgium, the Netherlands and even western Germany (Lower Saxony), shows a clear continuity in subsistence, lithic and organic industries, burial rites and settlement systems from the Late Mesolithic. The Swifterbant lithic inventory is still largely dominated by tools, such as trapezes and regular Montbani-like blades that have continuity with older technology. In addition, the raw material spectrum is still largely the same as that from the Late Mesolithic; a characteristic example is the continued use of Wommersom quartzite, an exotic raw material originating from central Belgium (Crombé et al. 2002; Gendel 1984). Tools (arrowheads, polished tools) and flint types (e.g., mined flint) associated with contemporaneous farming cultures only appear towards the end of the 5th millennium BC. Similarly, there is an important continuity in the use of organic tools, in particular in the use of perforated antler (beam) mattocks (Crombé et al. 1999). The settlement system, consisting of large base camps occupied for longer periods and/or during different seasons (e.g., P14, Brandwijk, Hardinxveld 'Polderweg') and short-term specialized campsites (e.g., Bergschenhoek) constitute a strong Mesolithic heritage (Louwe Kooijmans 1993). Subsistence still largely depended on the exploitation of wild resources such as hunting, fishing, fowling and gathering, although during the second half of the 5th millennium BC, small numbers of domesticated plants (emmer, naked barley, bread wheat) and animals (cattle, pig, sheep/goat) were introduced (Louwe Kooijmans 2003: 617–619; Raemaekers 1999: 112–115; Zeiler 1997). Whether these domesticates were locally cultivated still remains unclear (Bakels 1986; Crombé and Vanmontfort in press; Raemaekers 2003). Nearly all Swifterbant sites are situated in a wetland environment, on river dunes or levees; it is very questionable whether such environments would also have been suited for crop cultivation and cattle herding, but the possibility of small-scale agro-pastoral activities cannot be fully excluded. Initially, however, cereal grains and domesticated animals (or parts of them) may have been obtained in small quantities through exchange with contemporaneous agro-pastoral communities from the loess area. This idea is supported by the observation at Hardinxveld 'Polderweg' (Louwe Kooijmans 2003: 621) that the few bone fragments of domesticated animals (32 fragments), except for pig, all belong to limb bones, which have been interpreted as having been brought to the site around 4,600 BC as ready-to-cook quarters. In addition, it is assumed that these finds are connected with the supposed cultic deposition (offerings) of pots and red deer bones and antlers in adjacent pits rather than with the local subsistence.

The question of the acquisition of pottery manufacturing knowledge by these Swifterbant hunter-gatherers is still a subject of debate. In the 1970s, their origin was sought in the Ertebølle culture of southern Scandinavia and northern Germany. According to de Roever (1979), Swifterbant pottery is perhaps more variable, but it can nevertheless be attributed to the same ceramic tradition. This ultimately led to the interpretation of the

Swifterbant as the Dutch Ertebølle. In 1997, however, Raemaekers pointed out major differences between Swifterbant and Ertebølle pottery, suggesting that a direct affiliation was most improbable. One of the major differences is the total absence of clay lamps, typical of the Ertebølle. On the other hand, bowl-shaped pots, typical of the southern Swifterbant culture, do not occur within the Ertebølle inventory (Crombé et al. 2005: 57).

Based on morphological similarities, in particular the presence of pointed bases, others (Hogestijn and Peeters 1996: 111–112) have discussed a possible connection with La Hoguette pottery. This model too is hampered by numerous points of divergence between both pottery traditions, for example, in the decoration and tempering. Furthermore, a direct influence from La Hoguette pottery on Swifterbant pottery is difficult to explain, if the oldest dates of the latter are not accepted (see above). In addition, from a geographical point of view, an affiliation with La Hoguette pottery seems highly unlikely as there is no obvious spatial coincidence between both pottery traditions (see above). In fact one could rather expect a connection with the Limburg pottery, given its occurrence mainly between the Loire and the Rhine valleys. Here too, however, the lack of clear morpho-technological and decorative similarities (absence of S-shaped profiles and pointed bases, abundance of bone temper) totally excludes a direct affiliation with Swifterbant pottery. Except for the use of the coiling technique and the sporadic occurrence of rows of perforations underneath the rim (Constantin 1985: 94), there are no similarities at all between Limburg and Swifterbant pottery.

Today, most scholars believe that the knowledge of pottery production was passed from contemporaneous agro-pastoral cultures from the loess area through contact. According to Raemaekers and Gehasse (Raemaeckers 1999: 139–141), the pottery of the Rössen culture, dating approximately between 4,700 BC and 4,500 BC, was the most likely inspiration for Swifterbant pottery. De Roever (2004: 157) now considers Swifterbant pottery to be a mixture of Ertebølle (pointed bases) and Rössen/Bischheim (decoration) elements, while Crombé and colleagues (2002: 704, 2005: 57) have recently emphasized the similarities with the ceramics of the Blicquy/Villeneuve-Saint-Germain culture and the Grossgartach culture, dating approximately between 4,950 BC and 4,650 BC. These cultures—all of whom practised agro-pastoralism—provide good parallels for certain Swifterbant decoration methods, in particular for the *Randkerbung* and applied perforated as well as unperforated knobs, frequent on early Swifterbant pottery. While the former decoration technique is also frequently encountered on hunter-gatherer pottery (Timofeev 1998), the use of applied knobs seems to be a typical 'Neolithic' feature. In addition, the 'farmers' pottery offers close morphological similarities, for example with the S-shaped vessels (Rössen influence) and bowl-shaped vessels (Blicquy/VSG influence). Also, some technological features are to be found within Neolithic pottery traditions. Blicquy pottery, for example (Constantin 1985: 153–154, 196), contrary to LBK pottery, is also manufactured by means of the coiling technique and fired in a reduced-oxidized sequence, as is Swifterbant pottery.

However, no parallels for the pointed/conical bottoms, typical of the Swifterbant pottery, have been found within these traditions. Except for Blicquy pottery, 'farmer' ceramics are round- or occasionally flat-based. Only a small number of Blicquy pots have a slightly conical, thickened bottom; this, together with the dominant bone temper and the vertical decoration pattern, might be reminiscent of Limburg pottery (Constantin 1985: 198–201; van Berg 1990). As a whole the Blicquy/VSG culture seems to be a symbiosis of final LBK (architecture, some pottery forms and decoration patterns, flint industry, funeral rites), (Epi)Cardial (schist bracelets, poppy) and local elements (pottery temper and vertical decoration, projectile points) (Allard 2005: 213–232; Constantin and Ilett 1997; Hauzeur 2002). Whether the pointed bases (together with the coiling and firing technique) of the Swifterbant pottery are a legacy of Limburg pottery via the intermediary of Blicquy pottery remains to be investigated further. It is clear that the Blicquy culture maintained contact with the hunter-gatherers of the Swifterbant culture, as finds of Blicquy pottery have been reported on at least two Swifterbant sites, Hardinxveld 'Polderweg' (Louwe Kooijmans 2001a) and Melsele (Crombé and Vanmontfort in press; van Berg et al. 1992).

Still, the closest parallels for the pointed bases of Swifterbant pottery are to be found within hunter-gatherer contexts of north-northeastern Europe (see above). Whether we should interpret this as a proof of a direct derivation from these northern indigenous traditions remains at present difficult to verify. Contrary to other scholars (e.g., Dolukhanov et al. 2005; Hallgren 2004) we believe that the absolute chronology of most Baltic ceramic cultures, in particular the date of appearance of the first pottery, is still too uncertain as a result of dating problems related to the old wood effect in charcoal samples and the reservoir effect in food crust samples. As long as these problems are unresolved, it will be difficult to fully understand the relationship between these distant pottery areas.

Conclusion

It is clear from the above discussion that the debate concerning the origin of the earliest indigenous pottery in western Europe is still largely hampered by contextual and dating problems. At present, one can only state with confidence that the first hunter-gatherer pottery, named La Hoguette/Limburg or *Begleitkeramik*, appeared close to the (north)western limit of the LBK culture around 5,500–5,300 BC and had disappeared by around

5,000–4,900 BC (Figure 17.7). Echoes of these pottery traditions can be found in the ceramics of succeeding agro-pastoral traditions, for example in Blicquy/VSG pottery, probably indicating a further acculturation of local hunter-gatherer groups in the loess area. Simultaneously, during the 5th millennium BC in the sandy lowland alongside the coastal plain, a new indigenous type of pointed-based pottery named Swifterbant pottery emerged,

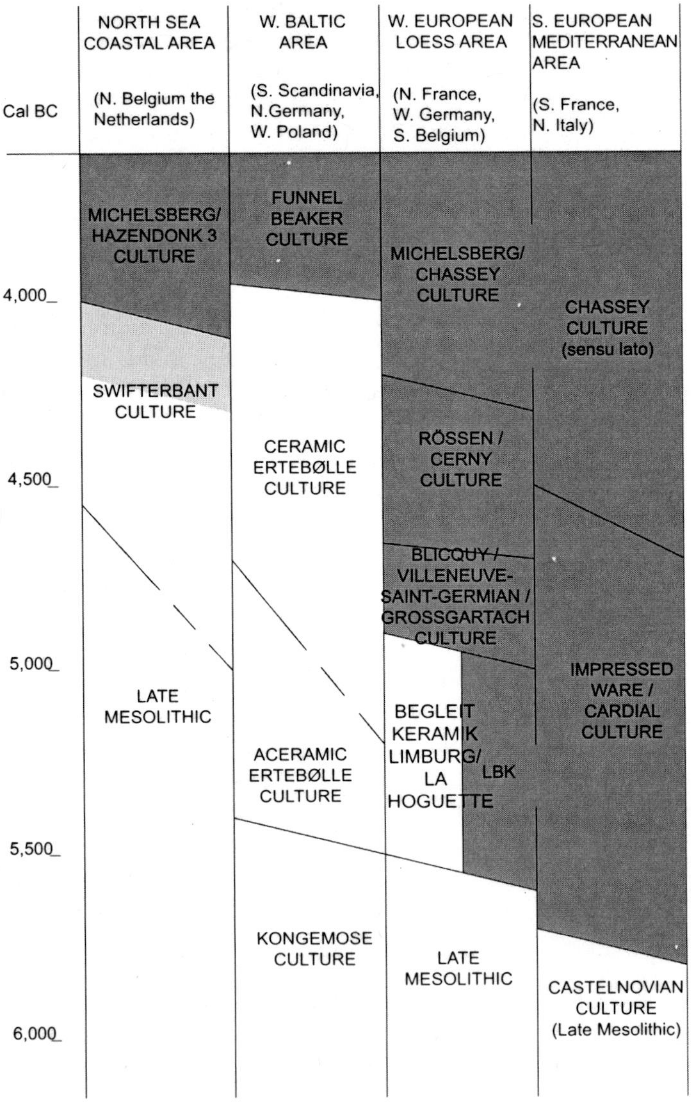

Figure 17.7 Chronological table by region of the main pottery traditions and cultures discussed in this chapter.

which was inspired by both north-northeastern (Baltic hunter-gatherers) and south-southeastern (farmer cultures in the loess area) pottery traditions. This pottery survived until this area too had fully adopted farming at the end of the 5th or the beginning of the 4th millennium BC, respectively south and north of the Rhine/Meuse.

Notes

1. Compared to La Hoguette and Limburg ceramics, little is known about the morphology of *Begleitkeramik*. Technically it is closely connected to both the former pottery styles, although the use of burnt bone fragments as temper seems to be less frequent.
2. For an overview of sites with La Hoguette and Limburg pottery, see Constantin 1985; Lüning et al. 1989; Manen 1997; van Berg 1990. For sites with *Begleitkeramik* we refer to Jeunesse 1994 and Brounen 1999.
3. However, in a recent study by Sénépart and Convertini (2003), the presence of bone temper in Cardial pottery has been reported.
4. In the 'Food crusts on Pottery Project', financed by the Belgian federal government, food crusts and absorbed fatty acids from Swifterbant potsherds in western Belgium and adjacent areas will be analyzed by means of various techniques, e.g., gas chromatography (combustion isotope ratio) mass spectrometry coupled with a conventional mass spectrometry (GC-G-IRMS-MS) and ^{14}C-accelerated mass spectrometry.

Acknowledgements

I would like to thank Michael Ilett and Pierre Allard, both from the University of Nanterre, France, and Paulien de Roever from the Groningen University, the Netherlands, for their critical reading and comments on this paper. Thanks also to Peter Jordan and Marek Zvelebil for the opportunity to publish this paper in their congress proceedings.

References

Aimé, G (1989) 'Les abris sous roche de Bavans (Doubs)', in Aimé, G and Thévenin, A (eds), *Epipaléolithique et Mésolithique entre Ardennes et Massif Alpin,* Table ronde de Besanson 26–27 Avril 1986, 2 Mémoire de la Société d'Agriculture, Lettres, Sciences et Arts de la haute-Saône, Archéologie, pp 89–98

Allard, P (2005) *L'industrie Lithique des Populations Rubanées du Nord-Est de la France et de la Belgique,* 86 Internationale Archäologie, Rahden/Westf.: Verlag Marie Leidorf

Arbogast, R-M (2001) 'Variabilité de la représentation des animaux sauvages et status de la chasse sur les sites du Rubané du Nord de la France', in Arbogast, R-M, Jeunesse, C and Schibler, J (eds), *Rôle et Status de la Chasse dans le Néolithique Ancien Danubien (5,500–4,900 av. J.-C.),* Internationale Archäologie, Arbeitsgemeinschaft, Symposium, Tagung, Kongress 1, Rahden/Westf.: Verlag Marie Leidorf, pp 77–90

Bakels, C C (1986) 'Akkerbouw in het moeras?', V *Rotterdam Papers*, 1–6

Bats, M, Crombé, P, Devriendt, I, Langohr, R, Mikkelsen, J, Ryssaert, C, Schreurs, J, Van de Water, A and Van Strydonck, M (2002) 'Onderzoek langs de omleiding N273 te Haelen (Provincie Limburg, Nederland): Vroeg-Mesolithicum en Vroeg-Neolithicum', 22 *Notae Praehistoricae*, 87–102

Brounen, F T S (1999) 'Vroegneolithische Begleitkeramik uit Limburg en Noord-Brabant', 82 *Archeologie in Limburg*, 59–64

Constantin, C (1985) *Fin du Rubané, Céramique du Limbourg et Post-Rubané*, British Archaeological Reports International Series, 273, Oxford: BAR

Constantin, C and Ilett, M (1997) 'Une étape terminale dans le Rubané Récent du bassin parisien', in Jeunesse, C (ed), *Le Néolithique Danubien et Ses Marges Entre Rhin et Seine, Actes du XXIIe Colloque Interrégional sur le Néolithique; Strasbourg, 27–29 Octobre 1997, 3 Cahiers de l'Association pour la Promotion de la Recherche Archéologique en Alsace*, Strasbourg, 281–300

Constantin, C and Vachard, D (2004) 'Anneaux d'origine méridionale dans le Rubané recent du Bassin parisien', 101(1) *Bulletin de la Société Préhistorique Française*, 75–83

Craig, O (2005) 'Organic analysis of "food crusts" from sites in the Schelde valley, Belgium: a preliminary evaluation', in Crombé, P (ed), *The Last Hunter-Gatherer-Fishermen in Sandy Flanders (NW Belgium); the Verrebroek and Doel Excavation Projects, Part 1: Palaeo-Environment, Chronology and Features*, Archaeological Reports 3, Ghent: Ghent University, pp 435–443

Craig, O E, Forster, M, Andersen, S H, Koch, E, Crombé, P, Milner, N J, Stern, B, Bailey, G N and Heron, C P (2007) 'Molecular and isotopic demonstration of the processing aquatic products in northern European prehistoric pottery', 49(1) *Archaeometry*, 135–152.

Crombé, P (ed) (2005) *The Last Hunter-Gatherer-Fishermen in Sandy Flanders (NW Belgium); the Verrebroek and Doel Excavation Projects, Part 1: Palaeo-Environment, Chronology and Features*, 3 Archaeological Reports, Ghent: Ghent University

Crombé, P (in press), 'Contact and interaction between early farmers and late hunter-gatherers in Belgium during the 6th and 5th millennium cal BC', in Gronenborn, D and Petrasch, J (eds), *The Spread of the Neolithic to Central Europe*, Mainz: Römisch Germanische Zentralmuseum Tagungen

Crombé, P, Boudin, M and Van Strydonck, M (in press) 'Swifterbant pottery in the Scheldt basin and the emergence of the earliest indigenous pottery in the sandy lowlands of Belgium', in v. Carnap-Bornheim, C, Sönke, H, Lüth, F and Terberger, T (eds), *The Earliest Pottery in the Baltic—Dating, Origin and Social Context*, Mainz: Berichte der Römisch-Germanische Kommission

Crombé, P, Perdaen, Y and Sergant, J (2005) 'La néolithisation de la Belgique: quelques réflexions', in Marchand, G and Tresset, A (eds), *Unité et Diversité des Processus de Néolithisation sur la Façade Atlantique de l'Europe (6e—4e Millénaires avant J.-C.)*, Table Ronde de Nantes, 26–27 Avril 2002, Mémoires de la Société Préhistorique Française 36, Paris: Société Préhistorique Française, pp 47–66

Crombé, P, Perdaen, Y and Sergant, S (in press) 'The Neolithisation of the Belgian lowlands: new evidence from the Scheldt basin', in *Proceedings of the International Meeting MESO 2005 Belfast*, Oxford: Oxbow Books

Crombé, P, Perdaen, Y, Sergant, J, Van Roeyen, J-P and Van Strydonck, M (2002) 'The Mesolithic-Neolithic transition in the sandy lowlands of Belgium: new evidence', 76 *Antiquity*, 699–706

Crombé, P and Vanmontfort, B (in press) 'The neolithisation of the Scheldt basin in western Belgium', in Whittle, A and Cummings, V (eds), *Going Over: The Mesolithic-Neolithic Transition in North-West Europe*, London: British Academy Press

Crombé, P, Van Strydonck, M and Hendrix, V (1999) 'AMS-dating of antler mattocks from the Schelde river in northern Belgium', 19 *Notae Praehistoricae*, 111–119

de Roever, J P (1979) 'The pottery from Swifterbant-Dutch Ertebølle?' 16 *Helinium*, 13–36

——— (2004) *Swifterbant-Aardewerk. Een Analyse van de Eolithische Nederzettingen bij Swifterbant, 5^{e} Millennium voor Christus,* Archaeological Studies 2, Eelde: Barkhuis

Dolukhanov, P, Shukurov, A, Gronenborn, D, Sokoloff, D, Timofeev, V and Zaitseva, G (2005) 'The chronology of Neolithic dispersal in Central and Eastern Europe', 32 *Journal of Archaeological Science*, 1441–1458

Fischer, A and Heinemeier, J (2003) 'Freshwater reservoir effect in 14C dates of food residue on pottery', 45(3) *Radiocarbon*, 449–466

Gendel, P (1984) *Mesolithic Social Territories in Northwestern Europe*, British Archaeological Reports International Series 218, Oxford: BAR

Gosselain, O P (1998) 'Social and technical identity in a clay crystal ball', in Stark, M T (ed), *The Archaeology of Social Boundaries*, Washington and London: Smithsonian Institution Press, pp 78–106

Gosselain, O P and Livingstone Smith, A (2005) 'The source. Clay selection and processing practices in sub-Saharan Africa', in Livingstone Smith, A, Bosquet, D and Martineau, R (eds), *Pottery Manufacturing Processes: Reconstruction and Interpretation*, British Archaeological Reports International Series 1349, Oxford: Archaeopress, pp 33–47

Gronenborn, D (1990) 'Eine Pfeilspitze vom ältestbandkeramischen Fundplatz Friedberg-Bruchenbrücken in der Wetterau', 68 *Germania*, 223–231

——— (1999) 'A variation on a basic theme: the transition to farming in southern central Europe', 13(2) *Journal of World Prehistory*, 123–210

Hachem, L (2000) 'New observations on the Bandkeramik house and social organization', 74 *Antiquity*, 308–312

Hallgren, F (2004) 'The introduction of ceramic technology around the Baltic Sea in the 6th millennium', in Knutsson, H (ed), *Coast to Coast—Arrival. Results and Reflections*, Uppsala: Uppsala University, pp 123–142

Hartz, S, Heinrich, D and Lubke, H (2002) 'Coastal farmers—the neolithisation of northernmost Germany', in Fischer, A and Kristiansen, K (eds), *The Neolithisation of Denmark: 150 years of Debate*, Sheffield: J R Collis Publications, pp 321–340

Hauzeur, A (2002) 'Affinités et influences dans le Néolithique ancient d'Europe occidentale: le Rubané de la moyenne vallée de la Moselle et la culture de Blicquy—Villeneuve-Saint-Germain', in Otte, M and Kozlowski, J K (eds), *Préhistoire de la Grande Plaine du Nord de l'Europe. Les Echanges Entre l'Est et l'Ouest dans les Sociétés Préhistoriques,* Actes du Colloque Chaire Francqui interuniversitaire Université de Liège 26 Juin 2001, 99 ERAUL, Liège: Université de Liège, pp 167–182

Hogestijn, J W H and Peeters, J H M (1996) 'De opgraving van de mesolithische en vroegneolithische bewoningsresten van de vindplaats "Hoge Vaart" bij Almere (prov. Fl.): een blik op een duistere periode van de Nederlandse prehistorie', 7 *Archeologie*, 80–113

Hogestijn, J W H and Peeters, J H M (eds) (2001) *De Mesolithische en Vroeg-Neolithische Vindplaats Hoge Vaart-A27 (Flevoland)*, 79 Rapportage Archeologische Monumentenzorg, Amersfoort

Jadin, I, Spier, F and Cauwe, N, (1991) 'Contribution à l'étude du Néolithique ancien de la Moselle: le village rubané de Weiler-la-Tour-Holzdreisch (Grand-Duché de Luxembourg)', 10 *Notae Praehistoricae* 61–67

Jeunesse, C (1986) 'Rapports avec le Néolithique ancient d'Alsace de la céramique "danubienne" de la Hoguette (à Fontenay-le-Marmion, Calvados)', supplément 1 *Revue Archéologique de l'Ouest*, 41–50

——— (1987) 'La céramique de la Hoguette: un nouvel 'élément non-rubané' du Néolithique ancien de l'Europe du Nord-Ouest', 30 *Cahiers Alsaciens d'Archéologie, d'Art et d'Histoire*, 3–33

——— (1994) 'Le Néolithique du sud de la plaine du Rhin supérieur', 69 *Prähistorische Zeitschrift*, 1–31

——— (2000) 'Les composantes autochtone et danubienne en Europe centrale ey occidentale entre 5500 et 4000 av. J.-C.: contacts, transferts, accumulations', in Richard, A, Cupillard, C, Richard, H and Thévenin, A (eds), *Les Derniers Chasseurs-Cueilleurs d'Europe Occidentale (13,000–5,500 av. J.-C.),* Actes du Colloque international de Besançon (Doubs, France), 23–25 Octobre 1998, Besançon: Presses universitaires de Franche-Comté, pp 361–378

——— (2002) 'Armatures asymétiques, regionalisation, acculturation, contribution à l'étude des relations entre le urbane et la composante autochtone dans l'ouest de la sphère danubienne', in Otte, M and Kozlowski, J K (eds), *Préhistoire de la Grande Plaine du Nord de l'Europe. Les Échanges Entre l'Est et l'Ouest dans les Sociétés Préhistoriques,* Actes du Colloque Chaire Francqui interuniversitaire Université de Liège 26 Juin 2001, 99 ERAUL, Liège: Université de Liège, pp 147–165

Jeunesse, C and Arbogast, R-M (1997) 'A propos du statut de la chasse au Néolithique moyen. La faune sauvage dans les déchets domestiques et dans les mobiliers funéraires', in Jeunesse, C (ed), *Le Néolithique Danubien et ses Marges Entre Rhin et Seine,* Actes du XXIIè Colloque Interrégional sur le Néolithique, 27–29 Octobre 1995, Supplément aux cahiers de l'Association pour la Promotion de la Recherche Archéologique en Alsace, 81–102

Kalis, A J, Meurers-Balke, J, van der Borg, K, von den Driesch, A, Rähle, W, Tegtmeier, U and Thiemeyer, H (2001) 'Der La-Hoguette-Fundhorizont in der Wilhelma von Stuttgart-Bad Cannstatt', in Gehlen, B, Heinen, M and Tillmann, A (eds), *Zeit-Räume. Gedenkschrift für Wolfgang Taute, Band 2,* 14 Archäologische Berichte, Bonn : Dr. Rudolf Habelt, 649–672

Lanting, J N and Van der Plicht, J (1999–2000) 'De 14C-chronologie van de Nederlandse pre- en protohistorie III: Neolithicum', 41/42 *Palaeohistoria*, 1–110

Lichardus-Itten, M (1986) 'Premières influences méditerranéennes dans le Néolithique du Bassin parisien: contribution au débat', in Demoule, J-P and Guilaine, J (eds), *Le Néolithique de la France. Hommage à G. Bailloud*, Paris: Picard, pp 147–160

Lodewijckx, M and Bakels, C (2000) 'The interaction between early farmers (Linearbandkeramik) and indigenous people in Central Belgium', in Henderson, J C (ed), *The Prehistory and Early History of Atlantic Europe,* Papers from a Session

Held at the E.A.A. Fourth Annual Meeting in Göteborg 1998, British Archaeological Reports International Series 861, Oxford: Archaeopress, pp 33–46

Löhr, H (1994) 'Linksflügler und Rechtsflügler in Mittel- un Westeuropa. Der Fortbestand der Verbreitungsgebiete asymmetrischer Pfeilspitzenformen als Kontinuitätsbeleg zwischen Meso- und Neolithikum', 57 *Trierer Zeitschrift*, 9–127

Louwe Kooijmans, L P (1993) 'Wetland exploitation and upland relations of prehistoric communities in the Netherlands', in Gardiner, J (ed), *Flatlands and Wetlands: Current Themes in East Anglian Archaeology*, 50 East Anglian Archaeology, pp 71–116

——— (ed) (2001a) *Hardinxveld-Giessendam Polderweg. Een Mesolithisch Jachtkamp in het Rivierengebied (5500–5000 v.Chr.),* Rapportage Archeologische Monumentenzorg 83, Utrecht

——— (2001b) *Hardinxveld-Giessendam De Bruin. Een Kampplaats uit het Laat-Mesolithicum en het begin van de Swifterbant-cultuur (5500–4450 v.Chr.),* Rapportage Archeologische Monumentenzorg 88, Utrecht

——— (2003) 'The Hardinxveld sites in the Rhine/Meuse Delta, the Netherlands, 5,500–4,500 cal. BC', in Larsson, L, Kindgren, H, Knutsson, K, Loeffler, D and Akerlund, A (eds), *Mesolithic on the Move.* Papers Presented at the 6th International Conference on the Mesolithic in Europe, Stockholm 2000, Oxford: Oxbow Books, pp 608–624

Lüning, J, Kloos, U and Albert, S (1989) 'Westliche Nachbarn der bandkeramischen Kultur: Die Keramikgruppen La Hoguette und Limburg', 67 *Germania*, 355–393

Manen, C (1997) *L'axe Rhodano-Jurassien dans le Problème des Relations Sud-Nord au Néolithique Ancient*, British Archaeological Reports International Series 665, Oxford: Hadrian Books

Manen, C and Mazurié de Keroualin, K (2003) 'Les concepts "La Hoguette" et "Limbourg" un bilan des données', in Besse, M, Stahl Gretsch, L-I and Curdy, P (eds), *ConstellaSion. Hommage à Alain Gallay,* 95 Cahiers d'archéologie romande, pp 115–145

Modderman, P J R (1970) *Linearbandkeramik aus Elsloo und Stein*, 3 Analecta Praehistorica Leidensia, Leiden: Leiden University

——— (1974) 'Die Limburger Keramik van Kesseleyk', 4 *Archäologisches Korrespondenzblatt*, 4–11

——— (1976) 'Theorie und Praxis bei der Erforschung des Frühneolithikums im Gebiet des Niederrheins und der Maas', 60 *Jahresschrift für Mitteldeutsche Vorgeschichte*, 49–60

Perrin, T (2002) 'La fin du Mésolithique dans l'arc jurassien: approche statistique des industries lithiques taillées', 99(3) *Bulletin de la Société Préhistorique Française*, 487–499

Raemaekers, D C M (1997) 'The history of the Ertebølle parallel in Dutch Neolithic studies and the spell of the point-based pottery', 1997(2) *Archaeological Dialogues*, 220–234

——— (1999) *The Articulation of a 'New Neolithic': The Meaning of the Swifterbant Culture for the Process of Neolithisation in the Western Part of the North European Plain (4,900–3,400 BC),* Archaeological Studies 3, Leiden: Leiden University

——— (2003) 'Cutting a long story short? The process of neolithization in the Dutch delta re-examined', 77 *Antiquity*, 740–748

Schut, P (1988) 'Poteries du Néolithique ancien à Ede (prov. De Gueldre)', 28 *Helinium*, 223–231

Sénépart, I and Convertini, F (2003) 'Essai de caractérisation des productions céramiques cardiales du Baratin à Courthézon (Vaucluse)', in Gasco, J, Gutherz, X and de Labriffe, P-A (eds), *Temps et Espaces Culturels du 6ᵉ au 2ᵉ Millénaire en France du Sud,* Actes des Quatrièmes Rencontres Méridionales de Préhistoire Récente, Nîmes 2000, pp 83–97

Sergant, J, Crombé, Ph and Perdaen, Y (2006) 'The sites of Doel "Deurganckdok" and the Mesolithic/Neolithic transition in the sandy lowlands of Belgium', in Guilaine J and van Berg, P-L (eds), *La Néolithisation/The Neolithisation Process (Commission XIV)*, Proceedings of the XIVth Congress of the 'Union International de Sciences Pré- et Protohistoriques', Liège, 2001, British Archaeological Reports International Series 1520, Oxford: Archaeopress, pp 53–60

Strien, H-C and Tillmann, A (2001) 'Die La-Hoguette-Fundstelle von Stuttgart-Bad Cannstatt: Archäologie', in Gehlen, B, Heinen, M and Tillmann, A (eds), *Zeit-Räum: Gedenkschrift für Wolfgang Taute, Band 2,* 14 Archäologische Berichte, Bonn : Dr. Rudolf Habelt, pp 673–681

Timofeev, V I (1998) 'The Beginning of the Neolithic in the eastern Baltic', in Zvelebil, M, Domanska, L and Dennell, R (eds), *Harvesting the Sea, Farming the Forest. The Emergence of Neolithic Societies in the Baltic Region,* Sheffield: Sheffield Academic Press, 225–236

van Berg, P-L (1990) 'La céramique néolithique ancienne non rubanée dans le nord-ouest de l'Europe', 12 *Bulletin de la Société Préhistorique Luxembourgeoise*, 107–124

——— (1997) 'La céramique et son décor en Eurasie', in Jeunesse, C (ed), *Le Néolithique Danubien et ses Marges Entre Rhin et Seine,* Actes du XXIIe Colloque Interrégional sur le Néolithique; Strasbourg, 27–29 Octobre 1997, 3 Cahiers de l'Association pour la Promotion de la Recherche Archéologique en Alsace, pp 223–264

van Berg, P-L, Keeley, L, Van Roeyen, J-P and Van Hove, R (1992) 'Le gisement mésolithique de Melsele (Flandre-Orientale, Belgique) et le subnéolithique en Europe occidentale', in Le Roux, C-T (ed), *Paysans et Bâtisseurs; L'émergence du Néolithique Atlantique et les Origines du Mégalithisme,* Actes du 17ème Colloque Interrégional sur le Néolithique, Vannes, 28–31 Octobre 1990, supplément 5 Revue Archéologique de l'Ouest, Rennes, pp 93–99

Van Strydonck, M and Crombé, P (2005) 'Features: radiocarbon dating', in Crombé, P (ed), *The Last Hunter-Gatherer-Fishermen in Sandy Flanders (NW Belgium); the Verrebroek and Doel Excavation Projects, Part 1: Palaeo-environment, Chronology and Features*, Archaeological Reports 3, Ghent: Ghent University, pp 180–212

Zeiler, J T (1997) 'Hunting, fowling and stock-herding at Neolithic sites in the western and central Netherlands; interaction between man and environment', 31 *Helinium*, 60–125

CHAPTER 18

CERAMIC TRAJECTORIES: FROM FIGURINES TO VESSELS

Mihael Budja

INTRODUCTION

The invention and the distribution of ceramic technology and of fired-clay vessel production in Levant, Anatolia and south-eastern Europe is far older than farming and it was already embedded in the agency of Eurasian hunter-gatherer societies. The appearance of ceramic technology in Europe was associated with female and animal figurine-making in the Gravettian, dated to about 26,000 BP. The invention of fired-clay vessels occurred first in hunter-gatherer contexts in eastern Eurasia, where it was associated with small-scale sedentary or semi-sedentary communities millennia before the advent of agriculture. The adoption of pottery making in the Levant seems to have been caught up in processes of social decentralization and community fragmentation which were part of the collapse of the 'ritual' elements in the Levantine Pre-Pottery Neolithic economy. Thus, domestic foodstuffs as luxury foods and the use of ceramic vessels as serving pots do not necessarily have to occur together, if at all, in the Levantine process of the transition to farming. In southeastern Europe, the adoption of pottery making seems to have been incorporated within the hunter-gatherers' social, domestic and mortuary structures. Within these contexts, it was the agency of hunter-gatherers which initiated and made possible both the later structural transformations in the economy, but also the almost simultaneous dispersal of a new form of ceramic materiality into the region.

The Invention of Ceramic Technology

The emergence of ceramic technology has been employed as a useful measure of socio-cultural progress ever since Morgan (1878) hypothesized that 'upper barbarism' could be distinguished from 'upper savagery' by the appearance of pottery. Childe also emphasized the importance of early pottery, arguing that it marked 'the earliest conscious utilization...of a chemical change', and formed a technological innovation that was a virtually universal characteristic of Neolithic communities, as well as an indicator of cultural identity (1936, 1951).

It was suggested recently that early pottery occurred first in hunter-gatherer contexts in east Asia, where it was associated with small semi-sedentary or sedentary communities and, presumably, the first cultivation of rice. The earliest dates are from southern China, where the direct dating of pottery at Miaoyan and Yuchanyan sites, based on insoluble residues, yield ^{14}C values of 17,200–16,300 cal BC (15,220±260 BP [BA94137b]) and 16,150–15,400 cal BC respectively (14,390±230 BP [BA95057b]) (Pearson 2005: 823; Zhao and Wu 2000: 236–237). In the Russian Far East, the earliest pottery at Gromatukha and Gasya is embedded between 14,560–13,070 cal BC (13,240±85 BP [AA-20939] and 14,160–12,530 cal BC (12,960±120 BP [LE-1781] (Kuz'min 2002: 41, Table 1; Zhushchikhovskaya 2005: 13, 17). Kuz'min, however, suggests that there was an almost simultaneous appearance of pottery in southern China around 13,700–13,300 BP, in Japan around 13,500 BP, and in the Russian Far East around 13,300 BP (Kuz'min 2006; see also Keally et al. 2004: 349).

It was hypothesized that the earliest pottery was fired at less than 250°C (Pearson 2005: 823) or at about 350°C (Zhushchikhovskaya 2005: 14–15). This seems unlikely, as vessels fired below 350°C are not appropriate for boiling food, the presence of which has been confirmed by organic residues. The earliest vessels are described as deep bowls, with flat or pointed bases, with walls up to 2 cm thick. The estimated volume of the pots is approximately 5.5 to 6 litres. The secondary burning, carbonized adhesion, soot and water lines seen on many fragments show that the basic functions of the pottery were for boiling water and food or other organic materials (δ^{13}C values are closest to C3 plants and herbivore meat) and extracting fish oils from salmonids (Keally et al. 2003: 5; Kuz'min 2002: 42; Zhushchikhovskaya 2005: 15, 29). These data correspond well with the concept of 'hearth-centred female activities' that became recognized with sedentism (Haaland 1997: 381).

It should be remembered, however, that ceramic technology had become 'inhabited' in the agency of Eurasian hunter-gatherers long before sedentary social structures appeared. It is not only that making ceramic figurines obviously predates pottery in Eurasia, but also that the production of the latter was not necessarily related to the technical and economic trajectories and dynamics of transition to farming (see below). The earliest European ceramic assemblages, which consist of anthropomorphic and zoomorphic figurines, pellets, 'earplugs' and 'structural ceramics', are assigned to the Pavlovian, a local variant of the eastern Gravettian techno-complex, and dated to about 26,000 BP (Verpoorte 2001: 86).

Why Ceramic Vessels?

To answer the question 'Why clay pots?' Brian Hayden (1995, 2001, 2003) has suggested an economically oriented socio-political scenario, based on the interconnected concepts of 'private ownership', 'economically based competition' and 'prestige technologies' which appeared among complex hunter-gatherers and early farmers. He hypothesized that ceramic vessels were first created to attract guests at communal feasts. Its shiny surfaces, forms and appearance made pottery a favoured prestige medium for serving food and drink at feasts.

The emergence of prestige technology is related to a critical economic distinction between 'generalized' and 'complex' hunter-gatherers. While sharing food is an absolute obligation among the former, the private ownership of stored food and the competitive use of food resources together with socio-economic inequalities are characteristic of the latter, trans-egalitarian societies. Hayden (1995: 258) believes that 'once technologies evolve to the point where private ownership and economically based competition can be sustained without immediate detrimental effects, it seems that some individuals in every community attempt to aggrandize themselves and to acquire material, social and political benefits for themselves and their families'. Their advantage was maintained through competitive feasting wherein they displayed wealth, power and prestige on the one hand and advertised their successes and thereby attracted people to participate in more labour-intensive production and debt. As feasts cause people to congregate, they offer ideal conditions for placing selected products, items and values in the focus of attention. It has been suggested repeatedly (Hayden 2001: 59, 2003: 458–469; see also van der Veen 2003: 405–427) that the desire to produce more exotic food (domestic animals and cereals) and serving bowls (ceramic pots) suitable for competitive feasting in the context of the Eurasian Early Neolithic appears to be one of the major motivating factors for the intensification of agriculture, pottery production and social ranking. The quantity, shape and size of pots associated with luxury foods were believed to have become status indicators. While the large vessels may point to the occurrence of communal feasts, a change from large communal pots and serving bowls to the use of individual plates and drinking vessels may point to a shift from eating from a communal bowl to eating individual portions.

The proposition that competitive feasting should be regarded as the driving force behind the transition to farming was based on ethnographic observation of trans-egalitarian societies which show the phenomenon 'that the primary force behind intensified subsistence production is not food shortage, but the desire to obtain social and political advantages—to obtain the most desirable mates, to create the most advantageous alliances, to wield the most political power' (Hayden 2003: 465). It is worth remembering, however, that domesticates such as luxury foods and ceramic vessels as serving pots are not necessarily correlative phenomena (if at all) in early competitive feasting. This is not because food preparation and consumption is the most difficult to reconstruct, but in an ethnographic sample of 862 non-sedentary and non-agricultural societies, 103 were pottery-making (cf. Rice 1999: 28).

Why the Absence of Ceramic Vessels in the West Asian 'Early' Neolithic?

The first occurrence of ceramic vessels in west Asia around 6,900–6,800 cal BC marks the transition from the Pre-Pottery Neolithic C (PPNC) to the Pottery (or Late) Neolithic. The pottery assemblages consist of coherent and homogeneous, coarse, plant-tempered, undecorated pots. It was not until several centuries after its first appearance that pottery may have received, with the evolution of elaborate painted styles and shapes, a wider significance in various social contexts (Akkermans et al. 2006: 123–156; Aurenche et al. 2001: 1197; Kozłowski and Aurenche 2005).

It remains unknown why ceramic technology was not applied to vessel production, although ceramic female and animal figurines had evidently been produced parallel with vessels made of 'white ware', sandstone, alabaster and marble since the beginning of the Pre-Pottery Neolithic (PPNA) around 10,200 cal BC onward.

The west Asian Early Neolithic is marked by dynamics and interrelated trajectories in subsistence practices and economic systems, in built environments, and in the formalization of ritual and religious practices. These culminate in the Middle, Late and Final Pre-Pottery Neolithic B (PPNC) periods around 9,800–7,300 cal BC. The estimated mean site size, which increased dramatically from 3 to 12 ha, shows shifts in the control of resources and centralization in large settlements, and population aggregation in regional centres. Differences in the built environment reflect two social realms. Monumental 'public structures' and 'ritual' architecture were placed inside and outside settlement boundaries. Most of these buildings were constructed with unique architectural features. They often appear to have been situated in highly visible locations and physically separated from residential areas. The residential area is marked by a shift from oval to small-scale sub-rectangular architecture to standardized rectangular multi-roomed buildings. Buildings were constructed, some with two floors, of stone and/or unfired mud brick in tandem with coloured lime-plaster floors which curved up the walls. The very small size of the cells suggests they were used for storage and other facilities. Geometrically shaped ceramic objects—counting tokens—were found in some sites at Ain Ghazal, Mureybet, Cheikh Hassan, Tell Aswad, Tepe Asiah and Tell Ramad (Iceland 2000; Schmandt-Besserat 1992: 168, 170–171).

While there is evidence for domesticated caprines at some Middle PPNB settlements, by the end of the Late PPNB domesticated species including goat and sheep, pig and cattle, provided the evident bulk of meat protein. Herding emerged in the context of crop cultivation, and it has been suggested that 'intensive mixed farming' was well suited to nuclear households. Culling patterns for ovicaprines reflect a strategy of keeping animals primarily for meat production rather than dairying. Parallel with small-scale herding and the close control of herds in the vicinity of settlements rather than extensive, loosely controlled herds, increasing household autonomy has been suggested (Bogaard 2005: 184–198; Byrd 2000: 63–98; Flannery 2002: 417–433; Wright 2000: 89–121).

Mortuary, ritual and cult practices, and the lack of obvious material differentiation among individual burials, are argued elsewhere as mechanisms linked to the intentional homogenization of community members at times of death, and designed to minimize differences and tensions within and between households and communities. These are believed to have been focused on primary and secondary mortuary rituals associated with skull removal and plastering. Beyond these, there is evidence of large plastered statues, often about half life-size, displayed by placement in upright positions within public structures (Goring-Morris 2000: 103–136; Kuijt 1996: 313–336, 2000: 75–102, 2001: 80–99; Kuijt and Goring-Morris 2002: 361–440). The longevity of 3,000 years of high-profile public ceremonies is surely indicative of the intensity and stability of a community's belief system and communal cohesion.

Social differentiation and the emergence of individual power, however, can be seen in post-mortem skull removal and plastering, as they have all been of young adult males and some display specific morphological (brachycephalic) tendencies, perhaps resulting from *in vivo* skull deformation. It is worth remembering that the individuals who practised skull plastering already had a knowledge of the 'pozzuolanic reaction', which is believed to be the domain of potters and metallurgists, in achieving very hard, long-lasting and resistant products (Goren et al. 2001: 671–690).

It is at best speculative as to whether the treatment of the skulls of deceased ancestors and plaster technology was practised only by certain individuals who also had additional roles, such as organizing planting, herding, harvesting and the building of monumental public and ritual structures. Hayden has hypothesized that they were 'successful aggrandizers', but they may have been 'shamans', a ritual elite with the necessary authority to maintain economic and social cohesion and collective identity, as suggested by Kuijt and Goring-Morris (2002: 361–440), Bischoff (2002) and Lewis-Williams (2004: 28–59; Lewis-Williams and Pearce 2005).

However, parallel with plaster and stone figurines, nearly a thousand small ceramic female, male and animal figurines have been found in west Asian Pre-Pottery Neolithic (Figures 18.1 and 18.2). In all the sites in the Levant and southeastern Anatolia, they were deposited in the residential architecture, and hypothetically connected with household cult practices. Human figurines were carefully handled but almost all were found in a broken state as heads or as bodies. Some were in ashy deposits within the roasting pits, and it was suggested they were intentionally exploded in an uncontrolled firing which might be a ritual death upon the decease of the figurine's owner (Rollefson 2000: 168; Voigt 2000: 265).

Many of the ceramic animal figurines appear to have been ritually killed by stabbing with pieces of flint before the clay was fully fired (Cauvin 2000: 105–120; Hauptmann 1999: 77, 2001; Kuijt and Goring-Morris 2002: 389; Rollefson 1986: 5–52).

In west Asia, pottery did not occur before the end of the Pre-Pottery Neolithic (c. 6,900–6,800 cal BC). Manufacture and use followed different trajectories; thus some communities remained 'aceramic', and others

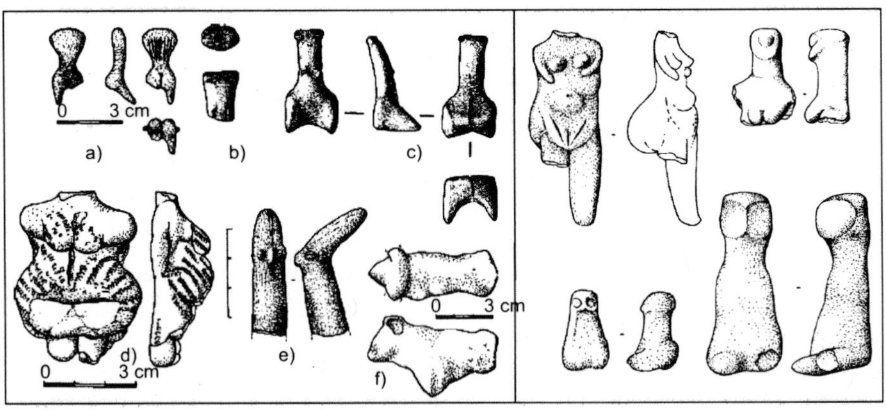

Figure 18.1 Anthropomorphic and zoomorphic ceramic figurines in West Asian Natufian and Pre-Pottery Neolithic (from Kuijt and Goring-Morris 2002, figure 5; Cauvin 2000, figure 8).

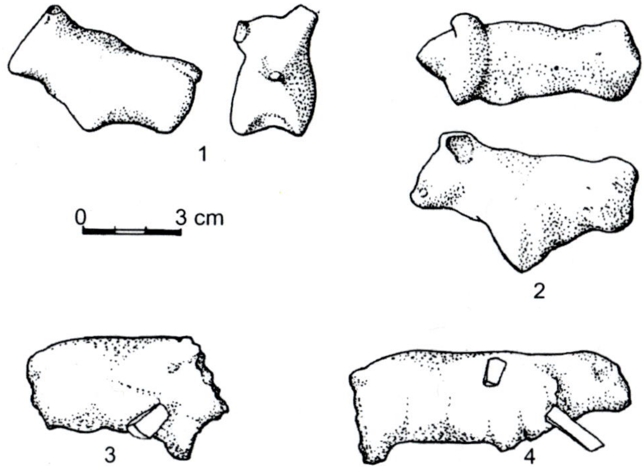

Figure 18.2 Levantine Pre-Pottery Neolithic ceramic animal figurines pierced by flint bladelets (from Cauvin 2000: 25, figure 38).

adopted pottery at much later dates and in its advanced rather than incipient form. At first, these seem to have been types of container, in addition to vessels made of white ware and basketry.

The appearance of pottery in west Asia correlates with the collapse of a 'ritual economy' and aggregation centres, the cessation of mortuary and ritual practices and the dispersal of peoples after the 'package' of crops and livestock had been composed (Johnson 1982: 322–333; Kuijt 2000: 75–102; see also Bandy 2004: 322–333). The transition to small Pottery Neolithic communities or even autonomous households seems to follow the 'scalar stress' scenario. The processes of decentralization and social fragmentation may thus have been linked to increasing population aggregation, population density, and conflicts between ritual and emerging economic elites or between new forms of community cooperation, individual lineages (households) and traditional kinship organization over rights and greater access to and control of resources and privileges, and in shaping social arrangements within communities in combination with regional environmental changes and local environmental degradation.

While in earlier periods of the Pre-Pottery Neolithic, food preparation was an arena for social interaction between households, food preparation facilities were later placed in hidden house spaces. Individual households seem to have controlled their own processing and storage. The 'privatisation' of storage, milling, cooking and eating that took place at the household level (Bogaard 2005: 177–196; Wright 2000: 89–121) may reflect the tensions mentioned above. It is possible therefore that the process of fragmentation and the cessation of ritual practices led to the secularization of ceramic technology. Within autonomous, economic and social household units, coarse and undecorated ceramic containers and vessels became associated with food processing and consumption. Easy access to ceramic technology certainly initiated craft specialization and the much later appearance of 'potters' marks' (Flannery 2002: 417–433).

On the other hand, we may hypothesize that in the context of a 'ritual economy', ceramic figurines were associated with ritual and mortuary practices carried out in hearth rooms, the most secluded spaces in houses in the private domestic area. Surprisingly little attention has been devoted to the resemblance with much later rituals at Çatal Höyük and Aşikli Höyük. The extent to which ceramic technology might have been controlled by ritual practitioners, for example, shamans, is of course speculative, although the making of early ceramics could have been not only a prestige, but also a sacred craft, controlled by privileged individuals or groups.

Demic Diffusion and the Ceramic Figurine Paradox

Jacques Cauvin (1978: 134, 2000: 22–29, 204–205, 207–208) postulated that Levantine ceramic female figurines were markers of new religion and ideology, a powerful force which made possible the transition to an agricultural way of life and 'very quickly revealed itself to be expansionist'. He thought he had found the reason villagers outside the Levant did not develop subsistence production for themselves. They supposedly did not adopt the 'humanization' of art and related new divinities that could have stimulated the necessary energy to develop a new type of palaeoeconomy. Europe, in this interpretative scenario, thus could not have become neolithicized until ceramic female figurines and the 'wave of advance' had reached the Balkans.

Ceramic female figurines have been recognized as the best 'genetic predictor' of molecular genetic (re)population dynamics in Europe. Roy King and Peter A Underhill (2002: 707–714) suggested that they are 'the best genetic predictor' of Neolithic farmer haplogroups in Eurasia and of 'demic diffusion' originating from southcentral Anatolia as their geographic distribution correlates with the southeast-northwest cline of frequencies of haplogroup Eu9 (J-M67* and J-M92 according to Semino et al. 2004: 1030) in current west Asian and European populations. This appreciation was based on the assumption that the 'European Neolithic package' consists of material and ideological content, and Y chromosome haplotype markers originated in west Asia exclusively. Indeed, it is broadly accepted that ceramic female figurines appeared along with the beginning of cereal cultivation in PPNA in the Levant, and all gender and symbolic attributes were visualized at that time. As such they were incorporated a millennia later in the 'new materiality' that defines the Balkan Neolithic (Bailey 2005; Biehl

1996: 153–175; Chapman 2000; Gimbutas 1989; Hansen 2005: 199–200; Marangou 1996: 176–202).

The introduction of ceramic female statuettes, animal figurines and structural ceramics was certainly not the domain of Levantine hunter-gatherers, and they did not appear on the 'eve of the appearance of an agricultural economy' exclusively, as Cauvin suggested (2000: 25). They can be traced from central Europe across the Russian Plain to southern Siberia and back to the Levant and northern Africa. They are well embedded in Eurasian hunter-gatherer social contexts, and chronologically clustered within a time span from 26,000 BP onwards (Budja 2004: 59–81, 2005: 53–72).

In central Europe, more than 16,000 fragments of anthropomorphic figurines, zoomorphic statuettes, pellets, 'earplugs', flat fragments and 'structural ceramic' were found at Dolní Věstonice, Pavlov, Petřkovice and Předmostí in Moravia. Ill-defined types of fired-clay fragments have been recorded at Krems-Wachtberg, Moravany-Lopata, Jarošov, and hypothetically at Kašov and Cejkov (Soffer and Vandiver 1997: 383–401; Verpoorte 2001) (Figures 18.3 and 18.4).

On the Russian Plain, low temperature fired clay was reported at Zaraisk and Kostenki Gravettian sites. At the latter, located on the banks of the River Don, more than 400 fragments were found, contextually associated with marl and ivory Venus figurines, and animal statuettes (Iakovleva 1999: 125–134; Soffer et al. 2000a: 511–537, 2000b: 814). The most easterly anthropomorphic ceramic figurine was found at an open air site at Maininskaia (Maina), on the left bank of the Yenisei River in Siberia (Sher 1999; Vasil'ev 1985: 193–196) (Figure 18.5).

The European assemblages are assigned to the Pavlovian, a local variant of the eastern Gravettian techno-complex, and dated to about 26,000 BP (Verpoorte 2001: 86). The ceramics at Kostenki are embedded in dates as early as 24,100 BP to as late as 18,000 BP (Soffer et al. 2000b: 814). Two dates are available for a ceramic figurine at Maininskaia: at 16,540±170BP

Site	Anthropomorphic	Figurative	'Ceramics'
Dolní Věstonice I	12 (note 1)	> 721	> 5,760
Dolní Věstonice II – north	–	≥ 2 (10?)	431
Dolní Věstonice II – west	–	1	7
Dolní Věstonice III	–	–	1?
Pavlov I	8 (note 1)	> 100	~ 10,000
Pavlov II	–	?	~ 135
Předmostí I	–	≥ 2	> 2
Jarošov II	–	–	1?
Krems-Wachtberg (note 2)	–	3	3

Figure 18.3 Statistics of ceramic assemblages of Upper Palaeolithic Pavlovian sites in Central Europe (after Verpoorte 2001, table 5.1).

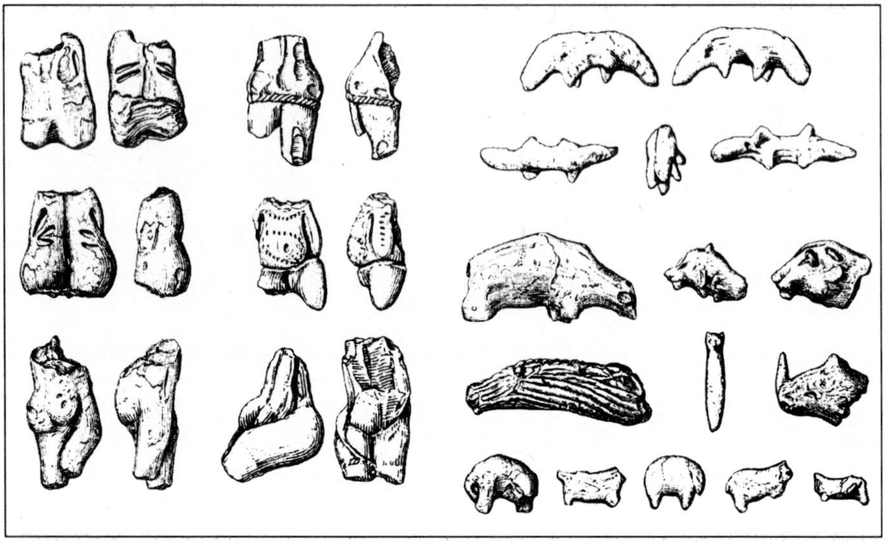

Figure 18.4 Anthropomorphic and zoomorphic ceramic figurines from the Upper Palaeolithic Pavlovian sites Dolní Věstonice, Pavlov and Předmostí (from Verpoorte 2001, figures 3.6, 7, 8, 9, 46, 3.73, 8.1 and 54).

Figure 18.5 Anthropomorphic ceramic figurine from Maina, Yenisei River basin in Siberia (from Vasil'ev 1985, figure 2).

and 16,176±180 BP (Vasil'ev 1985: 193–196; Vasil'ev et al. 2002: 526, table 1). A lesser-known ceramic fragment which was hypothesized to represent a Barbary Wild Sheep (*Ovis tragelaphus*) was found in an Iberomaurisian context, dated to 19,800±500 bp in Tamar Hat Cave in Algeria (Saxon 1976: 327–329).

It can be suggested, therefore, that ceramic technology had become 'inhabited' in the agency of Eurasian hunter-gatherers long before food production and agricultural social structures appeared. It is not only that the making of ceramic figurines predates pottery in Eurasia, but also that pottery was not necessarily associated with farming, as ceramic vessels appear before farming in eastern Asia, and afterwards in the Levant and Anatolia in southwestern Asia.

At Dolní Věstonice and Pavlov, Gravettian (Pavlovian) hunter-gatherer camps in central Europe, 16,000 ceramic objects have been found. Around and in the 'oven-like hearth' located in the middle of a hut, '2000 pieces of "ceramic", among which about 175 had traces of modelling' were dispersed (Verpoorte 2001: 56, 128). Some ceramics were deposited 'in the vicinity of a triple burial', and 'in the vicinity of a large hearth and a male burial' (Verpoorte 2001: 95). According to the available statistics almost all the figurines and statuettes are fragmented, but the pellets and balls which form a large part of the ceramic inventory were intact. This led Vandiver, Soffer, Klima and Svoboda (Soffer et al. 1993: 259–275; Soffer and Vandiver 1997: 383–401; Vandiver et al. 1989: 1002–1008) to reconstruct the entire process of ceramic production by examining the technological skills involved. They found that the local loess was suitable for shaping the female figurines, animal statuettes and pellets. Figurines and statuettes were made of several small pieces of clay stuck together. Heads, legs, feet, ears and tails were shaped separately and attached to the bodies. They were fired at temperatures between 500°C and 800°C. The most important finding was the evidence of thermal shock, an explosive reaction which shatters clay when it is being fired. It is believed the figurines, undried or fired at low temperatures, were rewetted to add some liquid, and then put into a hot fire where they loudly exploded, sending pieces flying in all directions. It seems the thermal shock was intentional, and the process of making and firing was therefore more important than achieving a lasting final product. All ivory objects and stone figurines, in contrast, survived in fairly complete states.

The locus of production located in the settlement may thus reflect a utilitarian, but controlled behaviour related to making, firing and the noticeable fragmenting of female figurines and animal statuettes. If we accept two basic premises: that the ceramics were not just kiln waste because the makers were 'awfully bad potters' and that the female figurines and animal statuettes had been intentionally fragmented in well visible and audible explosions, then this was not merely 'playing with fire', but well-controlled pyrotechnic manipulation with new media.

It may not be surprising that transubstantiation and fragmentation in the central European Upper Palaeolithic social context, whether formalized or not, were objectified with the help of Venuses, as they represent the principal component of the three-dimensional imagery of the Gravettian parthenon. But it is surprising that the *entoptics* were not attached to a new medium, although being broadly applied to ivory and bone imagery, and also stone figurines (Lewis-Williams 2002, 2004; Marshack 1991: 24; Svoboda et al. 1996: 167–168). Did the audio-visual effects of transubstantiation and fragmentation simply replace them, and the visual and audible magic of the new medium, which was not conditioned by the shaman's altered states of consciousness, become accessible to the all members of community?

When and Where Did the Earliest Pots in Southeastern Europe Appear?

Since Gordon Childe (1951: 76–77) put forward the idea that ceramic technology and pot making are virtually universal characteristics of Neolithic communities, as well as indicators of cultural identity, the appearance of pottery has been understood for decades as an exclusive marker of cultural discontinuity between the Late Mesolithic and Early Neolithic and between hunter-gatherers and farmers. The spatially restricted dispersals of selected ornaments attached to the pots reached paradigmatic status as the clusters of settlements producing painted pottery were thus believed to objectify the initial European Neolithic cultures and the confines of the region first settled by Anatolian migrants.

In the contexts of the orthodox 'centre and periphery' perception of the Eurasian Neolithic and the interpretation of processes of transition to farming, pottery maintains a central position in determining the direction of farmer movements, and in positioning the geographical boundaries between groups of hunter-gatherers and farmers. The determination of this structure is based on the assumption that Neolithic colonizers, when crossing the border between the Levant and Europe, brought in their most valuable objects, techniques, symbols and language(s).

It is noteworthy that when discussing the earliest ceramics and pottery making in the settlement contexts in Thessaly and Peloponnesus in southeastern Europe, whether assigned to the Initial or Early Neolithic, Claude Björk (1995), Karen Vitelli (1993, 1999), Paraskevi Yiouni (1996), and Catherine Perlès (2001) point out that potters produced few vessels, that the pots do not show evidence of use on fires, and that the vessel volumes are too small for adequate storage. Annual pottery production has been estimated at between 12 to 13 pots at Franchthi and 7 to 22 pots at Nea Nikomedei (Perlès 2001: 214). They suggest that the vessels should be considered as 'elite goods' with a special purpose value. Vitelli (1993: 213–219, 1999: 184–198) hypothesizes that the pottery was produced first for shamanistic purposes, because a number of saucers and plates show evidence of 'burning incense or narcotic substances in highly decorative containers, manufactured for special public occasions and ceremonies'. Women potters were shamans, she believes, as they had expert knowledge of the medical and narcotic properties of plants, and were able 'to bring about relief from pain and to bring on altered states of consciousness on the occasion of public ceremonies'. Pots and associated contexts are chronologically embedded in the period 7,050–6,450 cal BC (Perlès 2001: 84–92; Reingruber and Thissen 2005).

Locus, Agency and Habitus in the Late Mesolithic and Early Neolithic in Southeastern Europe

The appearance of pottery in hunter-gatherer and farming contexts in the northern Balkans seems to be more attractive. A series of mortuary and domestic palimpsests in the Danube Gorge provide a useful window into the agency of social and spiritual interactions, practices and ritual landscape. We have hypothesized already (Budja 2004, 2005) that pottery, a new medium adapted by complex hunter-gatherers, should be understood as a material correlate of dynamics in the reproductive trajectories of social and symbolic structures and relations. The earliest pottery in Mesolithic Lepenski Vir culture was well embedded in built structures and intentionally incorporated in the trans-generational communal perception of cosmological systems and associated rites, feasts and ceremonies.

The uniformity of the constructed spaces and internal structures at Lepenski Vir, Padina and Hajdučka Vodenica are evident. They consist of trapezoidal constructions with limestone mortared floors inside, and rectangular hearths made of vertical stone slabs positioned in the centre, boulders sculpted in complex designs and coloured in red and black, figurative stone statues and ornamented and unornamented mortar placed on the floors, primarily at the rear of the hearths. They were found associated with primary and secondary human and animal burials under the floors (Borić 1999, 2002a, 2002b, 2005; Radovanović 1996). At least 73 structures, termed 'houses' or 'shrines', have been reported from Lepenski Vir. Pottery was found in some, and it is worth pointing out two of them. In the centrally positioned built structure 54, a paradigmatic pot decorated with two plastic spirals placed on the two opposing sides of its body was associated with newborn and infant burials at the rear, the secondary burial of the mandible of a mature woman (within the hearth), and two carved and coloured boulders and mortar. In the second, peripherally positioned structure 28, pottery was contextually associated with three child burials, one with the skull absent, covered by a large stone plate with two flanking sculpted boulders at its sides and a huge red deer skull with antlers (Borić 2005: 59–60; Borić and Stefanović 2004: 539; Dimitrijević 2000: 113–114; Radovanović 1996: 183) (Figures 18.6 and 18.7). It should be kept in mind that there is no evidence of domesticates at the site and the first structure was dated to 6,375–5,733 cal BC (at 2σ)(Borić et al. 2004: 235; Whittle et al. 2002: 67).

Although there is no clear evidence of burials within the trapezoidal built structures at Padina, as 'the floors of these houses, made of a hard coating of burnt earth, do not show any noticeable damage or repairs corresponding to the position of the burial pit' (Jovanović 1972: 53) they are

Figure 18.6 In the centrally positioned trapezoidal built structure 54, a pot decorated with two plastic spirals placed on the two opposing sides of its body was associated with newborn and infant burials at the rear, the secondary burial of the mandible of a mature woman (within the hearth), and carved and coloured boulders and mortar (from Garašanin and Radovanović 2001, figures 1-3; Srejović and Babović 1983: 8, 18, 92-93, 167). The structure was dated around 6,430-5,980 cal BC (at 2σ)(Borić et al. 2004: 235; Whittle et al. 2002: 67).

Figure 18.7 In the trapezoidal built structure 28, pottery was contextually associated with three child burials, one with the skull absent, covered by a large stone plate with two flanking sculpted boulders at its sides and a huge red deer skull with antlers (from Srejović and Babović 1983: 69, 85, 107, but see also Borić 2005: 59–60; Dimitrijević 2000: 113–114; Radovanović 1996: 183; Stefanović and Borić 2004: 539).

stratigraphically, and to some extent chronologically, related to burials and mortuary structures at the site. While the palimpsest of the latter is chronologically embedded in a considerable 'time depth' from 9,965–9,275 cal BC to 5,780–5,560 cal BC (at 2σ), the sequence of ^{14}C dates of trapezoidal structures is available in fragments, as interest seemed to be focused on selected burial contexts and related individual diet and palaeodietary patterns. Only 3 of 21 trapezoidal structures and associated contexts are dated, and are fixed to between 6,410–6,090 cal BC and 5,990–5,720 cal BC (at 2σ) (Borić 1999: 47–53, 57–59; Borić et al. 2004: 221–248; Borić and Miracle 2004: 341–371; Whittle et al. 2002: 92).

It is worth emphasizing that 52 individuals have been found buried within the site, some incorporated into a stone burial mound of four levels, some below the trapezoidal structures, some in secondary burials (a skull with pebble and boulder cover), and four are in a sitting position with legs crossed (Jovanović 1969–1971: 32). Two of the seated bodies were walled into a conical stone structure up to the skull. A monumental boulder carved with a complex design was positioned 'at the upper level' of the stone burial mound. The stone constructions are spatially connected with the two trapezoidal built structures, but since there are no ^{14}C dates available for the latter, their chronological and structural interrelations remain speculative (Borić 1999: 49–70; Borić and Miracle 2004: figures 2–3; Jovanović 1969–1971: 24–38, 1972: 53, 1974: 1–22, 1987: 1–16; Roksandvić 2000: 26–27).

Similar to the Lepenski Vir context, complete vessels were found *in situ* on the floors within the trapezoidal built structures contextually associated with carved and painted pebbles and ornamented boulders. Perhaps the most interesting sets consist of small cups and bowls, a square dish on a low pedestal and a rhyton. They seem to be inappropriate, too small, and too unusual for routine domestic activities, and it has been suggested that they are 'prestigious serving dishes' (cf Borić 1999: 53; Radovanović 1996: 43). The second set of pots was found on the floor of trapezoidal structure 12, and associated with a decorated boulder, and covered 'by a layer of small fish and animal bones' (Jovanović 1969–1971: 30). Radiocarbon dates fix the pottery sets in structures 12, 17 and 18 within the period of 6,700–6,450,[1] 6,250–6,025 and 5,990–5,720 cal BC (at 2σ) (Borić and Miracle 2004: 347, 350) (Figure 18.8).

At Hajdučka Vodenica, pottery was reportedly linked to an elongated stone construction composed of a rectangular 'hearth', large plates, red burned flooring and a carved boulder. A few individual burials placed beside and covered by stone plates, and a group of 27 skeletons and a secondary skull burial embedded inside the 'grave chamber' suggest that this was a mortuary structure. The excavator points out that a stone axe and deer skull

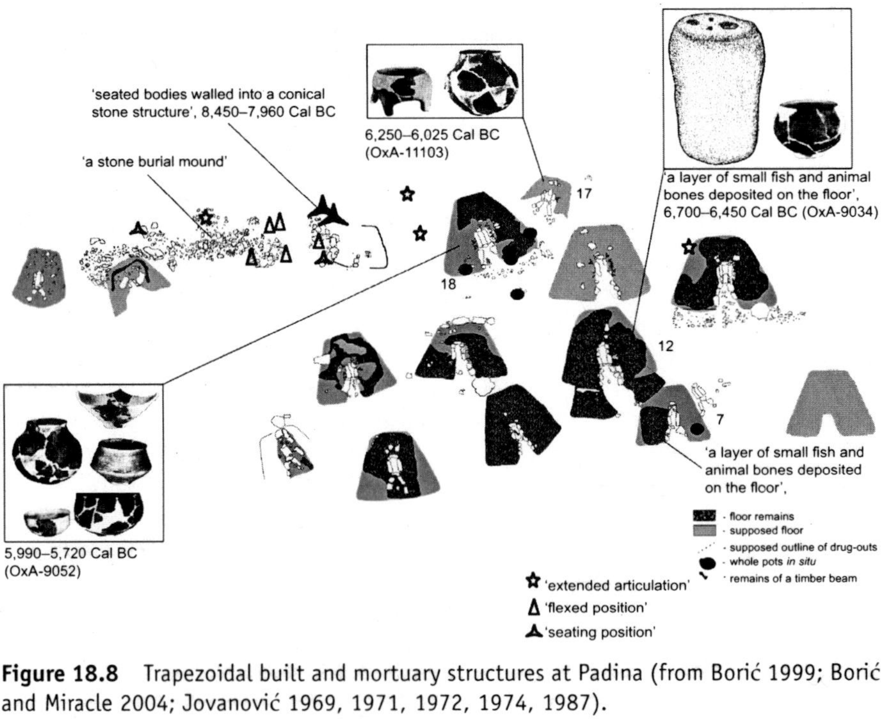

Figure 18.8 Trapezoidal built and mortuary structures at Padina (from Borić 1999; Borić and Miracle 2004; Jovanović 1969, 1971, 1972, 1974, 1987).

and antlers were placed beside the deceased inside the 'grave chamber'. The mortuary structure is dated around 6,500–6,090 cal BC (at 2σ) (Borić and Miracle 2004: 357–361; Radovanović 1996: 219–222).

When placed on the trapezoidal floors at Lepenski Vir and Padina, ceramic vessels become incorporated into hunter-gatherers' and farmers' mortuary and/or ritual complexes which seem to be objectified in long-term architectural structures, primary and secondary burials, burials in the sitting position, and associated 'artefacts' that consist of carved and painted boulders and mortars, the horned skulls of large bovids, and deer skulls with huge antlers. In the Balkans and Transylvania, the latter enjoyed a long tradition, as sculpted boulders were thus embedded in burials and within domestic contexts, some placed on ceramic pedestals and stone plates, in the Early Neolithic settlement at Gura Baciului (Lazarovici and Maxim 1995: 379; Vlassa 1972: 187–191). While children buried in the sitting position were incorporated in the Early Neolithic settlement contexts at Velesnica and Obre I, a large deer skull and infant burial were placed in a centrally positioned trapezoidal pit-dwelling structure at Blagotin-Poljna. A horned auroch skull was placed on top of a female body in a burial in a pit-dwelling at Golokut. Radiocarbon dates indicate that in farming contexts this tradition spanned around 6,230–6,030 cal BC (Blagotin), 5,620–5,470 cal BC (Golokut) and 5,730–5,520 cal BC (Obre)[2] 2δ (Benac 1973: 27–28, 35; Borić 1999; Gimbutas 1974: 19; Radovanović 1996: 223; Whittle et al. 2002: 66, 70–71).

Several readings and perceptions of settlement, mortuary and ritual complexes in the northern Balkans have become available in last decade. They focus on the dialectic opposition and symbolic confrontation of 'culture and nature' and 'the domestication of society' (Hodder 1990: 21–31, 41–43), or on 'apotropaic power and potency', and 'body metamorphosis and animality' (Borić 2002b: 46–74, 2005: 35–69). In Hodder's reading, the trapezoidal structures are seen as the 'locus, agency and habitus' where hunter-gatherers domesticated themselves, their ancestors and the 'wild' surroundings. They performed the process by the inhumation of ancestral bones and the dead, and by installing stone sculptures and statues which, being wild, depict fish-like ancestors and retained the form of natural boulders within the '*domus*'. Borić, however, suggests that the huge boulders commemorated particular individuals as much as 'the realm of the dead and ancestors' and the 'cognitive logic of animality' as practised by hunter-gatherers and farmers in northern Balkans.

More prosaic was the suggestion that the large stone boulders with sculpted representations of fish-human beings commemorate the '8,200 BP event', a global phase of climatic cooling that might have caused a rise in the Danube and frequent floods in the Danube Gorge. Bonsall and colleagues (2002–2003: 1–15) propose that 'neither the sculptures nor the mortar-floored buildings can be traced back before the catastrophic event, and when they appeared, they were intended to achieve dominance over the floods.

Little attention has been paid to the symbols and construals attached to the boulders. When discussing hunter-gatherer and farmer cosmologies in the Balkans, it seems appropriate to undertake a correlation with cave paintings in the Balkans and on the Adriatic and Ionian coast, where identical symbols and construals have been found in red and black on cave walls[3], and identifying them as entoptic motifs and iconography which might have been associated with altered states of consciousness, and shamanic beliefs, rituals and practices (Budja 2005: 63–65).

One of the most graphic narrative scenes is perfectly preserved in Cervi cave, near Porto Badisco at the Strait of Otranto. The iconography of the painting (Figure 18.9) on the left has a spiral motif and figures with bows who appear to be hunting deer. To the right are what appear to be headless figures riding deer and bovids, and dancing with raised arms. Above the scene, a painted anthropomorphic figure with headgear was placed (Graziosi 1996: 117–119, plates 12–17). A possible narrative reading of the scene is, however, closely associated with shamanistic complexes and practices related to altered consciousness. When interpreting a similar frieze in Çatal Höyük, Lewis-Williams (2004: 38–46; see also Lewis-Williams and Pearce 2005) suggests that in shamanistic societies the hunting of meat-producing

Figure 18.9 Examples of rock painting scenes inside Cervi cave (Porto Badisco, Otranto). The iconography of the painting consists of a spiral motif and figures with bows who appear to be hunting deer on the left, and the headless figures riding deer and bovids, and dancing with raised arms on the right. Above the scene, a painted anthropomorphic figure with head-gear was placed (from Graziosi 1996: plate 12).

animals often correlates with the acquisition of the animals' supernatural power, and that the headless figures may well represent the dead, although not simply concerned ancestors, but more probably ancestors who, at the lowest level of the cosmos, continued to be involved in the control and acquisition of animal power. The severing of skulls from bodies relates to the enactment of the ritual practice of the 'death' and 'rebirth' of a shaman and his ability to travel between the lower and the upper worlds. While the spiritual experiences of a decomposing and recomposing body in an altered state of consciousness and associated hallucinations is neurologically generated, the painted visualization, whether in built structures or in cave tunnels, may have been driven by the ritualized enactment of the experience in complex ritual circumstances (initiation into shamanism, perhaps), providing practitioners with respect, social status, and in some instances, political influence.

We may hypothesize that paradigmatic burials and associated material in the context of the trapezoidal built structures in the northern Balkans, depict a similar conception of the ritualized enactment of ritual practices: the skull of an 'old' woman, the horned skull of a large bovid and the skull of a deer with antlers were placed beside the body of a man (Figure 18.10); a headless child burial, covered by a large stone plate with two flanking sculpted boulders at its sides and a huge red deer skull with antlers on the floor; a skull of an older man placed beside the body of child in a sitting position, with legs crossed.

Figure 18.10 Lepenski Vir, burial No. 7a (from Radovanović 1996, figure 4.3; Srejović and Babović 1983: 18). The skull of an 'old' woman, the horned skull of a large bovid and the skull of a deer with antlers were placed beside the body of a man.

We may also hypothesize that, along with monumental boulders, the 'small cups and bowls, square dish on a low pedestal and rhyton' (see above), and 'a layer of small fish and animal bones'[4] deposited in these structures were maintaining an encoded agency, not only of 'bridging the world of the living and that of the dead', as suggested, but of feasting as an associated form of ritual activity in the locus. Like the shamanic ritual, the feast provides a site and medium for symbolic representations of social relations (Dietler 2001: 65–114; Jordan 2004: 117–134; Price 2001: 3–16) that, in the case of the northern Balkans, involved the control of rites and mediation between the community and the supernatural, as much as over the social hierarchy and structure, the intensity of social networks and the dynamics of the structural transformation of hunting and gathering communities.

The pottery and 'Neolithic' materiality in the Balkans, however, were contextualized within hunter-gatherers' domestic and mortuary structures, which were believed to have been dominated by a social and ritual hierarchy that monopolized power and prestige and maintained and controlled inter-regional networks and integrative mechanisms. It is suggested that this was the agency which initiated and made possible structural transformations and almost simultaneous dispersals of the new materiality in the region. This was certainly not the case in the Levant, where pottery making seems to correspond to the processes of decentralization and social fragmentation, and to the collapse of 'ritual' Early Neolithic economy. Pottery was contextualized within secular practices of food processing and consumption.

Conclusions

In recognizing that the appearance of ceramic technology and production of fired-clay vessels has an extended history, we must acknowledge that the emergence of this technology has not necessarily been related to the technical and economical trajectories and dynamics of the transition to farming, as had been broadly accepted. It is increasingly clear that diverse forms of ceramic technology had been 'inhabited' in the agency of Eurasian hunter-gatherers long before the emergence of sedentary social structures appeared. Moreover, these hunter-gatherer ceramic technologies included not only vessels, but also the much older practice of making figurines from fired clay. The earliest European ceramic assemblages consist of anthropomorphic and zoomorphic figurines rather than pottery vessels, and have been dated to the Pavlovian, a local variant of the eastern Gravettian techno-complex, dated to about 26,000 BP. In fact, there were two very different and indeed independent technological trajectories in western Eurasia: first the use of fired clay to make figurines and later vessels, and second the plant and animal domestication and rise of agro-pastoral farming.

This chapter has suggested that the ceramic vessel as serving pot is certainly not part of the 'competitive feasting' scenario for the emergence of agriculture in the Levant since domestic foodstuffs as luxury foods and the use of ceramic pots do not occur *together* in contexts of transition to farming in the region. The fired-clay technology in the Pre-Pottery Neolithic was restricted to the production of female, male and animal figurines and limited by the ideology of the 'ritual economy' which had close links to household rituals and mortuary practices. The earliest appearance of ceramic vessels there correlates with the collapse of aggregation centres, the cessation of previous mortuary and ritual practices and the dispersal of peoples after the 'package' of crops and livestock had been adopted. We may postulate that the collapse of a 'ritual economy' and the apparent decrease in the manufacture and use of figurines led to a shift away from the production of ceramics within a set of ritual acts and towards a 'secularization' of the ceramic craft. This then produced vessels in more utilitarian contexts, linked to the emerging economic elites, and associated with the new forms of community cooperation and household access to and control of food resources.

This scenario contrasts with southeastern Europe where the adoption of pottery making was closely associated with social and symbolic hunter-gatherers' practices. The pottery seems to have been a new form of materiality that was adapted by local complex hunter-gatherers, and as such should be understood as a material medium which was intentionally incorporated into rituals and feasting practices which upheld and reproduced across generations the communal perception of cosmological systems and social networks within which the structural transformation in the economy became verified. It was the agency of hunter-gatherers which initiated and made possible not only the transition to farming, but also the almost simultaneous dispersal of a new form of ceramic materiality into the region.

Notes

1. The date from the dog bone was not corrected for the freshwater reservoir effect, as suggested in Borić and Miracle (2004: 347, 350, table 4). If it is applied, the date is 200 years later, 6,500–6,230 cal BC (at 2σ), calibrated OxCal v. 3.10 (Bronk Ramsey 1995, 2001) and 220±23 years correction applied.
2. The date 5,730–5,520 cal BC 2δ, from Obre I (6,710±60 BP, UCLA 1605G), calibrated with OxCal v. 3.10 (Bronk Ramsey 1995, 2001), relates to the settlement context in which the burials were made.
3. In the Danube Gorge, paintings have been recorded in Climente II and Gaura Chindei (Boroneanţ 1977: 23–34, 1999), in Magourata Cave in the northern Balkans (Anati 1971) and in Cervi Cave, near Porto Badisco (Otranto) on the Adriatic coast (Graziosi 1996). For the painted anthropomorphic figure with headgear, see also Devlet (2001: 50–51).

4. Animal domesticates: *Bos tauros*, Ovis/capra and *Sus domesticus* have been reported from Padina (Clason 1980: 141–173).

References

Akkermans, M M G P, Cappers, R, Cavallo, C, Nieuwenhuyse, O, Nilhamn, B and Ott, N I (2006) 'Investigating the early pottery Neolithic of northern Syria: new evidence from Tell Sabi Abyad', 110(1) *American Journal of Archaeology*, 123–156

Anati, E (1971) 'Maggourata Cave, Bulgaria', 6 *Bollettino del Centro Camuno di Studi Preistorici*, 83–107

Aurenche, O, Galet, P, Régagnon-Caroline, E and Évin, J (2001) 'Proto-Neolithic and Neolithic cultures in the Middle East—the birth of agriculture, livestock raising, and ceramics: a calibrated ^{14}C chronology 12,500–5,500 cal bc', 43(3) *Radiocarbon*, 1191–1202

Bandy, S M (2004) 'Fissioning, scalar stress, and social evolution in early village societies', 106(2) *American Anthropologist*, 322–333

Bailey, D W (2005) *Prehistoric Figurines. Representation and Corporeality in the Neolithic*, London and New York: Routledge

Benac, A (1973) 'Obre I. 27–28', *Glasnik Zemaljskog muzeja Bosnei Hercegovine. N.S. Arheologija,* pp 5–171

Biehl, P F (1996) 'Symbolic communication system: symbols on anthropomorphic figurines of the Neolithic and Chalcolithic from south-eastern Europe', 4 *Journal of European Archaeology*, 153–176

Bischoff, D (2002) *Symbolic Worlds of Central and Southeast Anatolia in the Neolithic Age*, http://www.canew.org/lecbischoffbox.html

Björk, C (1995) *Early Pottery in Greece. A Technical and Functional Analysis of the Evidence from Neolithic Achilleion Thessaly*, Studies in Mediterranean Archaeology 115, Jonnsered: Paul Åströms Förlag

Bogaard, A (2005) '"Garden agriculture" and the nature of early farming in Europe and the Near East', 37(2) *World Archaeology*, 177–196

Bonsall, C, Macklin, G M, Payton, W R and Boroneanţ, A (2002–2003) 'Climate, floods and river gods: environmental change and the Meso–Neolithic transition in southeast Europe', 4(2) *Before Farming*, 1–15, http://www.waspress.co.uk/journals/beforefarming

Borić, D (1999) 'Places that created time in the Danube Gorges and beyond, c. 9,000–5,500 CalBC', 26 *Documenta Praehistorica*, 47–70

——— (2002a) 'The Lepenski Vir conundrum: reinterpretation of the Mesolithic and Neolithic sequences in the Danube Gorges', 76(294) *Antiquity*, 1026–1039

——— (2002b) '"Deep time" metaphor', 3(1) *Journal of Social Archaeology,* 46–74

——— (2005) 'Body metamorphosis and animality: volatile bodies and boulder artworks from Lepenski Vir', 15(1) *Cambridge Archaeological Journal*, 35–69

Borić, D, Grupe, G, Peters, J and Mikić, Ž (2004) 'Is the Mesolithic-Neolithic subsistence dichotomy real? New stable isotope evidence from the Danube Gorges', 7(3) *European Journal of Archaeology*, 221–248

Borić, D and Miracle, P (2004) 'Mesolithic and Neolithic (dis)continuities in the Danube Gorges: new AMS dates from Padina and Hajdučka Vodenica (Serbia)', 23(4) *Oxford Journal of Archaeology*, 341–371

Borić, D and Stefanović, S (2004) 'Birth and death: infant burials from Vlasac and Lepenski Vir', 78(301) *Antiquity*, 526–546

Boroneanţ, V (1977) 'Arta rupestra dis peştera Gaura Chindei, Comuna Pescari (Judetul Caraş-Severin)', 1 *Monumente Istorice şi de ArtăRevista Muzeelar şi Manumentelor*, 23–34

——— (1999) *The Mesolithic Habitation Complexes in the Balkans and the Danube Basin*, Living Past 1, http://cimec.cilea.it/livingpast/nr1/boroneant/mesolithic.htm

Bronk Ramsey, C (1995) 'Radiocarbon Calibration and Analysis of Stratigraphy: The OxCal Program', 37(2) *Radiocarbon,* 425–430

——— (2001) 'Development of the Radiocarbon Program OxCal', 43(2A) *Radiocarbon*, 355–363

Budja, M (2004) 'The transition to farming and the "revolution" of symbols in the Balkans. From ornament to entoptic and external symbolic storage', in Budja, M (ed), *11th Neolithic Studies. Documenta Praehistorica* 31, Ljubljana: Department of Archaeology, Faculty of Arts, Ljubljana University, pp 59–81

——— (2005) 'The process of Neolithisation in south-eastern Europe: from ceramic female figurines and cereal grains to entoptics and human nuclear DNA polymorphic markers', in Budja, M (ed), *12th Neolithic Studies. Documenta Praehistorica* 32, Ljubljana: Department of Archaeology, Faculty of Arts, Ljubljana University, pp 53–72

Byrd, B F (2000) 'Households in transition: Neolithic social organisation within southwest Asia', in Kuijt, I (ed), *Life in Neolithic Farming Communities: Social Organisation, Identity and Differentiation*. New York: Academic/Plenum, pp 63–98

Cauvin, J (1978) *Les Premiers Villages de Syrie-Palestine du IXe au VIIe Millénaire avant Jésus-Christ*, Travaux de La Maison de l'Orient 4, serie archéologique 3, Lyons: Maison de l'Orient

——— (2000) *The Birth of the Gods and the Origins of Agriculture*, Cambridge: Cambridge University Press

Chapman, J (2000) *Fragmentation in Archaeology,* London and New York: Routledge

Childe, V G (1936) *Man Makes Himself,* London: Watts

——— (1951) *Man Makes Himself,* London: Watts

Clason, A (1980) 'Padina and Starčevo: game, fish and cattle', 22 *Palaeohistoria*, 141–173

Devlet, E (2001) 'Rock art and the material culture of Siberian and Central Asian shamanism', in Price, N (ed), *The Archaeology of Shamanism,* London and New York: Routledge, pp 43–55

Dietler, M (2001) 'Theorizing the feast: rituals of consumption, commensal politics, and power in African contexts', in Dietler, M and Hayden, B (eds), *Feasts: Archaeological and Ethnographic Perspectives on Food, Politics, and Power,* Washington and London: Smithsonian Institution Press, pp 65–114

Dimitrijević, V (2000) 'The Lepenski Vir fauna: bones in houses and between houses', in Budja, M (ed), *7th Neolithic Studies. Documenta Praehistorica* 27, Ljubljana: Department of Archaeology, Faculty of Arts, Ljubljana University, pp 101–117

Flannery, K (2002) 'The origins of the village revisited: from nuclear to extended households', 67(3) *American Antiquity*, 417–433

Gimbutas, M (1974) 'Chronology of Obre I and Obre II', *Wissenschaftlichen Mitteilungen des Bosnisch-Hezegowinischen Landesmuseums 4. Archäologie,* 15–35

——— (1989) *The Language of the Goddess,* San Francisco: Harper and Row

Goren, Y, Goring-Morris, A N and Segal, I (2001) 'The technology of skull modelling in the Pre-Pottery Neolithic B (PPNB): regional variability, the relation of technology and iconography and their archaeological implications', 28(7) *Journal of Archaeological Science*, 671–690

Goring-Morris, A N (2000) 'The quick and the dead: the social context of aceramic Neolithic mortuary practices as seen from Kfar HaHoresh', in Kuijt, I (ed), *Life in Neolithic Farming Communities: Social Organization, Identity, and Differentiation,* New York: Kluwer Academic/Plenum, pp 103–136

Graziosi, P (1996) *The Prehistoric Paintings of the Porto Badiso Cave,* Firenze: Origines

Haaland, R (1997) 'Emergence of sedentism: new ways of living, new ways of symbolizing', 71(272) *Antiquity*, 374–385

Hansen, S (2005) 'Neolithic figurines east-west', in Lichter, C (ed), *How Did Farming Reach Europe? Anatolian-European Relations from the Second Half of the 7th through the First Half of the 6th millennium cal BC.* BYZAS 2, Istanbul: Veröffentlichungen des Deutschen Archäologischen Instituts Istanbul, pp 195–212

Hauptmann, H (1999) 'The Urfa region', in Özdogan, M and Basgelen, N (eds), *Neolithic in Turkey. The Cradle of Civilization, New Discoveries.* Istanbul: Arkeoloji ve Sanat Yayinlari, pp 65–86

——— (2001) 'Upper Mesopotamia in its regional context during the Early Neolithic', http://www.canew.org/siteplan.html

Hayden, B (1995) 'The emergence of prestige technologies and pottery', in Barnett, W and Hoopes, J (eds), *The Emergence of Pottery,* Washington DC: Smithsonian Institution Press, pp 257–266

——— (2001) 'Fabulous feasts: a prolegomenon to the importance of feasting', in Dietler, M and Hayden, B (eds), *Feasts: Were Luxury Foods the First Domesticates? Archaeological and Ethnographic Perspectives on Food, Politics, and Power.* Washington DC: Smithsonian Institution Press, pp 23–64

——— (2003) 'Were luxury foods the first domesticates? Ethnoarchaeological perspectives from Southeast Asia', 34(3) *World Archaeology*, 458–469

Hodder, I (1990) *The Domestication of Europe: Structure and Contingency in Neolithic Societies*, Oxford: Blackwell

Iakovleva, L (1999) 'The gravettian art of eastern Europe as exemplified in the figurative art of Kostenki 1', in Roebroeks, W, Mussi, M, Svoboda, J and Fennema, K (eds), *Hunters of the Golden Age: The Mid Upper Palaeolithic of Eurasia 30,000–20,000 BP*, Analecta Praehistorica Leidensia 31, Leiden: University of Leiden, pp 125–133

Iceland, H (2000) 'Token finds at Pre-Pottery Neolithic 'Ain Ghazal, Jordan: a formal and technological analysis', http://link.lanic.utexas.edu/menic/ghazal/Chap1/chapter1.htm

Johnson, G (1982) 'Organizational structure and scalar stress', in Renfrew, C, Rowlands, M and Segraves, B A (eds), *Theory and Explanation in Archaeology*, New York: Academic Press, pp 389–421

Jordan, P (2004) 'Examining the role of agency in hunter-gatherer cultural transmission', in Gardner, A (ed), *Agency Uncovered,* London: UCL Press, pp 117–134

Jovanović, B (1969–1971) 'Chronological frames of the Iron Gate Group of the Early Neolithic Period', 10 *Archaeologica Iugoslavica*, 23–38

——— (1972) 'The autochthonous and the migrational components of the Early Neolithic in the Iron Gates', 3 *Balcanica*, 49–58

——— (1974) 22 *Praistorija Gornjeg Djerdapa. Starinar N.S.*, 1–23

——— (1987) 'Die Architektur und Keramik der Siedlung Padina B am Eisernen Tor, Jugoslawien', 65(1) *Germania*, 1–16

Keally, T C, Taniguchi, Y and Kuz'min, V Y (2003) 'Understanding the beginnings of pottery technology in Japan and neighboring East Asia', in Kuz'min, Y V (ed), *The Nature of Transition from the Palaeolithic to the Neolithic in East Asia and the Pacific*, 24(2) *The Review of Archaeology*, 3–14

Keally, T C, Taniguchi, Y, Kuz'min, V Y and Shevkomud, I Y (2004) 'Chronology of the beginning of pottery manufacture in East Asia', 46(1) *Radiocarbon*, 345–351

King, R and Underhill, P A (2002) 'Congruent distribution of Neolithic painted pottery and ceramic figurines with Y-chromosome lineages', 76(293) *Antiquity*, 707–714

Kozłowski, K S and Aurenche, O (2005) *Territories, Boundaries and Cultures in the Neolithic Near East*, British Archaeological Reports International Series 1362, Oxford: Archaeopress

Kuijt, I (1996) 'Negotiating equality through ritual: a consideration of Late Natufian and Pre-Pottery Neolithic A Period mortuary practices', 15(4) *Journal of Anthropological Archaeology*, 313–336

——— (2000) 'People and space in early agricultural villages: exploring daily lives, community size, and architecture in the Late Pre-Pottery Neolithic', 19 *Journal of Anthropological Archaeology*, 75–102

——— (2001) 'Meaningful masks: place, death, and the transmission of social memory in early agricultural communities of the Near Eastern Pre-Pottery Neolithic', in Chesson, M S (ed), *Social Memory, Identity, and Death: Intradisciplinary Perspectives on Mortuary Rituals*, Vol. 10. Washington DC: American Anthropological Association, Archaeology Division, pp 80–99

Kuijt, I and Goring-Morris, N (2002) 'Foraging, farming, and social complexity in the Pre-Pottery Neolithic of the southern Levant: a review and synthesis', 16(4) *Journal of World Prehistory*, 361–440

Kuz'min, Y V (2002) 'The earliest centres of pottery origin in the Russian Far East and Siberia: review of chronology for the Oldest Neolithic cultures', in Budja, M (ed), *9th Neolithic Studies. Documenta Praehistorica* 29, Ljubljana: Department of Archaeology, Faculty of Arts, Ljubljana University, pp 37–46

——— (2006) 'Chronology of the earliest pottery in East Asia: progress and pitfalls', 80(308) *Antiquity*, 362–371

Lazarovici, G and Maxim, Z (1995) *Gura Baciuliu*, Cluj-Napoca: Biblioteca Musei Napocensis XI

Lewis-Williams, D (2002) *The Mind in Tomb: Consciousness and the Origins of Art*, London: Thames and Hudson

——— (2004) 'Constructing a cosmos: architecture, power and domestication at Çatalhöyük', 4(1) *Journal of Social Archaeology*, 28–59

Lewis-Williams, D and Pearce, D (2005) *Inside the Neolithic Mind: Consciousness, Cosmos and the Realm of the God*, London: Thames and Hudson

Marangou, C (1996) 'Assembling, displaying, and dissembling Neolithic and Eneolithic figurines and models', 4 *Journal of European Archaeology*, 177–202

Marshack, A (1991) 'The female image, a "time-factored" symbol: a study in style and modes of image use in the European Upper Palaeolithic', 57 *Proceedings of the Prehistoric Society*, 17–31

Morgan, L H (1878) *Ancient Society,* New York: World

Pearson, R (2005) 'The social context of early pottery in the Lingnan region of south China', 79(306) *Antiquity*, 819–828

Perlès, C (2001) *The Early Neolithic in Greece: The First Farming Communities in Europe,* Cambridge World Archaeology, Cambridge: Cambridge University Press

Price, S N (2001) 'An archaeology of altered states: shamanism and material culture studies', in Price, N (ed), *The Archaeology of Shamanism,* London and New York: Routledge, pp 3–16

Radovanović, I (1996) *The Iron Gates Mesolithic*, International Monographs in Prehistory. Archaeological Series 11, Michigan: Ann Arbor

Reingruber, A and Thissen, L (2005) 'CANeW 14C databases and 14C charts. Aegean Catchment. (E. Greece, S. Balkans and W. Turkey) 10,000–5,500 cal BC', www.canew.org/data.html

Rice, M P (1999) 'On the origins of pottery', 6(1) *Journal of Archaeological Method and Theory*, 1–54

Roksandvić, M (2000) 'Between foragers and farmers in the Iron Gates Gorge: physical anthropology perspective. Djerdap population in transition from Mesolithic to Neolithic', in Budja, M (ed), *7th Neolithic Studies. Documenta Praehistorica* 27, Ljubljana: Department of Archaeology, Faculty of Arts, Ljubljana University, pp 1–100

Rollefson, G O (1986) 'Neolithic 'Ain Ghazal (Jordan): ritua! and ceremony II', 12 *Paleorient*, 5–52

——— (2000) 'Ritual and social structure at Neolithic "Ain Ghazal", in Kuijt, I (ed), *Life in Neolithic Farming Comunities. Social Organization, Identity and Differentiation.* Fundamental Issues in Archaeology, New York: Kluwer Academic/Plenum, pp 165–190

Saxon, E C (1976) 'Pre-Neolithic Pottery: new evidence from North Africa', 42 *Proceedings of the Prehistoric Society*, 327–329

Schmandt-Besserat, D (1992) *Before Writing: From Counting to Cuneiform,* Volume 1, Austin: University of Texas Press

Semino, O, Magri, C, Benuzzi, G, Lin, A A, Al-Zahery, N, Battaglia, V, Maccioni, L, Triantaphyllidis, C, Shen, P, Oefner, J P, Zhivotovsky, A L, King, R, Torroni, A, Luca Cavalli-Sforza, L, Underhill, A P and Santachiara-Benerecetti, A S (2004) 'Origin, diffusion, and differentiation of Y-chromosome haplogroups E and J: inferences on the Neolithization of Europe and later migratory events in the Mediterranean area', 74 *American Journal of Human Genetics*, 1023–1034

Sher, A Y (1999) 'Palaeolithic Virtual Museum', http://www.vm.kemsu.ru/en/palaeolith/plastic/maina.html

Soffer, O, Adovasio, J M and Hyland, D C (2000a) 'The "Venus" figurines: textiles, basketry, gender, and status in the Upper Paleolithic', 41(4) *Current Anthropology*, 511–537

Soffer, O, Adovasio, J M, Illingworth, J S, Amirkhanov, H A, Praslov, N D and Street, M (2000b) 'Palaeolithic perishables made permanent', 74(286) *Antiquity*, 812–821

Soffer, O and Vandiver, P (1997) 'The Ceramics from Pavlov I—1957 excavation', in Svoboda, J and Škrdla, P (eds), *Pavlov I -Northwest. The Upper Paleolithic Burial and its Settlement Context*, The Dolni Věstonice Studies 4, Brno: Academy of Sciences of the Czech Republic, Institute of Archaeology, pp 383–401

Soffer, O, Vandiver, P, Klima, B and Svoboda, J (1993) 'The pyrotechnology of perfomance art: Moravian Venuse and Wolverines', in Knecht, H, Pike-Tay, A and White, R (eds), *Before Lascaux,* Boca Raton: CRC Press, pp 259–275

Svoboda, J, Ložek, V and Vlček, E (1996) *Hunters between East and West: The Paleolithic of Moravia*, New York and London: Plenum Press, pp 157–170

van der Veen, M (2003) 'When is food a luxury?' 34(3) *World Archaeology*, 405–427

Vandiver, B P, Soffer, O, Klima, B and Svoboda, J (1989) 'The origin of ceramic technology at Dolni Věstonice, Czechoslovakia', 246(4933) *Science*, 1002–1008

Vasil'ev, A S (1985) 'Une statuette d'argile paléolithique de Sibérie du Sud', 89(2) *L'Anthropologie*, 193–196

Vasil'ev, A S, Kuz'min, V Y, Orlova, A L and Dement'ev, N Y (2002) 'Radiocarbon-based chronology of the Paleolithic in Siberia and its relevance to the peopling of the New World', 44(2) *Radiocarbon*, 503–530

Verpoorte, A (2001) *Places of Art, Traces of Fire.* Archaeological Studies Leiden University 8 (Dolní Věstonice Studies 6), Leiden, Brno: Faculty of Archaeology, University of Leiden and Institute of Archaeology, Academy of Sciences of the Czech Republic

Vitelli, D K (1993) 'Franchthi Neolithic Pottery. Volume 1: Classification and Ceramic Phases 1 and 2', in Jacobsen, T W (ed), *Excavations in Franchthi Cave, Greece. Fascicle 8,* Bloomington and Indianapolis: Indiana University Press

——— (1999) '"Looking Up" at early ceramics in Greece', in Skibo, J M and Feinman, G M (eds), *Pottery and People: A Dynamic Interaction.* Foundations in Archaeological Inquiry, Salt Lake City: University of Utah Press, pp 184–198

Vlassa, N (1972) 'Eine frühneolitische Kultur mit bemalter Keramik der Vor-Starčevo-Körös±Zeit in Cluj-Gura Baciului, Siebenbürgen', 47 *Praehistorische Zeitschrift*, 174–197

Whittle, A, Bartosiewicz, L, Borić, D, Pettitt, P and Richards, M (2002) 'In the beginning: new radiocarbon dates for the Early Neolithic in northern Serbia and south-east Hungary', 25 *Antaeus*, 63–117

Voigt, M M (2000) 'Çatal Höyü in Context. Ritual at Early Neolithic Sites in Central and Eastern Turkey', in Kuijt, I (ed), *Life in Neolithic Farming Communities. Social Organization, Identity and Differentiation.* Fundamental Issues in Archaeology, New York: Kluwer Academic/Plenum, pp 253–293

Wright, K (2000) 'The social origins of cooking and dining in early villages of western Asia', 66 *Proceedings of the Prehistoric Society*, 89–121

Yiouni, P (1996) 'The Early Neolithic pottery: functional analysis', in Rodden, R J and Wardle, K A (eds), *Nea Nikomedeia I.* Supplementary Volume No. 25, Athens: British School at Athens, pp 181–191

Zhao, C and Wu, X (2000) 'The dating of Chinese Early Pottery and discussion of some related problems', in Budja, M (ed), *7th Neolithic Studies. Documenta Praehistorica* 27, Ljubljana: Department of Archaeology, Faculty of Arts, Ljubljana University, pp 233–239

Zhushchikhovskaya, S I (2005) *Prehistoric Pottery-making of the Russian Far East,* BAR International Series 1434, Oxford: Archaeopress

Chapter 19

Transregional Culture Contacts and the Neolithization Process in Northern Central Europe

Detlef Gronenborn

Introduction

In recent years, the evidence for culture contact between agriculturalists and hunter-gatherers along the 'frontier zone' of the classic Neolithic of the 'Danubian Tradition' has increased and the scenarios have become much more complex. Inter-group and inter-personal contacts appear to have resulted in a variety of mixed economies, with farming and pastoralism adopted to a certain extent in the west, while the hunter-gatherer population towards the north remained largely unaffected until the terminal 5th millennium cal BC, although long-term contacts had also existed across these economic barriers. The different economic—and cultural—traditions forming the neolithization process between the 7th and 5th millennium cal BC across Temperate Western Eurasia may be coarsely grouped into an Occidental-Mediterranean tradition of possibly partly African origins, the classic Danubian Tradition, and what may be termed the Hyperborean Tradition with its roots ultimately in the Russian steppe zones and possibly in the Russian Far East. These 'streams' are geographically correlated with geographical patterns of YDNA and mtDNA diversity and may be traced back to human re-colonization after the glacial maximum.

'Savage Tribes'

In 1859 Japetus Worsaae (2002), doyen of 19th-century Danish archaeology, separated the Danish Stone Age into two periods: the Early Stone

Age characterized by 'oyster heaps' as the classical site type with 'crude implements of stone and bone' 'together with very crude pottery', and the Late Stone Age with megalithic burials as the typical sites. Worsaae (2002 [1859]: 53) ascribed the 'oyster heaps' to 'several savage autochthonous tribes scattered over a large part of Scandinavia' and compared them to the 'races at the lowest rung of the cultural ladder, e.g., the Eskimos and many savage tribes in the South Sea Islands and in America'. Since these initial statements, scholars have speculated about the origins and cultural affiliation of the 'crude pottery' and its relation to the 'Late Stone Age', to the first Neolithic societies in the European North.

In his 1936 classic, Grahame Clark stated 'it is hard to think that pottery was invented separately and independently by the strand-loopers of the Litorina coast' (1977 [1936]: 156) and considers an origin from southern or southeastern Europe where 'Neolithic civilization' would have already been established. The latter statement is due to a chronological misconception as the Danubian (LBK and successor entities such as Rössen and Lengyel) was thought to date later and thus could not have been ancestral to Ertebølle (EBK) pottery; hence ceramic antecedents would have had to be searched for further south. Later in life, Clark saw farming as having spread from the Danubian but he considered pottery as 'an integral part of Ertebølle equipment', unrelated to the Danubian (Clark 1975: 195–196). Gordon Childe (1925: 15) saw the Ertebølle pottery tradition as 'due to cultural borrowing' and later he discussed possible stylistic antecedents from southern Spain and the Crimea (Childe 1950: 34). With this statement, the arena for possible connections had widened.

The Grand Stage: Afroeurasia

In 1960, Günter Smolla had published a by now largely forgotten booklet entitled 'Neolithic Culture Traits' (German: *Neolithische Kulturerscheinungen*). He further elaborated Clark's and Childe's arguments and seemed to favour a cultural affiliation of the Ertebølle pottery with the southeast: the Russian forest and steppe zones, the closest resemblances being finds from Zedmar A and D (Smolla 1960: 30). Apart from possible antecedents from the Russian steppe zones, Smolla also discusses western antecedents and looks as far as North Africa; lastly he abstains from a final decision but rules out an independent emergence in the north (Smolla 1960: 34).

A Ukrainian origin of Ertebølle pottery has then been suggested by Danilenko (1969) but his ideas had not crossed the Iron Curtain and, certainly unaware of Danilenko's concept, Schwabedissen (1967) explicitly ruled out any eastern contributions to the Baltic Neolithic and supported his statement with an alleged 'lack of preconditions'. Instead he discusses stylistic influences from the West Mediterranean Cardial and the later Rössen culture (Schwabedissen 1967: 459–461). Until the late 1990s, many western scholars continued to see Ertebølle and the stylistically related Swifterbant pottery in the Netherlands as results of east–west contacts on the northern continental European plain (e.g., Louwe Kooijmans 1998; Raemaekers 1999). An exception was Clark (1980: 83) who, in his final compendium on the Mesolithic, paralleled Ertebølle pottery with eastern Eurasian and North American wares and underlined that pottery technology was not restricted to farming societies.

During the 1980s, the picture became more complicated with the discovery of another non-LBK pottery tradition. Within LBK contexts, a curious ware of very different technological tradition appeared and was then labelled after its earliest findspot which, interestingly enough, lay far outward of its main distribution in eastern France and southwestern Germany: La Hoguette. This tradition had been seen in connection with the southern French Cardial (Jeunesse 1987; Lüning et al. 1989). Other non-LBK pottery traditions are known from northwestern continental Europe, namely Limburg (e.g., Brounen 1999; Modderman 1974, 1987) and the so-called *Begleitkeramik* often associated with La Hoguette (Jeunesse 1994: 15–16).

During the 1990s, Van Berg broadened the perspective and contrasted the pottery traditions of Eurasian hunter-gatherers—and, as we now know, horticulturalists and pastoralists—with those of the Neolithic farmers. He developed the term 'subnéolithique' for the western manifestations, Bliquy, Villeneuf-Saint-Germain, La Hoguette and Limburg, and sets it apart from the 'Neolithique céréalier', the Danubian traditions of southern Central Europe (van Berg 1990).[1] Later, Timofeev (1998) returned to Danilenko's idea of a spread of pottery from Ukraine and southeastern Russia to the Baltic coast. But Timofeev and Zaitseva (1998: 190) also suggested an independent emergence of pottery in the southeastern Russian steppe zones because the Elshan tradition in the Samara Valley appears to be earlier than surrounding traditions in western Central Asia (Mamonov 2000). Hence, while the considerable ancestry of eastern Siberian, Russian Far Eastern, Chinese and Japanese pottery (Habu 2004; Keally et al. 2004; Kuz'min 2002) would suggest a gradual westward spread of pottery technology towards Europe, the intermittent regions have not yet produced dates consistent with such a gradual expansion (Kuz'min and Orlova 2000). Elshan remains a phenomenon yet to be explained (Figure 19.1 a–d).

While Central and Eastern Asian roots remain uncertain, connections between the Russian steppes and the Baltic become increasingly evident, as the database of ^{14}C-dates from western Russia accumulates (Dolukhanov et al. 2005). Indeed, a number of western scholars now accept the possibility of an eastern origin of the pointed-base Ertebølle tradition (Hallgren 2004; Hartz 2000; Klassen 2000b).

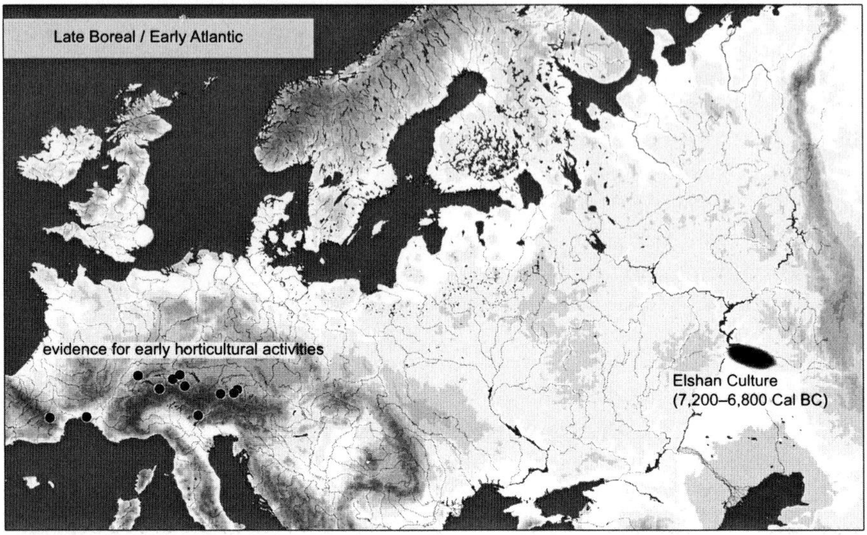
Late Boreal / Early Atlantic
evidence for early horticultural activities
Elshan Culture
(7,200–6,800 Cal BC)

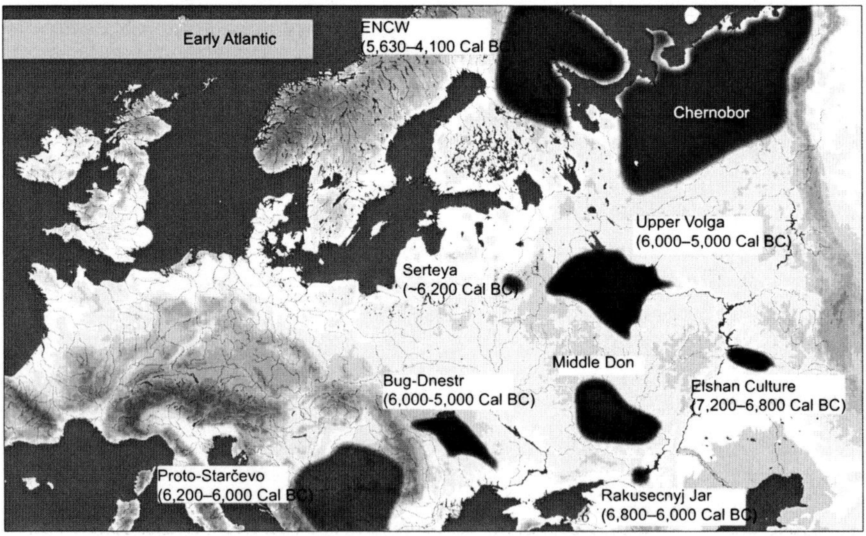
Early Atlantic
ENCW
(5,630–4,100 Cal BC)
Chernobor
Upper Volga
(6,000–5,000 Cal BC)
Serteya
(~6,200 Cal BC)
Middle Don
Bug-Dnestr
(6,000-5,000 Cal BC)
Elshan Culture
(7,200–6,800 Cal BC)
Proto-Starčevo
(6,200–6,000 Cal BC)
Rakusecnyj Jar
(6,800–6,000 Cal BC)

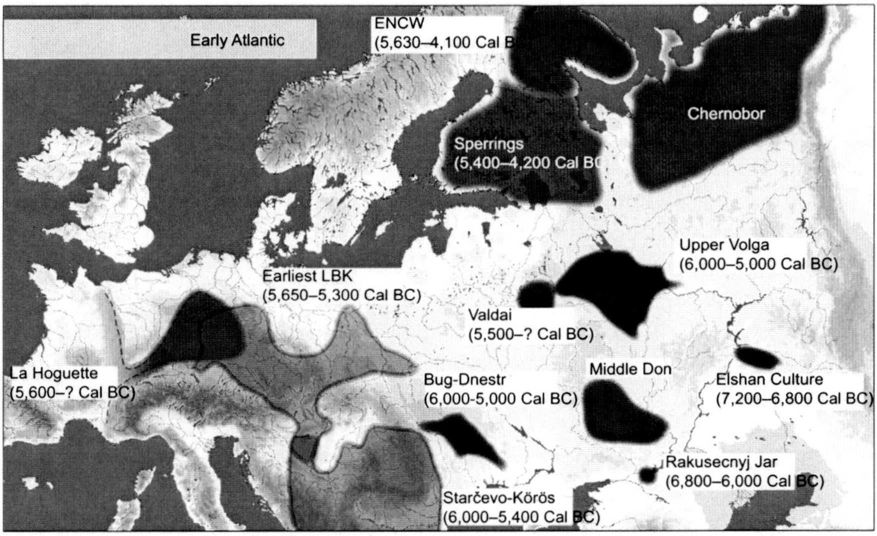

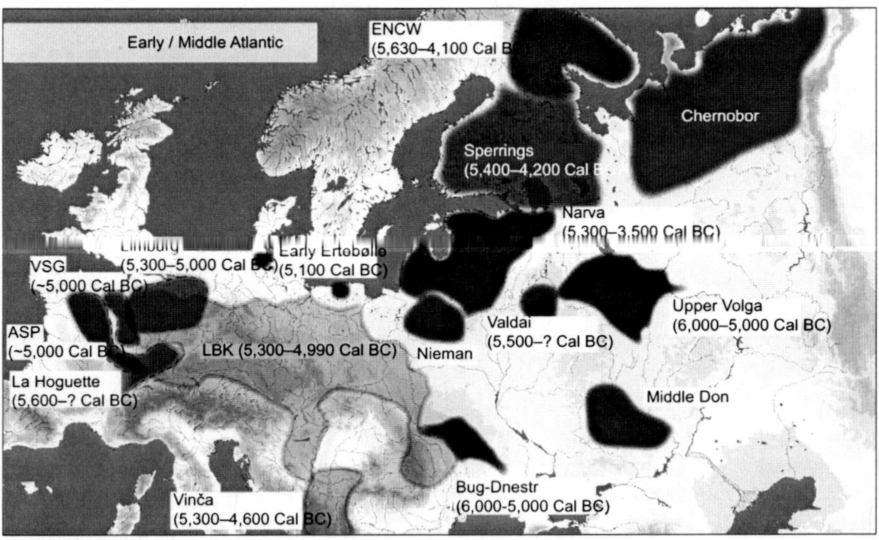

Figure 19.1(a–d) The spread of pottery and farming across Western Temperate Eurasia (after Gronenborn 2003a).

If Ertebølle and associated traditions were derived from the Russian steppes, and from regions even farther along the Pacific Coast, where could the western pottery of La Hoguette, with its Mediterranean stylistic connections, have originated? Unfortunately, mid-20th-century hypotheses on northwest African origins or contributions from the western Mediterranean Neolithic have become infamous in recent decades because earlier endeavours were swept away by uncomplicated and easy-to-digest scenarios of a rapid east–west coastline farmer expansion (e.g., Bernabò Brea 1950; Zilhão 2001). However, the question of an 'African connection' may have to be posed again.

African Saharan populations as contributors to the neolithization process of the western Mediterranean have recently been re-discussed by Winiger (1998) who sees population dispersals towards the north in the course of the early to mid-Holocene desiccation as the triggering force. While his coarse model may yet be criticized on archaeological as well as palaeoclimatological grounds (Linstädter 2004), northwest Africa and ultimately the early Holocene Saharan region remain good candidates for another independent centre for the 'Neolithic'. Indeed, the earliest Saharan pottery is dated to the 9th millennium cal BC (Roset 1987) and recent work on the Dogon Plateau in Mali at the present-day southern edge of the Sahara has produced evidence for a ceramic tradition which appears to date to the 8th millennium cal BC (Huysecom et al. 2004). In simulation studies of the emergence and expansion of farming based on mid-Holocene natural potentials, northwest Africa emerges as an independent centre (Wirtz and Lemmen 2003).

Despite these tempting indications, an African contribution to the early Neolithic of westernmost Eurasia is presently indicated only for the centuries around 5,000 cal BC (Manen et al. 2007). But generally the chronology of the western Mediterranean Early Neolithic and the connections between the different traditions in Iberia, southern France and western Central Europe are not well understood (van Willigen 2004). As new research along the southern French coast has shown, the earliest western Central European ware attributed to southern French influence—La Hoguette—has little or no stylistic affiliation to the contemporary western Mediterranean Impressa wares. The latter are nowadays believed to constitute the earliest 'Neolithic' pottery in southern France dating to 5,800–5,600 cal BC (van Willigen 2006). La Hoguette, according to the stratigraphy at Bavans rockshelter should date to 5,600–5,400 cal BC (Aimé 1991) and is well represented in earliest LBK context dating to 5,500–5,300 cal BC (Gronenborn 1999; Strien 2000; Strien and Gronenborn 2005). However, stylistically La Hoguette shows affiliations with the later Cardial wares which make their appearance no earlier than during the 5,400–5,300 cal BC (van Willigen 1997, 2004). This dating fault line has to be circumnavigated in order to understand the process of 'neolithization' in occidental Europe and adjacent regions. One recent suggestion is that an earlier as yet undiscovered ware may have been the stylistic antecedent in southern France (van Willigen 2006).

A Western Eurasian Passageway: Central Europe

Leaving the grand stage of Afroeurasia, we shall zoom into a narrower area within Western Temperate Eurasia where not only do eastern and western traditions meet but also where interactions between full scale farmers and pottery-using hunter-gatherers, pastoralists and agro-pastoralists can be studied in high resolution: Central Europe.

In recent years, a considerable body of data has been accumulated on interactions between Early Neolithic farmers of the Danubian world, the Linear Pottery culture (LBK) and native hunter-gatherers and pastoralists of the La Hoguette tradition along the western margins of the LBK ecumene. The picture now generally accepted is complex and shows that incorporation processes varied from region to region and throughout time. Moreover, because of the similarity of the material culture between LBK farmers and surrounding hunter-gatherers in the Carpathian Basin and adjacent regions, the visibility of such contacts is much lower—only in the western parts, along the Rhine, have scholars been able to formulate relatively robust and conclusive hypotheses.

Currently, the arrival of the first settlers at the western margin of the earliest LBK ecumene is dated to about 5,500 cal BC (Gronenborn 2003a; Strien and Gronenborn 2005). These settlers may have arrived from Franconia or the Neckar Valley, maybe both directions. Upon arrival, these farmers came into contact with pastoralists using La Hoguette pottery. This culture contact is archaeologically documented by La Hoguette vessels in LBK pits (Lüning et al. 1989), in the lithics—western European microliths and a non-earliest LBK lithic technology (Gronenborn 1990, 2003b)—and also in the unusual composition of the faunal spectrum with a number of wild species (Uerpmann and Uerpmann 1997). Another site where the complexity of interaction processes and the subsequent incorporation of local hunter-gatherer/pastoralists into LBK communities can be documented is Vaihingen upon Enz (Strien 2005). Contacts continue when the first step towards the Rhineland is taken around 5,400 cal BC (Heinen et al. 2004; Strien and Gronenborn 2005) where Limburg pottery makes its appearance in LBK contexts. To date, the Limburg pottery complex is an enigma not yet well understood as it is still unclear whether it represents a pottery tradition of an independent group or a specific ware used within LBK communities (Brounen 1999; Louwe Kooijmans 1998; van

Berg 1990). Furthermore, a third northwestern early ceramic tradition has recently been identified, the so-called *Begleitkeramik* (Jeunesse 1994: 15), often associated with La Hoguette ware (Brounen 1999; van Berg 1990). Along the Upper Rhine, in Alsace, La Hoguette is found in connection with later LBK, but vanishes from the sites east of the Rhine after earliest LBK around 5,400 cal BC (Jeunesse 1998–99; Strien 2000). While the pastoral component of the La Hoguette economy is nowadays largely agreed upon (Chaix 1997; Kalis et al. 2001), it is as yet unclear how the groups using Limburg and *Begleitkeramik* made their living—that is if these wares were socially independent of LBK. Hence, both along the Upper as well as Lower Rhine and the adjacent regions, a persistence of hunter-gatherer societies with and without pottery well into the Neolithic cannot be ruled out (e.g., Gronenborn 2007; Jeunesse 2000, 2001; Kind 1997). Since the archaeological recognition of the culturally diverse region with different economies co-existing, and by implication ethnicities, a number of physical anthropology and molecular archaeological studies have added to the hitherto only archaeologically derived picture (Bentley et al. 2002, 2003; Bentley and Knipper 2005; Haak et al. 2005).

Less intensively studied were the communication routes between LBK farming villages and the hunter-gatherer populations to the north, in the vast northern Central European lowlands between Flanders and the Polish Plains. Contacts between these regions had not emerged with the onset of the Neolithic; just as along the western margins of the LBK (Gronenborn

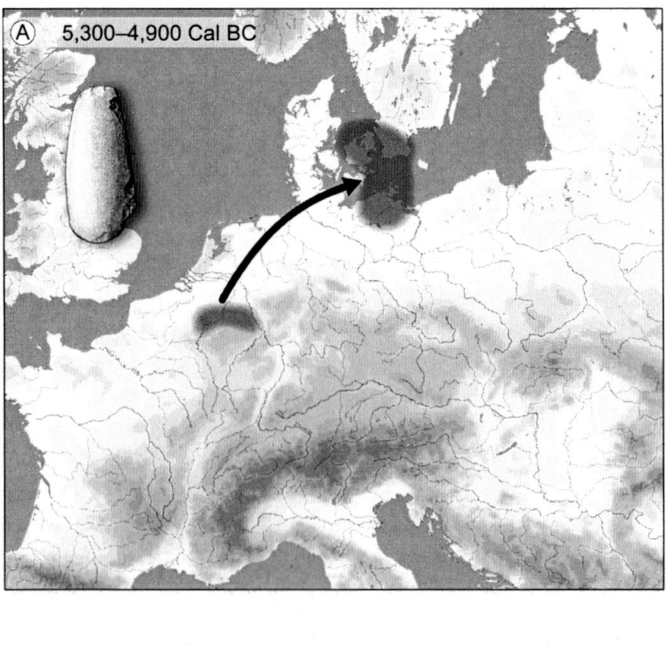

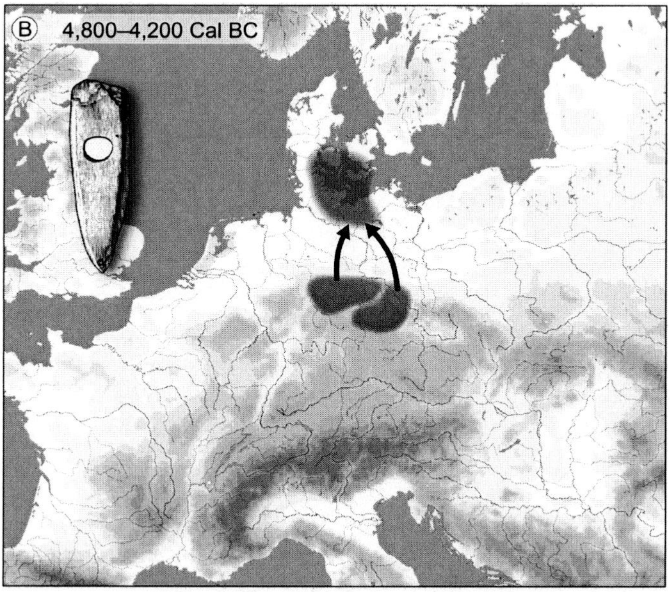

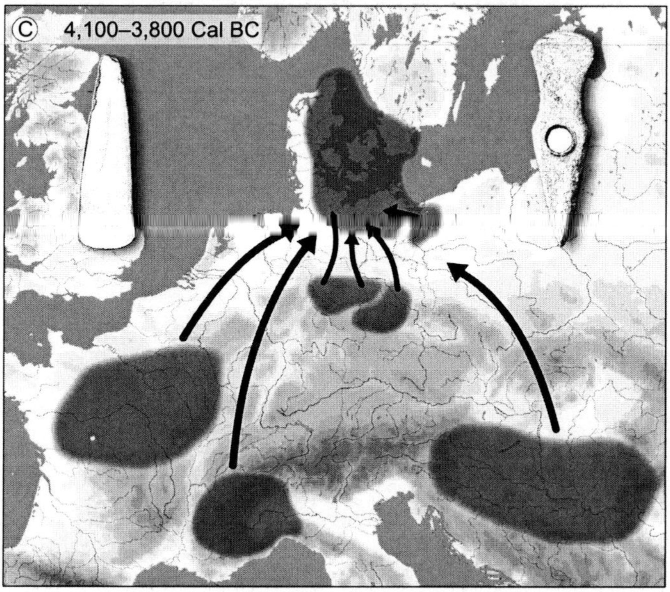

Figure 19.2(a–c) Shifting import regions for northern Central Europe and southern Scandinavia during the 5th millennium cal BC (inserts: (a) adze from Henriksholm-Bøgebakken, Denmark; (b) perforated shoe-last adze from Blankenberg, northern Germany; (c) copper axe from unknown location in Denmark, Jadeite celt from Zeeland, Denmark (all after or from Klassen 2004).

1999), these interactions have their roots in pre-Neolithic times. Already during the Late Palaeolithic and the Mesolithic, occasional objects of apparently northern European origin appear on sites in the south. Tanged points of Rhenish Lowland origin dating to the Younger Dryas are found on some southern German sites (Street et al. 2001: 396) and a reindeer metatarsus stems from the site of Siebenlinden 1 in the Upper Neckar Valley (Hahn et al. 1993; Kind 2003). Artefacts of northern lowland origin increase with the appearance of the LBK. An 'oblique transverse arrowhead' or 'broad trapeze' of southern Scandinavian origin dating to around 5,400 cal BC was found at the earliest LBK site of Bruchenbrücken (Gronenborn 2005). Occasionally, flake axes were found on LBK sites at the northern margin of their dispersion (Gronenborn 1997: 59–60; Kaczanowska 1990: 36, figure 7) and the T-shaped antler axes on LBK sites may also be of northern provenance (Hartz and Lübke 2005: 120; Klassen 2004: 124). In the south, the sites of Vaihingen upon Enz and Ditzingen near Stuttgart have produced antler fragments decorated with drilled pits typical for the circum-Baltic Late Mesolithic (Krause 2001; Terberger 2003). In the northwest the LBK well from Kückhoven, dating to the terminal 6th millennium cal BC, contained an amber pendant fragment similar to finds from southern Sweden (Klassen 2004: 314, figure 152; Weiner 1995a: 187) and a fragment of a longbow of the Holmegaard type (Weiner 1995b). Lastly the pointed-base vessel from the LBK site of Rosheim is to be mentioned, from a pit which produced a 14C-date between 5,200 cal BC and 4,800 cal BC (Jeunesse and Lefranc 1999: 44–47).

Taking these 'import finds' into account, the long-held view of a northward directed cultural drift with the hunter-gatherer populations in the lowlands playing the part of passive recipients has to be abandoned. With the acceptance of an eastern origin of Baltic pottery contact, scenarios for the Central European northern plains and southern Scandinavia had to be rethought. Nowadays, the situation appears much more complex. Judging from various ethnographic accounts on individual and group mobility (task groups) in boreal forest environments (e.g., Engelbrecht 2003; Hudson 1976), we may assume that at least some of the 'import objects' were brought to the LBK sites by individuals or groups which had travelled over considerable distances across the mid-Atlantic Central European landscape. Hence, people in the northern lowlands actively sought contact with the southern farmers and in the course of these contacts must have had the possibility of acquiring technical knowledge on farming; yet for over a millennium they had chosen not to apply this economy in their homelands or if so, only on a very limited scale in a restricted region, namely eastern Schleswig-Holstein.

Here pollen spectra indicate a gradual reduction in elm and an increase in ash after 5,000 cal BC which has been attributed to small-scale cattle-keeping. However, bovid remains from the site of Rosenhof, previously identified as domesticated and dated to the first half of the 5th millennium cal BC (Hartz et al. 2000: 136; Kalis and Meurers-Balke 1998: 13), have very recently been proved to come from female aurochsen (Scheu et al. 2008). The earliest cereal pollen appear between 4,800 cal BC and 4,500 cal BC, but only along the coast in eastern Holstein (Hartz et al. 2000: 135). However, Behre (2007: 208) has underlined the uncertain nature of these indications and concludes that 'neither the botanical nor the archaeological evidence' can be taken as a confirmation for farming before 4,100 cal BC. The image of eastern Schleswig-Holstein as the alleged 'bridge' for farming into southern Scandinavia is becoming blurred. Pottery of the Ertebølle tradition appears around 4,750 cal BC on coastal sites and may date as early as 5,100 cal BC at inland sites (Hartz et al. 2000: 140; Hartz and Lübke 2005: 125–127).

During the 5th millennium cal BC, the northern Central European lowlands and southern Scandinavia had been in intensive mutual contact and interaction with the more southerly regions inhabited by farmers of the post-LBK Großgartach and Stroke Ornamented Pottery traditions, later Rössen and Lengyel. A variety of shoe-last adzes, pecked stone adzes (German: *Walzenbeile*) and stone mace heads are taken as indications of northward directed contacts. Klassen (2004) has grouped these artefacts into chronologically differentiated 'import phases' and was able to show that the regions of origin of these artefacts shifted in the course of the 5th millennium cal BC (Figure 19.2 a–c). After 4,300 cal BC these networks had widened and intensified with more 'prestigious' artefacts being transported such as jadeite and copper celts of southern French and southeastern European origin (Klassen 2000a, 2004: 103–108). Klassen postulates that an alleged late Ertebølle socio-political intensification would have attracted these prestige items: 'This in fact speaks in favour of a considerable degree of social stratification in the western Baltic at the time in question, where some individuals in their strife for prestige obviously were able to control a group's surplus and subsequently to exchange this surplus' (Klassen 2000b: 45).

Apparently Scandinavian Late Mesolithic social archaeology had come a long way since Worsaae's 'lowest races' and Clark's 'strand-loopers'. Yet, Klassen's scenario may be on the extreme side of possible interpretations. Madsen (1991: 493–495) had already suggested that Late Ertebølle societies were segmented on a however 'minor scale' and he had also shown that 'probably warfare was the rule rather than the exception' particularly during the phase of culture change from Ertebølle to Early Neolithic Funnel Beaker around 4,100–4,000 cal BC. The latter conclusion, however, is meliorated by Fischer (2002: 372) who would not account for an increase in violence towards the EBK/TRB transition but he does support the idea that external contacts and the availability of exotic goods had stimulated political competition which finally would have led to the adoption of farming (Fischer 2002: 378–380). But where would the cultural triggering force have to be sought. In other words, what brought about the change in ceramic styles from EBK to TRB?

Previously, the emergence of TRB pottery had been discussed in connection with the northward movement of the Middle Neolithic Rössen culture. Several sites in the northern European plain do show stylistic affiliations such as Hüde I at Lake Dümmer (Kampffmeyer 1983) and Hamburg-Boberg (Schindler 1962). More recent studies, however, have shown that a chronological gap exists between latest Rössen pottery and earliest TRB pottery in the north. Thus, despite certain—albeit very general—similarities, Rössen as a source for TRB has to be ruled out (Madsen and Petersen 1984); nevertheless Ertebølle groups had been in continuous contacts with farming societies in the south.

While Ertebølle groups communicated with Rössen and Lengyel in the south, related Swifterbant and the *groupe de Melsele* in Belgium (Crombé et al. 2002; Van Neer et al. 2001) had been in contact with more westerly Neolithic societies, as is for instance indicated at the site of Hardinxveld in the Rhine-Meuse Delta in the Netherlands. Here, pottery of the Swifterbant-type made its appearance around 5,000 cal BC, but a vessel stylistically influenced by the *groupe de Blicquy* dates to around 4,800 cal BC (Raemaekers 2001). Cereals appear around 4,000 cal BC, however there is a hiatus between 4,400 cal BC and 4,000 cal BC so the exact initial date of farming is unknown (Louwe Kooijmans 2003).

While Swifterbant pottery remains in use in the Dutch sand zone up until 3,600 cal BC when it is replaced by the Michelsberg-influenced Hazendonk pottery (Louwe Kooijmans 1998), the Ertebølle tradition is succeeded by early Funnel Beakers (TRB) around 4,100–4,000 cal BC. At least this is suggested by the earliest dates from the site Wangels in eastern Schleswig-Holstein (Hartz et al. 2000: 143; Hartz and Lübke 2005: 128–129). Recently, the introduction of this new pottery tradition is seen in connection with the expansion of the Michelsberg culture and migration models are being discussed again (Jeunesse et al. 2004: 231–237):[2] some sites in Lower Saxony, of which one is dated to 4,000 cal BC, show affiliations to Michelsberg pottery and TRB pottery assemblages from northern Lower Saxony have Michelsberg-style influences (Hartz et al. 2000: 132–134). A typological stylistic detail indicating the eastward expansion of the Michelsberg tradition may be internally embossed knobs along vessel rims (French: *boutons au répoussé internes*; German: *innenständige Lochbuckel*) which appear in central and western France and then gradually spread eastward (Seidel and Jeunesse 2000), finally reaching the western Baltic (Figure 19.3).

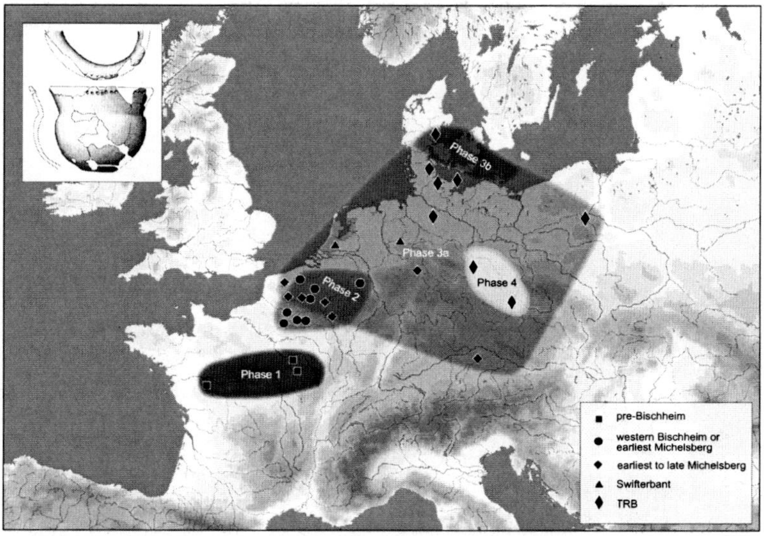

Figure 19.3 Eastward drift of internally embossed knobs as a possible indication of Michelsberg expansion (after Klassen 2004; insert from Søvind Mark, eastern Jutland, Denmark).

With the appearance of TRB the landscape was altered, notably along the coastline and particularly in eastern Holstein: birch and hazel were replaced by oak-dominated forests (Fischer 2002: 345–348; Jennbert 1988; Kalis and Meurers-Balke 1998), cereal horticulture was practised and sheep/goat were kept (Hartz et al. 2000: 138). From 4,100 cal BC, farming and animal husbandry had begun to change the north. However, in the Dutch sandy lowlands, hunting and gathering persisted to a considerable degree until the appearance of the TRB Drouwen style around 3,400 cal BC (Louwe Kooijmans 1998: 414–418, 2003: 622).

Remains ... Why?

While the scholarly interest in the process of neolithization has led to increasingly detailed scenarios for many regions in Europe, one major aspect still remains little understood: why? We know about the enormous contact networks which served as communication routes for the technical—and ritual—knowledge about farming. We know about the potential of these routes to channel prestigious artefacts which are supposed to have functioned as foundations for politically elevated positions. We may imagine charismatic personalities who saw an advantage in using farming products for their political careers. We know as well, or are at least fairly certain, that for a number of periods we have to account for population movements, for people migrating and bringing the new technologies to others. We know that out of migrant and local communities certain multi-traditional societies emerged which formed the basis for Neolithic social and political life. But, we still don't know what instigated all this.

Recently, however, a new parameter is increasingly being considered which might have influenced, if not occasionally triggered, the spread of farming, namely Holocene cooling cycles or ice-rafting-detritus events (Barber et al. 2004; Bond et al. 2001; Heiri et al. 2004). The idea of a climate-related emergence and spread of farming is not new, of course, and dates back to the earliest scholarly explanation models (Gronenborn 2005). But in the light of recent much more detailed Holocene palaeoclimatological data, a modest and differentiated 'neo' climate determinism is beginning to gain ground. Indeed, if looked at closely, the series of cooling events visible in a number of North Atlantic and continental European proxy data correlate with periods of a rapid advance of the Neolithic economy. This has been suggested for the now widely known so-called 6.2-event (Weninger et al. 2005) but also for the so-called Early Holocene Event and a number of other minor fluctuations (Gronenborn 2007).

What had happened? We may certainly not envisage huge masses of homeless people wandering across the European landscape, having been deprived of their belongings by environmental catastrophes. But we may search for certain key periods where climate change could have resulted in economic crises which again led to social friction, political instability and lastly social fissioning. Once groups or parts of groups decided to relocate, they would bring their technology to new regions. For the period in question, the Mesolithic-Neolithic transition in the northern lowlands, the Holocene ice-rafting-detritus event 4 (Bond et al. 2001; Heiri et al. 2004) may have had a considerable effect for over one thousand years up until the crisis years around 3,700 cal BC (Gronenborn 2007; Schibler et al. 1997).

Bonsall and colleagues (2002) have suggested a connection between the adoption of agriculture and drier conditions in the northern British Isles and southern Scandinavia. While they would not account for migrations under the interpretative influence of British post-processual archaeology, recent studies by Rowley-Conwy (2004) and Sheridan (2003) returned to these classic concepts. Apparently, now again, researchers believe that the first farmers in Britain may well have come from the continent. If migrations across the Irish Sea and the British Channel are acceptable, then an eastward expansion of Michelsberg in the course of migrations is not entirely unthinkable. Could this eastward expansion have been triggered by a peculiar anomaly around 4,350 cal BC which is visible as a growth depression in north-western European bog oak curves (Leuschner et al. 2002: 700)? This anomaly might have been an early effect of IRD 4 (Spurk et al.

2002). Too much rain, harvest failures, hunger, ...flight? Food for further thought.

Conclusion

Apart from the classic Danubian passageway, we are starting to acknowledge that material and economic aspects of what later constitutes Neolithic societies entered Central Europe from two other regions: one arrived through western France and has its origins somewhere in the western Mediterranean and—not entirely impossible—somewhere on the northern or north-western African continent. This sphere[3] was termed 'Occidental Tradition' (Gronenborn 2003a). The other direction from where Western Temperate Eurasia had been influenced lies far in the southeast of the European continent, in the Russian steppe zones, on the fringes of Central Asia. Apparently this is the region where pottery made its first appearance in Europe, solely in a hunter-gatherer environment. From the steppes it had spread to the Baltic and from there this technological innovation reached northwestern continental Europe. Referring to classical antiquity, we may baptize this vast network the 'Hyperborean Tradition' (Gronenborn 2003a). These three broad cultural 'streams' (Figure 19.4) contribute to the complex and intertwined process of neolithization in Central and northern Central Europe and thus

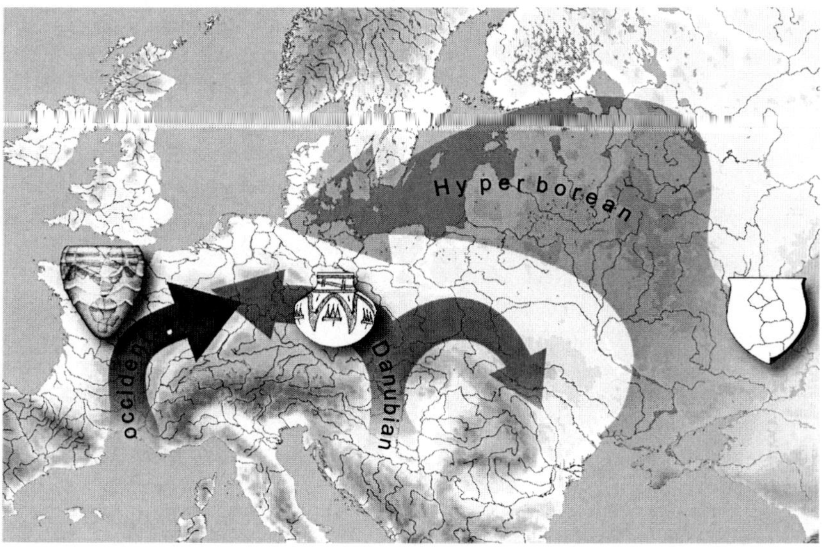

Figure 19.4 Three 'neolithization streams' into Central Europe (La Hoguette vessel from Alzey, Rhine-Hesse, Germany; LBK vessel from Frankfurt, Germany; Ertebølle-type vessel from Diepholz, Lower Saxony, Germany; Elshan vessel from Ivanovka, Samara Valley, Russia).

it is not surprising that they are coarsely reflected in Western Eurasian geographic patterns of mtDNA and YDNA diversity (e.g., Achilli et al. 2004; Richards 2003; Semino et al. 1996, 2000, 2004; Simoni et al. 2000). The 'streams' had followed networks in operation since Temperate Europe's repopulation after the glacial maximum.

Acknowledgements

I am indebted to Samuel van Willigen, Schweizerisches Landesmuseum, Zürich, Switzerland for informing me about the current state of research in the western Mediterranean, Hanny Piezonka, Deutsches Archäologisches Institut, Berlin, Germany for pointing out Smolla's early thoughts about the possible stylistic affiliations of EBK pottery and Ruth Bollongino, Palaeogenetics Group Mainz, for insights into the—at the time of writing—not yet publicly available information on the Rosenhof bovid aDNA analyses.

Notes

1. Recently Bliquy and Villeneuf-Saint-Germain have been rearranged chronologically and are now interpreted as Danubian-influenced local phenomena which adopted farming in the course of contact with the late LBK in the Paris Basin (Jeunesse 1998–99).
2. For a research history, see Klassen (1996–97, 2004).
3. I am yet hesitant to revive the term *Kulturkreis*; but consult e.g., Buttler (1938).

References

Achilli, A, Rengo, C, Magri, C, Battaglia, V, Olivieri, A, Scozzari, R, Cruciani, F, Zeviani, M, Briem, E, Carelli, V, Moral, P, Dugoujon, J-M, Roostalu, U, Loogväli, E-L, Kivisild, T, Bandelt, H-J, Richards, M, Villems, R, Santachiara-Benerecetti, A S, Semino, O and Torroni, A (2004) 'The molecular dissection of mtDNA haplogroup H confirms that the Franco-Cantabrian glacial refuge was a major source for the European gene pool', 75 *American Journal of Human Genetics*, 910–918

Aimé, G (1991) 'Les niveaux mésolithiques de Bavans dans le contexte jurassien', in Thévenin, A (ed), *Mésolithique et Néolithisation en France et dans les Régions Limitrophes. Actes du 113e Congrès National des Sociétés Savantes (Strasbourg, 5–9 avril 1988)*, Paris: Éditions du Comité des Travaux Historiques et Scientifiques, pp 323–345

Barber, K, Zolitschka, B, Tarasov, P and Lotter, A F (2004) 'Atlantic to Urals—The Holocene climatic record of mid-latitude Europe', in Battarbee, R W (ed), *Past Climate Variability through Europe and Africa*, Dordrecht: Kluwer, pp 417–442

Behre, K-E (2007) 'Evidence for Mesolithic agriculture in and around central Europe?', 16 *Vegetation History and Archaeobotany*, 203–219

Bentley, R A, Price, T D, Lüning, J, Gronenborn, D, Wahl, J and Fullagar, P D (2002) 'Prehistoric migration in Europe: Strontium isotope analysis of early Neolithic skeletons', 43(5) *Current Anthropology*, 799–804

Bentley, R A, Krause, R, Price, T D and Kaufmann, B (2003) 'Human mobility at the Early Neolithic settlement of Vaihingen, Germany: evidence from strontium isotope analysis', 45(3) *Archaeometry*, 471–486

Bentley, R A and Knipper, C (2005) 'Transhumance at the early Neolithic settlement at Vaihingen (Germany)', http://antiquity.ac.uk/ProjGall/bentley/index.html

Bernabó Brea, L (1950) 'Il Neolitico a ceramica Impressa e la sua diffusione nel Mediterraneo', 3 *Rivista di Studi Liguri*, 25–36

Bond, G, Kromer, B, Beer, J, Muscheler, R, Evans, M N, Showers, W, Hoffmann, S, Lotti-Bond, R, Hajdas, I and Bonani, G (2001) 'Persistent solar influence on North Atlantic climate during the Holocene', 294 *Science*, 2130–2136

Bonsall, C, Macklin, M G, Anderson, D E and Payton, R W (2002) 'Climate change and the adoption of agriculture in north-west Europe', 5 *European Journal of Archaeology*, 7–21

Buttler, W (1938) *Der Donauländische und der Westische Kulturkreis der Jüngeren Steinzeit. Handbuch der Urgeschichte Deutschlands* 2, Berlin: Walter DeGruyter

Brounen, F T S (1999) 'Vroegneolithische Begleitkeramiek uit Limburg en Noord-Brabant', 82 *Archeologie in Limburg*, 59–64

Chaix, L (1997) 'La transition Méso-Néolithique: quelques donnés de l'archéozoologie dans les Alpes du Nord et le Jura', in Jeunesse, C (ed), *Le Néolithique Danubien et ses Marges Entre Rhin et Seine. XXIIe Colloque Interrégional sur le Néolithique, Strasbourg 27–29 Octobre 1995*. Supplément n°3 Cahiers de l'Association pour la Promotion de la Recherche Archéologique en Alsace. Zimmersheim: Association pour la Promotion de la Recherche Archéologique en Alsace, pp 191–196

Childe, V G (1925) *The Dawn of European Civilization*, London: Routledge and Kegan Paul

——— (1950) *Prehistoric Migrations in Europe*, Oslo: Instituttet for Sammenlignende Kulturforskning Ser A Forelesninger no 20, H. Aschehoug and Co

Clark, J G D (1977 [1936]) *The Mesolithic Settlement of Northern Europe: A Study of the Food-Gathering Peoples of Northern Europe during the Early Post-Glacial Period*, Cambridge: Cambridge University Press

——— (1975) *The Earlier Stone Age Settlement of Scandinavia*, Cambridge: Cambridge University Press

——— (1980) *Mesolithic Prelude: The Palaeolithic-Neolithic Transition in Old World Prehistory*, Edinburgh: Edinburgh University Press

Cook, G T, Bonsall, C, Hedges, R E M, McSweeney, K, Boroneant, V, Bartosiewicz, L and Pettitt, P B (2001) 'Problems of dating human bones from the Iron Gates', 76 *Antiquity*, 77–85

Crombé, P, van Roeyen, J-P, Sergant, J, Perdean, Y, Cordemans, K and van Strydonck, M, (2000) 'Doel "Deurganckdok" (Flanders, Belgium): settlement traces from the Final Palaeolithic and the Early to Middle Neolithic', 20 *Notae Praehistoricae*, 111–119

Crombé, P, Perdaen, Y, Sergant, J, Van Royen, J-P and van Strydonck, M (2002) 'The Mesolithic-Neolithic transition in the sandy lowlands of Belgium: new evidence', 76 *Antiquity*, 699–706

Danilenko, V N (1969) *Neolit Ukrainy*, Kiev: Izdatel'stvo 'Naukova Dumka'

Dolukhanov, P, Shukurov, A, Gronenborn, D, Sokoloff, D, Timofeev, V and Zaitseva, G (2005) 'The chronology of Neolithic dispersal in Central and Eastern Europe', 32(10) *Journal of Archaeological Science*, 1441–1458

Engelbrecht, W (2003) *Iroquoia: The Development of a Native World*, Syracuse: Syracuse University Press

Fischer, A (2002) 'Food for feasting? An evaluation of explanations of the Neolithisation of Denmark and southern Sweden', in Fischer, A and Kristiansen, K H (eds), *The Neolithisation of Denmark: 150 Years of Debate*, Sheffield: Collis Publication, pp 343–393

Gronenborn, D (1990) 'Mesolithic-Neolithic interactions: the lithic industry of the Earliest Bandkeramik Culture site at Friedberg-Bruchenbrücken (West Germany)', in Vermeersch, P M and van Peer, P (eds), *Contributions to the Mesolithic in Europe. Papers Presented at the Fourth International Symposium 'The Mesolithic in Europe'*, Studia Praehist. Belgica 5, Leuven: Leuven University Press, pp 173–182

——— (1997) *Silexartefakte der Ältestbandkeramischen Kultur. Mit einem Beitrag von Jean-Paul Caspar*, Universitätsforsc hungen zur prähistorischen Archäologie 37, Bonn: Habelt

——— (1999) 'A variation on a basic theme: the transition to farming in southern Central Europe', 13 *Journal of World Prehistory*, 123–210

——— (2003a) 'Migration, acculturation, and culture change in Western Temperate Eurasia, 6500–5000 cal BC', in Budja, M (ed), *10th Neolithic Studies*. Documenta Praehistorica XXX. Ljubljana: Univerza v Ljubljani, Filosofska Fakulteta, Oddelek za Arheologijo, 79–91

——— (2003b) 'Lithic raw material distribution networks and the Neolithization of Central Europe', in Burnez-Lanotte, L (ed), *Production and Management of Lithic Materials in the European Linearbandkeramik/Gestion des Matériaux Lithiques dans le Rubané Européen*. Actes du XIVème Congrès UISPP, Université de Liège Belgique, 2–8 Septembre 2001, BAR International Series 1200. Oxford: Hadrian Books, pp 45–50

——— (2005) 'Eine Pfeilschneide aus Südskandinavien vom ältestbandkeramischen Fundplatz Friedberg-Bruchenbrücken in der Wetterau', 35 *Archäologisches Korrespondenzblatt*, 159–168

——— (2007) 'Beyond the models: "Neolithization" in Central Europe', in Whittle, A and Cummings, V (eds), *Going Over, the Mesolithic-Neolithic Transition in North-west Europe*, London: British Academy, pp 73–98

Haak, W, Forster, P, Bramanti, B, Matsumura, S, Brandt, G, Tänzer, M, Villems, R, Renfrew, C, Gronenborn, D, Alt, K W and Burger, J (2005) 'Ancient DNA from the first European farmers in 7500-year-old Neolithic sites', 310 *Science*, 1016–1018

Habu, J (2004) *Ancient Jomon of Japan*, Cambridge: Cambridge University Press

Hahn, J, Kind, C-J and Steppan, K (1993) 'Mesolithische Rentier-Jäger in Südwestdeutschland? Der mittelsteinzeitliche Freilandfundplatz Rottenburg "Siebenlinden I"', 18 *Fundberichte Baden-Württemberg*, 29–52

Hallgren, F (2004) 'The introduction of ceramic technology around the Baltic Sea in the 6th Millennium', in Knutsson, H (ed), *Coast to Coast—Arrival: Results and Reflections. Proceedings of the Final Coast to Coast Conference 1–5 October 2002 in Fallköping, Sweden*. Uppsala: Coast to Coast Project, pp 123–142

Hartz, S (2000) 'Austauschbeziehungen und Kulturkontakte während der Ertebøllekultur im südwestlichen Ostseegebiet', in Lüth, F and Schoknecht, U (eds), *Schutz des Kulturerbes unter Wasser: Veränderungen Europäischer Lebenskultur durch Fluß- und Seehandel. Beiträge zum Internationalen Kongreß für Unterwasserarchäologie (IKUWA '99) 18–21 Februar 1999 in Sassnitz auf Rügen*, 35 Beiträge zur Ur- und Frühgeschichte Mecklenburg-Vorpommerns, Lübstorf: Archäologisches Landesmuseum Mecklenburg-Vorpommern, pp 35–42

Hartz, S, Heinrich, D and Lübke, H (2000) 'Frühe Bauern an der Küste. Neue 14C-Daten und aktuelle Aspekte zum Neolithisierungsprozeß im norddeutschen Ostseeküstengebiet', 75 *Praehistorische Zeitschrift*, 129–52

Hartz, S, Heinrich, D and Lübke, H (2002) 'Coastal farmers—the neolithisation of northernmost Germany', in Fischer, A and Kristiansen, K H (eds), *The Neolithisation of Denmark. 150 Years of Debate*, Sheffield: Collis Publication, pp 321–340

Hartz, S and Lübke, H (2005) 'Zur chronostratigraphischen Gliederung der Ertebølle-Kultur und der frühesten Trichterbecherkultur in der südlichen Meckelnburger Bucht', 2004–52 *Jahrbuch Bodendenkmalpflege in Mecklenburg-Vorpommern*, 119–143

Heinen, M, Nehren, R and Weiner, J (2004) 'Älteste Bandkeramik und Flomborn ... Der Fundplatz Uckendorf wirft neue Fragen auf', in Kunow, J (ed), *Archäologie im Rheinland 2003*, Stuttgart: Konrad Theiss, pp 27–29

Heiri, O, Tinner, W and Lotter, A F (2004) 'Evidence for cooler European summers during periods of changing meltwater flux to the North Atlantic', 01/43 *PNAS*, 15285–15288

Hudson, C (1976) *The Southeastern Indians*, Knoxville: University of Tennessee Press

Huysecom, E, Ozainne, S, Raeli, F, Ballouche, A, Rasse, M and Stokes, S (2004) 'Ounjougou (Mali): A history of Holocene settlement at the southern edge of the Sahara', 78/301 *Antiquity*, 579–593

Jennbert, K (1988) 'Der Neolithisierungsprozeß in Südskandinavien', 63 *Praehistorische Zeitschrift*, 1–22

Jeunesse, C (1987) 'La céramique da La Hoguette. Un nouvel "élement non-rubané" du Néolithique ancien de l'Europe du Nord-Ouest', 30 *Cahiers Alsaciens*, 5–33

——— (1994) 'Le Néolithique du sud de la plaine du Rhin supérieur', 69 *Prähistorische Zeitschrift*, 1–31

——— (1998–99) 'La synchronisation des séquences culturelles des bassins du Rhin, de la Meuse et de la Seine et la chronologie du Bassin parisien au Néolithioque ancien et moyen (5,200–4,500 av. J.-C.)', 20–21 *Bulletin de la Societé Préhistorique Luxembourgeoise*, 337–392

——— (2000) 'Les composantes autochtone et danubienne en Europe Centrale et occidentale entre 5500 et 4000 av. J.-C.: contacts, transferts, acculturations', in *Les Derniers Chasseur-Cuilleurs d'Europe Occidentale, Actes du Colloque International de Besançon, Octobre 1998*. Besançon: Presses Universitaires Franc-Comtoises 361 à 378, Annales Littéraires 699, Série "Environnement, sociétés et archaéologie", pp 361–378

——— (2001) 'Armatures asymétriques, régionalisation, acculturation. Contribution à l'étude des relations entre la Rubané et la composante autochtone dans l'ouest de la sphère danubienne', in *Préhistoire de la Grande Plaine du Nord de l'Europe. Actes du Colloque Chaire Francqui Interuniversitaire au titre Étranger*, Liège: 99 ERAUL, pp 147–165

Jeunesse, C and Lefranc, P (1999) 'Rosheim "Sainte-Odile" (Bas-Rhin), un habitat rubané avec fossé d'enceinte. Première partie: les structures et la céramique', 15 *Cahiers de l'Association pour la Promotion de la Recherche Archéologique en Alsace*, 1–111

Jeunesse, C, Lefranc, P and Denaire, A (2004) 'Groupe de Bischheim, origine du Michelsberg, genèse du groupe d'Entzheim: La transition entre le Néolithique moyen et le Néolithique récent dans les régions rhénanes', 18/19 *Cahiers de l'Association pour la Promotion de la Recherche Archéologique en Alsace*. Zimersheim: Association pour la Promotion de la Recherche Archéologique en Alsace

Kaczanowska, M (1990) 'Die feuersteinindustrie der kultur mit linienbandkeramik von Eilsleben, Kr. Wanzleben', 73 *Jahresschrift für Mitteldeutsche Vorgeschichte*, 29–40

Kalis, A J and Meurers-Balke, J (1998) 'Die "Landnam"-Modelle von Iversen und Troels-Smith zur Neolithisierung des westlichen Ostseegebietes—ein Versuch zu ihrer Aktualisierung', 73 *Praehistorische Zeitschrift*, 1–144

Kalis, A J, Meurers-Balke, J, van der Borg, K, von den Driesch, A, Rähle, W, Tegtmeier, U and Thiemeyer, H (2001) 'Der La-Hoguette-Fundhorizont in der Wilhelma von Stuttgart-Bad Cannstatt. Anthrakologische, archäopalynologische, bodenkundliche, malakozoologische, radiometrische und säugetierkundliche Untersuchungen', in Gehlen, B, Heinen, M and Tillmann, A (eds), *Zeit-Räume. Gedenkschrift für Wolfgang Taute Band 2*. Bonn: Deutsche Gesellschaft für Ur- und Frühgeschichte / Rudolf Habelt, pp 649–672

Kampffmeyer, U (1983) 'Der neolithische Siedlungsplatz Hüde 1 am Dümmer', in Wegner, G (ed), *Frühe Bauernkulturen in Niedersachsen. Linienbandkeramik, Stichbandkeramik, Rössener Kultur*. Archäologische Mitteilungen aus Nordwestdeutschland Beiheft 1, Oldenburg: Staatliches Museum für Naturkunde und Vorgeschichte Oldenburg, pp 119–134

Keally, C T, Taniguchi, Y, Kuz'min, Y V and Shewkomud, I Y (2004) 'Chronology of the beginning of pottery manufacture in East Asia', 46 *Radiocarbon*, 345–351

Kind, C-J (1997) 'Die letzten Wildbeuter. Henauhof Nord II und das Endmesolithikum in Baden-Württemberg', 39 *Materialhefte zur Urgeschichte in Baden-Württemberg*, Stuttgart: Theiss

——— (2003) 'Das Mesolithikum in der Talaue des Neckars. Die Fundstelle von Rottenburg Siebenlinden 1 und 3', 88 *Forschungen und Berichte zur Vor- und Frühgeschichte in Baden-Württemberg*, Stuttgart: Theiss

Klassen, L (1996–97) 'The debate on the Mesolithic-Neolithic transition in the western Baltic: a central European perspective', 13 *Journal of Danish Archaeology*, 171–178

——— (2000a) 'The Ertebølle-Culture and the Neolithic continental Europe: traces of contact and interaction', in Fischer, A and Kristiansen, K (eds), *From Foraging to Farming: 150 Years of Neolithisation Debate in Southern Scandinavia*, Sheffield: Collis Publication, pp 303–318

——— (2000b) 'Waterborne exchange and late Ertebølle social structure', in Lüth, F and Schoknecht, U (eds), *Schutz des Kulturerbes unter Wasser. Veränderungen Europäischer Lebenskultur durch Fluß- und Seehandel. Beiträge zum Internationalen Kongreß für Unterwasserarchäologie (IKUWA '99) 18–21 Februar 1999 in Sassnitz auf Rügen*, 35 Beiträge zur Ur- und Frühgeschichte Mecklenburg-Vorpommerns, Lübstorf: Archäologisches Landesmuseum Mecklenburg-Vorpommern, pp 43–51

——— (2004) *Jade und Kupfer: Untersuchungen zum Neolithisierungsprozeß im Westlichen Ostseeraum unter Besonderer Berücksichtigung der Kulturentwicklung*

Europas 5,500–3,500 BC, 47 Jysk Arkæologisk Selskabs Skrifter, Aarhus: Aarhus University Press

Krause, R (2001) 'Stierkopf, Spondylus und verziertes Knochengerät: Neue Funde aus der bandkeramischen Siedlung von Vaihingen an der Enz, Kreis Ludwigsburg', *Archäologische Ausgrabungen in Baden-Württemberg 2000*, Stuttgart: Theiss, pp 23–27

Kuz'min, Y V (2002) 'The earliest centres of pottery origin in the Russian Far East and Siberia: review of chronology for the oldest Neolithic cultures', in Budja, M (ed), *9th Neolithic Seminar*. 29 Documenta Praehistoria, Ljubljana: Filosofska fakulteta, Oddelek za arheologijo, pp 37–46

Kuz'min, Y V and Orlova, L A (2000) 'The Neolithization of Siberia and the Russian Far East: radiocarbon evidence', 74 *Antiquity*, 356–365

Leuschner, H H, Sass-Klaassen, U, Jansma, E, Billie, M G L and Spurk, M (2002) 'Subfossil European bog oaks: population dynamics and long-term growth depressions as indicators of changes in the Holocene hydro-regime and climate', 12 *The Holocene*, 695–706

Linstädter, J (2004) *Zum Frühneolithikum des Westlichen Mittelmeerraums—Die Keramik der Fundstelle Hassi Ouenzga*. AVA-Forschungen Band 9. Bonn: Kommission für Allgemeine und Vergleichende Archäologie des deutschen Archäologischen Instituts

Louwe Kooijmans, L P (1993) 'The Mesolithic/Neolithic transformation in the lower Rhine Basin', in Bogucki, P (ed), *Case Studies in European Prehistory*. Boca Raton/Ann Arbor: CRC Press, pp 95–145

——— (1998) 'Understanding the Mesolithic/Neolithic frontier in the Lower Rhine Basin, 5300–4300 cal BC', in Edmonds, M and Richards, C (eds), *Social Life and Social Change: The Neolithic of North-Western Europe*. Glasgow: Cruithne Press, pp 407–427

——— (2003) 'The Hardinxveld sites in the Rhine/Meuse Delta, the Netherlands, 5,500–4,500 cal BC', in Kindgren, H, Knutsson, K, Loeffler, D and Åkerlund, A (eds), *Mesolithic on the Move. Papers presented at the Sixth International Conference on the Mesolithic in Europe, Stockholm 2000*, Oxford: Oxbow Books, pp 608–621

Lüning, J, Kloos, U and Albert, S (1989) 'Westliche Nachbarn der bandkeramischen Kultur: Die Keramikgruppen La Hoguette und Limburg', 67 *Germania*, 355–393

Madsen, T (1991) 'The social structure of Early Neolithic society in South Scandinavia' in Lichardus, J (ed), *Die Kupferzeit als Historische Epoche. Symposium Saarbrücken und Otzenhausen 6.-13.11.1988*. Saarbrücker Beiträge zur Altertumskunde 55. Bonn: Habelt 489–496

Madsen, T and Petersen, J E (1984) 'Tidligneolitiske Anlæg Ved Mosegården. Regionale or kronologiske forskelle i tidligneolithikum', 1982–83 *Kuml*, 61–120

Manen, C, Marchand, G and Carvalho, A F (2007) 'Le Néolithique ancien de la péninsule Ibérique: vers une nouvelle évaluation du mirage africain?', in Évin, J (dir.), *XXVIe Congrès Préhistorique de France, Congrès du Centenaire de la Société Préhistorique Française, Avignon 21–25 Septembre 2004. Vol. 3 "Aux conceptions d'aujourd'hui"*. Paris: Société Préhistorique Française, pp 133–151

Mamonov, A E (2000) 'Rannii Neolit. Elshanskaia Kul'tura', in Kabetov, P S, Vasil'ev, I B, Dubman, Z L, Smirnov, J H and Chramkov, L B (eds), *Istoriia Samarskogo Povolzh'ia. S Drevneishikh Vremen do Nashikh Dnei. Kamennyi Vek*. Samara: Rossiiskaia Akademiia Nauk. Samarskii Nauchnyi Tsentr. Tsentr "Integratsiia", pp 147–176

Meurers-Balke, J and Kalis, A J (2001) 'Früh-, alt- und jungneolithische Landnutzung—archäopalynologische Bearbeitung der Wilhelma-Travertine von Bad Cannstatt', in Gehlen, B, Heinen, M and Tillmann, A (eds), *Zeit-Räume. Gedenkschrift für Wolfgang Taute Band 2*. Bonn: Deutsche Gesellschaft für Ur- und Frühgeschichte/Rudolf Habelt, pp 631–648

Modderman, P (1974) 'Die Limburger Keramik von Kesseleyk', 4 *Archäologisches Korrespondenzblatt*, 4–11

——— (1987) 'Limburger aardewerk uit Sweikhuizen Gem. Schinnen Prov. Limburg', 20 *Analecta Praehistorica Leidensia*, Leiden: Leiden University, pp 87–94

Raemaekers, D C M (1999) *The Articulation of a "New Neolithic". The Meaning of the Swifterbant Culture for the Process of Neolithisation in the Western Part of the North European Plain (4,900–3,400 BC)*. Leiden: Archaeological Studies Leiden University. Faculty of Archaeology, University of Leiden

——— (2001) 'Aardewerk en verbrande klei', in Louwe Kooijmans, L P (ed), *Archeologie in de Betuweroute: Hardinxvel-Giessendam De Bruin. Een Kammplaats uit het Laat-Mesolithicum en het Begin van de Swifterbant-Cultuur (5500–4450 v. Chr.)*, Utrecht: ROB Rapportage Archeologische Monumentenzorg 88. NS Rilinfrabeheer B. V., pp 117–152

Richards, M (2003) 'The Neolithic Invasion of Europe', 32 *Annual Revue of Anthropology*, 135–162

Roset, J-P (1987) 'Paleoclimatic and cultural conditions of Neolithic development in the Early Holocene of Northern Niger (Aïr and Ténéré)', in Close, A E (ed), *Prehistory of Arid North Africa. Essays in Honor of Fred Wendorf*, Dallas: Southern Methodist University Press, pp 211–234

Rowley-Conwy, P (2004) 'How the West was lost. A reconsideration of agricultural origins in Britain, Ireland, and Southern Scandinavia', 45/S4 *Current Anthropology*, S83–S113

Scheu, A, Hartz, S, Schmölcke, U, Tresset, A, Burger, J and Bollongino, R (2008) 'Ancient DNA provides no evidence for independent domestication of cattle in Rosenhof, Northernmost Germany', 35(5) *Journal of Archaeological Science*, 1257–1264

Schibler, J, Jacomet, S, Hüster-Plogmann, H and Brombacher, C (1997) 'Economic crash in the 37th and 36th centuries cal. BC in Neolithic lake shore sites in Switzerland', 25/26 *Anthropozoologica*, 553–570

Schindler, R (1962) 'Rössener elemente im Boberger Neolithikum', 40 *Germania*, 245–255

Schwabedissen, H (1967) 'Ein horizontierter "Breitkeil" aus Satrup und die mannigfachen Kulturverbindungen des beginnenden Neolithikums im Norden und Nordwesten', 12 *Palaeohistoria*, 409–468

Seidel, U and Jeunesse, C (2000) 'À propos d'un tesson du Néolithique récent de la vallée du Neckar. La technique du bouton au repussé et la question de la diffusion du Michelsberg', 97 *Bulletin de la Société Préhistorique Française*, 229–237

Semino, O, Passarino, G, Brega, A, Fellous, M and Santachiara-Benerecetti, S (1996) 'A view on the Neolithic demic diffusion in Europe through two Y chromosome-specific markers', 59 *American Journal of Human Genetics*, 964–968

Semino, O, Passarino, G, Oefner, P J, Lin, A A, Arbuzova, S and Beckman, L E (2000) 'The genetic legacy of Paleolithic Homo sapiens sapiens in extant Europeans: a Y chromosome perspective', 290 *Science*, 1155–1159

Semino, O, Magri, C, Benuzzi, G, Lin, A, Al-Zahery, N, Battaglia, V, Maccioni, L, Triantaphyllidis, C, Shen, P, Oefner, P, Zhivotovsky, L A, King, R, Torroni, A, Cavalli-Sforza, L L, Underhill, P A and Santachiara-Benerecetti, A S (2004) 'Origin, diffusion, and differentiation of Y-Chromosome haplogroups E and J: inferences on the neolithization of Europe and later migratory events in the Mediterranean area', 74 *American Journal of Human Genetics*, 1023–1034

Simoni, L, Calafell, F, Pettener, D, Bertranpetit, J and Barbujani, G (2000) 'Geographic patterns of mtDNA diversity in Europe', 66 *American Journal of Human Genetics*, 262–278

Sheridan, A (2003) 'French Connections I: spreading of the *marmites* thinly', in Armit, I, Murphy, E, Nelis, E and Simpson, D (eds), *Neolithic Settlement in Ireland and Western Britain*, Oxford: Prehistoric Society, Oxbow Books, Queen's University Belfast, pp 3–17

Smolla, G (1960) *Neolithische Kulturerscheinungen. Studien zur Frage ihrer Herausbildung*, Bonn: Rudolf Habelt

Spurk, M, Leuschner, H H, Baillie, M G L, Briffa, K R and Friedrich, M (2002) 'Depositional frequency of German subfossil oaks: climatically and non-climatically induced fluctuations in the Holocene', 12 *The Holocene*, 707–715

Street, M, Baales, M, Cziesla, E, Hartz, S, Heinen, M, Jöris, O, Pasda, C, Terberger, T and Vollbrecht, J (2001) 'Final Palaeolithic and Mesolithic research in reunified Germany', 15 *Journal of World Prehistory*, 365–453

Strien, H-C (2000) *Untersuchungen zur Bandkeramik in Württemberg*, 69 Universitätsforschungen zur Prähistorischen Archäologie, Bonn: Habelt

——— (2005) 'Familientraditionen in der bandkeramischen Siedlung bei Vaihingen/Enz', in Lüning, J, Frirdich, C and Zimmermann, A (eds), *Die Bandkeramik im 21. Jahrhundert. Symposium in der Abtei Brauweiler bei Köln vom 16.9–19.9.2002*, 7 Internationale Archäologie Arbeitsgemeinschaft, Symposium, Tagung, Kongress Band, Rahden/Westf.: Marie Leidorf, pp 189–198

Strien, H-C and Gronenborn, D (2005) 'Klima- und Kulturwandel während des mitteleuropäischen Altneolithikums (58./57.-51./50. Jahrhundert v. Chr.)', in Gronenborn, D (Hrsg), *Klimaveräanderung und Kulturwandel in Neolithischen Gesellschaften Mitteleuropas, 6700–2200 v. Chr.* RGZM—Tagungen 1. Mainz: Verlag des Römisch-Germanischen Zentralmuseums, pp 131–150

Terberger, T (2003) 'Decorated objects of the older Mesolithic from the northern lowlands' in Kindgren, H, Knutsson, K, Loeffler, D and Åkerlund, A (eds), *Mesolithic on the Move: Papers presented at the Sixth International Conference on the Mesolithic in Europe*, Stockholm: Oxbow Books, pp 547–557

Timofeev, V I (1998) 'The east-west relations in the Late Mesolithic and Neolithic in the Baltic Region' in Domańska, L and Jacobs, K (eds), *Beyond Balkanization.* 5 Baltic-Pontic Studies, Poznań: Adam Mickiewicz University, pp 44–58

Timofeev, V I and Zaitseva, G (1998) 'On the problem of the Neolithisation of eastern Europe and the position of the South Russian Area in this process', in Évin, J and Oberlin, C (eds), *Actes du Colloque "C14 Archéologie": 3ème Congrès International, Lyon, 6–10 Avril 1998*. Paris: 26 Mémoires de la Société Préhistorique Française, pp 189–192

Uerpmann, M and Uerpmann, H P (1997) 'Remarks on the faunal remains of some early farming communities in Central Europe', 25/26 *Anthropozoologica*, 571–578 van Berg, P-L (1990) 'La ceramique Néolithique Ancienne non rubanée dans le nord-ouest de l'Europe', 12 *Bulletin de la Societé Préhistorique du Luxembourg,* 107–124

van Berg, P-L (1997) 'La céramique et son décor en Eurasie' in Jeunesse, C (ed), *Le Néolithique Danubien et ses Marges Entre Rhin et Seine. XXIIe Colloque Interrégional sur le Néolithique, Strasbourg 27–29 Octobre 1995,* Supplément n°3, Zimmersheim: Cahiers de l'Association pour la Promotion de la Recherche Archéologique en Alsace, pp 223–264

Van Neer, W, Ervynck, A, Lentacker, A, Crombé, P, Sergant, J, Perdaen, Y, Van Strydonck, M and Van Roeyen, J-P (2001) 'Dierenresten uit een vroege Swifterbant-nederzetting te Doel-Deurganckdok (Vlaanderen, België): jachtweld, maar vooral veel vis', 21 *Notae Praehistoricae*, 85–96

van Willigen, S (1997) 'Zur zeitlichen und räumlichen Differenzierung des südfranzösischen Néolithique ancien', 75 *Germania*, 423–442

——— (2004) 'Aspects culturels de la néolithisation en Méditerranée occidentale: le Cardial et l'Épicardial', 101 *Bulletin de la Société Préhistorique Française*, 463–495

——— (2006) *Die Neolithisierung im Nordwestlichen Mittelmeerraum*, Mainz: Philipp von Zabern

Weiner, J (1995a) 'Eine zimmermannstechnische Glanzleistung: der 7000 Jahre alte Eichenholzbrunnen aus Erkelenz-Kückhoven', in Horn, H G, Hellenkemper, H, Koschik, H and Trier, B (eds), *Ein Land Macht Geschichte: Archäologie in Nordrhein-Westfallen.* Mainz: Philipp von Zabern, pp 179–187

——— (1995b) 'Bogenstab- und Pfeilschaftfragmente aus dem altneolithischen Brunnen von Erkelenz-Kückhofen. Ein Beitrag zur Bogenwaffe der Bandkeramik', 25 *Archäologisches Korrespondenzblatt,* 355–372

Weninger, B, Alram-Stern, E, Bauer, E, Clare, L, Danzeglocke, U, Jöris, O, Kubatzki, C, Rollefson, G and Todorova, H (2005) 'Die Neolithisierung von Südosteuropa als Folge des abrupten Klimawandels um 8200 cal BP', in Gronenborn, D (Hrsg), *Klimaveränderung und Kulturwandel in Neolithischen Gesellschaften Mitteleuropas, 6700–2200 v. Chr.* RGZM—Tagungen 1. Mainz: Verlag des Römisch-Germanischen Zentralmuseums, pp 75–118

Winiger, J (1998) *Ethnoarchäologische Studien zum Neolithikum Südwesteuropas*, BAR International Series 701, Oxford: Archaeopress

Wirtz, K and Lemmen, C (2003) 'A Global Dynamic Model for the Neolithic Transition', *59 Climatic Change*, 333–367

Worsaae, J J A (2002 [1859]) 'Concerning a new division of the Stone and Bronze Ages, and a strange find a Engestofte on Lolland from the early Stone Age', in Fischer, A and Kristiansen, K (eds), *The Neolithisation of Denmark. 150 Years of Debate*, Sheffield: Collis Publication, pp 47–56

Zilhão, J (2001) 'Radiocarbon evidence for maritime pioneer colonization at the origins of farming in west Mediterranean Europe', 98 *Proceedings of the National Academy of Sciences of the United States of America*, 1480–1485

Part 4

Ceramic Dispersals in World Perspective

CHAPTER 20

DISCUSSION: LIVING AT THE EDGE

William K Barnett

INTRODUCTION

Sometimes, advances in archaeology have been led by new programmes of empirical information collection, analysis and integration. At other times, the discipline has been transformed by the development of fresh theoretical approaches which challenge established perspectives. In a few cases, the discipline has been affected by external factors that have nothing to do with the practice of archaeology. As we build political forces into our explanatory models for past societies, we must also admit their effect on our practice.

The contributions we see in this volume were in some degree made possible by two events on the political stage. The first was China opening itself to the West, through notable events such as Richard Nixon's 1972 visit, and China's 'Open Door' policy since 2000. The second was Mikhail Gorbachev's 'Glasnost', which opened up the Soviet Union beginning in 1985 and which played an important role in its ultimate dissolution.

These events also ultimately spelled the end of Maoist and Soviet archaeological doctrines. As a new generation of Eurasian archaeologists began to engage in archaeological research, two things happened. New and decidedly neither western nor communist academic traditions began to form. These scholars could pick and choose their preferred orientation and approach without being beholden to an established paradigm. It also opened up Eurasian academies, publications and sites to the rest of the world. These vast regions had long been archaeologically invisible to the western world. The 2003 European Archaeological Association Annual Meeting in Saint

Petersburg, Russia, from which many of these chapters originate, was a notable event in this archaeological 'coming out'.

This is a potentially landmark event for the study of Old World prehistory. It is significant that the editors have recognized the potential impact of this archaeological 'Glasnost' and have further built upon it by encouraging discussions from the Balkans, eastern Europe, northern Europe and hints of Africa. It builds on a research tradition that Marek Zvelebil has developed in bringing to light research on early agricultural societies in eastern Europe and on Peter Jordan's work in northern Eurasia that introduces new perspectives on pottery-using hunter-gatherers. These topics have been considered peripheral to traditional interpretations of the first pottery-using societies primarily in British, French and German research on western Europe and the Near East.

Consider a map of the world from the tip of Portugal to the Bering Straits. Compared to the rest of the contiguous hemisphere, western Europe is a small triangle appended to the western side of Eurasia. Eurasia, traditionally not addressed in classic Old World prehistories, and therefore peripheral, is 54,000,000 square kilometres, representing over 10% of the world's landmass. This 'edge area' dwarfs that of western Europe and the Near East at 5,000,000 square kilometres. Despite the richness of prehistoric cultures known from the west, the implications of uncovering the archaeology of the remaining 90% of the Eurasian landmass should be met with anticipation and openness of our own.

Obviously, synthesizing information from such a vast expanse is challenging. This volume attempts a portion of that task, covering Russia, Japan, Korea, Scandinavia, and eastern and western Europe. The potential of the new data demonstrated in this volume is significant and the voices of eastern scholars that contribute represent a refreshing new look at an 'old world'. This discussion can hope to do no more than brush that surface, and I hope the themes I raise can help provide an entrée to further archaeological research and interpretation across this vast area.

Challenges to Tradition

Has there been western chauvinism in the development of culture histories? Perhaps my assertion that this is the case would be a *mea culpa* for my own limited archaeological upbringing. In western Europe, where I have done my research, pottery is the identifier of the Neolithic. The term 'Neolithic', traditionally associated with early pottery-producing agricultural societies and still seen in force in this volume, comes from early western archaeology. In many cases, and throughout this volume, the term is also associated with the independent rise of pottery, not agriculture. So, we have a terminological challenge to sort out. It turns out that if Neolithic means agriculture, however, it cannot mean pottery. As discussed below, many of the authors examine extensive prehistoric periods of pottery use in foraging contexts, leaving the term with little analytical clarity. Furthermore, as evidence for ceramic vessels now stretches back into the Pleistocene, it brings pottery use well into the previously unassailable realm of the Upper Palaeolithic as well, further challenging established models of periodicity.

Of course there are still areas where these terms have currency, principally western Europe where they were invented. For Eurasia, however, it is a semantic trap we find ourselves in without an easy way out. In the past, when Eurasia was largely invisible and therefore peripheral to Europe, we could force fit these terms. Now that Europe is looking more like a small province of Eurasia, we may need to think about new ways of describing these cultural categories. Part of the challenge is that the socioeconomic formulations discussed throughout this book do not fit neatly into any established culture evolutionary schema, making it difficult to create a rubric.

This challenge is intensified by the tradition of using ceramic typologies as cultural or functional units. The problems with this, and potential alternative solutions, are discussed by several of the authors. Budja brings socioeconomic models to a broad discussion of the development of ceramics as prestige technologies across a broad swath of Eurasia and West Asia. Larsson proposes the Pitted Ware Culture of Sweden as a system of material communication and the result of human agency. Kaner also uses social agency in his discussion of the Jomon. These approaches successfully challenge the notion of ceramic typologies, attempting to understand the material record from the perspective of the actor rather than the archaeologist. In a similar vein, Skandfer presents Early Northern Comb Ware as a manifestation of technological choice and style. Kashina discusses anthropomorphic figurines in terms of their relationships to manufacture and use, particularly in terms of gender and domestic spaces. It is through these tools that we will be able to make sense of the past rather than just categorize it in space and time.

Many of the authors in this volume, however, still struggle to escape from their culture history shackles. In discussing the archaeological literature, there is too much of a tradition of ceramic-defined culture histories from which it is difficult to escape if one wants to use the work of archaeologists that have come before, a necessary condition in this discipline. We know we must reject the concept of the 'Pitted Ware Culture', but are hesitant to replace it with the Pitted Ware Exchange Sphere, Economy, or System of Material Communication (with apologies to Larsson). Many of the authors here do grapple with the issue, but without better models to understand stylistic stability across large areas and long time periods, this particular challenge will not easily be met.

Pottery and Foragers

Certainly, all archaeologists now acknowledge that pottery-using foraging societies existed in the past, particularly at the agricultural transition. One of the lingering themes in Old World prehistory, however, is that these societies were generally considered transitional—that the production and use of ceramics somehow placed scheduling or other pressures on small-scale foragers that moved them towards agricultural villages or made ceramics a logistical challenge. What we see throughout the volume is that nothing appears to have been further from the truth in many areas.

Kaner presents 16,000 BP pottery in Japan, yet agriculture was not established until 3,000 BP. Cho and Ko cite evidence for pottery as early as 10,000 BP with agriculture arriving at 4,000 BC. In the Russian Far East, Zhushchikhovskaya documents the use of pottery by foragers from 13,000 BP until the agriculture of the palaeometal period of around 2,000 BC. In eastern Siberia, McKenzie likewise documents ceramics in a foraging context dating back to 10,000 BP. Pottery use by foragers dates back to 8,000 BP in western Siberia according to Chairkina and Kosinskaia, and in eastern Europe, Dolukhanov and colleagues have evidence of foragers using pottery by at least 7,000 BC on the eastern steppe. Farther west, foragers with pottery are noted by German in Karelia and in Sweden by Stilborg and Holm at around 6,500 BP, in Finnmark at 5,600 BC and very far to the north in Åland by Hallgren.

What conclusions are we to draw from such a vast geographic and temporal establishment of ceramics by foragers across Eurasia? Looking more carefully at the contexts of use (as we are encouraged to do by Kaner and Larsson), it is possible to make a broad and overreaching generalization that much of the pottery use occurred in fishing contexts. No doubt, the consumption, storage or exchange of these products had something to do with the popularity of ceramics and with the persistent use of this technology for millennia and over many thousands of miles. In the case of Japan, pottery was used in a foraging context for 13,000 years and, since the adoption of agriculture in the first millennium BC, only 3,000 years have passed to the present day. This pattern is seen, although not in this extreme, across much of Eurasia. We therefore not only need to make a place in our prehistorical stories for vast traditions of pottery-using foragers, but to consider it the normative mode of Holocene human existence in the Old World.

What is more interesting, however, is to consider the possibility that the predominant motivation for the invention and spread of ceramics in the Old World had little to do with domesticated cereals, long considered the *raison d'être* for pottery. The earliest ceramics, as noted here by Budja, may have originated at three separate centres: southern China (also Jordan and Zvelebil), Japan (also Kaner), and the Russian Far East (also Zhushchikhovskaya). Each of these purported centres is associated with an extremely long foraging tradition. Budja proposes that these early ceramics were primarily used for cooking herbivores, C3 plants and salmonids. Similar traditions were established in eastern (McKenzie) and western (Chairkina and Kosinskaia) Siberia, Korea (Cho and Ko), Poland (Nowak), Karelia (German), Finnmark (Skandfer), Finland (Pesonen and Leskinen) and Sweden (Hallgren, Papmehl-Dufay, Stilborg and Holm). In Sweden, Stilborg and Holm note that some of these vessels have capacities of 45, and in some cases, 70 litres. These were not casual ceramic experiments by peripheral groups, but longstanding sustainable traditions of significant socioeconomic investment.

Implications of Early Dates

Not only is the long tradition of forager pottery notable in the East, but also the extraordinarily early dates. As mentioned above, there are now several dates associated with separate areas that extend back to 16,000 BP. Even if the earliest dates are rejected, there are still more that come in at 13,000 BP. As much as we might be enamoured by the prospect of Pleistocene pottery, I think it important to consider broader implications of these dates.

First, however, I would like to register my on-going frustration at the obvious disjunction among dates expressed as BP, bp, BC, or bc throughout the chapters. This has primarily to do with two factors, the research tradition of Pleistocene versus Holocene scholars, and the use of radiocarbon calibration curves. Having worked with Palaeolithic archaeologists, I understand their disdain for the Gregorian Calendar and appreciate that it has little impact on their remote dates. On the other hand, I understand prehistorians of the Holocene who, in addition to having somewhat reliable radiocarbon calibration curves against which to plot their dates, also aspire to tie their prehistoric pasts into an historical trajectory that extends into the present day. Both types of scholarly traditions are represented throughout this volume, and frequently within the same chapter. For the readers' sake, I would have preferred adherence to one system in the service of communicability to the reader even if at the expense of a little accuracy. That being said, I appreciate Hommel's pioneering attempts to develop and map a single chronology that brings a great deal of clarity to the dating of early ceramics across Eurasia (see Figures 21.1 and 21. 2; and see also Jordan and Zvelebil's preliminary model for the periodization of early pottery in northern Eurasia, Figure 1.5, this volume).

What is impressive in this volume is not only the fact that the dates are shockingly early, but that these early dates are so pervasive across such a vast expanse. As noted above, China, Japan and the Russian Far East come in at 13,000 BP. This is still comfortably in the Pleistocene and, as Zhushchikhovskaya notes, during a period of climatic amelioration. Early pottery makes it to Korea by 10,000 BP (Cho and Ko), eastern Siberia by 10,000 BP (McKenzie), western Siberia by 8,000 BP (Chairkina and Kosinskaia), the eastern steppe at around 9,000 BP (7,000 BC), the Bug-Dniester region by 8,000 BP (6,000 BC), the Russian lake district at 7,900 BP (6,300 BC) (Dolukhanov et al.), and the Balkans (Lepenski Vir) by 8,500 BP (6,500 BC) (Budja). To the north, we see early pottery in Sweden (Stilborg and Holm) and Karelia (German) at around 6,500 BP, in the Finnmark region at 5,600 BP (Skandfer), and in the arctic island environments of Åland by 7,000 BP (Hallgren).

This independent trajectory of ceramics is tracked by other contributors in this volume farther to the west, in either exclusively forager contexts or in a frontier situation in interaction with agriculturalists. Gronenborn argues for three separate streams of ceramic diffusion, one of which, the Hyperborean, describes a Ukranian origin for Ertebølle pottery, which also influenced the enigmatic La Hoguette and Limburg wares. Crombé, in his study of western European early pottery, also argues for an eastern European origin for La Hoguette, Limburg, and *Begleitkeramik* wares. Nowak argues that the users of Stroke Band Pottery in Poland connect to a broader trend of pottery-using foragers to the east even though they interacted with LBK farmers in nearby Loess zones. Dolukhanov, Mazurkevich and Shukurov note that eastern European ceramics appeared 1,500 years before the appearance of pottery in the Near East.

What we see in the basic consideration of the data from these chapters is a pervasive shift in our knowledge about early pottery across Eurasia. This shift is geographic and temporal and indicates a vast development of early ceramics across Eurasia that precedes the traditional 'out of the Near East' scenario for the development of pottery. Several of the chapters (Budja, Crombé, Dolukhanov et al. and Gronenborn) all explicitly discuss the implications of pottery adjacent to and earlier than ceramics traditionally associated with the spread of agriculture from the Near East across Europe.

It is well known that there were periods of settled village agriculture in the Near East long before the appearance of pottery and subsequent rapid spread of the agricultural package of domesticated cereals, domesticated herbivores, ceramics and ground stone tools. The Near East has long been viewed as the source of agricultural-based civilizations that laid the groundwork for western civilization. I think it is quite possible, and the chapters in this volume provide interesting insight for this supposition, that as pottery spread from Asia across the continent, it served as a technological catalyst. Ceramic cooking and storage capabilities were external innovations that were adopted into the pre-existing Near Eastern socioeconomic context, and went on to trigger the subsequent spread of cereal agriculture across western Europe.

I think this proposal, that the arrival of pottery from the east into the Near East was the catalyst for the spread of agriculture, is one of the more significant potential outcomes of this volume. It truly links east and west. Consider first that pottery is now known to have an antiquity far beyond the establishment of settled village agriculture in the Near East. Consider also, as do Dolukhanov, Mazurkevich and Shukurov, that pottery in the southern Urals precedes pottery in the Near East by 1,500 years. These areas are separated by a distance of approximately 2,000 km. Even relatively slow demic diffusion models can easily account for the spread of pottery from the Urals to the Near East.

Looking at it from the other side, there was settled village agriculture in the Near East for millennia without pottery yet with other forms of thermal technologies such as plastering. It was only after the appearance of pottery that agriculture spread into adjoining regions and the great ancient civilizations began to take shape. I would propose that ceramics were spread into the Near East, into societies with a potential for producing a surplus yet without a technology to unleash that productive capacity. They would have been the source of a second agricultural revolution, based on this new technology. So, the further spread of agriculture across western Europe was triggered by the arrival of pottery technology from further east, and it was not a product of purely local cultural evolution.

Let us also look backwards (and eastwards) to the origins of Old World ceramics along the Pacific coast of Asia at around 13,000 BP (at least for now!). If we now cast our gaze across the Bering Straits, we see that some of these early pottery areas, are not too far from Alaska and the new world. It is possible also to wonder what impact ceramic vessels might have played in the peopling of the Americas. If the advantages of pottery were so universal that it as a tool found purchase in so many diverse societies across Eurasia, it is also possible that it could have played a role in enabling at least one wave of peoples across the Bering Straits. I would be interested in how scholars of the peopling of the new world interpret the implications of this volume.

There are several questions that still lie before us. What was the genesis of pottery in the Old World? What gave rise to its success and its spread in a foraging economy? Ceramic figurines are known from the Palaeolithic so the knowledge of hardening earth with fire was not difficult to discern.

McKenzie proposes that, as the first pottery in eastern Siberia is net-impressed, nets may have been the functional model for these early ceramics. Liliana Janik, in a paper presented at another session of the Saint Petersburg EAA (European Association of Archaeologists) conference in 2003, proposed that ceramics and cordage spread together across Eurasia and were linked technologies. The data are still quite thin, but I am encouraged by the contributors to this volume and by their synthesis of vast amounts of data, innovative presentation of new perspectives, and above all, by their fearless challenge of conventional wisdoms.

Chapter 21

Hunter Gatherer Pottery: An Emerging 14C Chronology

Peter Hommel

Early Pottery Dispersals in Northern Eurasia

The invention and dispersal of ceramic vessel technology has become the focus of significant academic attention in recent years, and improvements in 14C measurement techniques have allowed archaeologists to push back the oldest pottery dates by many thousands of years. The secure dating of very early hunter-gatherer pottery in North Africa and eastern Asia has made it increasingly clear that the traditional models of a Near Eastern and 'farming' origin for ceramic technology is not supported by the available data (Close 1995; Kuz'min 2006; Jordan and Zvelebil, this volume).

This chapter presents preliminary analysis of 1,105 published radiocarbon dates from 82 sites associated with early pottery technology from across the Old World. The results enable provisional mapping of potential early pottery dispersal routes (see Figure 1.5, Jordan and Zvelebil, this volume), and the full dataset will be made available at: http://www.cecd.ucl.ac.uk/home/

Structure of the Database and Map

The data set was produced in Microsoft© Excel 2004 drawing on a range of published sources (in print and online). These include academic publications, radiocarbon laboratory date lists and other relevant databases (all sources are cited in full in the database and full references are listed in the

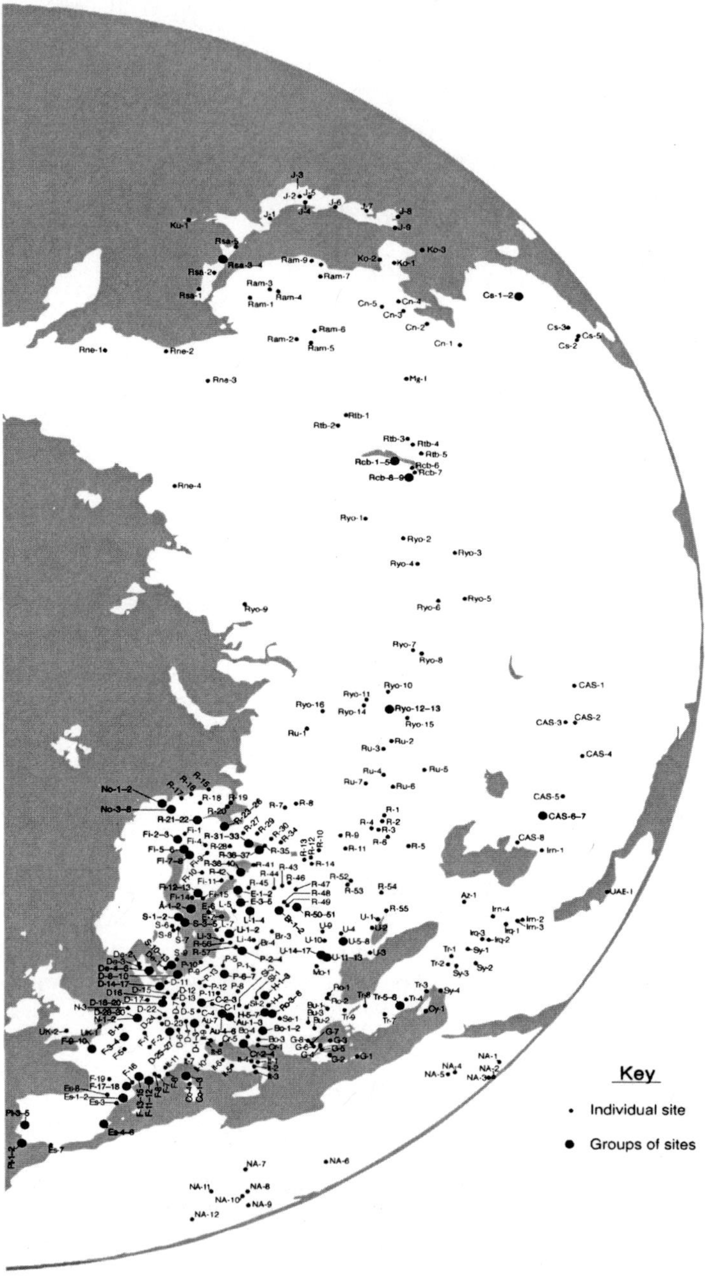

Figure 21.1 Geographic locations of early pottery sites (for explanation and details of radiocarbon database, see text).

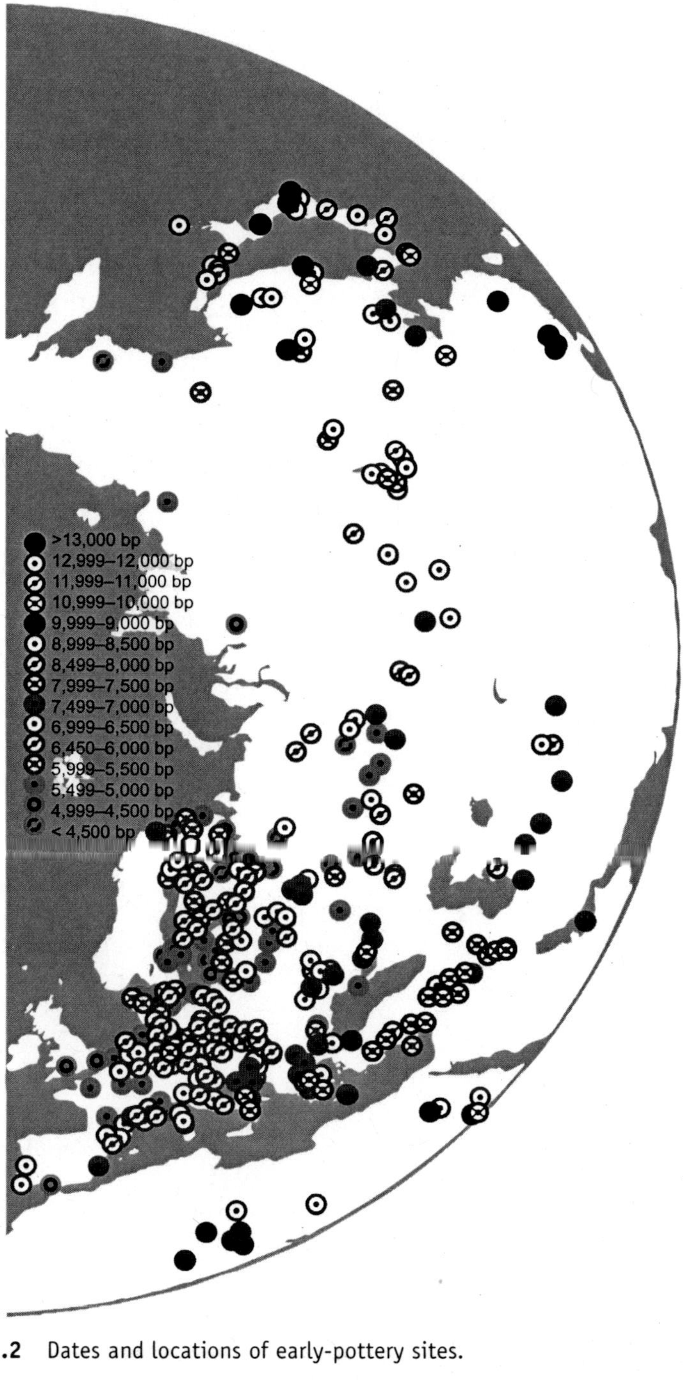

Figure 21.2 Dates and locations of early-pottery sites.

bibliography). For each date, the features recorded were: the name of the site, stratigraphic details, the material dated, uncalibrated radiocarbon date bp, cultural associations and source reference. The calibrated radiocarbon date BC was then calculated for every date using CalPal2005_SFCP, the Cologne Radiocarbon and Palaeoclimatic research package CalPalOnline (Danzeglocke et al. 2007).

Where information about a published date was found to be incomplete, further details were often obtained by searching for further site information elsewhere in the literature. For brevity, the listed reference reflects only the most complete source except in those cases where two or more sources were required to obtain full information for a date or where different sources provided conflicting information.

In most instances, it was possible to confirm that early dates refer to layers containing pottery; however, in some cases original sources and site reports were not available. Where sufficient contextual information was available, only the dates from the earliest layers to contain pottery fragments were included in the database. Where no contextual information was available or the direct association with pottery could not be confirmed, all available dates were retained.

As noted above, the individual dates were also assigned an alphanumeric code and plotted on a map encompassing northern Eurasia and North Africa (see Figure 21.1 and 21.2, and see Figure 1.5, Jordan and Zvelebil, this volume). Many dates were published with associated grid references and most other site locations could be found elsewhere in the literature. However, in some cases, the location of sites was less precise and so the dates were assigned to the centre of the nearest known village/town. This strategy enabled a larger number of sites to be included in the map and database, and was deemed acceptable given the requirement to identify very large-scale and coarse-grained patterning in the chronology of early pottery dispersals.

Bibliography

Åkerlund, A (1996) *Human Responses to Shore Displacement: Living by the Sea in Eastern Middle Sweden During the Stone Age. Arkeologiska undersökningar och skrifter 16*, Stockholm: Studier Från uv Stockholm

Andric, M (2001) 'The Holocene vegetation dynamics and the formation of Neolithic and present-day Slovenian landscape', 23 *Documenta Praehistoria*, 133–175

Anthony, D W and Brown, D R (2000) 'Eneolithic horse exploitation in the Eurasian steppes: diet, ritual and riding', 74 *Antiquity*, 75–86

BaNaDoRa Database (2007) *BAnque NAtionale de DOnnées Radiocarbone pour l'Europe et le Proche Orient* (online): http://www.archeometrie.mom.fr/banadora/index.php

Barker, H and Mackey, C J (1963) 'British Museum natural radiocarbon measurements IV', 5 *Radiocarbon*, 104–108

Barton, L W, Brantingham, P J and Ji, D X (2007) 'Late Pleistocene climate change and Paleolithic cultural evolution in northern China: implications from the last glacial maximum', in Madsen, D B, Chen, F and Gao, X (eds), *Human Adaptation to Climate Change in Arid China. Developments in Quaternary Science Series*, Vol. 9, Amsterdam: Elsevier, pp 105–128

Böhner, U and Schyle, D (2004) *Radiocarbon CONTEXT Database* (online): http://context-database.uni-koeln.de/download/NEAR%20EAST_1.3%20(Testversion).zip

Bond, C J (2003) 'The coming of the earlier Neolithic, pottery and people in the Somerset Levels', in Gibson, A (ed), *Prehistoric Pottery: People, Pattern and Purpose,* BAR International Series 1156, Oxford: Archaeopress, pp 1–27

Buvit, I, Waters, M R, Konstantinov, M V and Konstantinov, A V (2003) 'Geoarchaeological investigations at Studenoe, an Upper Paleolithic site in the Transbaikal Region, Russia', 18 *Geoarchaeology*, 649–673

Chapman, J and Müller, J (1990) 'Early farmers in the Mediterranean Basin: the Dalmatian evidence', 64 *Antiquity*, 127–134

Chi, Z (2002) 'The discovery of early pottery in China', 29 *Documenta Praehistorica*, 29–35

Close, A E (1995) 'Few and far between: early ceramics in North Africa', in Barnett, W K and Hoopes, J W (eds), *The Emergence of Pottery: Technology and Innovation in Ancient Societies*, Washington DC: Smithsonian Institution Press, pp 23–38

Courdaget, J and Le Run, J (1966) 'Gif-sur-Yvette natural radiocarbon measurements', 8 *Radiocarbon*, 128–141

Danzeglocke, U, Jöris, O and Weninger, B (2007) CalPal-2007online. http://www.calpal-online.de, accessed March–April 2007

Delabrias, G and Guillier, M T (1988) 'Gif natural radiocarbon measurements XI', 30 *Radiocarbon*, 61–124

Derevianko, A P, Kuz'min, Y V, Burr, G S, Jull, A J T and Kim, J C (2004) 'AMS 14C age of the earliest pottery from the Russian Far East: 1996–2002 results', 223–224 *Nuclear Instruments and Methods in Physics Research B*, 735–739

Dolukhanov, P, Romanova, Y N and Sementsov, A A (1970) 'Radiocarbon dates of the Institute of Archaeology II (1 January–31 July 1967)', 12 *Radiocarbon*, 130–155

Dolukhanov, P, Shukurov, A, Gronenborn, D, Sokoloff, D, Timofeev, T and Zaitseva, G (2005) 'The chronology of Neolithic dispersal in central and eastern Europe', 32 *Journal of Archaeological Science*, 1441–1458

Evin, J, Longin, G, Marien, G and Pachiaudi, C (1971) 'Lyon natural radiocarbon measurements', 13 *Radiocarbon*, 52–73

Fukai, S and Matsutami, T (1981) *Telul eth-Thalatat—The Excavation of Tell I., The Fifth Season 1976, vol. 4, report 17,* Tokyo: The Institute of Oriental Culture

Geyh, M A (1967) 'Hannover radiocarbon measurements IV', 9 *Radiocarbon*, 198–217

Gronenbron, D (2003) 'Migration, acculturation and culture change in Western Temperate Europe', 30 *Documenta Praehistorca*, 79–91

Grygiel, R and Bogucki, P (1986) 'EN Sites at Brześć Kujawski, Poland: preliminary report on the 1980–1984 excavations', 13 *Journal of Field Archaeology*, 121–137

Guilaine, J (1979) 'The earliest Neolithic in the West Mediterranean: a new appraisal', 53 *Antiquity*, 22–30

Håkansson, S (1983) 'University of Lund radiocarbon dates XVI', 25 *Radiocarbon*, 875–891

——— (1986) 'University of Lund radiocarbon dates XIX', 28 *Radiocarbon*, 1111–1152

——— (1987) 'University of Lund radiocarbon dates XX', 29 *Radicoarbon*, 353–379

Hallgren, F (2002) 'The introduction of ceramic technology around the Baltic Sea in the 6th millennium BC', in Knutsson, H (ed), *Coast to Coast——Arrival: Results and Reflections, Proceedings of the Final Coast to Coast Conference 1–5 October 2002 in Falköping, Sweden*, Uppsala: Uppsala University, pp 123–142

Harris, D R, Gosden, C and Charles, M P (1996) 'Jeitun: excavations at an Early Neolithic site in southern Turkmenistan', 62 *Proceedings of the Prehistoric Society*, 423–442

Hartz, S, Heinrich, D and Lübke, H (2002) 'Coastal farmers—the neolithisation of northernmost Germany', in Fischer, A and Kristiansen, K (eds), *The Neolithisation of Denmark,* Sheffield: J R Collis Publications, pp 319–340

Hartz, S and Lübke, H (2006) 'New evidence for a chronostratigraphic division of the Ertebølle Culture and the earliest Funnel Beaker Culture on the southern Mecklenburgian Bay', in Kind, C J (ed), *After the Ice Age. Settlements, Subsistence and Social Development in the Mesolithic of Central Europe,* Materialhefte zur Archäologie in Baden-Württemberg 78, Stuttgart: Theiss, pp 61–77

Housley, R A (1994) 'Eastern Mediterranean chronologies: the Oxford AMS contribution', 36 *Radiocarbon*, 55–73

Improta, S and Pessina, A (1998) 'La Neolitizzazione dell'Italia settentrionale. Il nuovo quadro chronologico', in Pessina, A and Muscio, G (eds), *Settemila Anni fa il Primo Pane, Ambienti e Culture delle Societé Neolitiche*, Udine: Museo Friulano di Storia Naturale, pp 107–115

Keally, C T, Taniguchi, Y, Kuz'min, Y V and Shevkomud, I Y (2004) 'Chronology of the beginning of pottery manufacture in East Asia', 46 *Radiocarbon*, 345–351

Kihlstedt, B, Larsson, M and Nordqvist, B (1997) 'Neolitiseringen i Syd-Vast-och Mellansverige', in Larsson, M and Olsson, E (eds), *Regionalt och Interregionalt. Stenildersundersokningar i Syd-och Mellansverige*, RAÅ Skrifter, 23, pp 85–133

Kohl, G and Quitta, H (1964) 'Berlin radiocarbon measurements I', 6 *Radiocarbon*, 308–317

——— (1970) 'Berlin radiocarbon measurements IV', 12 *Radiocarbon*, 400–420

Kriiska, A, (2001) *Stone Age Settlement and Economic Processes in the Estonian Coastal Area and Islands*, unpublished academic dissertation, Helsinki: University of Helsinki (Online) http://ethesis.helsinki.fi/julkaisut/hum/kultt/vk/kriiska/

Kuz'min, Y V (1994) 'Prehistoric colonization of northeastern Siberia and migration to America: radiocarbon evidences', 36 *Radiocarbon*, 367–376

——— (2002) 'The earliest centers of pottery origin in the Russian Far East and Siberia: review of chronology for the oldest Neolithic cultures', 29 *Documenta Praehistorica*, 37–46

——— (2004) 'The beginnings of ancient ceramics in East Asia (geoarchaeological aspect)', 2 *Rossiiskaia Arkheologiia*, 79–86 (in Russian)

——— (2006) 'Chronology of the earliest pottery in the Far East: progress and pitfalls', 80 *Antiquity*, 362–371

Kuz'min, Y V and Jull, A J T (1997) 'AMS radiocarbon dating of the Paleolithic-Neolithic transition in the Russian Far East', 14 *Current Research in the Pleistocene,* 46–48

Kuz'min, Y V, Jull, A J T, Orlova, L A and Sulerzhitskii, L D (1998) '14C chronology of Stone Age cultures in the Russian Far East', 40 *Radiocarbon*, 675–686

Kuz'min, Y V and Keally, C T (2001) 'Radiocarbon chronology of the earliest Neolithic sites in East Asia', 43 *Radiocarbon*, 1121–1128

Kuz'min, Y V and Orlova, L A (2000) 'The neolithization of Siberia and the Russian Far East', 74 *Antiquity*, 356–364

Lenneis, E and Stadler, P (1995) 'Zur Absolutchronologie der Linearbandkeramik aufgrund von 14C-Daten', 6 *Archäologie Österreichs*, 4–12

Liiva, A and Loze, I (1993) 'Mesolithic and Neolithic habitation of the eastern Baltic', 35 *Radiocarbon*, 503–506

Lübke, H and Terberger, T (2002) 'New evidence on the Ertebølle culture on Rügen and neighbouring areas', 27 *Greifswalder Geographische Arbeiten*, 47–53

Martin-Socas, D, Cámalich, M D, Buxó, R, Chávez, E, Echallier, J C, González, P, Goñi, A, Mañosa, M, Orozco, T, Paz, M A, Rodríguez, M O, Tusell, M and Watson, J P N (2004) 'Cueva de El Toro (Antequera, Malaga-Spain): a Neolithic stockbreeding community in the Andalusian region, between the 6th and 3th millennia BC', 31 *Documenta Praehistorica*, 163–181

Matiushin, G N (1986) 'The Mesolithic and Neolithic in the Southern Urals and Central Asia', in Zvelebil, M (ed), *Hunters in Transition: Mesolithic Societies of Temperate Eurasia and their Transition to Farming*, Cambridge: Cambridge University Press, pp 133–150

Nelson, S (1990) 'The Neolithic of northeastern China and Korea', 64 *Antiquity*, 234–248

Neustupnyi, E F and Veselyi, K (1977) 'Most radiocarbon dates I', 19 *Radiocarbon*, 183–187

Nuñez, M (1990) 'On subneolithic pottery and its adoption in Late Mesolithic Finland', 7 *Fennoscandia Archaeologica*, 27–52

Oshibkina, S V, Spiridonova, E A and Sulerzhitskii, L D (2006) 'Dynamics of natural environment and people in the Holocene (on material from the site of Luk[illegible])', 4 *Rossiiskaia Arkheologiia*, 5–17 (in Russian)

Pazdur, M F, Awsiuk, R, Goslar, T, Pazdur, A, Walanus, A and Zastawny, A (1994) 'Glwice Radiocarbon Dates XI', 36 *Radiocarbon*, 257–279

Reingruber, A and Thissen, L (2005) 'CANeW 14C databases and 14C charts Aegean Catchment (E GREECE, S BALKANS and W TURKEY) 10,000–5500 cal BC', (online) http://www.canew.org/files/CANeW%20Aegean%20C14%20dbase%20(June2005).pdf

Rimantiené, R (1992) 'Neolithic hunter-gatherers at Sventoji in Lithuania', 66 *Antiquity*, 367–376

Rubin, M and Alexander, C (1960) 'US geological survey radiocarbon dates V', 2 *Radiocarbon*, 129–185

Sato, J, Sato, T, Otomori, Y and Suzuki, H (1969) 'University of Tokyo radiocarbon measurements II', 11 *Radiocarbon*, 509–514

Séfériadès, M L (2004) 'Neolithisation in Mongolia: the Mesolithic-Neolithic site of Tamsagbulag (Dornod district)', 31 *Documenta Praehistorica*, 139–149

Sementsov, A A, Dolukhanov, P M, Romanova, Y N and Timofeev, V I (1972) 'Radiocarbon dates of the Institute of Archaeology III (1 July 1967–1 July 1969)', 14 *Radiocarbon*, 336–367

Serizawa, C (1979) 'Cave sites in Japan', 10 *World Archaeology*, 340–349

Shemiakin, U P and Karacharov, K G (2002) 'Drevniaia Istoriia Surgutskogo Priob'ia', in Wiget, A (ed), *Ocherki Istorii Traditsionnogo Zemlepol'zovaniia (Materialy k Atlasu)*, 2nd edition, Ekaterinburg: Tezic

Skandfer, M (2005) 'Early Northern Comb Ware in Finnmark: the concept of Säräisniemi 1 reconsidered', 23 *Fennoscandia Archaeologica*, 3–27

Skeates, R (1994) 'Towards an absolute chronology for the Neolithic in central Italy', in Skeates, R and Whitehouse, R (eds), *Radiocarbon Dating and Italian Prehistory*, BSR Archaeological Reports 8, London: British School at Rome, pp 61–288

Spataro, M (2006) 'Pottery typology versus technological choices: an Early Neolithic case study from Banat (Romania)', 14 *Analele Banatului*, 63–78

Stäuble, H (1995) 'Radiocarbon dates of the earliest Neolithic in central Europe', 37 *Radiocarbon*, 227–237

Stuckenrath, R (1967) 'University of Pennsylvania radiocarbon dates X', 9 *Radiocarbon*, 333–345

——— (1963) 'University of Pennsylvania radiocarbon dates VI', 5 *Radiocarbon*, 82–103

Sulgostowska, Z (1998) 'Continuity, change and transition: the case of northern-eastern Poland during the Stone Age', in Zvelebil, M, Domańska, L and Dennell, R (eds), *Harvesting the Sea, Farming the Forest: The Emergence of Neolithic Societies in the Baltic Region*, Sheffield: Sheffield Academic Press, pp 87–94

Tauber, H (1968) 'Copenhagen radiocarbon dates IX', 10 *Radiocarbon*, 295–327

——— (1973) 'Copenhagen radiocarbon dates', 15 *Radiocarbon*, 86–112

Telegin, D Y, Lillie, M, Potekhina, I D and Kovaliukh, M M (2003) 'Settlement and economy in Neolithic Ukraine: a new chronology', 77 *Antiquity*, 456–470

Timofeev, V I (1984) 'On the Early Neolithic links between the East Baltic area and Fenno-Scandia', in Edgren, H (ed), *Suomen Muinaismuistoyhdistys Finska Fornminnesföreningen,* ISKOS 4, Helsinki: Helsingfors, pp 36–41

——— (1988) 'On the problem of the Early Neolithic of the East Baltic area', 58 *Acta Archaeologica,* 207–212

Timofeev, V I and Zaitseva, G I (1996) 'Problemy Absoliutnoi Khronologii', in Oshibkina, S V (ed), *Neolit Severnoi Evrazii, Chast' II: Neolit Lesnoi Zony*, Moscow: Nauka, pp 330–348 (in Russian)

Vetrov, V M (1995) 'Early ceramics of the Stone Age cultures of Upper Vitim', in Kajiwara, H (ed), *The Origin of Ceramics in the East Asia and the Far East. Abstracts of International Symposium, Tohoku Fukushi University, Sendai, Japan, September 29–October 5, 1995*, Sendai: Tohoku Fukushi University Press, pp 31–35

——— (2006) 'Problems with similarity in preparation techniques and ornamentation of vessels of early ceramic complexes of northern Eurasia', in Derevianko, A P and Molodin, V I (eds), *Modern Problems in Russian Archaeology*, Novosibirsk: Institute of Archaeology and Ethnography Press, pp 173–176 (in Russian)

Vogel, J C and Waterbolk, H T (1963) 'Groningen radiocarbon dates IV', 5 *Radiocarbon*, 163–202

——— (1964) 'Groningen radiocarbon dates V', 6 *Radiocarbon*, 349–369

——— (1967) 'Groningen radiocarbon dates VII', 9 *Radiocarbon*, 107–155

Weber, A W (1995) 'The Neolithic and Early Bronze Age of the Lake Baikal region, Siberia: a review of recent research', 9 *Journal of World Prehistory*, 99–165

Whittle, A (1990) 'Radiocarbon dating of the Linear Pottery Culture: the contribution of cereal and bone samples', 64 *Antiquity*, 297–302

Wu, X and Zhao, C (2003) 'Chronology of the transition from Palaeolithic to Neolithic in China', 24 *The Review of Archaeology*, 15–20

Yuan, S, Zhou, G, Guo, Z, Zhang, Z, Gao, S, Li, K, Wang, J, Liu, K, Li, B and Lu, X (1995) '14C AMS dating the transition from the Paleolithic to the Neolithic in South China', 37 *Radiocarbon*, 245–249

Zaitseva, G I and van Geel (2005) 'The occupation history of the southern Eurasian steppe during the Holocene: chronology, the calibration curve and methodological problems of the Scythian chronology', in Scott, E M, Alekseev, A Y and Zaitseva, G (eds), *Impact of the Environment on Human Migration in Eurasia. NATO Science Series, vol. 42,* Netherlands: Springer, pp 63–82

Zaitseva, G I, Popov, S G, Krylov, A P, Knozorov, U V and Spevakovskii, A B (1993) 'Radiocarbon chronology of the archaeological sites of the Kurile islands', 35 *Radiocarbon*, 507–510

Zhang, C (1999) 'The Mesolithic and the Neolithic in China', 26 *Documenta Praehistorica*, 1–13

Zhao, C and Wu, X (2000) 'The dating of Chinese early pottery and a discussion of some related problems', 27 *Documenta Praehistorica*, 233–239

Zilhão, J (2001) 'Radiocarbon evidence for maritime pioneer colonization at the origins of farming in West Mediterranean Europe', 98 *Proceedings of the National Academy of Science*, 14180–14185

Zvelebil, M and Rowley-Conwy, P (1986) 'Foragers and farmers in Atlantic Europe', in Zvelebil, M (ed), *Hunters in Transition: Mesolithic Societies of Temperate Eurasia and their Transition to Farming*, Cambridge: Cambridge University Press, pp 67–95

Index

C

D

G

H

I

J

K

R

S

T

U

V

W

Y

Z

Publications of the
Institute of Archaeology, University College London

Series Editor: Ruth Whitehouse
Director of the Institute: Stephen Shennan
Founding Series Editor: Peter J. Ucko

Encounters with Ancient Egypt Subseries, Peter J. Ucko (ed.)
Jean-Marcel Humbert and Clifford Price (eds.), ***Imhotep Today***
David Jeffreys (ed.), ***Views of Ancient Egypt since Napoleon Bonaparte***
Sally MacDonald and Michael Rice (eds.), ***Consuming Ancient Egypt***
Roger Matthews and Cornelia Roemer (eds.), ***Ancient Perspectives on Egypt***
David O'Connor and Andrew Reid (eds.), ***Ancient Egypt in Africa***
John Tait (ed.), ***'Never Had the Like Occurred': Egypt's View of its Past***
David O'Connor and Stephen Quirke (eds.), ***Mysterious Lands***
Peter Ucko and Timothy Champion (eds.), ***The Wisdom of Egypt***

Critical Perspectives on Cultural Heritage Subseries, Beverley Butler (ed.)
Beverley Butler, ***Return to Alexandria***
Ferdinand de Jong and Michael Rowlands (eds.), ***Reclaiming Heritage***
Dean Sully (ed.), ***Decolonizing Conservation***
Mingming Wang, ***Empire and Local Worlds***

Other Titles
Andrew Gardner (ed.), ***Agency Uncovered***
Okasha El-Daly, ***Egyptology, The Missing Millennium***
Ruth Mace, Clare J. Holden, and Stephen Shennan (eds.), ***Evolution of Cultural Diversity***
Arkadiusz Marciniak, ***Placing Animals in the Neolithic***
Robert Layton, Stephen Shennan, and Peter Stone (eds.), ***A Future for Archaeology***
Joost Fontein, ***The Silence of Great Zimbabwe***
Gabriele Puschnigg, ***Ceramics of the Merv Oasis***
James Graham-Campbell and Gareth Williams (eds.), ***Silver Economy in the Viking Age***
Barbara Bender, Sue Hamilton, and Chris Tilley, ***Stone World***
Andrew Gardner, ***An Archaeology of Identity***
Sue Hamilton, Ruth Whitehouse, and Katherine I. Wright (eds.), ***Archaeology and Women***
Gustavo Politis, ***Nukak***
Sue Colledge and James Conolly (eds.), ***The Origins and Spread of Domestic Plants in Southwest Asia and Europe***
Timothy Clack and Marcus Brittain (eds.), ***Archaeology and the Media***
Janet Picton, Stephen Quirke, and Paul C. Roberts (eds.), ***Living Images***
Tony Waldron, ***Paleoepidemiology***
Eleni Asouti and Dorian Q. Fuller, ***Trees and Woodlands of South India***
Russell McDougall and Iain Davidson (eds.), ***The Roth Family, Anthropology, and Colonial Administration***
Elizabeth Pye (ed.), ***The Power of Touch***
Miriam Davis, ***Dame Kathleen Kenyon***

Marcos Martinón-Torres and Thilo Rehren (eds.), *Archaeology, History, and Science*
John Tait, *Why the Egyptians Wrote Books*
Peter Jordan and Marek Zvelebil (eds.), *Ceramics Before Farming*
Peter J. Ucko and Rachael Thyrza Sparks (eds.), *A Future for the Past*
Ruth D. Whitehouse (ed.), *Gender and Italian Archaeology*
Simon Hillson, *Mammal Bones and Teeth*
Nick Merriman, *Beyond the Glass Case*
Paulette M. McManus (ed.), *Archaeological Displays and the Public*
Sally-Ann Ashton, *Petrie's Ptolemaic and Roman Memphis*
Norah Moloney and Michael J. Shott (eds.), *Lithic Analysis at the Millennium*
Mark Nesbitt, *Identification Guide for Near Eastern Grass Seeds*
Christopher Tilley, *The Dolmens and Passage Graves of Sweden*
Sarah L.R. Mason and Jon G. Hather (eds.), *Hunter-Gatherer Archaeology*
Jon G. Hather, *The Identification of the Northern European Woods*
Jon G. Hather, *Archaeological Parenchyma*
D.F. Clark, M.M. Roxan and J.J. Wilkes (eds.), *The Later Roman Empire Today*
Nicholas Balaam and James Rackham (eds.), *Issues in Environmental Archaeology*
Rob Sands, *Prehistoric Woodworking*
David R. Harris and Kenneth D. Thomas (eds.), *Modelling Ecological Change*
Shmuel Ahituv and Eliezer D. Oren (eds.), *The Origin of Early Israel—Current Debate*
A.J. Ammerman (ed.), *The Acconia Survey*
Hans-Gert Bachmann, *The Identification of Slags from Archaeological Sites*
Anthony J. Barham and Richard I. Macphail (eds.), *Archaeological Sediments and Soils*
D.R. Brothwell, K.D. Thomas & Juliet Clutton-Brock (eds.), *Research Problems in Zooarchaeology*
Beatrice de Cardi, *Archaeological Surveys in Baluchistan, 1948 and 1957*
J.C. Mann and M.M. Roxan (eds.), *Legionary Recruitment and Veteran Settlement During the Principate*
Martin Millett (ed.), *Pottery and the Archaeologist*
D. Price Williams, *The Tombs of the Middle Bronze Age II Period from the "500" Cemetery at Tell Fara (South)*
Richard Reece, *Excavations in Iona 1964 to 1974*
Jonathan N. Tubb (ed.), *Palestine in the Bronze and Iron Ages*
Paula J. Turner, *Roman Coins from India*
Margaret M. Roxan, *Roman Military Diplomas 1985 to 1993*
Margaret M. Roxan, *Roman Military Diplomas 1978 to 1984*
Margaret M. Roxan, *Roman Military Diplomas 1954 to 1977*

About the Editors and Contributors

Editors

Peter Jordan: Peter Jordan is Senior Lecturer in the Department of Archaeology, University of Aberdeen, and he is also a co-PI of the AHRC Centre for the Evolution of Cultural Diversity (CECD). His PhD (University of Sheffield) focused on the ethnoarchaeology of Siberian hunter-gatherers, and was subsequently published as *Material Culture and Sacred Landscape: The Anthropology of the Siberian Khanty,* by Rowman and Littlefield Press (2003). He is currently researching cultural transmission amongst hunter-gatherer groups in western North America and Siberia, which will form the basis of a new book, and he also works on forager landscape enculturation and worldview, prehistoric hunter-gatherer rock art and the ethno-history of Siberia.

Marek Zvelebil: Marek Zvelebil is a world-leading authority on the archaeology and anthropology of hunter-gatherer societies and is based in the Department of Archaeology, University of Sheffield. In addition to numerous authoritative journal papers on wide-ranging hunter-gatherer themes, he has produced a suite of edited volumes which have synthesized key debates in hunter-gatherer studies, and signalled new frontiers for research. He also directs several research projects, which include investigations of prehistoric hunter-gatherer diet, health and mobility and bio-social perspectives on forager farmer-interactions in prehistoric Europe.

Contributors

Barnett, William K: William K Barnett studies the transition to agricultural societies in the western Mediterranean. He specializes in the reconstruction of socio-economic aspects of the transition through the production, distribution and technology of early pottery. He has a BA from the College of William and Mary and an MA and PhD from Boston University.

Budja, Mihael: Mihael Budja is based at the University of Ljubljana, Slovenia. His research and publications focus on cultural trajectories, population dynamics and climatic anomalies in Eurasian prehistory. He is Chief Editor of the archaeology journal *Documenta Praehistorica* and organizes the conference series *Neolithic Seminars,* which are held annually at the University of Ljubljana.

Chairkina, Natal'ia M: Natal'ia M. Chairkina is Head of Department of Archaeology of Stone Age at the Institute of History and Archaeology, Urals Branch of the Russian Academy of Science. Her research and publications address a range of topics pertaining to wetland archaeology as well as archaeological cultures of the Neolithic-Eneolithic Age of the Urals and northern Eurasia.

Cho, Daeyoun: Daeyoun Cho is Lecturer in Archaeology at Chonbuk National University, South Korea. His research focuses on early Korean ceramics, with particular emphasis on pottery production and consumption.

Crombé, Philippe: Philippe Crombé is Senior Lecturer in European Prehistory at Ghent University. He is also a senior researcher specializing in final Late Glacial and early Post-Glacial archaeology and is a director of major excavation and survey projects in northern Belgium (Flanders).

Dolukhanov, Pavel M: Pavel M. Dolukhanov is a leading specialist on the prehistory of Eastern Europe with special reference to the Mesolithic-Neolithic transition. His recent publications include *Le Néolithique en Europe* (co-authored, 2007); *The Black Sea Flood Question* (co-edited, 2007) and several articles across a range of journals.

German, Konstantin: Konstantin German is Head of Departament of Archaeology at the Open-Air Museum 'Kizhi', Republic of Karelia, Russia. His research is devoted to the Early Neolithic Cultures of Karelia.

Gronenborn, Detlef: Detlef Gronenborn is Curator of Prehistoric Archaeology at the Römisch-Germanisches Zentralmuseum and teaches at the Johann Gutenberg University in Mainz, Germany. His research focuses on the Early Neolithic in western Eurasia, with a particular focus on central Europe.

Hallgren, Fredrik: Fredrik Hallgren graduated from the University of Uppsala in 2008. His doctoral dissertation focused on the earliest Neolithic horticultural societies of central Scandinavia. He is currently researching the 'Neolithic' hunter-gatherers of central and northern Scandinavia, with a special focus on interaction between hunter-gatherers and horticulturalists.

Hayden, Brian: Brian Hayden is Professor of Archaeology at Simon Fraser University in British Colombia. He has carried out extensive ethnoarchaeological research on hunter-gatherers, feasting and prestige technologies.

Holm, Lena: Lena Holm is Reader in Archaeology at Umeå University, Sweden. Her research focuses on the northern Stone Age, with a special emphasis on ceramics, and on archaeological cultural heritage.

Hommel, Peter: Peter Hommel is a post-graduate research student at the University of Sheffield under the supervision of Peter Day, Patrick Quinn and Peter Jordan. His interests focus on the invention, adoption and use of ceramic and high-temperature technology among prehistoric hunter-gatherer and nomadic societies.

Kaner, Simon: Simon Kaner is Assistant Director of the Sainsbury Institute for the Study of Japanese Cultures and has taught and published on many aspects of Japanese and East Asian archaeology. Recent publications include editing Tatsuo Kobayashi's *Jomon Reflections* (2004).

Kashina, Ekaterina: Ekaterina Kashina is Researcher and Curator of the Urals Neolithic-Eneolithic Collections, State Historical Museum, Moscow. Her doctoral thesis was entitled 'Mobile art of the European Russian Forest Zone in the Neolithic-Eneolithic' (Moscow State University, 2005).

Ko, Ilhong: Ilhong Ko is Research Fellow at the Institute of Humanities, Seoul National University, South Korea. She recently graduated from the University of Sheffield, completing a doctoral dissertation on burial practices and social reproduction in the Korean Bronze Age. Her research interests include the burials, rituals and rock art of prehistoric Korea.

Kosinskaia, Lubov' L: Lubov L. Kosinskaia is based at the Laboratory for Archaeological Research, Urals State University, Russia.

Larsson, Mats: Mats Larsson is Professor of Archaeology at the University of Kalmar, Sweden. His research interests include the transition to farming in southern Scandinavia, the Middle Neolithic in southern Sweden and especially, the Pitted Ware Culture and its relations with the other Middle Neolithic groups.

Leskinen, Sirpa: Sirpa Leskinen has worked as Field Archaeologist at Stone Age sites in Finland. Her main research interests include the subsistence economy and pit houses.

Mazurkevich, Andrei M: Andrei M. Mazurkevich is Curator of the Department of Archaeology of Eastern Europe at the State Hermitage Museum of St Petersburg, and Director of the North-West Expedition. He has authored numerous publications on the Mesolithic and Neolithic of northwest Russia.

McKenzie, Hugh G: Hugh G. McKenzie is Instructor at Grant MacEwan College, Canada and is Co-Investigator of the long-standing Baikal Archaeology Project (http://baikal.arts.ualberta.ca). For the past 10 years, he has been exploring the archaeology and bioarchaeology of northern hunter-gatherers in eastern Siberia, Inner and North East Asia.

Nowak, Marek: Marek Nowak is Senior Lecturer in Archaeology at the Jagiellonian University of Kraków. He writes and teaches on the Mesolithic-Neolithic transition and on the Neolithic of central Europe. His research interests also include environmental archaeology. Recent publications include *Mesolithic/Neolithic Interactions in the Balkans and in the Middle Danube Basin* (co-edited, 2007).

Papmehl-Dufay, Ludvig:: Ludvig Papmehl-Dufay is currently working in contract archaeology at Kalmar County Museum, Kalmar, Sweden. Parallel to his excavation work, he continues to research the Neolithic of Öland Island, including ceramics of the Pitted Ware Culture.

Pesonen, Petro: Petro Pesonen has worked as field archaeologist in several excavations and field projects, especially in eastern Finland. He is a specialist in Stone Age and Early Metal period archaeology, particularly in ceramics and cultural ecology.

Shukurov, Anvar M: Anvar M. Shukurov is currently Professor of Astrophysical Fluid Dynamics in the School of Mathematics and Statistics at Newcastle University, United Kingdom. Apart from mathematical modelling for archaeology, his research focuses on the astrophysical magnetic fields, dynamo theory, and the physics of the interstellar medium.

Skandfer, Marianne: Marianne Skandfer is Project Leader at the Archaeolgical Institute, University of Tromsø, Norway. Since gaining her PhD in

2003, she has undertaken two post-doctoral projects on Stone Age and Early Metal Age archaeology in northern Fennoscandia, and has led various excavations.

Stilborg, Ole: Ole Stilborg obtained a PhD in 1997, focusing on Roman Iron Age settlement and grave pottery from eastern Funen, Denmark. His recent research also covers Ertebølle, Middle Neolithic and Iron Age pottery in Scania as well as technical ceramics (crucibles, furnace walls) from Bronze and Iron Age in Sweden.

Zhushchikhovskaya, Irina S: Irina S. Zhushchikhovskaya is Senior Researcher at the Insititute of History, Archaeology and Ethnology of the Russian Far East, Far Eastern Branch of the Russian Academy of Sciences. Her research focuses on the archaeology of ceramics and the prehistoric and ancient historic pottery-making traditions of Russian Far East and east Asia.